APPLIED ATOMIC COLLISION PHYSICS

Volume 2

Plasmas

APPLIED ATOMIC COLLISION PHYSICS
A Treatise in Five Volumes

Edited by

H. S. W. MASSEY
E. W. McDANIEL
B. BEDERSON

Volume 1 Atmospheric Physics and Chemistry

Volume 2 Plasmas

Volume 3 Gas Lasers

Volume 4 Condensed Matter

Volume 5 Special Topics

Appendixes in this volume list sources of information.

This is Volume 43-2 in
PURE AND APPLIED PHYSICS
A Series of Monographs and Textbooks
Consulting Editors: H. S. W. MASSEY AND KEITH A. BRUECKNER
A complete list of titles in this series appears at the end of this volume.

APPLIED ATOMIC COLLISION PHYSICS

Volume 2

Plasmas

Volume Editors

C. F. BARNETT
Physics Division
Oak Ridge
National Laboratory
Oak Ridge, Tennessee

M. F. A. HARRISON
Culham Laboratory
Abingdon, Oxfordshire
England

ACADEMIC PRESS, INC.   1984
(Harcourt Brace Jovanovich, Publishers)

Orlando   San Diego   New York   London
Toronto   Montreal   Sydney   Tokyo

COPYRIGHT © 1984, BY ACADEMIC PRESS, INC.
ALL RIGHTS RESERVED.
NO PART OF THIS PUBLICATION MAY BE REPRODUCED OR
TRANSMITTED IN ANY FORM OR BY ANY MEANS, ELECTRONIC
OR MECHANICAL, INCLUDING PHOTOCOPY, RECORDING, OR ANY
INFORMATION STORAGE AND RETRIEVAL SYSTEM, WITHOUT
PERMISSION IN WRITING FROM THE PUBLISHER.

ACADEMIC PRESS, INC.
Orlando, Florida 32887

United Kingdom Edition published by
ACADEMIC PRESS, INC. (LONDON) LTD.
24/28 Oval Road, London NW1 7DX

Library of Congress Cataloging in Publication Data
Main entry under title:

Applied atomic collision physics.

  (Pure and applied physics ; v. 43)
  Includes bibliographies and indexes.
  Contents: v. 1. Atmospheric physics and chemistry
-- v. 2. Plasmas -- v. 3. Gas lasers -- [etc.]
  1. Collisions (Nuclear physics) I. Massey, Harrie
Stewart Wilson, Sir. II. McDaniel, Earl Wadsworth,
Date. III. Bederson, Benjamin.
QC794.6.C6A65   1983   539.7′54   82-4114
ISBN 0–12–478802–5 (v. 2)

PRINTED IN THE UNITED STATES OF AMERICA

84 85 86 87   9 8 7 6 5 4 3 2 1

Contents

List of Contributors	ix
Treatise Preface	xi
Preface	xiii

1 Introduction
C. F. Barnett

I.	Evolution of Atomic Physics in Fusion Research	1
II.	Approaches to Fusion	5
III.	Atomic Physics in Fusion	21
	References	23

2 Basic Concepts of Fusion Research
M. F. A. Harrison

I.	Introduction	27
II.	The Principles of a D–T Fusion Reactor	28
III.	Magnetic Confinement of Fusion Plasmas	32
IV.	Energy Balance Conditions in a Magnetically Confined Fusion Plasma	36
V.	Auxiliary Heating	42
VI.	Inertial Confinement	46
	References	49

3 Atomic Radiation from Low Density Plasma
R. W. P. McWhirter and H. P. Summers

I.	Preliminary Discussion	52
II.	The Boltzmann Equation	55
III.	Components of the Statistical Balance Equations	60
IV.	The Collective Viewpoint of Ionization and Recombination	71
V.	The Distribution among the Stages of Ionization	81
VI.	Spectral Line Intensities	85
VII.	Radiated Power Loss	97
	References	108

4 Properties of Magnetically Confined Plasmas in Tokamaks
John T. Hogan

I.	Introduction	114
II.	Magnetic Configuration	116
III.	Moment Equations	122

IV.	Particle Balance	126
V.	Energy Balance	130
VI.	Impurity Transport	136
	References	138

5 Diagnostics

5A Diagnostics Based on Emission Spectra

N. J. Peacock

I.	Introduction	143
II.	Ionization Equilibrium	144
III.	Atomic Level Populations	162
IV.	Spectral Features and Their Diagnostic Application to Fusion Plasmas	169
V.	Neutral-Beam Spectroscopy	181
VI.	Basic Atomic Physics	183
	References	186

5B Laser Diagnostics

D. E. Evans

I.	Introduction	192
II.	Faraday Rotation	192
III.	Interferometry	199
IV.	Thomson Scattering	211
	References	224

5C Plasma Diagnostics Using Electron Cyclotron Emission

D. A. Boyd

I.	Introduction	227
II.	The Theory of Electron Cyclotron Emission	229
III.	Instrumentation	233
IV.	Applications	240
V.	Concluding Remarks	244
	References	246

5D Particle Plasma Diagnostics

C. F. Barnett

I.	Introduction	249
II.	Particle Diagnostic Atomic Physics	251
III.	Neutral-Particle Spectrometers Used in Determining Ion Temperatures	257
IV.	Plasma Ion Density and Effective Charge by Neutral-Beam Attenuation	278
V.	Beam Scattering Diagnostics	282
VI.	Impurity Ion Density	286
VII.	Magnetic Field Measurements	290

VIII.	Heavy-Ion Beam Probe	296
	References	303

5E The Electron Bremsstrahlung Spectrum from Neutral Atoms and Ions

R. H. Pratt and I. J. Feng

I.	Introduction	307
II.	Bremsstrahlung in a Plasma: Observables and Assumptions	308
III.	Coulomb Spectrum	311
IV.	Atomic Electron Screening Effects for an Isolated Atom or Ion	313
V.	End Points of the Spectrum: Elastic Scattering and Direct Radiative Recombination	315
VI.	Angular Distributions and Polarization Correlations	317
VII.	Bremsstrahlung Emission in Hot Dense Plasmas	318
	References	319

6 Heating of Plasma by Energetic Particles

6A Introduction   325

M. F. A. Harrison

	Reference	326

6B Trapping and Thermalization of Fast Ions

J. G. Cordey

I.	Introduction	327
II.	Fast-Ion Deposition	328
III.	The Slowing Down of the Fast Ions	332
IV.	Energy and Momentum Transfer Rates	334
V.	Effect on Plasma Temperature, Current, and Rotation	335
	References	338

6C Neutral-Beam Formation and Transport

T. S. Green

I.	Introduction	339
II.	Ion Beam Extraction and Acceleration	341
III.	Plasma Sources for Positive Ions	352
IV.	Beam Transport in a Gas Neutralizer	363
V.	Negative-Ion Beams	372
	References	378

6D Alpha-Particle Heating

D. E. Post

I.	Alpha-Particle Production and Heating	381
II.	Fast-Alpha-Particle Diagnostics	387
III.	Alpha-Particle Ash	390
	References	393

7 Boundary Plasma
M. F. A. Harrison

I.	Description of the Boundary Region	395
II.	The Boundary of a Toroidal Device	399
III.	The Sheath and Long-Range Electric Field Regions	413
IV.	Particle–Surface Interactions	416
V.	Atomic Processes in the Boundary Plasma	423
VI.	Significance of the Boundary Plasma	436
	References	437

8 Atomic Phenomena in Hot Dense Plasmas
Jon C. Weisheit

I.	Introduction	441
II.	The Plasma Environment	443
III.	Perturbations of Atomic Structure	450
IV.	Perturbations of Atomic Collisions	460
V.	Formation of Spectral Lines	467
VI.	Dielectronic Recombination	479
	References	482

Index   487

List of Contributors

Numbers in parentheses indicate the pages on which the authors' contributions begin.

C. F. *Barnett* (1, 249), Physics Division, Oak Ridge National Laboratory, Oak Ridge, Tennessee 37830

D. A. *Boyd* (227), Laboratory for Plasma and Fusion Energy Studies, University of Maryland, College Park, Maryland 20742

J. G. *Cordey* (327), Joint European Tokamak, Culham Laboratory, Abingdon, Oxfordshire OX14 3DB, England

D. E. *Evans* (191), Euratom/UKAEA Fusion Association, Culham Laboratory, Abingdon, Oxfordshire OX14 3DB, England

I. J. *Feng*[*] (307), Department of Physics and Astronomy, University of Pittsburgh, Pittsburgh, Pennsylvania 15260

T. S. *Green* (339), Euratom/UKAEA Fusion Association, Culham Laboratory, Abingdon, Oxfordshire OX14 3DB, England

M. F. A. *Harrison* (27, 325, 395), Culham Laboratory, Abingdon, Oxfordshire OX14 3DB, England

John T. *Hogan* (113), Fusion Energy Division, Oak Ridge National Laboratory, Oak Ridge, Tennessee 37830

R. W. P. *McWhirter* (51), Space and Astrophysics Division, Rutherford Appleton Laboratory, Chilton, Didcot, Oxfordshire OX11 0QX, England

N. J. *Peacock* (143), Culham Laboratory, Abingdon, Oxfordshire OX14 3DB, England

D. E. *Post* (381), Plasma Physics Laboratory, Princeton University, Princeton, New Jersey 08544

R. H. *Pratt* (307), Department of Physics and Astronomy, University of Pittsburgh, Pittsburgh, Pennsylvania 15260

H. P. *Summers* (51), Department of Natural Philosophy, The University of Strathclyde, Glasgow G4 0NG, Scotland

Jon C. *Weisheit*[†] (441), Plasma Physics Laboratory, Princeton University, Princeton, New Jersey 08544

---

[*] Present address: AT&T Bell Laboratories, Murray Hill, New Jersey 07974.

[†] Present address: Theoretical Physics Division, Lawrence Livermore National Laboratory, Livermore, California 94550.

Treatise Preface

Research in atomic physics and especially in the physics of atomic collisions has developed at an explosive rate since the Second World War. The high rate of increase of knowledge of atomic collision processes has been of great value in many applications to pure and applied physics and chemistry. For the full understanding of the physics of planetary and stellar atmospheres, including those of the earth and the sun, detailed knowledge is required of the rates of a great variety of atomic and molecular reactions. Gas lasers depend for their operation on atomic collision processes of many kinds, and a knowledge of the corresponding reaction rates is important for laser design. The release of energy by controlled nuclear fusion offers a possibility of an effectively infinite source of power in the future. Many aspects of the complex techniques involved are affected by atomic reactions. Again, there are many applications of collision physics to the study of condensed matter.

These major activities have expanded rapidly at a rate which has been accelerated by the availability of data and understanding from atomic collision physics. There are many smaller areas which depend on this subject.

In these five volumes we planned to give an account of the wide range of applications which are now being made, as well as the additional requirements for further applications. Volume 1 deals with applications to atmospheric and astrophysics, Volume 2 to controlled fusion, Volume 3 to laser physics, and Volume 4 to condensed matter. Volume 5 includes various special applications.

In all cases the emphasis is on the discussion of these applications and the atomic physics involved therein. However, sufficient background is provided to make clear what has been achieved and what remains to be done through further research in collision physics.

We are much indebted to Academic Press for the ready assistance they have afforded us at all times.

H. S. W. MASSEY
E. W. MCDANIEL
B. BEDERSON

As we go to press with this volume word has reached us that Sir Harrie Massey has passed away in his 75th year.

Sir Harrie's professional life, through no coincidence, spanned the field of atomic and molecular collisions since in large part he helped create and guide its course. This he did through his seminal and amazingly perceptive formulation of atomic collision theory, his role as teacher of generations of atomic and molecular theorists, and his interests in atmospheric and space science and many other areas of fundamental and practical importance.

This series of volumes was Sir Harrie's concept in the first place and thus appropriately takes its place among the many contributions he has made to science.

We, the co-editors of Sir Harrie, add our voices to those who mark his passing with a feeling of loss and sadness.

BENJAMIN BEDERSON
EARL W. MCDANIEL

*Preface**

Since the beginning of controlled fusion energy research in 1951–1952, atomic and molecular physics processes have played a significant role in the heating, cooling, loss, diagnostics, and modeling of high-temperature plasmas. The probability or cross section for an atomic collision to occur is approximately 12 orders of magnitude greater than that for the D–T thermonuclear reaction. Thus, a great effort has been expended over the past 30 years to minimize or eliminate those atomic processes that have detrimental effects on the plasma, e.g., excessive cooling of a tokamak plasma by radiation from multicharged impurity tungsten ions. This volume of "Applied Atomic Collision Physics" deals with those atomic processes that have been important in fusion research developments during the past 30 years.

Two distinct approaches have been utilized toward the goal of achieving a fusion reactor—magnetic confinement and inertial confinement. Since inertial confinement research was initiated several years after magnetic confinement, studies of atomic processes in high-density, high-temperature plasmas (10^{22}–10^{26} cm^{-3}) have been rather limited. For this reason most of the chapters in this volume are devoted to studies concerning magnetically confined plasmas. Chapter 1 is a historical summary of the history of fusion research along with a brief description of the various approaches and efforts being pursued in both magnetically and inertially confined plasmas. A general discussion of the basic concepts and properties in confinement and heating of a plasma is presented in Chapters 2 and 4. In Chapter 3 the theory of atomic collisions that result in excited quantum states, particularly highly ionized impurity atoms, is introduced. The greatest contribution that atomic physics has made to fusion research is in the area of diagnostics. Without the use of techniques and knowledge that have been developed in atomic physics, our understanding of the physics of high-temperature plasmas would still be in its infancy. Such diverse diagnostic topics as emission spectra, laser scattering, electron cyclotron emission, particle beams, and bremsstrahlung are treated in Chapter 5.

Probably, the greatest problem encountered in obtaining a stable, long-duration, high-temperature plasma is the coupling of more energy into the plasma than that lost by radiation and particle transport. The most successful means to date has been plasma heating by neutral beams whose formation, transport, and thermalization are described in Chapter 6. Sustaining a

---

* Research sponsored by the Office of Fusion Energy, U.S. Department of Energy under contract W-7405-eng-26 with the Union Carbide Corporation.

burning D–T plasma must rely on α-particle heating also discussed in Chapter 6. In the past few years the importance of the plasma boundary-wall region on plasma stability has been recognized. The boundary or edge plasma, along with particle–surface interactions, is discussed in Chapter 7. Finally, in the last chapter the role of atomic physics in hot dense plasmas is shown to be remarkably different than in hot tenuous plasmas found in magnetically confined devices.

Oak Ridge National Laboratory   C. F. BARNETT
Oak Ridge, Tennessee

1
Introduction[†]

C. F. Barnett

Physics Division
Oak Ridge National Laboratory
Oak Ridge, Tennessee

    I. Evolution of Atomic Physics in Fusion Research . 1
   II. Approaches to Fusion 5
       A. Toroidal Geometries. 6
       B. Open-Ended Geometries 14
       C. Inertial Confinement. 20
  III. Atomic Physics in Fusion 21
       References 23

I. Evolution of Atomic Physics in Fusion Research

Since the beginning of an active effort in magnetically confined fusion research in 1951–1952, atomic physics has played a significant role in the heating, cooling, modeling, and diagnostics of high temperature plasmas. The importance of atomic processes in plasmas can best be illustrated by considering the cross sections for energy production and loss in a 10-keV deuterium–tritium (D–T) plasma. The cross section for the D–T nuclear reaction is $\sim 10^{-27}$ cm^2, whereas the atomic cross section for energy loss is of the order of 10^{-15} cm^2. Because atomic cross sections are approximately 12 orders of magnitude greater than the relevant nuclear cross sections, extreme care must be used to minimize energy and particle loss mechanisms.

The history of the fusion research effort is replete with accounts of the use of atomic physics processes to create a high temperature plasma, only to have similar processes limit the plasma density and temperature. In the 1950s most atomic physics problems centered on the production and confinement of a plasma in magnetic mirror geometries, with little attention placed on the role of impurity atoms entering the plasma from surrounding walls. Following the declassification of the controlled fusion program in 1958, Teller (1959), at the Second Geneva Conference on Peaceful Uses of Atomic

---

[†] Research sponsored by the Office of Fusion Energy, U.S. Department of Energy under Contract W-7405-eng-26 with the Union Carbide Corporation.

Energy, warned that impurities could have a catastrophic effect on plasma properties. Not only would the presence of impurities lead to increased bremsstrahlung and line radiation, resulting in rapid energy loss, but another mechanism could also be important: if neutral metallic atoms, sputtered from the plasma wall, enter the plasma, they will undergo charge exchange collisions with plasma ions. The fast neutral atoms formed in the collision will cross the magnetic field lines and impact the wall—sputtering additional atoms and producing an avalanche effect on the plasma. To overcome problems such as this Spitzer (1958) had suggested the use of a divertor to prevent impurities from the wall entering the plasma. In this concept a cylindrical outer shell of a magnetic flux surface that surrounds the confined plasma is diverted by a set of appropriately placed auxiliary magnetic coils to a region outside the plasma edge. Impurity atoms from the surrounding surfaces enter the diverted flux surface where they are immediately ionized. The resulting ions flow along the magnetic flux lines to a target plate located in an external vacuum chamber. Thus, the number of particles striking the plasma wall is greatly reduced. Currently an active divertor program is being pursued in toroidal confinement experiments. The divertor is discussed in more detail in Chapter 7.

Most of the effort in mirror confinement in the 1950s was directed toward injecting energetic hydrogen molecular ions or neutral particles into the containment volume, trapping ions through an irreversible collision, and containing the ions a sufficiently long time for thermalization to take place. In the Oak Ridge DCX-1 experiment 600-keV H_2^+ ions were injected into a simple mirror configuration and dissociated by a carbon arc with a density of $\sim 10^{14}$ cm^{-3}. Approximately 30% of the H_2^+ ions were dissociated on a single pass through the arc. Theory had predicted that charge exchange losses of H^+ in the highly ionized carbon arc would be very small. However, transmission measurements suggested that the cross section for $H^+ + C^{3+} \rightarrow H^0 + C^{4+}$ was of order 10^{-17}–10^{-18} cm^2 with the consequence that multiple passes of H^+ through the arc did lead to excessive particle loss. Subsequent experiments involved dissociating H_2^+ on residual gas and a D_2 arc. The U.S.S.R. OGRA experiments used both residual gas and low temperature gaseous discharges to trap H_2^+. In the ALICE experiment at the U.S. Lawrence Livermore National Laboratory and the PHOENIX experiment at the U.K. Harwell Laboratory, H^0 or D^0 was injected into the mirror volume and trapping was accomplished by the Lorentz field ($v \times H$) ionization of those atoms in highly excited n levels or Rydberg states. All of the injection experiments showed that if the injected current exceeded a critical value I_c, the residual gas would become fully ionized. Losses due to charge exchange would approach zero, and the plasma density would increase exponentially with time. The critical current was given by

$$I_c = \sigma_{ex} v V / 4 \sigma_T^2 \lambda^2,$$

1. Introduction

where σ_{ex} is the charge exchange cross section, σ_T is the dissociation or ionization cross section, v is the injected particle velocity, V is the plasma volume, and λ is the mean free path of the trapped ion before it is lost. Although particle containment times of several seconds were achieved, all the experiments were limited to plasma densities in the range of 10^9–10^{11} cm^{-3}. These maximum density limits were the result of particle loss either by atomic collisions or plasma instabilities.

In the toroidally confined experiments including tokamaks, stellarators, and pinches, atomic physic processes were of less importance than were plasma stability studies and the "anomalous" loss of particles and energy from the plasma. Impurities from the walls in the U.K. ZETA toroidal pinch experiment resulted in a high plasma resistivity. Impurity levels as high as 30% of the plasma density were reported for a He plasma (Butt *et al.*, 1959). Highly luminous spots were observed on the plasma walls, leading to intense flashes of radiation from the plasma. These spots resulted from plasma interaction with the walls and are known today as unipolar arcs. High-energy x rays were observed which originated from "runaway" electrons. Classically, as the plasma electron temperature increases, the Coulomb collision cross section for energy transfer to the plasma ion decreases or the plasma resistivity decreases since it is proportional to $T_e^{-3/2}$. In such geometries an electric field is imposed longitudinally, and each time an electron traverses the torus it is accelerated by the electric field. As the collision cross section decreases the electron energy rapidly increases. In many toroidal experiments the electron energy approached several mega-electron-volts. Present-day experiments minimize runaway conditions by careful control of plasma parameters. In early toroidal plasma experiments application of atomic physics processes and techniques was indispensable to determining plasma parameters or properties: bremsstrahlung radiation and ratios of emission lines were used to obtain electron temperatures, and line emission spectra were used to determine impurity densities and ion temperatures from thermal Doppler broadening and electron densities.

Pessimism can best describe the status of plasma research in the early 1960s. Plasmas were plagued with seemingly universal instabilities, and many investigators believed that the instability problem was not amendable to solution. This gloomy attitude was dispelled in 1967 when Artsimovich reported at the Third International Conference on Plasma Physics and Controlled Nuclear Fusion that a stable, high temperature plasma had been obtained in the T-3 tokamak at the Soviet Kurchatov Institute (Artsimovich *et al.*, 1969). The plasma resistivity was "anomalous," and was later shown to be the result of excessive impurities originating at the plasma limiter and wall.

At plasma temperatures of 1–2 keV, high-Z atoms are only partially stripped, and are thus an important source of energy loss by radiation. The deleterious effect of high-Z ions on plasma properties was dramatically illustrated in the Oak Ridge ORMAK tokamak. Plasma measurements indicated that 40–80% of the radiant power originated from tungsten ions with charge states 29–34 (Isler *et al.*, 1977). During this same period it was observed that the electron temperature profile of the Princeton PLT plasma was depressed in the plasma center rather than exhibiting the normal peaked condition (Hinnov and Mattioli, 1978). This depression was attributed to plasma cooling by tungsten radiation from charge states 19–34. Computations using an average-ion model indicated that a concentration of tungsten as low as 5×10^{-4} of the electron density would prevent a thermonuclear reactor from igniting (Mead, 1974). Since the experiments in ORMAK and PLT, emphasis has been placed in preventing all high-Z materials from coming into contact with a high temperature plasma.

Plasmas in tokamaks, stellarators, and reversed-field pinches have traditionally been heated by ohmic heating in which a current flows through the plasma. At electron plasma temperatures of 1–2 keV, ohmic heating becomes inefficient, and other means must be used to increase the plasma temperature to the desired level. Multimegawatt H^0 and D^0 neutral-beam systems have been developed to supply the auxiliary heating. By injecting 2.4 MW of neutral power into a PLT tokamak plasma of 5×10^{13} cm^{-3} density and free of tungsten impurity, the ion temperature was increased to 6.5 keV. Neutral-beam injection heating has opened up a new regime of relevant atomic processes in fusion research. These processes include low-energy collisions in positive and negative ion sources, ion neutralization, transport of intense beams of neutral particles, beam penetration into the plasma, and beam thermalization through Coulomb collisions.

In 1974 the U.S. Office of Controlled Thermonuclear Research (ERDA, 1974) appointed a panel to identify those atomic physics data needed in fusion research and to recommend steps toward obtaining them. The panel recommended a tenfold increase in the funding of atomic physics in areas of atomic structure (wavelengths, energy levels, and transition probabilities) of highly ionized atoms; cross-section data relevant to plasma heating, cooling, modeling, and diagnostics; fundamental data on plasma interaction with surfaces; and evaluating and compiling atomic data relevant to fusion research. The impact of the report was immediate. Atomic physics has subsequently received tremendous impetus as indicated by a survey of the open, published literature in this field. The total number of relevant papers published during 1975–1980 was ~2400, increasing from ~160 in 1975 to ~530 in 1980.

Unlike the magnetically confined fusion research program, the inertial confinement program has been in existence only during the past decade. Because the plasma density approaches a few grams per cubic centimeter at temperatures of ~1 keV, it is not surprising that some of the atomic physics problems are unique (Hauer and Burns, 1982). Much of the atomic physics is concentrated in the generation and transport of intense laser, ion, or electron

1. Introduction

beams, which act as drivers for pellet compression. During the past ten years excellent progress has been made in understanding the atomic physics processes involved in coupling the driver energy to the pellet during the compression stage. Probably of most importance is the role of atomic physics in the diagnostics of high-density plasmas. Optical spectroscopy of line radiation from transitions in highly stripped ions and bremsstrahlung radiation has been instrumental in determining plasma properties.

Atomic physics processes will continue to have a prominent role in plasma physics as the next generation of large plasma experiments becomes operable.

II. Approaches to Fusion

A thorough discussion of the role of atomic physics in fusion research requires a working knowledge of basic concepts and nomenclature, including the endless list of acronyms used in various fusion research facilities. This section will present only a brief account of the past and present approaches directed toward the ultimate goal of a power-producing fusion reactor. For a detailed summary of the history, facilities, and problems of fusion research, the reader is referred to Part 1 of the recent two-part volume "Fusion," edited by Teller (1981).

During the past 30 years two principal concepts have evolved in efforts to develop a working fusion reactor: magnetic confinement and inertial confinement. In the magnetic confinement concept a magnetic field restricts the high temperature plasma to a specific volume by the magnetic pressure $B^2/8\pi$ acting against the material plasma pressure $2nkT$, where n and T are the plasma density and temperature and k is the Boltzmann constant. The ratio of the material pressure to the magnetic pressure in the vacuum field is known as β. Obviously, the maximum value of β is unity. Because it is proportional to the plasma power density, large efforts have been expended in designing magnetic configurations to maximize β. (For further discussion of β, see Harrison, Chapter 2, this volume.)

The basic problems encountered in all magnetic confinement devices include plasma heating, confinement, fueling, impurity control, equilibrium, and stability. Solutions to problems in each of these areas are necessary before a stable D–T reactor plasma with an electron temperature of 10 keV, a plasma density of 10^{14} cm^{-3}, and containment times of 1 s can be realized. So far, the principal limitation in achieving these plasma parameters has been plasma instabilities either of a fluid (MHD) or kinetic type.

In the inertial confinement scheme, a pulse of energy in the form of laser, electron, or ion beams is incident on a D–T pellet. The pellet is rapidly compressed ($\sim 10^{-9}$ s) by ablation and, during compression, is heated to thermonuclear temperatures. Two problems are dominant in inertial confinement: absorption of energy by the pellet and the development of lasers or particle sources with sufficient power to heat the pellet. Energy must be absorbed rapidly in the pellet periphery to prevent preheating of the interior and the subsequent dispersion before ignition temperatures are reached. The beam energy needed to produce the required plasma parameters in a reactor has been estimated at 10^{14} W.

Magnetically confined plasmas fall into two categories. In one category the magnetic lines of force close upon themselves within the confinement region. Best examples of a closed system are tokamaks, stellarators, reversed-field pinches, and bumpy tori. In the other class are the open-ended systems in which the field lines escape out the ends of the confinement region (e.g., magnetic mirrors and theta pinches). Soviet scientists in the late 1960s succeeded in containing a high temperature, stable, low-β plasma in a tokamak. Consequently, today in laboratories worldwide, major emphasis is placed on tokamak geometries.

A. Toroidal Geometries

1. Tokamaks

The basic ideas of tokamak confinement were established by Tamm and Sakharov and by Spitzer in 1950, with the first experimental tokamak being put into operation in 1956 at the Soviet Kurchatov Institute (see Bezbatchenko *et al.*, 1960; Spitzer, 1952; Tamm and Sakharov, 1961). This first tokamak was designed with an insulating porcelain inner wall or liner. Interaction of the plasma with the wall resulted in a large influx of oxygen impurity into the plasma. Later, attempts were made in the T-1 and T-2 tokamaks to reduce the impurity level by installing a bakable stainless-steel liner and a plasma limiter—a mechanical obstruction made of a high temperature material that protruded 1–2 cm from the walls. Ideally, the limiter prevents the plasma from interacting with the wall. At the Novosibirsk Third Conference on Plasma Physics and Controlled Fusion Research, Artsimovich announced that electron temperatures of ~1000 eV, ion temperatures of 400–500 eV, and average plasma densities of 4×10^{13} cm^{-3} had been achieved in the Kurchatov T-3 tokamak (Artsimovich *et al.*, 1969). Throughout the plasma physics community skepticism was expressed over the reported electron temperature. A large discrepancy existed in the reported temperature as determined from the plasma diamagnetism and the plasma resistivity. The plasma resistivity was termed "anomalous" in that the resistivity was not proportional to $T_e^{-3/2}$. Uncertainties in the measurements were resolved by the joint efforts of a team from the U.K. Culham Laboratory and the Soviet Kurchatov Institute (Peacock *et al.*, 1969). Making use of the atomic physics technique of Thomson scattering of a laser beam, they confirmed the

1. Introduction

1000-eV temperature. Later, it was shown that the diamagnetic signal could be distorted by the runaway electron distribution and that the anomalous resistivity results from impurities in the plasma. The success of the T-3A and TM-3 tokamaks prompted a rapid conversion of the Princeton model C stellarator to the ST tokamak, which confirmed the TM-3 results. Shortly thereafter, a proliferation of tokamak experiments occurred throughout the world.

Probably the most important advance in tokamak plasma research during the past decade was the application of intense neutral beams as a means of auxiliary plasma heating. Initial experiments with neutral-beam heating in the Oak Ridge ORMAK, Princeton ATC, and Fontenay TFR-400 tokamaks indicated that the energetic trapped ions were confined for times sufficient to permit transfer of energy to the plasma particles and that heating plasma ions to ignition temperatures was indeed feasible. In the late 1970s 2.5 MW of neutral beams was injected into the Princeton PLT tokamak, raising the ion temperature to a record 7.1 keV and the electron temperature to 3.5 keV (Eubank *et al.*, 1979).

The obvious closed magnetic field line confinement system is the simple torus with nested magnetic surfaces. However, such a system is unstable for two reasons. First, the magnetic field of a torus decreases as the major radius R increases. Stated differently, the magnetic field is greater on the inside of the minor diameter than on the outside. Thus, the magnetic pressure on the inside of the torus tends to expand the plasma toward the outside at greater R. Second, plasma electrons and ions spiral in opposite directions, with a tendency for electrons to accumulate on the top or bottom of the minor cross section and on the opposite side from the positive ion accumulation. This separation of charge creates an electric field with the resulting **E** × **B** drift moving the ions to the plasma wall. If a twist in the magnetic field (a rotational transform) is provided, the electrons can flow along the field lines to the positive charge region, thereby shorting out the electric field. Rotational transforms are used in both tokamak and stellarator geometries to stabilize the plasma.

Comprehensive reviews of tokamak research have been published by Artsimovich (1972), Furth (1975, 1981), and Bodin and Keen (1977). A schematic diagram of a tokamak is shown in Fig. 1. The plasma is confined by a strong toroidal magnetic field B_T produced by the toroidal field coils. A plasma current I_p induced by the transformer creates a poloidal magnetic field B_P, which when superimposed on the toroidal field results in magnetic field lines with helical paths. This scheme provides the rotational transform illustrated in Fig. 2. A third field B_v in the vertical direction (not shown in Fig. 1) must be provided to center the plasma in the vacuum vessel. In the early tokamaks the vertical field was produced by the image current in a highly conducting shell surrounding the plasma. All present-day tokamaks

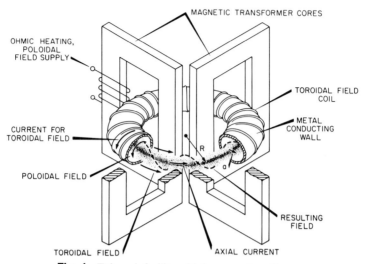

Fig. 1. Tokamak facility with iron core transformer.

have an external set of coils to produce the stabilizing and centering field. Stability in tokamak plasmas is expressed by requiring the quantity $q = B_T a / B_P R > 1$, where a is the plasma radius.

Not only does the plasma current I_p flowing through the plasma produce the required rotational transform, but it also heats the plasma through ohmic or Joule heating. With only ohmic heating plasma temperatures are limited to 1–2 keV. Thus, auxiliary heating must be supplied, usually by neutral-beam injection and trapping. Intensive efforts are now under way to provide auxiliary heating employing RF wave heating in the form of Alfvén waves, ion cyclotron harmonics, lower hybrid waves, and electron cyclotron harmonics. Other heating methods that offer some promise are adiabatic compression, turbulent heating, and of course α-particle heating after ignition temperatures have been obtained.

Because the plasma current is driven by transformer action, the tokamak is inherently a non-steady-state device, a severe handicap in designing a fusion reactor. To overcome the pulsed nature of the machine, attempts are being made to drive the plasma current after a stable plasma has been established by either creating a current flow through neutral beams injected in the toroidal direction or by coupling RF power to the high-energy tail of the plasma electron distribution. Some success has been achieved with each of these two methods.

During the past 25 years tremendous progress has been made in understanding the physics of tokamak plasmas. From the first tokamak with minor radius 13 cm, major radius 80 cm, $I_p = 100$ kA, $B_T = 1.5$ T, and electron temperatures of a few tens of electron volts, the devices have grown to the size of the present Princeton TFTR experiment with minor radius 1.1 m,

1. Introduction

major radius 2.7 m, $I_p = 2.5$ MA, $B_T = 5$ T, and ion temperatures that may approach 10 keV with neutral-beam heating. Other facilities of comparable size which will begin operation in the mid-1980s are the JET (Joint European Tokamak) at Culham, JT-60 in Japan, and T-15 in the Soviet Union. It is hoped that problems of disruptive instabilities, impurities, and non-steady-state operation can be solved in this generation of experiments.

2. Stellarators

During the decade of the 1950s the stellarator concept of magnetic toroidal confinement was developed under Spitzer's leadership at Princeton (Spitzer, 1958). Five experimental facilities were constructed starting with the B series of stellarators and culminating with the C stellarator in 1968. The early B-stellarator experiments confirmed theoretical predictions on individual particle confinement. The period of the 1960s saw an expansion of stellarator facilities in a large number of countries including the L-1 and TOR-1 at the Soviet Lebedev Institute, Uragan-1 at the Soviet Khar'kov Institute, Wendelstein I and II at West Germany's Garching Laboratory, and Proto-Cleo at the U.K. Culham Laboratory. Plasmas formed in these experiments were characterized as having an anomalous loss of plasma to the walls caused either by plasma instabilities or charge exchange collisions with impurities. Results obtained from the Princeton C stellarator have been summarized by Young (1974). Miyamoto (1978) and Shohet (1981) have published reviews of stellarator research and facilities. A compendium of papers in a special issue of *IEEE Transactions on Plasma Science* (IEEE, 1981) suveyed stellarator research—past, present, and future plans. At the present time stellarator research facilities include Wendelstein VIIA—West Germany; Heliotron E—Kyoto University, Japan; Uragan III—Khar'kov; JIPPT II—Nagoya, Japan; and the L-2—Lebedev Institute. Results

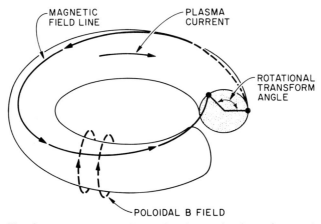

Fig. 2. Tokamak plasma illustrating the rotational transform angle.

obtained from these facilities indicate that confinement in stellarators is equal or superior to that found in comparable-sized tokamaks.

Unlike the tokamak, which depends on internal current flow through the plasma to provide the stabilizing field, the stellarator derives its rotational transform from the magnetic field produced by current flowing in a set of external helical windings wrapped around the vacuum vessel. Stellarator field configurations can best be visualized by considering the field produced by a current flowing in a single conductor wrapped in a helical path around the plasma. Such a configuration produces toroidal (B_T), poloidal (B_P), and vertical field (B_V) components. If a second conductor is installed with the current flowing in the opposite direction from the first one, the B_T and B_V components will cancel, leaving only the B_P component. An additional set of circular coils must be placed around the torus to provide a confining toroidal field. Designs of conventional stellarators include multiple sets of continuous helical windings with currents flowing in opposite directions to produce the required rotational transform. A schematic diagram of a hexapole field formed by six current-carrying conductors is shown in Fig. 3. Not shown are the circular coils needed to produce B_T. The symmetry of the magnetic surface formed by the external field coils is characterized by the l number. As viewed in the minor plasma cross section, the coils in Fig. 3 will form a triangular-shaped plasma with $l = 3$.

An alternative method of forming the rotational transform is to twist the torus out of a planar geometry, forming a figure-eight configuration as used in the Princeton B-1 stellarator (Coor *et al.*, 1958).

If, instead of permitting the adjacent helical coil current to flow in the opposite direction, the current is reversed so that the current flows in the same direction in all the helical coils, the resulting field has B_T, B_P, and B_V components. A separate set of windings must be added to cancel out the unwanted B_V component. Plasma machines with geometries of this type are known as torsatrons. By modulating the pitch in the helical windings, the vertical field can be canceled, leading to the term "ultimate" torsatron. One

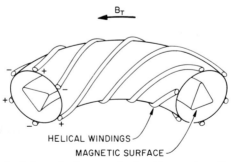

Fig. 3. Schematic diagram of a $l = 3$ hexapole stellarator.

1. Introduction

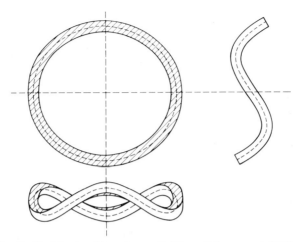

Fig. 4. Modular $l = 3$ torsatron coil. [From Grieger et al. (1981).]

other term used in describing one of the stellarator geometries is the heliotron—a torsatron with the addition of a toroidal field.

A distinct advantage of stellarator or torsatron geometry is that the field windings can be modular, a concept which not only eliminates the large inward-directed mechanical forces arising from the coils, but also facilitates remote-controlled repairs. A modular coil can be formed by twisting and bending a circular coil out of its planar geometry as shown in three views of Fig. 4. When installed around the torus, a set of coils such as these produces an $l = 3$ plasma similar to that shown in Fig. 3. In modular coils no net toroidal current and hence no vertical field exists. Modular coil geometries have been designed into the IMS torsatron at the University of Wisconsin (Anderson et al., 1981).

Stellarators have no net current flow in the plasma, so disruptive instabilities or other current-driven instabilities are eliminated. Early stellarators relied on ohmic heating (which results in net current flow) to heat the temperature to modest temperatures. Typical plasma properties obtained were 5×10^{12}–10^{13} cm^{-3} density, 300–750 eV electron temperature, and 50–250 eV ion temperature. To obtain maximum benefits from stellarators, a nonohmically heated, low temperature plasma needs to be produced and subsequently heated by either neutral-beam injection or RF wave heating. Several experiments have used these methods with good results. Typical equilibrium plasma properties in the Heliotron-E facility, when heated by 200 kW of electron cyclotron resonant heating (ECRH) power, were $T_e \sim 200$ eV, $n_e = 4 \times 10^{12}$ cm^{-3}, and $T_i \sim 100$ eV. Injecting neutral beams into the Wendelstein VII A plasma has produced a 5×10^{13} cm^{-3} average density, a T_e of ~600 eV, and a T_i of ~800 eV (Rostagni, 1981). However, the plasma properties were limited by the influx of impurities. Stellarators should have an advantage over tokamaks in controlling impurities in that the shape of the stellarator field permits the easy installation of divertors. Modular stellarators are in fact modular divertors.

Although stellarators have many advantages over tokamaks, there are some disadvantages. The plasma is nonsymmetric, causing difficulty in theoretical computations. Also, the magnetic field is not efficiently utilized and hence is expensive. Experience indicates that stellarators are a viable alternative for fusion reactor design.

3. *Reversed-Field Pinches (RFPs)*

A linear pinch plasma can be established between two electrodes in a gas by applying an electric field and causing a large current to flow between the electrodes. The current induces an azimuthal or poloidal magnetic field that pinches the discharge from the surrounding vessel walls. The pinch takes place only if the magnetic field is above a critical value. With linear geometry such as this, plasma energy is lost to the electrodes, lowering the plasma electron temperature. Also, sputtering of the electrodes by the plasma injects metallic impurity ions and atoms into the plasma. Placing the pinch discharge in a toroidal magnetic field eliminates end losses to the electrodes.

In reversed-field pinch experiments, a geometrical configuration like that described previously for a tokamak has been used. The toroidal field is maintained by a set of external coils. A very fast-rising plasma current is induced to flow around the torus by discharging a fast capacitor bank into the primary of a transformer. The plasma acts as the transformer secondary. The implosion and compression from the fast-rising poloidal field pinches and heats the plasma. Shortly after the current in the plasma is induced and the plasma pinches, the direction of the current in the external toroidal field is reversed on a time scale that is short in term of instability growth time. Consequently, the external field is in an opposite direction to the magnetic field trapped in the plasma. Stability is produced by the rotational transform that results from combining B_T and B_P and also by the high magnetic shear produced in the magnetic field configuration. Equilibrium is provided by a highly conducting metal shell surrounding the plasma. Plasma confinement time is limited by the time it takes for the trapped magnetic flux to leak or diffuse out of the confined volume. A theoretical review of RFP confinement has been published by Christiansen and Roberts (1978). Baker and Quinn (1981) and Bodin and Keen (1977) have written general reviews.

RFP plasmas are similar to tokamak plasmas in that stability is provided in both experiments by a plasma current producing a rotational transform in the magnetic field while equilibrium is supplied by a conducting shell or external vertical field. In tokamaks $B_T \gg B_P$, while in RFP experiments $B_P \gtrsim B_T$. Tokamak stability requires the plasma current to be less than a critical value; RFP plasmas do not have constraints imposed on the plasma current, thereby permitting large ohmic heating currents. Theory predicts that it may be possible to heat RFP plasmas to ignition by ohmic, implosive, and compression heating without the use of auxiliary heating methods. RFP plasmas are stable with β's of 0.5 or even higher.

The first RFP experiment was the stabilized ZETA pinch at the U.K. Harwell Laboratory (Butt et al., 1959). The early operation was plagued with instabilities and excessive impurity levels in the plasma. Several years after the initial operation of ZETA the discovery was made that, if the H_2 filling pressure was increased to 1–3 mTorr, a magnetic field developed such that the initial turbulent plasma reverted to a quiescent plasma. This quiescent plasma was characterized by a much lower level of plasma fluctuations and a factor of 2 increase in plasma electron temperature. In addition, the plasma resistivity decreased, which implied a lower impurity content. Later, the change from an unstable to a stable state was found to be the result of the plasma generating a reversed toroidal field at the plasma edge. This phenomena has been termed "self-reversal" in contrast to the programmed reversal used in present-day RFP facilities.

Several small ($a < 10$ cm) RFP experiments have advanced the state of knowledge of pinch discharges over the past decade. These include the HBTX-1 at Culham, ZT-1 and ZT-2 at Los Alamos, Eta-Beta 1 at Padua, and ET-TPEI in Tokyo. In these various experiments the plasma electron temperature was limited by impurity radiation. Intermediate size ($a > 10$ cm) reversed-field pinches are currently in operation at Los Alamos (ZT-40), Padua (Eta-Beta II), and Culham (HBTX-1A). A variation of the reversed-pinch configuration is embodied in the OHTE experiment at the General Atomic Laboratory. A helical winding is installed around the torus in order to obtain a more controllable rotational transform. With the possibility of high β the RFP plasmas are attractive as a fusion reactor candidate.

4. *Elmo Bumpy Torus*

One solution to plasma loss out the ends of magnetic mirrors is to join together several simple mirrors to form a torus. Single particles leaking out one mirror enter the adjoining mirror and under ideal conditions would be lost only by diffusing radially across the magnetic field lines. However, such a geometrical configuration is unstable to simple interchange-type perturbations. In the late 1960s a series of experiments with electron cyclotron heating a simple mirror showed that a hot-electron annulus could be formed in the mirror midplane. The electron rings were stable with an electron temperature of ~1 MeV and a plasma β of order 0.5. Such a ring produced a minimum B field or magnetic well whose positive magnetic field gradient provided the conditions necessary for a stable plasma.

The next step in the series of developments was to experimentally verify that a stable ring and plasma could be formed in a canted mirror. With the mirror coils canted at an angle of 15°, a stable ring was formed, indicating that a multiple-ring, mirror plasma would be stable in a torus configuration. Dandl and Guest (1981) have reviewed the theory and development of the EBT concept and experiments. An EBT-1 bumpy torus facility was constructed at Oak Ridge by joining 24 mirror coils to form a torus. The maximum magnetic field was 1.3 T with an axial mirror ratio of 2. To heat and confine the hot-electron ring, each mirror volume was enclosed by a microwave cavity fed by a microwave source with a frequency of 10.6 GHz corresponding to the resonant electron cyclotron frequency. The toroidal plasma component was heated by electron collisions in the ring and also by off-resonant RF power at 18 GHz. The resulting plasma was stable at a density of $\sim 10^{12}$ cm^{-3} and an electron temperature of 100–300 eV. EBT-1 has been upgraded into EBT-S by operating at higher field strengths and using 200 kW of 28-GHz RF power for plasma heating. EBT reactors produce steady-state plasmas; they are the only large plasma devices with this characteristic. A second bumpy torus, NBT, has been fabricated and experiments are being conducted at Nagoya, Japan.

B. Open-Ended Geometries

1. *Magnetic Mirrors*

Experiments in plasma confinement using open-ended magnetic mirror geometries were initiated in the early 1950s. Comprehensive reviews of both theoretical and experimental developments have been written by Baldwin and Logan (1977), Fowler (1981), and Post (1981). In the simple mirror trap two magnetic coils are separated by a distance of at least the coil diameter. The field configuration is as shown in Fig. 5 where the central or midplane field B_C is less than the field in the coil throat B_M. As the plasma electrons and ions spiral along the field lines from the weaker central field toward the stronger field, the gyroradius becomes smaller. Since the magnetic moment

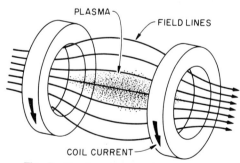

Fig. 5. Simple magnetic mirror geometry.

1. Introduction

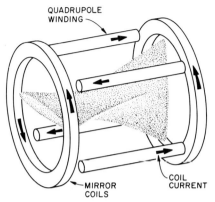

Fig. 6. Simple mirror with the addition of Ioffe bars.

μ, defined as $mv_\perp^2/B$, is constant or adiabatically invariant, the perpendicular velocity $v_\perp$ increases while the parallel velocity decreases. At some point all the parallel velocity will be changed into $v_\perp$, and the particle will be reflected. One can show that the smallest angle that a confined particle makes with the magnetic axis is given by

$$\sin^2 \Theta_m = B_C/B_M = 1/R_M,$$

where R_M is termed the mirror ratio. Depending on the mirror ratio Θ_m defines a loss cone, and all particles that make an angle less than Θ_m with the axis will be lost out the ends. Examples of early simple mirror experiments are the DCX in Oak Ridge, OGRA in the U.S.S.R., and Table Top I at Livermore.

Geometries of the simple mirror type are inherently unstable. At all points between the coils the magnetic field lines are convex. As the plasma pressure increases there is a tendency for the field lines to bow out or stretch until the plasma is lost radially to the walls. Also, the negative outward gradient in the magnetic field may give rise to the so-called flute or interchange instability. If the plasma is displaced slightly from the magnetic axis, ions and electrons will drift in the opposite direction, leading to a charge separation. The resulting electric field when crossed with the B field causes the plasma to drift outward to the wall in the case of a simple mirror field with negative gradient.

The radial loss of plasma through instabilities has been partially eliminated by superimposing a quadrapole or higher multipole field on the symmetric mirror field. In the PR-6 simple mirror experiment at Moscow a set of Ioffe bars or conductors were placed parallel to the magnetic axis and azimuthally symmetric about the plasma (Baiborodov *et al.*, 1963) as shown in Fig. 6. The current in adjacent bars flows in opposite directions. A field arrangement of this type results in a positive gradient in the radial direction

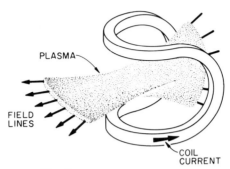

Fig. 7. Baseball minimum-B coil.

and is known as a minimum B field. Energizing the Ioffe bars in the PR-6 experiment increased the particle containment time by a factor of 5.

Combining the simple mirror coils with the Ioffe bars into a single coil configuration resulted in the baseball coil as shown in Fig. 7. (The terminology derives from the fact that the geometric shape of the coil is the same as that of the seams of a baseball.) In attempts to better utilize the minimum B field yin–yang coils were invented and are shown schematically in Fig. 8. The advantage of these coils over the baseball coil is that the coils can be individually adjusted. Experiments in which minimum B field geometries have been used to increase the plasma containment time and suppress instabilities include Phoenix II and MTSE II at Culham; PR-6 and 7 at Moscow; Decca II at Fontenay; and ALICE, Table Top II, Toy Top, and 2X at Livermore. For all the experiments, in the absence of instabilities, the particle confinement is limited by Coulomb collisions which scatter the ions and electrons into the loss cone. One of the key problems in the minimum-B geometry is the elimination of the end losses.

To decrease unacceptable end losses the concept of the tandem mirror has evolved over the past ten years. In a mirror-contained plasma the elec-

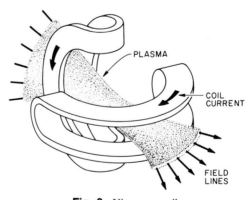

Fig. 8. Yin–yang coil.

1. Introduction

trons have a much higher velocity than the plasma ions. Thus, electrons will diffuse out of the confinement region faster, resulting in a net positive space potential termed the ambipolar potential. In the ambipolar tandem mirror a long central solenoid plasma is contained at the ends by baseball winding end plugs as shown in Fig. 9 (Fowler and Logan, 1977). The basic idea is to create a plasma in the end plugs whose density n_p is much greater than the solenoid density n_c, or central cell plasma density. By virtue of the greater end plug density the ambipolar potential will be higher than that of the central cell. As ions diffuse out to the end plugs they encounter an increased positive potential and are electrostatically reflected. A typical tandem mirror geometry and plasma are shown in Fig. 9. By injecting H^0 currents of 500 A into the end plugs of the Livermore tandem TMX experiment, an electrostatic potential difference of 300 V was obtained. In the same experiment the particle confinement time was increased by an order of magnitude over that found without beam injection into the end plugs. In mirror devices impurity problems are less severe than in toroidal devices. There are two main reasons for this happy state of affairs: (1) impurity ions with the same energy as hydrogen ions scatter into the loss cone much faster, and (2) the addition of an ambipolar plasma potential readily scatters out the low-energy impurity

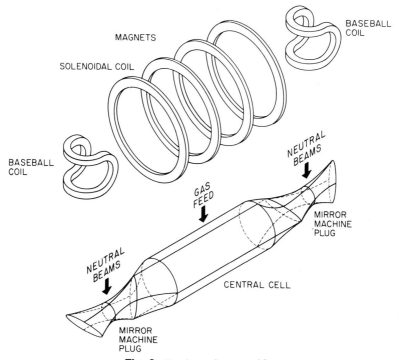

Fig. 9. Tandem mirror machine.

ions. In the TMX experiment the impurity content was low, with only 10% of the plasma energy being lost by impurity radiation. Currently two facilities have tested and proved the theory of increased plasma containment by ambipolar potentials—the TMX as described and the GAMMA-6 experiment located at Tsukuba, Japan.

In the end plugs the ambipolar potential not only depends on ion density but also increases with the electron temperature. In the tandem mirror geometry depicted in Fig. 9, electrons flow freely between the central cell and the end plug with only modest potential differences possible between the two regions. Ideally, the solution would be to insulate the end plugs from the central cell. Hence, the thermal barrier concept has been introduced, which involves the addition of a strong mirror field and a weak field region between the central cell and end plug (Baldwin and Logan, 1977). High end plug ambipolar potentials would be created by heating the electrons with RF, ECRH. Also, intense neutral beams are injected to maintain the required high density. With high electron temperatures n_p can be reduced by scaling down the energy and intensity of the neutral beam. If a way can be found to "pump out" or decrease the plasma density in the mirror region, the plasma potential will decrease, thereby trapping the electrons and thermally isolating the end plugs from the central cell. To test the thermal barrier concept the TMX-U (Livermore), GAMMA-10 (Tsukuba), and TARA (MIT) machines are being fabricated and put into operation. In the TMX-U facility the thermal barrier is produced by injecting neutral beams at 45° to the magnetic axis. The trapped ions reflect or "slosh" between the B field maximum and cause a density decrease in the midplane. To prevent slow positive ions from filling in the density or potential hole, neutral beams are injected at 15° to the axis. Slow ions in this region are lost by charge exchange collisions, with the resulting energetic positive ions from the collisions being readily lost out the mirror ends. Innovative and ingenious ideas have been and will continue to be used in the magnetic mirror confinement program to eliminate or minimize end losses without sacrificing plasma stability.

Brief mention should be made of efforts to minimize mirror end losses using the field-reversed mirror concept (Fowler and Post, 1977). Containment in the axial direction is obtained by the magnetic field induced by current flow in the plasma. In the early 1960s attempts were made in the Astron experiment at Livermore to inject intense electron currents into the plasma setting up the field reversal. Currents can also be established by injecting neutral particles and trapping the atoms by collisions with a target plasma. The trapped ions circulate in a ring and generate a self-magnetic field. If this self-field can be made larger than the mirror field, the resulting field is reversed and a volume is created in which field lines close upon themselves to form an elliptical torus. Field reversal was attempted in the 2XIIB experiment at Livermore without success. This facility has been converted to the Beta II experiment to verify the reversed-field principle.

1. Introduction

2. *Theta Pinches*

A schematic diagram of a θ-pinch is shown in Fig. 10. A capacitor bank is discharged into a single-turn coil which generates a rapidly rising longitudinal magnetic field to drive the plasma column inward and form the pinched plasma. The theta-pinch plasma formed is stable with densities as high as 10^{17} cm^{-3}. Temperatures of several kilo-electron-volts have been obtained by initial shock heating followed by compression heating. However, the plasma containment time has been only a few microseconds, determined principally by end losses. Although various schemes have been devised to eliminate end losses—increasing the cell length to several meters, putting the pinch in a mirror or multiple mirror field, RF stoppering, electrostatic trapping, using end plugs of gas or metal walls, or forming the linear pinch into a torus—an insurmountable problem arises from the large collisional scattering rate at densities of 10^{16}–10^{17} cm^{-3}. At these densities most methods are inefficient.

Early experiments with theta pinches were conducted on the SCYLLA series of long theta pinches at Los Alamos. Although stable, high-density,

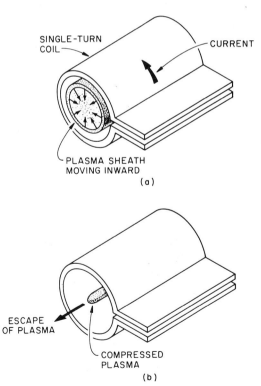

Fig. 10. Theta-pinch experiment: (a) plasma cross section at the beginning of the compression stage, (b) cigar-shaped pinched plasma.

high temperature plasmas could be formed, the containment time was always low. Other linear theta-pinch research has been conducted in laboratories throughout the world.

The most ambitious attempt to develop a theta pinch in a toroidal configuration was the SCYLLAC experiment at Los Alamos. However, a toroidal instabilities prevent the attainment of a high temperature, stable plasma. Currently, the theta pinch is not highly regarded as a reactor concept.

C. Inertial Confinement

Basic principles involved in the physics of forming a high temperature plasma by inertial confinement are discussed in Chapters 2 and 8 of this volume. Only a cursory account of inertial confinement will be presented in this section to acquaint the reader with a few of the problems and the status of research in this field. Several reviews have been written on the technological developments in laser drivers (Holzrichter et al., 1982) and pulsed-ion-beam drivers (Humphries, 1980). The status of inertial confinement research as of 1982 has been summarized by Seigel (1982).

In the inertial confinement method of producing fusionlike plasmas a pellet of D–T is compressed to a high density by ablating the outer surface. Usually the design of the pellet consists of an outer layer of plastic or glass, an intermediate layer of a heavy element, and a central core of the D–T fuel. When high-power beams are incident on the spherical pellet, the outer surface is heated. Ablation occurs, which drives the intermediate layer inward and compresses the pellet to densities of 10^{25}–10^{26} cm^{-3}. After compression the D–T reaction time as determined by the compressed plasma diameter, temperature, and density must be greater than the time for the plasma to dissociate.

Careful attention must be paid to how the energy (in the form of a laser, ion, or electron beam) is coupled to the plasma. The spherical pellet must be uniformly illuminated, which requires many beams impacting from various directions. As in magnetic confinement, instabilities can develop as the pellet is being compressed. These instabilities accelerate plasma ions and electrons to high energies, resulting in preheating of the D–T fuel. Preheating causes the D–T volume to expand, negating the compression of the fuel. In laser beam inertial confinement research, preheat has been minimized by controlling the wavelength of the laser, tailoring the shape of the laser pulse in time, and changing the structure of the pellet outer shells. During the past decade a major fraction of research has gone into developing the technology of the power drivers. Intense focused laser, ion, and electron beams have been developed during this period. According to estimates, power levels of 10^{14} W (3–5 MJ in 10^{-8} s) are required to produce net power.

Most of the laser drivers have been either Nd–glass lasers (1.05 μm) or

1. Introduction

CO_2 gas lasers (10.6 μm). The first lasers at Livermore were the Argus and Shiva Nd–glass laser systems. The Shiva system consisted of 20 beams with an output power of 10–20 kJ. This was followed by the Nova Nd–glass laser consisting of 40 beams with 200–300 kJ energy and power levels $>2 \times 10^{14}$ W. Experimental data indicated that the power absorption by the target pellet increased as the laser wavelength decreased. Thus, the 2-beam Novette laser system was constructed with frequency doubling to operate at a wavelength of 0.53 μm.

At Los Alamos emphasis has been placed on developing high-power, pulsed CO_2 lasers. The first laser was the 8-beam Helois, followed by the 72-beam Antares 100-kJ system. In the early 1980s the 6-beam Nd–glass Beta laser system was put into service at the University of Rochester. This system was expanded into the 24-beam Omega system. In Japan Nd–glass lasers are being used at the University of Osaka, which started with the Gekko-4 system operating at a wavelength of either 1.05 or 0.53 μm. Other lasers include the GM-II and the 8-beam Lekko-8. Their most powerful laser at the present time is the 12-beam, 4×10^{13} W Gekko-12 system. High-power Nd–glass lasers are also being used at the Lebedev Institute in Moscow.

Several facilities have been built to investigate further the effect of wavelength on absorption. At the Garching Laboratory the Austerix III iodine laser (1.3 μm) is being operated with one beam at 2 kJ. The United Kingdom's Rutherford Laboratory is concentrating on the development of rare-gas halide lasers operating in the ultraviolet region.

Electron and ion beam drivers have been developed primarily in the United States at the Sandia Laboratory, Naval Research Laboratory, and Cornell University. In Europe the work has been concentrated at Karlsruhe, West Germany. Early research indicated that electron beam energy could be coupled to the pellet efficiently. However, since the range of electrons in matter is large, the electrons penetrated the pellet outer layer and preheated the D–T. One of the first experiments at Sandia was the electron Proto I facility in 1975. This experiment was later upgraded to the 12-beam Proto II source in 1977. Plans were made for a 36-beam EBFA machine, but before construction was completed the design was changed to the 36-light-ion-beam PBFA experiment capable of delivering 1 MA of H$^+$ with a power of 3×10^{13} W. Computations and designs have been made at Berkeley and Brookhaven for accelerating heavy ions (e.g., Hg^{n+}) by a linear accelerator or a synchrotron to very high energies before impacting on a pellet. So far, however, the cost of a prototype facility such as this has been prohibitive.

III. Atomic Physics in Fusion

A successful fusion program requires an understanding of the limitations and characteristics of present-day magnetically and inertially confined plasmas so that scaled-up future machines can be based on sound plasma dynamics. Analysis of plasma heating and cooling mechanisms and particle transport, as well as the application and development of diagnostic techniques, requires a large amount of atomic and molecular data. In essence the understanding of plasma behavior depends to a great extent on a knowledge of atomic processes.

With all the charge states of light and heavy impurity ions and atoms in the plasma, the number of cross sections or reaction rates required is enormous. New experimental techniques, apparatuses, and computational methods have been developed and implemented to satisfy these needs. At one time most of the demands for atomic data concerned only the total cross section, which could be measured relatively easily by projecting an ion or atom beam through a static gas and measuring either the beam attenuation or species formation. With the demand for electron ionization, excitation, and recombination cross sections or rates for multicharged ions, crossed beam-apparatuses had to be updated. Ion sources have been invented and fabricated to produce copious quantities of multicharged ions. Scattering chambers were upgraded to operate at higher vacuum conditions. To obtain information for higher charge states, high temperature plasmas as found in tokamaks and theta-pinch plasmas have been used, requiring detailed information on spatial and temporal distribution of the plasma properties. Optical spectroscopy has been extremely beneficial in providing information as to plasma properties. Use of high-power laser-produced plasmas, vacuum sparks, and accelerator beam–foil excitation has permitted high-resolution spectroscopy not only to identify allowed and forbidden transitions but also to measure transition probabilities or oscillator strengths of highly ionized ions.

During the past ten years a highly reliable data base has been accumulated for both atomic structure and collision cross sections or rates. Many lacunae still exist, but work is proceeding rapidly to provide a complete data base for impurity ions normally found in the plasma and for the multitude of problems that limit our understanding of the detailed behavior of hydrogen in magnetically and inertially confined plasmas. The objectives of this book are (1) to discuss atomic data needs in fusion research, and (2) to describe how the data are used in understanding high temperature plasma behavior. Chapters 2–7 are devoted to magnetically confined plasmas, and Chapter 8 discusses the atomic physics of inertial confinement fusion.

The fundamentals of high temperature plasma heating and confinement and the production of fusion energy are explored in Chapter 2. Chapter 3 emphasises the atomic properties of low-density, optically thin plasmas. The theory that has been developed for both equilibrium and nonequilibrium plasmas is summarized. Atomic binary collisions of individual processes are presented as they affect the total radiant power lost by the plasma. The dynamics of impurities and their effect on particle and energy balance is treated in Chapter 4.

1. Introduction

Diagnostics based on atomic physics plays a most important role in understanding plasma properties and dynamics. Throughout the history of fusion research resonance emission spectroscopy has been used to identify impurity ion species and their concentration. The application of optical diagnostics to determine the effect impurities have on energy balance and stability is discussed in Subchapter 5A from a basic atomic physics viewpoint. Thomson scattering from laser beams has established a quantitative standard in the measurement of electron temperature. The application of lasers to the all-important measurements of both electron and ion temperatures, electron density by interferometry, and magnetic field strengths is presented in Subchapter 5B. Measurements of spatial and temporal electron temperature by the nonperturbing electron cyclotron emission technique have confirmed laser scattering measurements and are discussed in Subchapter 5C. Particle diagnostics, covered in Subchapter 5D, have been instrumental in determining spatial plasma ion temperature and density, impurity ion density, and plasma potential, and work is proceeding on measuring the plasma current via the poloidal magnetic field producing Zeeman splitting in a transmitted beam of energetic ions or atoms. Finally, in Subchapter 5E the theoretical basis of bremsstrahlung radiation is established.

Both tokamak and mirror-confined plasmas rely at the present time on neutral-beam auxiliary heating. Chapter 6 discusses the atomic physics of the formation of positive and negative ions in sources, their neutralization, trapping, and thermalization. All magnetically confined fusion plasmas will depend on the 3.5-MeV α particles produced in the D–T reaction to sustain and heat the plasma once the ignition temperature is reached (Subchapter 6C).

During the past few years the importance of the edge plasma in controlling impurity influx has been realized and studied intensively. Also, the problem exists for the removal of He particles after thermalization of α particles (the "He-ash problem"). Chapter 7 describes the atomic physics processes occurring in the plasma boundary and the development of pump limiters and divertors as a means of impurity and He-ash control.

In Chapter 8 Weisheit develops the basic atomic physics of inertially confined plasmas whose densities are in the range 10^{22}–10^{26} cm^{-3}. In these dense plasmas the normal electron states of the ions are perturbed and the continuum is lowered. This property effects the interpretation of the optical (including x rays) diagnostics.

References

Anderson, D. T., Derr, J. A., and Shohet, J. L. (1981). *IEEE Trans. Plasma Sci.* **PS-9,** 212–220.
Artsimovich, L. A. (1972). *Nucl. Fusion* **12,** 215–252.
Artsimovich, L. A., Bobrovsky, G., Gorbunov, E. P., Ivanov, D. P., Kirillov, V. D., Kunwraoc, E. I., Mirnov, S. V., Petrov, M. P., Razumova, K. A., Strelkov, V. S., and Shekeglov,

D. A. (1969). *Plasma Phys. Controlled Nucl. Fusion Res., Proc. Int. Conf., 3rd, Novosibirsk, 1968* **1**, 157; Engl. transl.: *Nucl. Fusion* **17**, Suppl., 17–24 (1969).
Baiborodov, Y. I., Ioffe, M. S., Petrov, V. M., and Sobolev, R. I. (1963). *J. Nucl. Energy, Part C* **5**, 409–410.
Baker, D. A., and Quinn, W. E. (1981). *In* "Fusion" Vol. 1, Part A, (E. Teller, ed.), pp. 437–475. Academic Press, New York.
Baldwin, D. E., and Logan, B. G. (1977). *Rev. Mod. Phys.* **49**, 317–339.
Baldwin, D. E., and Logan, B. G. (1979). *Phys. Rev. Lett.* **43**, 1318–1321.
Bezbatchenko, A. L., Golovin, I. N., Kozlov, P. I., Strelkov, V. S., and Yavlinskii, N. A. (1960). *In* "Plasma Physics and the Problem of Controlled Thermonuclear Reactions" (M. A. Leontovich, ed.), Vol. 4, pp. 135–156. Pergamon, Oxford.
Bodin, H. A. B., and Keen, B. E. (1977). *Rep. Prog. Phys.* **40**, 1415–1565.
Butt, E. P., Carruthers, R., Mitchell, J. T. D., Pease, R. S., Thoneman, P. C., Bird, M. A., Blears, J., and Hartill, E. R. (1959). *In* "Plasma Physics and Thermonuclear Research" (C. Longmire, J. L. Tuck, and W. B. Thompson, eds.), Vol. 1, pp. 281–317. Pergamon, New York.
Christiansen, J. P., and Roberts, K. V. (1978). *Nucl. Fusion* **18**, 181–197.
Coor, T., Cunningham, S. P., Ellis, R. A., Heald, M. A., and Kranz, A. Z. (1958). *Phys. Fluids* **1**, 411–420.
Dandl, R. A., and Guest, G. E. (1981). *In* "Fusion" (E. Teller, ed.), Vol. 1, Part B, pp. 79–101. Academic Press, New York.
Energy Research and Development Administration (ERDA) (1974). "The 1974 Review of the CTR Research Program," ERDA Rep. No. 39, pp. 143–157. Washington, D.C.
Eubank, H. P., *et al.* (1979). *Plasma Phys. Controlled Nucl. Fusion Res., 7th, Innsbruck, 1978* **1**, 167–198.
Fowler, T. K. (1981). *In* "Fusion" (E. Teller, ed.), Vol. 1, Part A, pp. 291–333. Academic Press, New York.
Fowler, T. K., and Logan, B. G. (1977). *Comments Plasma Phys. Controlled Fusion* **2**, 167–172.
Fowler, T. K., and Post, R. F. (1977). *Fiz. Plazmy (Moscow)* **3**, 1408–1417; *Sov. J. Plasma Phys. (Engl. Transl.)* 787–797.
Furth, H. P. (1975). *Nucl. Fusion* **15**, 487–534.
Furth, H. P. (1981). *In* "Fusion" (E. Teller, ed.), Vol. 1, Part A, pp. 124–242. Academic Press, New York.
Grieger, G., Dove, W. F., Johnson, J. J., Lees, D. J., Politzer, P. A., Shohet, J. L., Wobig, H., and Rau, F. (1981). *Max-Planck-Inst. Plasmaphys. [Ber.] IPP* **IPP 2/254.**
Hauer, A., and Burns, E. J. T. (1982). *In* "Physics of Electronic and Atomic Collisions" (S. Datz, ed.), pp. 797–810. North-Holland Publ., Amsterdam.
Hinnov, E., and Mattioli, M. (1978). *Phys. Lett. A* **66A**, 109–111.
Holzrichter, J. F., Eimerl, D., George, E. V., Trenholme, J. B., Simmons, W. W., and Hunt, J. T. (1982). *J. Fusion Energy* **2**, 5–45.
Humphries, S., Jr. (1980). *Nucl. Fusion* **20**, 1549–1612.
Isler, R. C., Neidigh, R. V., and Cowan, R. D. (1977). *Phys. Lett. A* **63A**, 295–297.
IEEE (1981). *IEEE Trans. Plasma Sci.* **PS-10**, No. 4.
Mead, D. M. (1974). *Nucl. Fusion* **14**, 289–291.
Miyamoto, K. (1978). *Nucl. Fusion* **18**, 243–284.
Peacock, N. J., Robinson, D. C., Forrest, M. J., Wilcock, P. D., and Sannikov, V. V. (1969). *Nature (London)* **224**, 488–490.
Post, R. F. (1981). *In* "Fusion" (E. Teller, ed.), Vol. 1, Part A, pp. 357–435. Academic Press, New York.
Rostagni, G. (1981). *Nucl. Fusion* **21**, 1673–1682.
Seigel, R. (1982). *Nucl. Fusion* **22**, 665–669.

1. Introduction

Shohet, J. L. (1981). *In* "Fusion" (E. Teller, ed.), Vol. 1, Part A, pp. 243–289. Academic Press, New York.
Spitzer, L., Jr. (1952). *USAEC Rep.* No. 115, p. 12.
Spitzer, L., Jr. (1958). *Phys. Fluids* **1,** 253–264.
Tamm, I. E., and Sakharov, A. D. (1961). *In* "Plasma Physics and the Problem of Controlled Thermonuclear Reactions" (M. A. Leontovich, ed.), Vol. 1, pp. 1–47. Pergamon, Oxford.
Teller, E. (1959). *In* "Plasma Physics and Thermonuclear Research" (C. Longmire, J. L. Tuck, and W. B. Thompson, eds.), Vol. 1, pp. 55–56. Pergamon, New York.
Teller, E., ed. (1981). "Fusion," Vol. 1, Parts A and B. Academic Press, New York.
Young, K. M. (1974). *J. Nucl. Energy, Part C* **16,** 119–152.

2
Basic Concepts of Fusion Research

M. F. A. Harrison

Culham Laboratory
Abingdon, Oxfordshire
England

I.	Introduction	27
II.	The Principles of a D–T Fusion Reactor	28
III.	Magnetic Confinement of Fusion Plasmas	32
IV.	Energy Balance Conditions in a Magnetically Confined Fusion Plasma	36
V.	Auxiliary Heating	42
	A. Neutral-Beam Injection Heating	42
	B. Radio-Frequency Heating	44
VI.	Inertial Confinement	46
	References	49

I. Introduction

Research into the controlled release of energy from nuclear fusion covers many fields of pure and applied science as well as technology and engineering. This present brief account surveys the more salient features with the objective of providing a perspective against which the diverse role of atomic processes may be viewed. The discussion is general in nature and the reader is directed to more specialized literature for details and for references to individual contributions. General reviews such as Post (1956), Glasstone and Lovberg (1960), Rose and Clark (1961), and Artsimovich (1964) present details of the basic problems involved and describe some of the earlier approaches, whereas useful sources of information concerning more recent developments can be found in Kammash (1975), Hagler and Kristiansen (1977), Motz (1979), and Stacey (1981). A vivid appreciation of the present status of research and the breadth of practical problems involved can be gleaned from the report of the International Atomic Energy Agency for INTOR (International Tokamak Reactor, 1980, 1982), which sets out the requirements and objectives of the next likely step towards a fusion reactor.

However, the conceptual approaches to a fusion reactor are many and diverse and valuable surveys can be found in Ribe (1975, 1977).

In all cases the basic source of energy is nuclear and so is dependent upon reduction of mass in a system of two colliding nuclei (i.e., 931 MeV/amu). If the collision partners are light nuclei, then loss of mass occurs due to nuclear fusion, and potentially usable amounts of energy are released in collisions between the heavy isotopes of hydrogen and in reactions of protons or deuterons with isotopes of helium, lithium, and boron. However, the initial partners of all fusion reactions consist of two charged nuclei, and so the probability of reaction is low unless the colliding nuclei have sufficient kinetic energy to overcome the powerful force of their Coulomb repulsion. Considerations of attaining the necessary collision energies and of the availability of fuel have concentrated interest upon reactions involving deuterium. This isotope has a natural abundance of about one part in 6500 and it can be readily extracted from water. It has been conservatively estimated that sea water could provide sufficient deuterium to meet the predicted energy needs of the world for at least 10^9 yr; moreover, the reactor systems envisaged are unlikely to pose serious problems of hazardous waste.

The most significant reactions involving deuterium are

$$D + D \nearrow (T + 1.01 \text{ MeV}) + (p + 3.03 \text{ MeV}), \quad \text{(i)}$$
$$\searrow (^3He + 0.82 \text{ MeV}) + (n + 2.45 \text{ MeV}) \quad \text{(ii)}$$

$$D + {}^3He \rightarrow ({}^4He + 3.67 \text{ MeV}) + (p + 14.67 \text{ MeV}), \quad \text{(iii)}$$

$$D + T \rightarrow ({}^4He + 3.52 \text{ MeV}) + (n + 14.06 \text{ MeV}), \quad \text{(iv)}$$

and the D–D reactions occur with almost equal probability. Cross sections for these reactions are shown as a function of deuteron energy in Fig. 1 and it is apparent that the peak value for D–T is some 50 times greater than that of the total for D–D [i.e., (i) plus (ii)] and it occurs at about 100 keV, which is very much less than the corresponding energies for both D–D and D^3–He. The energy released in the D–T reaction (17.58 MeV) is about five times the energy released by either of the D–D processes, and the combination of these properties favors the use of a D–T mixture as the fuel for a fusion reactor despite the fact that tritium is radioactive and not available from natural sources. (The half-life of tritium is 12.4 yr and it emits β particles of 5.4 keV average energy.)

II. The Principles of a D–T Fusion Reactor

A tritium-burning reactor must incorporate a facility for breeding a sufficient supply of tritium and it is envisaged that this can be achieved by absorp-

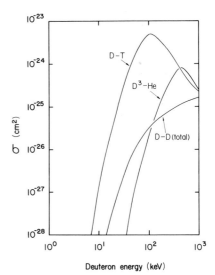

Fig. 1. Cross sections for deuterium fusion reactions plotted as a function of deuteron energy. The curve labeled D–D (total) shows the sum of reactions (i) and (ii) described in the text.

tion of the fusion neutrons within a blanket of lithium. Inelastic scattering of high-energy neutrons by the more abundant ^{7}Li isotope gives rise to the breeding reaction,

$$^7\text{Li} + n \rightarrow {}^4\text{He} + \text{T} + n - 2.47 \quad \text{MeV},$$

and slow neutrons can subsequently be captured by the reaction,

$$^6\text{Li} + n \rightarrow {}^4\text{He} + \text{T} + 4.8 \text{ MeV}.$$

Net yields somewhat in excess of one triton per each fusion neutron are predicted even when allowance is made for leakage of neutrons and for their absorption in the structure of the blanket. The overall breeding process is exothermic and provides energy in addition to that released by each D–T fusion collision so that the total yield of energy $Q_{\text{D-T}}$ per event can be taken as about 20.1 MeV.

The instantaneous rate at which fusion energy is released in a specific device is dependent upon the distribution of velocities among the fuel particles and also upon the distribution of density throughout the reaction region. Nevertheless, the basic concepts can be appreciated by considering a homogeneous and hot fuel mixture wherein $n_\text{D} = n_\text{T}$ and the distribution of particle velocities is Maxwellian. The rate K_F at which collisions occur within a unit volume of D–T fuel is then given by,

$$K_\text{F} = n_\text{D} n_\text{T} \langle \sigma v \rangle_{\text{D-T}} \quad \text{cm}^{-3} \text{ s}^{-1}, \tag{1}$$

where $\langle \sigma v \rangle_{\text{D-T}}$ is the rate coefficient, i.e., the product of cross section and collision velocity, averaged over the Maxwellian velocity distribution corresponding to temperature T. The rate coefficients for D–T are shown in Fig.

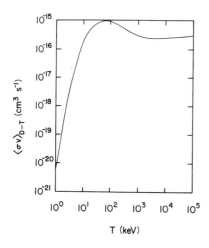

Fig. 2. Rate coefficient $\langle \sigma v \rangle_{\text{D-T}}$ of the D–T fusion reaction plotted as a function of temperature T.

2, and it is apparent that the rate of fusion reactions decreases dramatically when $T \lesssim 10^4$ eV.[†]

Fuel that is sufficiently energetic to initiate fusion must also be fully ionized because the collision rate for atomic ionization processes is typically $\sim 10^7$ times greater than that for D–T fusion. Each fusion event produces an He^{2+} ion with an energy $Q_\alpha = 3.5$ MeV, and these ions can dissipate their kinetic energy by elastic scattering due to Coulomb collisions within the ionized fuel (i.e., among the D^+ and T^+ ions together with their free electrons). This situation leads to the concept of α-particle "ignition" whereby energy lost from the energetic fuel particles to the walls of the reactor vessel is balanced by energy given to the fuel by the energetic α-particles. Fusion reactions then become self-sustaining and a reactor can "burn" continuously provided that sufficient D–T fuel is supplied and nonreactive ^4He is exhausted from the reaction region. Alternatively, the reaction might be terminated when a limited amount of fuel present in the reactor has burnt to such a degree that the reaction rate cannot compensate for energy losses; this gives rise to a pulsed burn cycle somewhat analogous to that of the internal combustion engine.

The nuclear cross sections associated with the stopping of neutrons within the fuel are negligibly small so that the fusion neutrons leave the fuel and can enter a separate region which houses the lithium breeder blanket. Absorption of the energy carried by fusion neutrons, together with the exothermic nature of the breeding process, raises the temperature of this blanket, whose heat output forms the main source of external energy from a D–T reactor.

[†] Throughout Chapter 2 the symbol T is used whenever kT is expressed in electron volts and in general Gaussian units are used for other parameters. The symbols c, e, k, etc., have their conventional meaning. The electron and ion masses are m_e and m_i and for D–T, $m_i \approx 2.5\, m_i$ (proton); Z is the charge state of the ion.

2. Basic Concepts of Fusion Research

The preceding principles are embodied in the reactor system that is illustrated in a highly simplified form in Fig. 3. A mixture of approximately equal quantities of deuterium and tritium is fed into the reaction vessel where sufficient energy is introduced for fusion reactions to take place. The reaction region is enclosed by a "first wall" that is permeable to neutrons but not fuel. Energy deposited in the blanket, which completely surrounds the reaction region, is extracted by a heat exchanger system and used to raise steam and thereby drive electrical generators. Not all of the fuel can be burnt, so an exhaust system must be provided to extract the fuel and separate the helium. The processed exhaust gas is then recirculated to the fuel input where it is joined by tritium that has been extracted from the blanket. The correct composition of fuel mixture is maintained by adding fresh deuterium.

It is evident that a preeminent issue must be the provision of a practical environment in which to house the very energetic fuel. An appreciation of the magnitude of this problem can be gained from the following assessment. The characteristic time τ_F for a fusion reaction is

$$\tau_F = (n_D + n_T)/n_D n_T \langle \sigma v \rangle_{D-T} \quad \text{s} \tag{2}$$

and, if the reactor is to be efficient, the energetic fuel particles must retain their energy for a time τ_E which is comparable to τ_F. Assume that this energy containment time can be crudely expressed as the average time for a fuel particle to travel to the first wall, namely,

$$\tau_E = r/v_t,$$

where r is a characteristic dimension of the reactor and $v_t = (8kT/\pi m_i)^{1/2}$ is the thermal velocity of the fuel. Thus $\tau_E \sim 10^{-6}$ s when 100 cm is taken as the dimension r and T is about 10^4 eV. Substitution of $\tau_E = \tau_F$ into Eq. (2) yields $n_D + n_T \approx 2 \times 10^{22}$ cm s^{-1}. If the fully ionized fuel is considered to be a hot

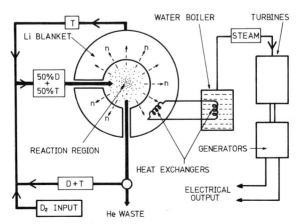

Fig. 3. Simplified concept of a D–T fusion reactor.

ideal gas comprising free electrons and free ions, then its pressure can be expressed as

$$p = n_e k T_e + n_i k T_i = 2nkT, \qquad (3)$$

where $n = n_e = n_i = n_D + n_T$ are the number densities of electrons and ions and the corresponding temperatures are T_e and T_i. When $T_e = T_i = T$, the pressure is $2nkT \approx 6 \times 10^{14}$ dyn cm^{-2} (i.e., 6×10^8 atm). This exceeds by many orders of magnitude the maximum steady state pressure that can be sustained by a practical structure. Even if such pressures could be contained, the heat flux Γ_t carried to the first wall by the impact of energetic D$^+$ and T$^+$ ions traveling with thermal velocity, is

$$\Gamma_t \sim \tfrac{1}{4}(n_D + n_T) v_t \tfrac{3}{2} kT \sim 10^{15} \quad \text{W cm}^{-2},$$

and this would rapidly cause the wall to disintegrate because acceptable steady state power loadings lie in the range 10^2–10^3 W cm^{-2}.

The wall is bombarded by fuel particles, α-particles, neutrons, and radiation and its ability to survive greatly influences the volume and lifetime of the device and hence the economics of fusion power. Moreover, erosion of the wall introduces impurity elements into the D–T fuel and thereby gives rise to energy losses due to atomic collision processes. Solutions to these problems are sought by invoking "confinement" mechanisms that reduce the physical interaction between the hot fuel and the reactor vessel. Two basic approaches are envisaged, "magnetic confinement" and "inertial confinement." In the first approach the transport of charged fuel particles towards the wall is impeded by imposing a magnetic field so oriented that the lines of force lie parallel to the wall and, as a consequence, the confinement time τ_E is enhanced by several orders. In the second, fusion energy is released by imploding a small fuel pellet that remains at a substantial distance from the wall throughout its brief burn time.

III. Magnetic Confinement of Fusion Plasmas

The relatively long containment time enables a magnetic confinement device to operate in a low-density regime, $\sim 10^{14}$ cm^{-3}, where fuel pressure lies in the range 10–50 atm and can therefore be sustained in steady state. In this regime the fully ionized fuel is a "plasma" and its motion is dominated by collective electromagnetic interactions. A plasma can be defined as being electrically neutral overall, so on average $n_e = n_i = n$ (where n is the plasma density). Another characteristic is that the mobile electrons screen the Coulomb field of the less mobile plasma ions, and the distance λ_D at which this field is reduced to a negligible value must be smaller than the dimensions of the ionized gas. This Debye shielding length is expressed as

2. Basic Concepts of Fusion Research

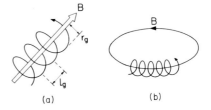

Fig. 4. (a) Trajectory of a charged particle gyrating around a magnetic field line. The gyroradius r_g and pitch l_g are described in the text. (b) Gyration around a closed field line.

$$\lambda_D = (kT_e/4\pi n_e e^2)^{1/2} = 7.43 \times 10^2 (T_e/n_e)^{1/2} \text{ cm}, \quad (4)$$

and for typical fusion conditions $\lambda_D \ll 1$ cm.

Consider a plasma formed by ionization of neutral atoms or molecules within a region of uniform magnetic field of strength B. A particle of mass m and unit electronic charge e moving normal to the field lines is constrained to move along a circular trajectory whose radius r_g (i.e., the gyromagnetic radius) is

$$r_g = mv_\perp c/eB. \quad (5)$$

Here $v_\perp$ is the particle velocity perpendicular to the field. Charged particles within a plasma have a random distribution of velocities, so movement perpendicular to the direction of the field is constrained but motion parallel to the field is not affected. The trajectories therefore form spirals of radius r_g (and pitch $l_g = v_\parallel 2\pi/\omega_c$ where $v_\parallel$ is the parallel velocity) that gyrate around the field lines as shown in Fig. 4a. Here ω_c is the gyromagnetic or cyclotron frequency,

$$\omega_c = eB/mc. \quad (6)$$

To achieve plasma confinement it is therefore necessary either to impede this free flow parallel to the field or else to bend the field lines back upon themselves so that they become "closed" and thereby allow particles to gyrate around the resultant ring of magnetic field in a manner illustrated in Fig. 4b.

One approach to confinement based on the linear concept is that of magnetic mirrors. Here regions of increased magnetic field are located at each end of the device and the localized increase in magnetic field reflects both ions and electrons by the principle of conservation of the magnetic moment of the gyrating particles. However, some "end losses" of particles with $v_\parallel \gg v_\perp$ are unavoidable and the energy containment time tends to be short. Consequently most experiments and reactor concepts are based upon configurations of closed, or effectively closed, lines of force and such devices are toroidal in form (i.e., shaped like a doughnut).†

In effect, the magnetic field exerts a pressure which can be shown to be equivalent to the energy density $B^2/8\pi$ of the field and so the maximum

---

† Magnetic confinement devices are surveyed in Barnett (Chapter 1, this volume).

plasma pressure that can be confined by a field of intensity B_0 (external to the plasma) can be expressed as

$$n_i k T_i + n_e k T_e = 2nkT = B_0^2/8\pi, \tag{7}$$

where it is assumed that $T_e = T_i = T$. However, the field may not be utilized effectively, and the parameter

$$\beta = \frac{2nkT}{B_0^2/8\pi} \tag{8}$$

is customarily employed to define its confinement efficiency. The magnitude of β depends upon details of both the field and confined plasma and maximum values are predicted to lie in the range 5–30%. The majority of present-day reactor concepts (i.e., the tokamak approach) invoke plasma parameters that are typically $\beta \approx 0.05$, $n = 2 \times 10^{14}$ cm^{-3}, and T $= 10^4$ eV; Eqs. (7) and (8) therefore indicate that such a plasma could be confined when $B_0 \approx 40$ kG.

The characteristic fusion reaction time τ_F for the above example is about 100 s and during this period the free electrons and ions within the plasma suffer many Coulomb scattering collisions. Such scattering is not dominated by single events but by the accumulation of a large number of collisions which eventually cause scattering through an angle equal to 90° where it can be considered that the initial energy of the incident particle has been shared with its numerous collision partners. Scattering through 90° in the presence of a magnetic field also implies that the incident particle has moved transversely across the field by a distance $\Delta r \sim r_g$. Within a hydrogenic plasma (i.e., $Z = 1$) the frequencies of electron–electron collisions ν_{ee} and electron–ion collisions ν_{ei} are comparable in magnitude, and Spitzer (1962) shows that ν_{ee} for a Maxwellian distribution of velocities is given by

$$\nu_{ee} = \frac{4(2)^{1/2} \pi n_i Z^2 e^4 \ln \Lambda}{m_e^{1/2}(3kT_e)^{3/2}} = \frac{3.0 \times 10^{-6} n_i Z^2 \ln \Lambda}{T_e^{3/2}} \quad \text{s}^{-1}. \tag{9}$$

Here the "Coulomb logarithm" $\ln \Lambda$ is determined by

$$\Lambda = \lambda_D/b_0,$$

where b_0 is the smallest impact parameter at which small-angle scattering can occur. The choice of b_0 is not critical and $\ln \Lambda \approx 10$ to 20 for plasmas of interest here. For the preceding fusion plasma $\nu_{ee} = 6 \times 10^3$ s^{-1} and so the rate for Coulomb scattering collisions is $\sim 10^6$ times that of fusion collisions.

Scattering causes diffusion of particles across the magnetic field and thereby sets up a gradient in density dn/dr in the direction r normal to the magnetic axis. The "classical" diffusion coefficient can be expressed as

$$D_\perp = \Delta r^2/\Delta t,$$

where the mean time step Δt can be approximated to $\Delta t \approx \nu_{ei}^{-1}$ so that

$$D_\perp \sim r_g^2 \nu_{ei} \quad \text{cm}^2 \text{ s}^{-1}. \tag{10}$$

2. Basic Concepts of Fusion Research

Since r_g is greatest for the ions and $\nu_{ei} \approx \nu_{ee}$ it follows that ions tend to diffuse at a faster rate than the electrons. However, when allowance is made for the need to maintain ambipolar conditions, e.g., in which electron and ion diffusion rates are coupled, it is found, for the case in which $T_i \sim T_e$, that

$$D_{\text{amb}\perp} \approx 2D_{e\perp}.$$

The plasma flows across the magnetic field with a velocity

$$v_{p\perp} = \frac{D_\perp}{n}\frac{dn}{dr},$$

so that the particle confinement time in a closed-line toroidal device can be expressed in the form $\tau_n \sim r_w/v_{p\perp}$ where r_w can be approximated to the distance of the wall from the magnetic axis. Assuming that the gradient $dn/dr \sim n/r_w$ results in the expression

$$\tau_{n(\text{classical})} \sim r_w^2/D_\perp,$$

and $\tau_{n(\text{classical})} \sim 10$ s when the preceding plasma parameters are assumed.

Diffusion in a practical device is far more complex;[†] the magnetic field is neither uniform nor axisymmetric with respect to r_w. The plasma is not quiescent but subject to many forms of instabilities that enhance transport across the magnetic field. Gradients exist in temperature as well as density so that thermal conduction, particularly that due to anomalous conduction by electrons, causes losses that exceed the convection of energy due to the diffusion of energetic particles and so the energy containment time τ_E is less than τ_n. The gradients in n and T are not linear but are affected by boundary conditions that involve atomic and surface processes such as recycling of plasma particles and the release of impurity elements. Present predictions of the energy confinement time for reactor-sized toroidal devices are typically a few seconds although considerable uncertainty exists owing to the complex and interactive nature of the processes that govern both particle and energy confinement.

Electron motion parallel to the magnetic field is impeded only by scattering collisions and so the plasma is an excellent electrical conductor; its electrical resistivity can be expressed in the form

$$\eta \approx c^2 m_e n_i \nu_{ei}/e^2 n_e,$$

and Spitzer and Harm (1953) quote, for a plasma where $Z = 1$,

$$\eta = [(5.24 \times 10^{-3})/T_e^{3/2}] \ln \Lambda \quad \Omega \text{ cm}. \tag{11}$$

The plasma can be heated when a current is induced to flow in the direction of the confining field, and this method of "ohmic" heating is a powerful

---

[†] Plasma transport across the magnetic field of a tokamak is discussed in Hogan (Chapter 4, this volume).

adjunct to magnetic confinement. However, $\eta \propto T^{-3/2}$, and if the plasma current density j is constant then the ohmic heating power density

$$P_\Omega = j^2 \eta \tag{12}$$

reduces with increasing plasma temperature.[†] Current flowing in the plasma generates a magnetic field and so, depending particularly upon the topography of the external field, some constraints must be imposed upon j. These in turn limit the temperature at which ohmic heating is effective. The maximum effective temperature for many toroidal devices is about 1 keV, which is below that required for fusion; therefore such systems need some form of "auxiliary heating."

IV. Energy Balance Conditions in a Magnetically Confined Fusion Plasma

The effective power density generated by fusion reactions in a homogeneous D–T plasma wherein $Z = 1$, $n_D = n_T = \tfrac{1}{2}n_i = \tfrac{1}{2}n_e$, and $T = T_e = T_i$ can be expressed as

$$P_F(T_i) = \tfrac{1}{4}n_i^2 \langle \sigma v \rangle_{D-T} Q_{D-T} = 8.05 \times 10^{-13} n_i^2 \langle \sigma v \rangle_{D-T} \quad \text{W cm}^{-3}, \tag{13}$$

where $Q_{D-T} \approx 20.1$ MeV per event is typical of the combined yield of the plasma and breeding blanket. The corresponding power density distributed by Coulomb scattering of 3.5-MeV α-particles[‡] with electrons and ions of the plasma is

$$P_\alpha(T_i) = \tfrac{1}{4}n_i^2 \langle \sigma v \rangle_{D-T} Q_\alpha = 1.41 \times 10^{-13} n_i^2 \langle \sigma v \rangle_{D-T} \quad \text{W cm}^{-3}. \tag{14}$$

It is evident that these power sources must exceed all other forms of energy loss if a fusion reactor is to provide a net gain in energy. Coulomb scattering in electron–electron, electron–ion, and ion–ion collisions causes energy losses due to transport of plasma particles and energy across the field (i.e., convection and conduction) but free–free collisions between electrons and ions also give rise to bremsstrahlung radiation which is emitted during acceleration of the electron in the unscreened field of an ion. For a plasma comprising fully stripped ions the power density radiated over all wavelengths is given by

$$P_{br}(T_e) = 1.69 \times 10^{-32} n_e T_e^{1/2} \sum (n_z Z^2) \quad \text{W cm}^{-3}, \tag{15}$$

where n_z is the ion density in each fully stripped charge state Z. Electromagnetic radiation of frequency ω greater than a critical value ω_x can pass

---

[†] The influence of atomic processes upon η and the consequent effects upon plasma properties are considered in Hogan (Chapter 4, Section II).

[‡] Details of α-particle heating are presented in Post (Subchapter 6D, Section I).

2. Basic Concepts of Fusion Research

through a plasma, and for the present discussion ω_x can be regarded as the electron plasma frequency

$$\omega_{pe} = (4\pi n_e e^2/m_e)^{1/2}, \tag{16}$$

so that a cutoff wavelength can be expressed as

$$\lambda_x \approx 3.4 \times 10^6 n_e^{-1/2} \text{ cm}. \tag{17}$$

The value of λ_x is about 3 mm when $n_e \sim 10^{14}$ cm^{-3}, so that the plasma is transparent to visible and shorter-wavelength radiation. Bremsstrahlung radiation is emitted by the hot plasma in the wavelength range below 10 Å and it therefore gives rise to unavoidable loss of energy to the walls of the containment vessel.

Charged particles also emit cyclotron (or synchrotron) radiation due to the centripetal acceleration associated with their gyration around the magnetic field lines. Ion velocities are relatively low and so ion cyclotron radiation can be neglected, but cyclotron radiation by electrons can be significant. It is not possible to give a precise estimate of the power loss in simple terms[†] because (i) the radiation is distributed over a range of harmonics of the fundamental frequency ω_{ce} which lie in the infrared and μ-wave regions, so that the fundamental and lower harmonics may be adsorbed within the plasma; (ii) the walls of the vessel can be powerful reflectors; (iii) the radiation is polarized and also emitted into a narrow cone in the direction of the electron motion; (iv) the magnetic field is not uniform throughout the plasma. Nevertheless, an indication of the effects of electron cyclotron radiation can be gained from the following description: the power radiated in a unit volume of homogeneous plasma by electrons with a Maxwellian distribution of velocities can be expressed as

$$P_{cy} \approx (4e^4 k/3 m_e^3 c^5) B^2 n_e T_e,$$

and substitution for B^2 from Eq. (8) yields, under conditions in which $n_e = n_i$ and $T_e = T_i$,

$$P_{cy} \approx 5 \times 10^{-38}[(1 - Y_{cy})(1 - R_w)/\beta] n_e^2 T_e^2 \text{ W cm}^{-3}. \tag{18}$$

Here, Y_{cy} and R_w are respectively coefficients that describe absorption in the plasma and reflection at the wall. Comparison with Eq. (15) shows that losses due to cyclotron radiation will probably exceed those due to bremsstrahlung at high temperature.

The balance of power between fusion and radiative processes can be seen from the power density functions $P_\alpha(T)/n^2$, $P_F(T)/n^2$, $P_{br}(T)/n^2$, and $P_{cy}(T)/n^2$ that are plotted as a function of temperature in Fig. 5.[‡] It is apparent that

---

[†] The theory of electron cyclotron emission is discussed in Boyd (Subchapter 5C, Section II).

[‡] It is arbitrarily assumed in Fig. 5 that $\beta = 0.05$, $Y_{cy} = 0.9$, and $R_w = 0.9$.

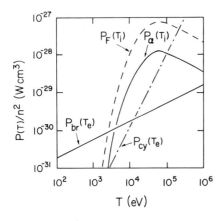

Fig. 5. Power density functions for a pure D–T fusion plasma plotted against plasma temperature T. The functions $P_F(T_i)/n^2$ and $P_\alpha(T_i)/n^2$ are given by Eqs. (13) and (14), respectively. $P_{br}(T_e)/n^2$ is given by Eq. (15) and $P_{cy}(T_e)/n^2$ is determined from Eq. (18), in which it is arbitrarily assumed that $\beta = 5 \times 10^{-2}$, $R_w = 0.9$, and $Y_{cy} = 0.9$.

the plasma temperature must be in excess of about 5 keV before energy deposited by α-particles exceeds that lost by bremsstrahlung radiation. This temperature is indicative of ignition conditions for a pure D–T plasma but in practice the losses due to plasma transport of energy must also be taken into account. After ignition, the plasma would, in principle, rise in temperature until $P_\alpha(T)/n^2 \rightarrow P_{cy}(T)/n^2$ but such excursion must be limited to the regime within which plasma pressure can be opposed by the confining magnetic field, and the range 10–15 keV is characteristic of present objectives (it should, however, be noted that cyclotron radiation may be an impediment to the longer-term attainment of controlled D–D reactions).

These idealized conditions are appreciably changed if the plasma contains impurity elements; not only do these enhance bremsstrahlung power losses, which are dependent upon Z^2, but such ions may not be fully stripped and the consequent presence of bound electrons gives rise to atomic line-radiation losses.

A magnetically confined fusion plasma is optically thin to the most significant components of atomic radiation from impurity ions, and moreover the radiation generally arises from atomic levels that decay back to the ground state of the ion during the time interval between electron–ion collisions. Therefore the plasma is not in thermodynamic equilibrium but, in the present simple context,[†] it is reasonable to assume that the population of ion charge states and the population of excited levels of each charge state are in equilibrium with the associated electron–ion collision rates. Models based upon this assumption of "local thermal equilibrium" (sometimes referred to as "coronal equilibrium") have been used to determine the temperature dependence of the power losses due to line radiation, two-body recombination,

---

[†] Atomic radiation losses from a magnetically confined plasma are considered in McWhirter and Summers (Chapter 3, this volume) where limitations of the present simplified treatment are discussed in detail.

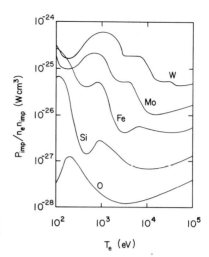

Fig. 6. Radiated power density function for common impurity elements plotted against electron temperature T_e. The function $P_{imp}/n_e n_{imp}$ is taken from data of Jensen et al. (1977), which are based upon a coronal equilibrium model.

and dielectronic recombination.[†] The total radiated power density function $P_{imp}(T_e) n_e n_{imp}$ from one such calculation (Jensen et al., 1977) is shown for some typical impurity elements in Fig. 6. In general the total radiative loss peaks at about 10^2 eV (due to a maximum in the balance between the ion charge state, the electron excitation rate, and the photon energy). At higher temperatures the number of bound electrons is reduced and there is a decrease in the power loss from line radiation; eventually bremsstrahlung becomes the dominant source of radiation and the $Z^2 T^{1/2}$ dependence is evident. The total radiative power losses increase with increasing atomic number of the impurity element, and moreover ions of the heavier elements are less readily stripped of their electrons so that the dominance of line radiation is maintained throughout the temperature regime of present interest.

The significance of small concentrations of impurities, e.g., 1% of carbon and 0.1% of iron, which have been added to a D–T plasma is shown in Fig. 7, where the expression

$$\frac{P_r(T_e)}{n^2} = \frac{P_{br}(T_e)}{n^2} + \frac{C_{imp} P_{imp}(T_e)}{n^2} \tag{19}$$

is plotted as a function of electron temperature. Here, $C_{imp} = n_{imp}/n_i$ is the concentration of impurities and $C_{imp} \ll 1$. It is apparent that while modest concentrations of elements such as carbon (or oxygen) do not significantly influence the power balance of the ignited fusion plasma, they nevertheless greatly increase the amount of energy radiated from a cold plasma and hence the energy that must be fed into the plasma in order to heat it to ignition

---

[†] See McWhirter and Summers (Chapter 3, Section VII).

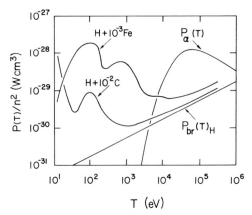

Fig. 7. Power density functions for a D-T plasma that contains a small concentration of impurity elements. Curves for 1% of carbon and 0.1% of iron were determined using Eq. (19) and $P_{br}(T)_H$ refers to a pure D-T plasma.

conditions. Radiative losses due to heavy elements are greater and they extend into the higher temperature regime; this behavior is evident from the substantial effect of only 0.1% of iron.

An assessment of energy balance within a fusion plasma must account not only for the bremsstrahlung losses but also for the energy input needed to sustain the thermal energy content of the hot plasma. The average energy of each particle within an idealized hot plasma is $\frac{3}{2}kT$ so that the amount of energy required to raise the electrons and ions in a unit volume of the cold plasma to temperature T is

$$E_{ei} = \tfrac{3}{2}(n_e kT_e + n_i kT_i) = 3nkT; \qquad (20)$$

this can be regarded as the energy density within the plasma. Suppose that all of this thermal energy is lost from the plasma in an energy confinement time τ_E and moreover that the heating system is capable of raising the cold plasma to a temperature T in a time that is considerably less than τ_E. Then the energy balance needed to sustain ignition by α-particle heating alone is given by,

$$\tau_E P_\alpha(T) = 3nkT + \tau_E P_r(T),$$

which can be expressed as

$$n\tau_E = \frac{3kT}{\left[P_\alpha(T) - \left(P_{br}(T) + \sum C_{imp} P_{imp}(T)\right)\right]n^{-2}} \quad \text{cm}^{-3}\,\text{s}, \qquad (21)$$

where the summation accounts for all species of impurities present in the plasma. This $n\tau_E$ criterion is shown in Fig. 8, and for pure D-T it has a minimum of about 2×10^{14} cm^{-3} s in the region of about 1.5×10^4 eV. A favorable energy balance can be attained at a somewhat lower temperature but it should be noted that, if τ_E is independent of temperature, the energy density, which can be expressed in the form

$$E_{ei} = 3(n\tau_E)kT/\tau_E,$$

also has a minimum around 15 keV, which can thereby be taken as a criterion for minimum expenditure of heating energy. This simple analysis omits many important details but nevertheless provides a reasonable indication of the necessary objectives. Accepting that $\tau_E \sim 1$ s indicates that the plasma density must be in excess of $\sim 10^{14}$ cm^{-3} to achieve ignition from α-particle heating.

The balance of energy is deleteriously affected by the presence of even small concentrations of radiating impurity elements and Fig. 8 also shows the consequences of adding 1% and 0.1% of iron to pure D–T. The effects of the lower concentration are predominantly evident in the increased values of $n\tau_E$ needed at lower temperature, but 1% of iron drives the complete ignition characteristic into a high-temperature regime that is incompatible with magnetic confinement.

The preceding conditions may be regarded as somewhat extreme because no allowance has yet been made for the neutron energy produced in D–T fusion. Some fraction of the energy produced in the reactor blanket could be reintroduced into the plasma and thereby used to complement α-particle heating. Suppose that the fusion energy, the plasma thermal energy, and the radiated energy are all collected within the reactor and are there converted with an efficiency η_t into electrical energy. Assume also that this electrical energy can be converted into plasma heating and containment mechanisms with an efficiency η_D and that all of the electrical energy is recirculated for a

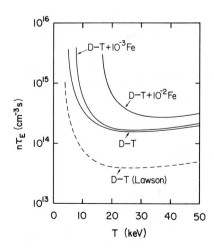

Fig. 8. The $n\tau_E$ criterion for ignition plotted as a function of plasma temperature T. Data for α-particle ignition were determined using Eq. (21) for a pure D–T plasma and for contamination by 0.1 and 1.0% of iron. The dashed curve shows the Lawson (1959) break-even criterion given by Eq. (22), in which it is assumed that $\eta_t\eta_D = 0.4$.

time τ_L during which no electrical power is extracted from the reactor. The energy density balance for an idealized pure D–T plasma is therefore given by

$$\eta_t \eta_D [\tau_L P_F(T) + \tau_L P_{br}(T) + 3nkT] = [\tau_L P_{br}(T) + 3nkT].$$

Such a reactor can provide a net output of electrical power if $\tau_E > \tau_L$, the "breakeven" condition being $\tau_E = \tau_L$. Rearrangement yields

$$n\tau_E \geq \frac{3kT}{[\eta_t \eta_D (1 - \eta_t \eta_D)^{-1} P_F(T) - P_{br}(T)] n^{-2}} \quad \text{cm}^{-3} \text{ s}, \tag{22}$$

where the equality corresponds to break even. This criterion was first identified by Lawson (1959) and is illustrated for $\eta_t \eta_D = 0.4$ by the dashed curve in Fig. 8. The minimum value of $n\tau_E$ is about 4×10^{13} cm^{-3} s and occurs at a temperature of about 2.5×10^4 eV but, since this temperature is somewhat excessive for magnetic containment, the fusion objective is usually taken at 10^4 eV where $n\tau_E \approx 10^{14}$ cm^{-3} s.

V. Auxiliary Heating

The ability to attain ignition conditions by means of the ohmic heating process is limited by the temperature dependence of plasma resistivity [$\eta \propto T^{-3/2}$; see Eq. (11)] and by restrictions that must be applied to the plasma current density in order to comply with criteria for confinement. Equality between P_Ω and P_{br} for a pure D–T plasma in a toroidal device is generally in the temperature range below about 3 keV and additional or "auxiliary" heating methods must thus be used in order to raise the plasma temperature to the ignition regime. Two approaches which offer particular advantages are the injection of intense beams of energetic atoms and the injection of powerful sources of radio-frequency electromagnetic radiation.

A. Neutral-Beam Injection Heating

To heat the plasma using an external source of energetic particles it is necessary to meet the following requirements:

(a) Each injected particle must be appreciably more energetic than the electrons and ions of the plasma.

(b) The particles must be able to pass through the strong magnetic field that confines the plasma.

(c) When inside the plasma, the particles must be able to travel for a distance sufficient for there to be an efficient equipartition of energy.

2. Basic Concepts of Fusion Research

Neutral-beam injection has, in some measure, been successful in meeting all of these requirements. The technique is based upon the following sequence of events: (i) the production of ions in an external source; (ii) acceleration of these ions in an electrostatic field; (iii) subsequent passage of the energetic ion beam through a neutralizer gas cell where some of the ions are neutralized (generally by charge capture); (iv) injection of the resultant neutral beam into the plasma where many of its atoms are ionized and therefore confined within the magnetic field; and (v) the subsequent transfer of energy to the plasma due to Coulomb collisions during the confinement time of the energetic ions.

Intense beams of $\sim 10^2$ A are required to achieve ignition and the direction of the beam relative to that of the flow of ohmic heating current is also important. The production and transmission of neutral beams is described in Green (Subchapter 6C) whereas the deposition and subsequent slowing down of its energetic ions is discussed in Cordey (Subchapter 6B).

Energetic ions from the beam can share their energy with both the ions and electrons of the plasma and the relative rates of energy transfer, which can be inferred from Eq. (35) in Section VI, give rise to the expression

$$\frac{\text{rate of energy transferred to ions}}{\text{rate of energy transferred to electrons}} \approx \frac{56}{A_p} \left(\frac{A_i T_e}{E_i}\right)^{3/2}, \qquad (23)$$

where E_i and A_i are respectively the beam energy (in eV) and the atomic weight of the trapped ions. A_p is the atomic weight of the plasma ions (i.e., 2.5 for D–T). Disproportionately large transfer of energy to the plasma electrons is deleterious to the balance between fusion power and radiative losses, and so the ratio in Eq. (23) should be as large as is compatible with the needs of beam formation and of magnetic confinement of energetic ions. The obvious choice for the injected beam species is D atoms, and for these equality in the energy transfer rates to electrons and the ions is given by

$$E_{in} \approx 16 T_e.$$

Beam energies in the range 100–200 keV are therefore desirable to attain a plasma temperature ~ 10 keV. These energies are also necessary to obtain adequate penetration of the plasma prior to ionization which, in a pure D–T plasma, occurs largely through "proton"[†] impact; i.e.,

$$H + \overline{H^+} \rightarrow H^+ + e + \overline{H^+},$$

where the overbars denote the plasma particle. At the lower beam energies presently encountered in laboratory-scale experiments (e.g., $v_{\text{beam}} < 2 \times 10^8$ cm s^{-1}) beam deposition is predominantly due to charge exchange,

$$H + \overline{H^+} \rightarrow H^+ + \overline{H},$$

[†] Quotes are used because it is not necessary for this discussion to distinguish between the isotopes of hydrogen.

and it should be noted that each event gives rise to an atom whose energy corresponds to that of a plasma ion. Such atoms are not confined by the magnetic field; they may be lost to the wall of the containment vessel and thereby cause the release of impurities.

The magnitude of power deposition needed for ignition (~100 MW) is unlikely to be attained except at high injection energy. Positively charged ion beams can be produced with adequate intensity but, at high injection energy, the efficiency for conversion to atoms in the neutralizer cell is low because the probability of charge stripping is greater than the probability of electron capture. This is a fundamental disadvantage of neutral injection systems based upon positively charged ions and a solution is sought through the route of negative ion production. Such ions can be neutralized by stripping reactions of the type

$$H^- + H \rightarrow H + e + H,$$

which have relatively large cross sections at high energy. The problems of negative ion formation and the subsequent neutralization of the beam are discussed in Green (Chapter 6C, Section V).

B. Radio-Frequency Heating

The relatively low electrical efficiency and the size and complex technology of beam heating systems for high injection energy has stimulated interest in the alternative technique of heating by means of externally launched electromagnetic waves. While this approach has major significance in the overall context of fusion research, it has little direct bearing upon atomic collision mechanisms and so will not be discussed in detail [the theory of radio-frequency heating has been recently reviewed by Fielding (1981) and the status of experiments surveyed by Riviere (1981)]. Large amounts of radio-frequency power can be efficiently generated and transmitted to a wave launcher located on the containment vessel. In this respect the wave heating approach is superior to that of neutral-beam injection but the mechanisms by which such power is coupled to the plasma and subsequently converted into thermal energy are much more complex.

An electromagnetic wave whose electric field vector lies parallel to the direction of the magnetic field cannot penetrate the plasma if the wave frequency ω_W is less than the electron plasma frequency ω_{pe}, which [see Eq. (16)] is typically 100 GHz in a fusion relevant plasma. Wave penetration can occur at lower frequencies but only if a component of its electric field lies perpendicular to the magnetic field; this lower-frequency regime is employed in many forms of radio-frequency heating.

A magnetically confined plasma is not homogeneous, so that the launched wave can be converted into one or more modes of oscillation that are natural to different regions of the plasma. There are many possible modes, and their accessibility depends not only upon the parameters of the plasma and magnetic field but also upon the direction of the phase velocity of the launched wave relative to the direction of the magnetic field. In order that an oscillatory field may irreversibly transfer energy to plasma particles it is necessary for the wave frequency to exceed the lowest frequency for relaxation of plasma energy, typically that for Coulomb collisions between the plasma ions, i.e.,

$$\omega_W > \nu_{ii} = Z^4(m_e/m_i)(T_e/T_i)^{3/2}\nu_{ee}, \tag{24}$$

where ν_{ee} is given by Eq. (9). Waves at a frequency that can heat the plasma are damped by a variety of collision processes involving ions and electrons of the bulk plasma but there are certain groups of plasma particles that can damp the wave by collisionless transfer of energy. Despite the oscillatory nature of the wave, these groups of plasma particles experience a static electric field within their rest frame and they can be conveniently identified by means of the following expression:

$$\omega_W - k_\parallel v_\parallel - N\omega_c = 0. \tag{25}$$

Here N is an integer and $k_\parallel$ is the wave propagation vector in the direction parallel to the magnetic field. If N is taken to be zero, then Eq. (25) shows that the wave sets up a static electric field that acts upon those plasma particles whose parallel velocity $v_\parallel$ is close to the phase velocity of the wave. This is the mechanism of Landau damping commonly encountered in plasmas. When $N = \pm 1$ or $|N| \geq 2$ Eq. (25) describes resonant damping by excitation of the cyclotron frequency ω_c of the plasma particles. These resonance conditions can occur for either electrons or ions and also for minority ions such as helium. Wave damping by resonant processes provides an irreversible deposition of energy to the bulk plasma only if the directed motion imparted to these groups of particles is subsequently randomized by scattering collisions.

A wide range of frequencies can be employed for plasma heating; typical of the lower regime (i.e., a few megahertz) are the hydromagnetic (or Alfvén) waves, whereas the practical upper limit is bounded by the electron cyclotron frequency

$$\omega_{ce} = eB/M_e c, \tag{26}$$

which lies in the region 15–300 GHz. An obvious candidate in the intermediate regime is the ion cyclotron resonance

$$\omega_{ci} = (m_e/m_i)\omega_{ce}, \tag{27}$$

which lies around 10–100 MHz and there is also a "lower hybrid resonance"

$$\omega_{LH} = \omega_{pi}(1 + \omega_{pe}^2/\omega_{ce}^2)^{-1/2} \tag{28}$$

in the range 0.5–2.5 GHz. [Here ω_{pi} is the ion plasma frequency $\omega_{pe} = (4\pi n_i Z^2 e^2/m_i)^{1/2}$.]

The present status of radio-frequency heating is encouraging and it may eventually supplant neutral-beam heating.

VI. Inertial Confinement

The concept of inertial confinement relies upon the attainment of a regime of extremely high density, $n \sim 10^{26}$ cm^{-3}, wherein the mean free path for D–T fusion reactions is $\sim 10^{-2}$ cm. A worthwhile yield of fusion energy might therefore be obtained from a small volume of fuel, namely a pellet of frozen D–T gas. The particle density of cold, solid D–T is about 5×10^{21} cm^{-3}, so, to achieve the preceding criteria, the pellet must be compressed as well as heated to ignition temperature prior to the fusion burn. It is conceived that the frozen pellet be launched into a large (~ 10-m radius) vacuum chamber and while in free flight be subjected to an intense pulse of energy uniformly distributed over its surface. This "driving" energy may be delivered either from a distributed array of laser beams or from beams of high-energy particles. The material of the pellet surface is rapidly heated and ablates at high velocity so that reaction against this ablation causes an implosion of the inner region, thereby compressing the D–T core and heating it to thermonuclear temperature.

The ablating material is ionized and the incident energy is coupled to the electrons, which, as the density increases, become degenerate and thereby attain superthermal energies. Ideally the fuel should be adiabatically compressed and heated by an inward-moving shock wave, but the superthermal electrons penetrate to the core faster than the shock and generate a pressure that opposes compression. To overcome this problem of "preheat," pellets are constructed with a central core of fuel encased in several concentric shells; the outer made from lightweight material serve as an energy absorber and an ablator whereas the inner, constructed from heavy elements, act as an absorber for the preheat electrons and as a pusher for the D–T core.

The principles for ignition are similar to those for magnetic confinement; D–T temperature should be about 10 keV but n and τ_E must be matched to the available driver energy, predicted to be 1–10 MJ per pulse. It is convenient to define a disassembly time τ_d related to the radius r_F of a spherical pellet which has just been ignited. The ablating cloud expands with a velocity comparable to the sound speed, and if nondegeneracy is assumed,

$$c_s \approx (2kT_i/m_i)^{1/2},$$

when $T_e = T_i$. The transit time of an acoustic wave across this pellet is r_F/c_s

2. Basic Concepts of Fusion Research

and so may be regarded as equivalent to the fuel confinement time. However, the pellet expands during this time and, because of its spherical geometry, the density reduction is greatest in the outer regions; when allowance is made for these effects (see, e.g., Ribe, 1975),

$$\tau_d \approx \tau_E \approx r_F/4c_s. \tag{29}$$

Criteria for $n\tau_E$ can thus be converted into the form

$$m_i 4 c_s n \tau_E = \rho r_F, \tag{30}$$

where ρ is the mass density and $m_i = 4.2 \times 10^{-24}$ g for D–T fuel. The reactor must be pulsed and the repetition rate is limited by technological constraints of the driving system; ~ 10 Hz is typical. To achieve worthwhile efficiency it is thus essential to extract the maximum practicable amount of fusion energy from each pellet. This in turn demands that the burnup fraction f_B of the fuel be high. If the burn time is τ_d and the amount of fuel is constant, then

$$f_B = \frac{\tau_d/2\tau_F}{1 + \tau_d/2\tau_F}, \tag{31}$$

where τ_F is the D–T reaction time and the factor 2 arises because one D^+ and one T^+ ion are involved in each fusion collision. The reaction time given by Eq. (2) is expressed here as

$$\tau_r = 1/n\langle\sigma v\rangle_{D-T} = m_i/\rho\langle\sigma v\rangle_{D-T},$$

so that

$$f_B = \frac{\rho r_F}{(8 m_i c_s/\langle\sigma v\rangle_{D-T}) + \rho r_F}. \tag{32}$$

In practice the burnup fraction should be limited to about 30% to avoid an unproductive buildup of helium ash; substitution of $f_B = 0.3$ in Eq. (32) and evaluating the other parameters at $T = 10^4$ eV gives $\rho r_F \approx 3$ g cm^{-2}. This value can thus be considered as a reasonable criterion for an inertial confinement reactor.

Compression of the pellet is due to conversion of thermal energy into compressive forces and these must be matched to available driver energy. The total amount of thermal energy E_{ei} of the electrons and ions in a pellet of radius r_F can be expressed as

$$E_{ei} = \frac{4}{3} \pi r_F^3 E_{ei} = \frac{4}{3} \pi r_F^3 3nkT = 4\pi \frac{(\rho r_F)^3}{\rho^2} \frac{kT}{m_i}, \tag{33}$$

and, since the value of ρr_F has been prescribed at the ignition temperature, it is apparent that the thermal energy of the pellet is dependent upon ρ^{-2}. If the density of solid D–T (i.e., 0.213 g cm^{-3}) is substituted into Eq. (33) then $E_{ei} \approx 2.7 \times 10^{12}$ J, which is incompatible with any envisaged driver system.

Compressing the fuel by about 10^4 so that $\rho \approx 10^3$ g cm^{-3} yields $E_{ei} \approx 1.2 \times 10^5$ J, which could be compatible with a driver pulse of 10^6 J provided that the conversion efficiency of driver energy to thermal energy were about 0.1. The mass of this particular pellet is about 100 μg and so its radius at cryogenic temperature is 0.5 mm, its compressed radius r_F is about 0.03 mm, and its disassembly time is 1.4×10^{-10} s. The D–T reaction time is about 10^{-11} s. Fusion energy produced by the burnup of each pellet is given by

$$E_F = \frac{4}{3}\pi r_F^3 \frac{\rho}{m_i} \frac{f_B}{2} Q_{D-T} \tag{34}$$

and so, when $Q_{D-T} = 20$ MeV is assumed, $E_F \approx 12.7$ MJ. Some appreciation of the problems of constructing a suitable reactor vessel can be gained from the fact that the preceding value of E_F is equivalent to the explosion of about 1 kg of TNT.

Two other properties of the pellet are worthy of consideration. First, the range of 3.5-MeV α-particles should be less than r_F and this is well satisfied when $\rho = 10^3$ g cm^{-3}. Second, the time t_{eq} required for electrons to transfer energy to the ions must be appreciably less than τ_d. Spitzer (1962) has derived an expression for the equipartition time for a group of "test particles" moving through a group of "field particles" denoted by the subscript f, namely,

$$t_{eq} = \frac{3mm_f k^{3/2}}{8(2\pi)^{1/2} n_f Z^2 Z_f^2 e^4 \ln \Lambda} \left(\frac{T}{m} + \frac{T_f}{m_f}\right)^{3/2}. \tag{35}$$

If electrons are the field particles and $Z = Z_f = 1$, then Eq. (35) yields $t_{eq} \approx 3 \times 10^{-12}$ s for $n \approx 2 \times 10^{26}$ cm^{-3} (i.e., 10^3 g cm^{-3}), $kT = 0$, $kT_f = 10^4$ eV, and $\ln \Lambda = 10$. If it is accepted that the ablating cloud has the properties of a plasma, then Eq. (35) indicates that there is ample time for the ions to become heated before the pellet disassembles.

In the preceding example E_F/E_{ei} is about 100 but the overall gain in energy is much less because the electrical energy E_D fed to the driver appreciably exceeds E_{ei}. If the efficiency for beam production and focusing is η_D then $E_F = \eta_D G E_D$ where G is defined as the gain of the pellet. This gain is predicted to increase with the total energy delivered in the driving pulse but, because it is dependent upon the detailed mechanism of energy absorption and conversion within the pellet, it is also sensitive to the nature of the driving beams. Laser beams suffer substantial reflection in the absorber unless their wavelength is less than about $\frac{1}{4}$ μm [See Eq. (17)] and the kinetic energy of electrons and light ions should not exceed a few million electron volts; otherwise these particles penetrate deeply into the pellet and cause preheating. By contrast heavy ions can be stopped in a thin absorption region even when the particle velocity corresponds to several billion electron volts. It is predicted (see, e.g., Lawson, 1980), that η_D is likely to range

2. Basic Concepts of Fusion Research

from 5% for short-wavelength lasers such as KrF to 25% for heavy ions and that a reasonably efficient reactor will require $\eta_D G = 10\text{--}15$.

Acknowledgment

All the figures in this chapter have been reproduced courtesy of Culham Laboratory.

References

Artsimovich, L. A. (1964). "Controlled Thermonuclear Reactions." Gordon & Breach, New York.
Fielding, P. J. (1981). *In* "Plasma Physics and Nuclear Fusion Research" (R. D. Gill, ed.), p. 477. Academic Press, New York.
Glasstone, S., and Lovberg, R. H. (1960). "Controlled Thermonuclear Reactions." Van Nostrand, Princeton, New Jersey.
Hagler, M. O., and Kristiansen, M. (1977). "An Introduction to Controlled Thermonuclear Fusion." Lexington Books, New York.
International Tokamak Reactor—Zero Phase (1980). Report of the International Tokamak Reactor Workshop, 1979. IAEA, Vienna; also *Nucl. Fusion* **20,** 349 (1980).
International Tokamak Reactor—Phase One (1982). Report of the International Tokamak Reactor Workshop, 1980 to 1981. IAEA, Vienna.
Jensen, R. V., Post, D. E., Grasberger, W. H., and Tarter, C. M. (1977). *Nucl. Fusion* **17,** 1187.
Kammash, T. (1975). "Fusion Reactor Physics." Ann Arbor Sci. Publ., Ann Arbor, Michigan.
Lawson, J. D. (1959). *Proc. Phys. Soc., London, Sect. A* **106,** Suppl. 2, 173.
Lawson, J. D. (1980). *Fusion Technol., Proc. Symp., 11th, Oxford, Engl.* **1,** 89.
Motz, H. (1979). "The Physics of Laser Fusion." Academic Press, New York.
Post, R. F. (1956). *Rev. Mod. Phys.* **28,** 338.
Ribe, F. L. (1975). *Rev. Mod. Phys.* **47,** 7.
Ribe, F. L. (1977). *Nucl. Technol.* **34,** 179.
Riviere, A. C. (1981). In "Plasma Physics and Nuclear Fusion Research" (R. D. Gill, ed.), p. 501. Academic Press, New York.
Rose, D. J., and Clark, M. (1961). "Plasmas and Controlled Fusion." MIT Press, Cambridge, Massachusetts.
Spitzer, L. (1962). "Physics of Fully Ionized Gases." Wiley, New York.
Spitzer, L., and Harm, R. (1953). *Phys. Rev. A* **16,** 1811.
Stacey, W. M. (1981). "Fusion Plasma Analysis." Wiley, New York.

3
Atomic Radiation from Low Density Plasma

R. W. P. McWhirter

Space and Astrophysics Division
Rutherford Appleton Laboratory
Chilton, Didcot, Oxfordshire
England

and

H. P. Summers

Department of Natural Philosophy
The University of Strathclyde
Glasgow, Scotland

I.	Preliminary Discussion	52
II.	The Boltzmann Equation	55
III.	Components of the Statistical Balance Equations	60
	A. Collisional Ionization by Electron Impact	62
	B. Collisional Excitation by Electron Impact	64
	C. Spontaneous Emission of Photons	66
	D. Radiative Recombination	67
	E. Dielectronic Recombination	68
	F. Scaling of the Coefficients with Ionic Charge	69
IV.	The Collective Viewpoint of Ionization and Recombination	71
V.	The Distribution among the Stages of Ionization	81
VI.	Spectral Line Intensities	85
	A. Hydrogenlike Ions	86
	B. Heliumlike Ions	87
	C. Satellite Lines	90
	D. Lithiumlike Ions	91
	E. Berylliumlike Ions	93
	F. Boronlike Ions	95
VII.	Radiated Power Loss	97
	References	108

I. Preliminary Discussion

In this chapter the theory of the atomic processes that give rise to the spectra emitted by hot tenuous plasmas is introduced. Upon this theory is based the calculation of the distribution of the atomic species among their various stages of ionization and the intensities of spectral line and continuum radiation emitted by the plasma. These calculations are important in making estimates of the radiated power loss and the plasma resistivity as well as providing the theoretical basis for the spectroscopic determination of plasma parameters such as the electron density and temperature. In the presentation given here the primary emphasis is on fusion plasmas, but much of the theory is equally relevant to astronomical spectra.

Typical of the spectra of fusion plasmas is Fig. 1, which shows a microdensitometer trace of a small portion of the soft x-ray spectrum emitted by the DITE Tokamak at the Culham Laboratory. The radiation is composed almost entirely of emission spectral lines from impurity ions which contaminate the hydrogen plasma. In spite of the concentrations of these impurities being small (less than a few percent) their radiation dominates the overall power loss from the plasma. It is one of the purposes of this chapter to explain the basis of this power loss mechanism.

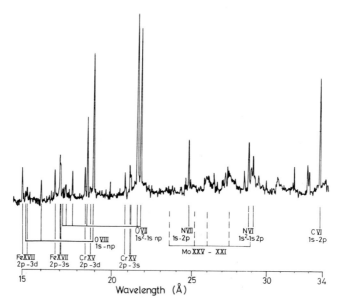

Fig. 1. XUV emission spectra from DITE (Tokamak Experiment) at Culham Laboratory, UKAEA. Emission from the hydrogenlike and heliumlike ionization stages of carbon, oxygen, and nitrogen is particularly prominent. Spectral lines from heavy elements are also abundant. [From Peacock et al. (1979).]

3. Atomic Radiation from Low Density Plasma

The broad mechanism is simple. Thermal kinetic energy of the free electrons in the plasma is transferred by collisions to the internal energy of impurity ions:

$$\mathcal{A} + e \rightarrow \mathcal{A}^* + e,$$

where $\mathcal{A}^*$ denotes an excited state of the impurity ion $\mathcal{A}$. This energy is then radiated as photons which escape from the plasma volume

$$\mathcal{A}^* \rightarrow \mathcal{A} + h\nu.$$

A detailed quantitative description is complicated because of the need to evaluate individually the many controlling collisional and radiative processes, a task which is compounded by the variety of atoms and ions which participate.

The plasmas of interest are in general evolving in time and space so that the ions that give rise to the spectra may be moving through regions of varying temperature and density. Thus in fusion devices there are regions such as the zones near the walls of the containing vessel where impurity recycling is pronounced. Equally there are examples in astrophysics where plasma motion and turbulence have a dramatic effect on the distribution of the ions. In addition, and in both cases, there are generally regions where gradients of temperature, density, or concentration give rise to diffusion effects which can have a strong influence on the distributions of ionic species. Although this chapter addresses only the atomic processes, the importance of plasma processes should not be underestimated as they influence the details of the emitted spectra. Plasma processes are considered in detail in other chapters of this book but some mention of them is made here in Section VII.

Powerful magnetic fields are present in many fusion experiments as a means of confining the plasma and thereby reducing contact with the walls of the containment vessel. In principle such a field can influence the spectrum through its effect on the details of the atomic processes. In practice, the main effect of the field is to inhibit free-electron thermalization transverse to the field so that different temperatures can exist in different zones of the plasma. In general for fusion plasmas the influence of magnetic fields on the structure of the excited levels of ions is masked by the relatively high frequency of electron collisions.

Cyclotron radiation (or at relativistic particle velocities: synchrotron radiation) is produced when charged particles move across a magnetic field. In the fusion domain the contribution to this radiation from positive ions is negligible compared with electrons so only the latter need be considered. The radiation is not isotropic but related to the magnetic field direction and the direction of instantaneous electron motion. For thermonuclear plasmas the wavelength of the radiation is in the far infrared where laboratory size plasmas (~ 1 m) are optically thick not only to cyclotron radiation but also to bremsstrahlung. Where practical calculations have been done for current laboratory plasmas it turns out that the power loss due to cyclotron radiation is less than that due to bremsstrahlung. However, because cyclotron radiation power rises steeply with temperature it may represent an important loss mechanism for a fusion reactor. The relevant physics does not fall conveniently within the scope of this chapter.

The main emphasis here is placed upon low-density fusion plasmas where even resonance radiation, once created, escapes from the plasma volume without being reabsorbed; i.e., the theory of the atomic processes is developed in the optically thin approximation. However, it is worth remarking that many of the results are still applicable under conditions of moderate opacity. The complete theoretical treatment of opacity requires the simultaneous solution of the equation of radiative transfer with the equations describing the atomic processes. The complexities of this approach are well known and elaboration in this direction is not justified in the circumstances of modest or small opacity that apply to most plasmas of interest here. Instead it will suffice to draw attention to some of the physical consequences of opacity as they apply to low-density plasmas. Resonance radiation is most susceptible to the effects of opacity, which usually causes the intensity to be reduced although there may be significant directional effects when the physical shape of the plasma is far from spherical. In some cases resonance radiation may be able to decay by an alternative transition at a different wavelength, in which case the alternative line intensity is enhanced by opacity in the resonance line. Solutions of the equation of radiative transfer are quite common in astrophysics, but rare in magnetically confined fusion plasmas.

Having described the areas of work which lie outside the scope of this chapter it remains to describe the many features of the theoretical plasma model that will be treated. Both astrophysical and fusion plasmas consist primarily of hydrogen with small admixtures of heavier elements and in both cases interest is in plasmas where the hydrogen is more or less fully ionized. The distinction between ordinary hydrogen and deuterium or tritium is usually unimportant in the theory treated here. Thus for brevity the word "proton" will be used to indicate a positive hydrogen ion of any isotope. Where there is need to do so reference will be made to the specific isotope. Thus the theory will be developed for a plasma composed of electrons and protons as the major species and relatively small concentrations of hydrogen atoms and heavier elements in various stages of ionization. The latter will be referred to as minor species. Usually the concentration of the (heavy) minor species is less than a few percent of the hydrogen concentration (atoms plus protons). In this connection it is worth noting that with the exception of helium the cosmological abundances of the elements are one part in ten thousand or less (compared to hydrogen). Despite these small concentrations the atomic radiation emitted by these (heavy) minor species dominates that from hydrogen and may be the greatest single factor in the power loss

3. Atomic Radiation from Low Density Plasma

from a plasma. It will be shown later that, for example, 10% of carbon impurity in a fusion reactor would be sufficient to quench the nuclear reaction and that 0.01% of molybdenum would have the same effect.

The development of the theory is presented in the following pages and is organized into six sections. In Section II a nonequilibrium model is introduced by way of the Boltzmann equation. This is used to treat binary collisions between major species and hence to outline the derivation of the equation describing the Maxwellian velocity distribution. Binary collisions between major and minor species are then examined to show how the Boltzmann equation may be specialized to a form more suitable for the description of the excited levels of ions. Finally, in this section there is a discussion of local thermal equilibrium and complete thermodynamic equilibrium. This rather formal introduction to the theory creates a framework to which the more complex parts developed later may be related and establishes the physical basis of some of the concepts.

In Section III the individual atomic processes are examined. These are then discussed separately with some care taken to allow the reader to gain access to the relevant literature in order to perform practical calculations.

In Section IV the statistical balance equations are introduced in a general form and the collective effects of the individual atomic processes are treated. As a consequence the collisional–radiative and collisional–dielectronic coefficients of recombination, ionization, and excitation are defined. The justification is made in this section for separating the equations of statistical balance into (a) those establishing the stage of ionization and (b) those describing excitation leading to the emission of radiation.

Section V takes a practical situation and applies the equations of time-dependent ionization balance to make predictions of the time development of the state of ionization and then makes comparison with experiment. Steady-state solutions of ionization balance equations are discussed.

Section VI treats the solution of equations of statistical balance as they apply to excitation. This is done by treating a selection of ionic species in turn in order to illustrate different effects.

Finally, in Section VII the calculation of the total radiated power is discussed for steady-state plasmas. The importance of limiting the impurity level in fusion plasma is established.

II. The Boltzmann Equation

As the intention is to describe nonequilibrium situations, an appropriate initial point for analysis is the Boltzmann equation for the set of interacting particle types comprising the plasma. The objective is to provide an outline of the factors involved in the reduction of the Boltzmann equation to the statistical balance equations which form the usual starting point for the theory and also to derive some important relations and distributions. These rather abstract considerations, from a spectroscopic viewpoint, can be bypassed by moving directly to the final summary paragraph of this section.

For a particular class of particle, the time rate of change of its distribution function f is given by (Chapman and Cowling, 1970)

$$\frac{df}{dt} = \left(\frac{\partial f}{\partial t}\right)_{\text{int}}. \tag{1}$$

The right-hand side of this equation symbolizes the rate of change of f due to interaction with other classes of particles (including photons). The left-hand side is the rate of change of f in phase space under the influence of external fields. With the interaction term zero, Eq. (1) asserts the constancy of f, traveling with the particles in phase space.

Important classes of particles which must be identified here include free electrons, free protons, and ions of different species in distinct internal quantum states. The distribution function for photons should also be examined. However, as described in Section I the radiation field density may be taken to be low in the optically thin plasmas of interest here, the only coupling to the radiation field being via spontaneous emission. For that reason, attention is initially concentrated on the particle distributions. Also, since the geometry of the contained plasma is not of concern in this chapter it may be supposed that the particle distributions are uniform in space, depending only on velocity and time.

Since separation into major and minor species is possible for fusion plasma the important contributions to $(\partial f/\partial t)_{\text{int}}$ are those for which at least one interacting particle belongs to a major species. Consider first the electron and proton distributions for which e–e, p–p, and e–p collisions are relevant. As is well known, owing to the long range of the Coulomb interaction, distant encounters are most effective in changing the distribution functions. For isotropic distributions of particle velocities, the kinetic-energy redistribution time scales τ_{e-e}, τ_{p-p}, and τ_{e-p} are significant parameters. It may be shown that, because of the mass factors, $1/\tau_{e-e} \simeq 43/\tau_{p-p} \simeq 1849/\tau_{e-p}$ (with appropriate factors for the heavier isotopes of hydrogen). Time scales of processes leading to the emission of radiation are much longer. Therefore the interaction term for the electron distribution is determined primarily by e–e collisions and likewise for the proton distribution by p–p collisions. Also these time scales indicate that a situation can occur, and is in fact common in fusion plasmas, in which equipartition of energy has occurred for electrons and protons separately but without equipartition between protons and electrons.

The interaction term for the electron distribution $f_e(\mathbf{u})$, where $\mathbf{u}$ is the electron velocity, takes the form

3. Atomic Radiation from Low Density Plasma

$$\left(\frac{\partial f_e(\mathbf{u})}{\partial t}\right)_{int} \simeq -\int f_e(\mathbf{u})f_e(\mathbf{v})|\mathbf{u} - \mathbf{v}|\sigma(\mathbf{u}, \mathbf{v} \to \mathbf{u}', \mathbf{v}')\, du'\, dv'\, dv$$
$$+ \int f_e(\mathbf{u}')f_e(\mathbf{v}')|\mathbf{u}' - \mathbf{v}'|\sigma(\mathbf{u}', \mathbf{v}' \to \mathbf{u}, \mathbf{v})\, du'\, dv'\, dv, \quad (2)$$

with the constraints that total energy and momentum are conserved in each binary collision; σ describes the collisional transition for collisions between pairs of electrons of initial velocities $\mathbf{u}$ and $\mathbf{v}$ and final velocities $\mathbf{u}'$ and $\mathbf{v}'$. The invariance of the dynamical equations under time reversal implies that

$$\sigma(\mathbf{u}, \mathbf{v} \to \mathbf{u}', \mathbf{v}') = \sigma(\mathbf{u}', \mathbf{v}' \to \mathbf{u}, \mathbf{v}) \quad (3)$$

and so the usual form for the electron collision term is obtained, namely,

$$\left(\frac{\partial f_e(\mathbf{u})}{\partial t}\right)_{int} \simeq \int [f_e(\mathbf{u}')f_e(\mathbf{v}') - f_e(\mathbf{u})f_e(\mathbf{v})]|\mathbf{u} - \mathbf{v}|$$
$$\times \sigma(\mathbf{u}, \mathbf{v} \to \mathbf{u}', \mathbf{v}')\, du'\, dv'\, dv. \quad (4)$$

The isotropic distribution for which the above integral vanishes identically is that for which $f_e(\mathbf{u}')f_e(\mathbf{v}') = f_e(\mathbf{u})f_e(\mathbf{v})$ when $\mathbf{u}$, $\mathbf{v}$, $\mathbf{u}'$, and $\mathbf{v}'$ satisfy the conservation constraints. This is the Maxwellian distribution, namely,

$$f_e(\mathbf{u}) = n_e \left(\frac{m_e}{2\pi kT_e}\right)^{3/2} \exp\left(\frac{-m_e u^2}{2kT_e}\right); \quad (5)$$

T_e is the free-electron kinetic temperature, m_e the electron mass, and n_e the electron density. This distribution is independent of the particular form of σ and is usually deduced from general thermodynamic considerations. The distributions for other mass particles (neglecting internal structure) follow immediately. If there is more than one kind of particle present, in mutual equilibrium they will take up the same kinetic temperatures and their mean speeds will be related inversely as the square roots of their masses. In an electron–proton plasma in equilibrium, the electrons have, on average, velocities 43 times greater than the protons. This has important consequences. Useful expressions are given by Spitzer (1962) for the self-collision time τ_e characterizing the time required for the kinetic-energy distribution of identical particles to approach a Maxwellian distribution, and for the equipartition time τ_{eq} characterizing the time for equipartition of energy of particles of type 1 with particles of type 2. These are

$$\tau_e = 0.12 \frac{1}{\alpha c a_0^2} \left(\frac{m}{m_e}\right)^{1/2} \left(\frac{kT}{I_H}\right)^{3/2} \frac{1}{nz^4 \ln \Lambda} \quad \text{(s)}, \quad (6)$$

$$\tau_{eq} = 0.0529 \frac{1}{\alpha c a_0^2} \frac{m_1 m_2}{m_e m_e} \left(\frac{kT_1}{I_H}\frac{m_e}{m_1} + \frac{kT_2}{I_H}\frac{m_e}{m_2}\right)^{3/2} \frac{1}{n^2 z_1^2 z_2^2 \ln \Lambda} \quad \text{(s)}, \quad (7)$$

where m denotes particle mass, n number density (in cm^{-3}), T temperature, I_H the ionization potential of hydrogen, and z the ion charge number; ln Λ (~ 20) is the Coulomb logarithm, α is the fine-structure constant, and a_0 is the first Bohr radius. Since the self-collision times for electrons and protons are short compared with times scales for changes in the internal ion state distributions, in examining the latter it is usual to assume Maxwellian distributions for the electrons and protons. The main factor in the justification of this assumption is the rate at which energy is imparted to the particles in question.

Consider now the distribution of a minor-species ion among its excited quantum states. First, some extension of notation is required. For an element $\mathcal{A}$ of nuclear charge z_0, denote the abundance or population (per unit volume) of ions of charge z in the excited state p, of all velocities, by $n(\mathcal{A}, z, p)$. The particular ion is denoted by $\mathcal{A}(z, p)$ and the distribution function by $f(\mathcal{A}, z, p, \mathbf{u})$. The significant contributions to the interaction term for the ions in state p are the various encounters of such ions with electrons, protons, and occasionally other ions when they have a major-species role, together with the coupling to the radiation field. It is emphasized that this last factor may not be neglected here in general as it was for the free-particle distributions. Typical events then include conversion of ions in state p to higher excited states q by electron or proton collision, creation of ions in state p from ions in state q similarly, conversion to an excited state of a different ionization stage by electron collision, spontaneous emission of radiation involving capture of a free electron into the state p, and so on. As in Eq. (2) the various components, excluding radiative emission processes, can be grouped into forward and reverse reaction pairs. However, since radiation trapping is ignored in this work, the inverse and complementary reactions of photoabsorption and stimulated emission which group with spontaneous emission reactions are not present in the interaction term here. The variation of the relative magnitudes of the collision processes and spontaneous emission processes in the interaction term for excited-state distribution functions are responsible for the subtleties of observed spectra.

The procedure for further reduction of the Boltzmann equation to a more useful form for atomic physics can be exemplified by focusing attention on the pair of events that lead to electron collisional excitation and deexcitation. Because of the small mass of the electron relative to the ion nucleus, conservation of total momentum and energy in these collisions can be approximated respectively to conservation of nuclear velocity and of the sum of electron kinetic and internal energy. If it is supposed that the ion velocities are small and equal to a constant, then the interaction term takes the form

3. Atomic Radiation from Low Density Plasma

$$\int \left(\frac{\partial f(\mathcal{A}, z, p, \mathbf{u})}{\partial t}\right)_{\text{int}} d\mathbf{u} = -\sum_q n_e n(\mathcal{A}, z, p) X^e(p \rightarrow q)$$
$$+ \sum_q n_e n(\mathcal{A}, z, q) X^e(q \rightarrow p) + \cdots, \quad (8)$$

where

$$X^e(p \rightarrow q) = \frac{1}{n_e} \frac{8\pi}{m_e^2} \int f_e(\mathbf{v}) E Q_{p \rightarrow q}(E) \, dE \quad (9)$$

is the electron collisional excitation rate coefficient; $E(=\frac{1}{2}m_e v^2)$ is the colliding electron energy, Q is the collision cross section, and $\Delta E(p, q)$ is the excess internal energy of level q over that of level p. So $\Delta E(p, q) = E(p) - E(q)$, where $E(p)$ is the ionization potential of an ion in state p. f_e is given by Eq. (5). The terms not explicitly written down in the above equation comprise similar coefficients and factors for ionizing collisions, collisions with protons, spontaneous emission, etc. Restricting discussion now to a particular element, the element label $\mathcal{A}$ may be dropped from the various expressions, so, for example, writing $n(z, p)$ for $n(\mathcal{A}, z, p)$.

It is convenient at this point to note that various ion states may have the same internal energy and, in an isotropic plasma, be bundled together. Denote the statistical weight of the bundled state p by $\omega(z, p)$. Time-reversal invariance now implies that

$$X^e(p \rightarrow q) = \exp\left(-\frac{\Delta E(p, q)}{kT_e}\right) \frac{\omega(z, q)}{\omega(z, p)} X^e(q \rightarrow p). \quad (10)$$

This detailed balance relationship enables excitation rate coefficients to be deduced immediately from the corresponding deexcitation rate coefficients or vice versa. Similar detailed balance relations exist between collisional ionization and three-body recombination (see Section II).

In Eq. (8) the interaction term consists of collisional parts occurring in detailed balance pairs and spontaneous emission parts for which the inverse processes are not present. In many laboratory and astrophysical plasmas the collisional parts are much larger than the spontaneous emission parts. This is true of the interaction term for highly excited ion state distributions and generally when the electron density becomes large. With spontaneous parts negligible, an equilibrium exists with the overall interaction term zero, called "local thermal equilibrium" or simply LTE. Solution for the populations in LTE yields

$$\frac{n(z, p)}{n(z, q)} = \frac{\omega(z, p)}{\omega(z, q)} \exp\left(+\frac{\Delta E(p, q)}{kT_e}\right), \quad (11)$$

relating populations of the same ionization stage, and

$$\frac{n(z, \text{p})}{n_e n(z+1, 1)} = \left(\frac{h^2}{2\pi m k T_e}\right)^{3/2} \frac{\omega(z, \text{p})}{2\omega(z+1, 1)} \exp\left(+\frac{E(\text{p})}{kT_e}\right), \qquad (12)$$

relating populations of neighboring ionization stages. Formula (11) is called the Boltzmann distribution of populations and formula (12) the Saha–Boltzmann distribution. For simplicity, 1 in Eq. (12) denotes the ground level.

The Saha–Boltzmann and Boltzmann population distribution limits are often approached in practice. It is useful therefore to introduce b factors which denote the deviations of the populations from LTE values. They are defined by

$$\frac{n(z, \text{p})}{n_e n(z+1, 1)} = \left(\frac{h^2}{2\pi m k T_e}\right)^{3/2} \frac{\omega(z, \text{p})}{2\omega(z+1, 1)} \exp\left(+\frac{E(\text{p})}{kT_e}\right) b(\text{p}). \qquad (13)$$

For completeness, mention should be made of the situation with contained radiation. The inverse radiative processes are then present, the coefficients satisfying the familiar Einstein relations. In complete equilibrium between particles and radiation, that is, "thermodynamic equilibrium," the distributions of particle states again follow formulas (11) and (12), while the distribution of radiation is the Planck distribution. For the photon energy density at frequency ν in the frequency interval $d\nu$ this takes the form

$$u(\nu)\, d\nu = \frac{8\pi h \nu^3\, d\nu/c^3}{\exp(h\nu/kT_r) - 1}, \qquad (14)$$

where T_r is the Planck temperature of the radiation field and $T_r = T_e$.

To summarize, on time scales appropriate to the development of excited-state populations, free-particle major species can generally be assumed to follow Maxwellian velocity distributions [Eq. (5)]. The Boltzmann equation for the distribution function of ions in a particular state can be simplified to an equation for the rate of change of the local population density of ions in that state expressed as a sum of terms, each being a product of a rate coefficient and the population densities of species entering the reaction. This special form of the Boltzmann equation is called the set of statistical balance equations. Important detailed balance relations exist between rate coefficients of processes and their inverses exemplified by Eq. (10). The equilibrium population densities when the electron density is high or radiation emission is small approach Boltzmann and Saha–Boltzmann values given by Eqs. (11) and (12).

III. Components of the Statistical Balance Equations

The densities of the majority of plasmas with which this chapter is concerned are too low for the concepts of thermodynamic equilibrium or even

3. Atomic Radiation from Low Density Plasma

local thermal equilibrium to be applicable within an adequate degree of approximation. It becomes necessary to seek solutions of the statistical balance equations where the dominant terms are not the inverse of each other. It will be seen very often that the dominant processes are such that the internal energy of an ion is raised as a result of a collision process whereas it decays to a lower energy by the emission of radiation. It is necessary in this situation to include specific terms in the equation to account for an enormous number of possible atomic processes. The range of the specific terms to be included will now be discussed.

In principle any atomic process that results in an ion being transferred from one quantum state to another should be included. In practice a selection is made of those processes considered to be sufficiently significant for the level of approximation in mind. In making this selection the following factors are generally taken into account. As has already been discussed, the plasma is made up of the major constituents electrons and protons and minor constituents atomic hydrogen and atoms and ions of heavier elements, and so the first major simplification is achieved by neglecting collisions between minor constituents on the grounds that they are rare compared with collisions involving the major constituents (although special reference will be made to the possibility of charge exchange between hydrogen atoms and some ions). Thus for collision processes involving the minor constituents only electrons and protons need be considered. Because in conditions of approximate equipartition of kinetic energy between electrons and protons the electrons have much greater velocities and since the rate coefficients are the product of cross section with velocity, electrons are usually much more effective than protons in causing collisional transitions among the states of the minor constituents. There are, however, special circumstances where the proton collision cross sections are sufficiently large to overcome the disadvantage of low velocity and make these processes important. Of particular importance in this connection are collisions which redistribute populations between fine-structure levels. Despite this rather severe restriction in the collision processes to be taken into account there remain a very large number of possible collision processes. The next step is to dismiss processes which cause the ionic charge to change by more than one unit of charge. There may be important circumstances where ionization collisions cause the ejection of more than one electron, but it appears that no practical calculation has included this possibility. For multiple recombination to be effective a many-body collision would be required and this may be dismissed as unlikely. It will be seen that authors of the calculations to be discussed have adopted further restrictions in the range of collision processes included in the solution of their particular version of the statistical balance equations. However, the situation may be summarized as follows:

Collision processes included in solutions of the statistical balance equations:

$\mathcal{A}(z, p) + e \rightleftharpoons \mathcal{A}(z, q) + e$	(excitation and deexcitation by electrons).
$\mathcal{A}(z, p) + p \rightleftharpoons \mathcal{A}(z, q) + p$	(excitation and deexcitation by protons).
$\mathcal{A}(z, p) + e \rightleftharpoons \mathcal{A}(z + 1, q) + 2e$	(ionization and three-body recombination by electrons).
$\mathcal{A}(z, p) + H \rightarrow \mathcal{A}(z - 1, q) + p$	(charge exchange with hydrogen atoms).

In addition the following radiation processes are usually included:

$\mathcal{A}(z + 1, p) + e \rightarrow \mathcal{A}(z, q) + h\nu$	(radiative and dielectronic recombination).
$\mathcal{A}(z, p) \rightarrow \mathcal{A}(z, q) + h\nu$	(spontaneous radiative decay).

Having identified the individual processes it remains to discuss the evaluation of the coefficients used to describe them in the statistical balance equations. The remainder of this section is devoted to this purpose.

The parametric dependence of the various coefficients on the ion charge z frequently involves the factor $z + 1$. It is convenient to introduce $\xi_e = (z + 1)^2 I_H / kT_e$.

A. *Collisional Ionization by Electron Impact*

Accurate quantum-mechanical calculation of this reaction is complex and so, for general application, recourse has been made to simpler estimates. These rely upon the collision of the incident electron with the target ion approximating to a classical binary encounter between the incident and initially bound electrons. The rate coefficient may be written

$$S(p) = 16\sqrt{\pi}\alpha c a_0^2 \frac{1}{(z + 1)^3 \xi_e^{1/2}} \nu^4(p) F \exp\left(-\frac{\xi_e}{\nu^2(p)}\right), \qquad (15)$$

where z is the charge of the initial ion and $\nu(p)$ is the effective principal quantum number of the level p from which ionization takes place. The factor $F = 1$ for the classical rate (Thomson, 1912) when the mean electron energy is much less than the ionization potential. This formula has been improved semiempirically by a number of authors by modifying F to give better correspondence with experiment or quantum-mechanical theory near threshold and by introducing an improved high temperature asymptotic form. Seaton (1964) gives the semiempirical formula (SEF). Lotz (1967a) gives further adjustments which, however, do not change the threshold behavior significantly from that of Seaton. [Lotz (1967b, 1968, 1970) gives more elaborate parametric forms for neutral ionization rates.] Burgess (1964a; see also Burgess and Percival, 1968) gives a set of formulas which unite a symmetrized

3. Atomic Radiation from Low Density Plasma

binary encounter approach for close collisions with a semiclassical impact parameter approach for distant encounters. In its most common usage it is termed the ECIP approximation (Burgess and Summers, 1976; Summers, 1979). These various formulas have been compared with experiment in the near-threshold region by Burgess et al. (1977). The approximate formulas can be substantially in error (standard deviation ~90%), with no formula being strongly favored. More refined Coulomb–Born calculations are available for some ions (Moores, 1972, 1978, 1979; Frank, 1980). Agreement with experiment in the threshold region is not markedly improved although there is indication of reduced variance. More recently Sampson and co-workers have applied Coulomb–Born calculations, computed for infinite z and then applied for finite z using scaling laws, to some ionization cross sections (see Sampson, 1978; Sampson and Golden, 1978, 1979; Golden and Sampson, 1977, 1980; Golden et al., 1978; Moores et al., 1980). Typically their results fall between ECIP and SEF and are convenient to use. Even more recently Burgess and Chidichimo (1983) have published an analysis of available ionization cross sections, both measured and calculated. They derive a general formula for the total ionization cross section (i.e., including inner-shell excitation and autoionization as discussed below) that represents the available experimental data to within ±23% rms deviation.

It is important to take ionization from inner shells and ionization occurring by excitation to autoionizing levels properly into account when calculating ionization from the ground level of an ion. Thus the ionization rate from the ground level (denoted by 1 here for simplicity) is modified to

$$S(1) = \sum_i \zeta(i)S(i, 1) + \sum_{\substack{r \\ E(r)<0}} X^e(1 \to r) \qquad (16)$$

where the summations are over shells i and autoionizing levels r. The $S(i, 1)$ are the direct ionization coefficients of the form (15) from the shell i and the X^e are excitation rate coefficients to autoionizing levels; $\zeta(i)$ is the number of equivalent electrons in shell i. This expression assumes that excitation to autoionizing levels certainly leads to ionization. When the autoionizing levels lie densely through the outer electron ionization threshold, the combined inner-shell and autoionization contribution can be obtained more simply by omitting the second sum and the inner-shell direct ionization and increasing ζ for the outer shell by the number of inner-shell electrons that have the same principal quantum number as the outer shell (see Burgess et al., 1977; Burgess and Chidichimo, 1983). Figure 2 shows experimental ionization cross section data for C^{3+}. The autoionization thresholds are evident. Theoretical approximations are also shown for comparison.

The inverse process of three-body recombination is obtained from detailed balance

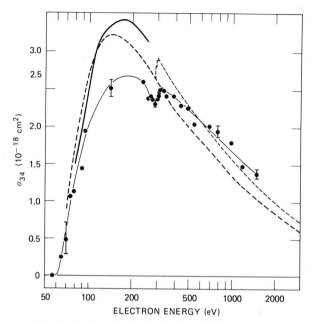

Fig. 2. Cross section for electron impact ionization of C^{3+} measured by the crossed-beams method. [From Crandall *et al.* (1979).] Experimental points are shown, with a best fitting curve. Autoionization enhancement is evident above the 1s ionization threshold arising from excitation to resonances of the form 1s2s2*l*. Coulomb–Born calculations of Moores (1978) (heavy solid line) and infinite-*z* Coulomb–Born calculations of Golden and Sampson (1977) (dashed curves) are shown for comparison. The latter show the calculated effect of autoionization. [Copyright 1979 The Institute of Physics.]

$$\alpha_3(p) = 2^6 \pi^2 \alpha c a_0^5 \frac{\xi_e}{(z+1)^6} \frac{\nu^4(p)\omega(z, 1)}{\omega(z+1, 1)} F; \tag{17}$$

z is the charge of the recombined ion.

B. Collisional Excitation by Electron Impact

A useful review has been written by Seaton (1975). This updates and extends earlier reviews by Gabriel and Jordan (1972), Bely and Van Regemorter (1970), and Kunze (1972). A large number of methods and elaborate computer codes are now available for the calculation of collisional excitation cross sections and rates. In relation to fusion plasmas, a number of areas of application with different demands can be distinguished.

Provided the excitation energy of the transition is fairly small compared with the incident electron energy, electric dipole excitation rates are generally the largest. Recognizing that distant encounters are of greatest importance, a simple approximation for the excitation rate can be obtained:

3. Atomic Radiation from Low Density Plasma

$$X^e(p \to q) = 8\pi\sqrt{3} \, \frac{2\sqrt{\pi}\alpha ca_0^2}{3} \, \frac{\xi_e^{1/2}}{(z+1)^3} \, \frac{(z+1)^2 I_H}{\Delta E(p, q)}$$

$$\times f(p, q) \exp\left(-\frac{\Delta E(p, q)}{kT_e}\right) \int_0^\infty g^{III}(x) \exp(-w) \, dw. \quad (18)$$

Here $f(p \to q)$ is the absorption oscillator strength and $g^{III}(x)$ is the free–free Gaunt factor for the incident electron in the ion Coulomb field. It is conveniently expressed in terms of the parameter $x = [wkT_e/\Delta E(p, q)]^{1/2}$, where wkT_e is the electron's final energy. In practice additional factors must be taken into account. The transition probability may become so large that conservation of probability is violated, the atomic levels being strongly coupled. Also the cross-section threshold behavior is different for atoms and ions, the ion threshold being finite. It is usual in consequence to replace g^{III} by an adjustable function $\bar{g}$, the mean Gaunt factor, chosen to represent accurate cross sections as well as possible. Seaton (1962a) chooses $\bar{g} \simeq 0.2$ (constant) for ions. Van Regemorter (1962) tabulates $\bar{g}$ and its integral P defined by

$$P\left(\frac{\Delta E(p, q)}{kT_e}\right) = \int_0^\nabla \bar{g}(x) \exp(-w) \, dw \quad (19)$$

separately for atoms and ions. Semiclassical expressions for P are given by Seaton (1962b) and Burgess (1964a; cf. Burgess and Summers, 1976). These approximate methods are useful for calculating highly excited level populations. Important additional formulas based on the correspondence principle have also been given for this region in an extended series of papers by Percival and Richards (see Percival and Richards, 1975).

The above formulas are too inaccurate for detailed application to low levels; also, nondipole excitations play a major role in determining metastable level populations. Figure 3 shows the calculated electron impact collision strength for the $2s^2$ ^1S–$2s2p$ ^3P transition in O^{4+} and illustrates well the complex resonance structure. Collision strength is a dimensionless quantity related to collision cross section and its behavior would be reflected in the rate coefficients.

Useful results are provided by the infinite-z Coulomb–Born approximation of Sampson and co-workers based on the earlier work of Burgess et al. (1970). They are contained in an extended collection of papers originating with Sampson (1974) and Sampson and Parks (1974; see Magee et al., 1977, 1980, for further references). However, the principal source of reliable data for such transitions are the large computer codes implementing "distorted-wave" and "close-coupling" methods. There is a very extensive literature which cannot reasonably by surveyed here. Valuable compilations of theoretical data have been assembled by Magee et al. (1977, 1980). There is also a recent review by Henry (1981).

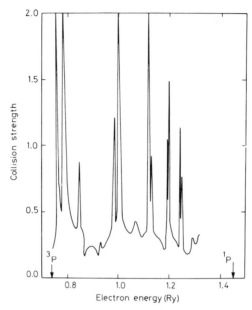

Fig. 3. Electron impact collision strengths for the $2s^2$ ^1S–$2s2p$ ^3P transitions in O^{4+} calculated in the close-coupling approximation. [From Berrington et al. (1977). Copyright 1977 The Institute of Physics.] Elaborate resonance structure is evident in the $2s^2$ ^1S–$2s2p$ ^3P collision strength between the ^3P and ^1P thresholds. The collision strengths are finite at threshold.

The deexcitation rate coefficient is obtained from the detailed balance relationship:

$$X^e(q \rightarrow p) = \frac{\omega(p)}{\omega(q)} \exp\left(\frac{\Delta E(p, q)}{kT_e}\right) X^e(p \rightarrow q). \tag{20}$$

Proton impact is important for causing transitions between closely neighboring levels, such as highly excited degenerate l states (Pengelly and Seaton, 1964) and in the quadrupole mixing of J sublevels. Semiclassical and close-coupling methods are available. The semiclassical methods are extensions of the nuclear Coulomb excitation work of Alder et al. (1956) to the atomic case. There is again a substantial literature (see Bahcall and Wolf, 1968; Kastner and Bhatia, 1979; Landman and Brown, 1979; Sahal-Brechot, 1974; Bely and Faucher, 1970; Faucher, 1975).

C. Spontaneous Emission of Photons

The values of the A coefficients depend on the details of the atomic structure of atoms and ions, and their calculation and measurement are an important part of atomic physics. Compilations and bibliographic reviews of

3. Atomic Radiation from Low Density Plasma

transition probabilities are maintained by the National Bureau of Standards in Washington. The main compilations are Wiese *et al.* (1966, 1969). Bibliographies include NBS Special Publications 320 (three parts) and 505. Large computer codes are available for the calculation of transition probabilities implementing Hartree–Fock and multiconfiguration direct diagonalization methods (see Froese-Fischer, 1977; Cowan, 1980; Eissner *et al.*, 1974). These generally incorporate relativistic corrections.

For electric dipole transitions, the method of Bates and Damgaard (1949) is of fairly good accuracy and simply applied. In certain sensitive transitions incorporation of a correction for rapidly changing quantum defect (Seaton, 1958) improves reliability. Wiese and Younger (1976) give useful isoelectronic sequence extrapolations of oscillator strengths. For dipole transitions the A coefficient may be written as

$$A(p \rightarrow q) = \frac{16\alpha^4 c}{3\sqrt{3}\pi a_0} \frac{(z+1)^4}{\omega(z, p)} \frac{g^I(p, q)}{\nu(p)\nu(q)[\nu^2(q) - \nu^2(p)]} \quad (21)$$

g^I is the bound–bound Gaunt factor, and is of order unity. Electric quadrupole, magnetic dipole, and electric dipole transitions occurring through LS coupling breakdown have different z dependences. $\nu(p)$, etc., denote effective principal quantum numbers.

D. Radiative Recombination

Recombination rates are required to all levels p of the ion. The energies of highly excited levels are approximately hydrogenic and so hydrogenic theory for recombination coefficients to them is appropriate. The calculation of radiative recombination coefficients for the formation of hydrogenlike ions has been considered by Menzel and Peckeris (1935) and corrected by Burgess (1958; see also Seaton, 1959; Burgess, 1964b; Burgess and Summers, 1976). The recombination coefficient may be expressed as

$$\alpha_r(p) = \frac{2^7}{3} \sqrt{\frac{\pi}{3}} \alpha^4 c a_0^2 \xi_e^{3/2} \frac{(z+1)}{\omega(z, p)} \frac{1}{\nu(p)}$$
$$\times \exp\left(\frac{E(p)}{kT_e}\right) \int_{E(p)/kT_e}^{\infty} \frac{g^{II} \exp(-w)}{w} dw \quad (22)$$

where g^{II} is the bound–free Gaunt factor and is of order unity near threshold; w is the recombining electron energy divided by kT_e. For nonhydrogenic ions, a useful approximate method has been given by Burgess and Seaton (1960). This is an extension into the continuum of the method of Bates and Damgaard (1949) for bound–bound transitions. Computer codes are available for more detailed calculations of low-level recombination rates. The associated processes of photoionization and stimulated recombination are not considered here.

E. Dielectronic Recombination

This process occurs in two stages, resonant capture followed by radiative stabilization:

$$\mathcal{A}(z+1, i) + e \rightleftharpoons \mathcal{A}(z, j, p),$$
$$\mathcal{A}(z, j, p) \to \mathcal{A}(z, k, p) + h\nu, \qquad (23)$$

where i, j, and k label the parent or core levels. The incident electron excites the parent ion in level i to level j, the incident electron losing sufficient energy to enter a bound level denoted by p. This condition is transient and may break up in an Auger transition, shown as the left-pointing arrow in the first expression. The following outline of the process is based on the work of Burgess (1965a). It is assumed that no other effects interrupt the resonant level and that the electron in level p is passive in the stabilization phase. Usually the most important contributions come from levels for which i → j is optically allowed, k belongs to the ground configuration, and the level p corresponds to large principal and angular momentum quantum numbers. In these circumstances the incident electron energy E is

$$\frac{E}{I_H} = (z+2)^2 \Delta\varepsilon(i,j) - \frac{(z+1)^2}{\nu^2(p)}, \qquad \Delta\varepsilon(i,j) = \frac{1}{\nu^2(i)} - \frac{1}{\nu^2(j)}. \qquad (24)$$

The transition of the parent ion i → j is termed the "core transition."

Total and partial dielectronic recombination coefficients at low electron density may then be written from detailed and statistical balancing arguments (Burgess, 1965a). It is convenient to introduce a dielectronic recombination "cross section." Then

$$\tilde{Q}(i,j,p) = \int_{\Delta E(p)} Q \, dE = \frac{2\pi^2 a_0^3 I_H}{\alpha c} \frac{I_H}{E}$$
$$\times \left(\frac{\omega(z,j,p)}{\omega(z+1,i)} \frac{\sum_{\varepsilon'} A_a(j, p \to i, \varepsilon') \sum_k A_r(j, p \to k, p)}{\sum_k [A_r(j, p \to k, p) + \sum_{\varepsilon''} A_a(j, p \to k, \varepsilon'')]} \right). \qquad (25)$$

In this expression A_a denotes Auger rates and A_r radiative rates; ε' and ε'' denote free electron states of energy E. The cross section is integrated over an interval $\Delta E(p)$ of energy equal to the spacing between energy levels, approximately,

$$\Delta E(p)/I_H = 2(z+1)^2/\nu^3(p). \qquad (26)$$

The dielectronic recombination coefficient to level p is then

$$\alpha_d(p) = \sum_j \alpha_d(i,j,p) = \sum_j \frac{f_e'(v)}{n_e m_e} \tilde{Q}(i,j,p), \qquad (27)$$

3. Atomic Radiation from Low Density Plasma

where $f'_e(v)$ is the Maxwellian electron speed distribution [$=4\pi v^2 f_e(\mathbf{v})$ in Eq. (5)] with $\frac{1}{2}m_e v^2 = E$. Letting k = i for simplicity, we have

$$\frac{1}{(z+1)^2}\bar{Q}(i,j,p) = 2\pi^2 a_0^2 \alpha^3 I_H \frac{1}{[(z+2)/(z+1)]^2 \Delta\varepsilon(i,j) - 1/\nu^2(p)}$$

$$\times \left(\frac{z+2}{z+1}\right)^4 \Delta\varepsilon^2(i,j) f_{i\to j} \frac{\omega(p)\sum A_a}{A_r + \sum A_a} \quad (28)$$

with $f_{i\to j}$ the absorption oscillator strength. The last term corresponds to the term in large parentheses in Eq. (25). If there were no z dependence in the factors on the right-hand side, the dielectronic recombination coefficient would depend linearly on $z + 1$, as does the radiative recombination coefficient. The dependence of $\bar{Q}/(z+1)^2$ on z and $\nu(p)$ is illustrated in Figs. 4 and 5. Seaton and Storey (1976) have given a valuable review of the details of dielectronic recombination calculations.

A very useful general formula for the total low-density dielectronic rate coefficient, summed over all levels, is given by Burgess (1965b). Simplifications made in obtaining this formula lead to erroneous results in certain circumstances. Recombination to heliumlike ions is a special case and has been examined by Burgess and Tworkowski (1976). The opening of alternative Auger channels leads to substantial reduction of the dielectronic rate over that given by the general formula (Jacobs et al., 1977a). Also, the fine structure of the parent ion state can permit a secondary autoionization event (Blaha, 1972). Elaborate calculations have been made of dielectronic rates to low levels of three-electron and two-electron systems (see Bely-Dubau et al., 1979a,b; Dubau et al., 1980), the work being directed toward interpretation of the associated satellite photons emitted in stabilization. This is considered further in Section VI.C.

F. Scaling of the Coefficients with Ionic Charge

It is important to emphasize the dependence of the various rate coefficients above on z. If "reduced" electron temperatures and densities defined by

$$\theta_e = T_e/(z+1)^2, \quad \eta_e = n_e/(z+1)^7 \quad (29)$$

are introduced, it is evident that $A/(z+1)^4$, $X^{(e)}(z+1)^3$, $S(z+1)^3$, $\alpha_r/(z+1)$, and $\alpha_3(z+1)^6$ are functions of θ_e and η_e and only weakly dependent on z. This is provided only electric dipole A's corresponding to a change in principal quantum number are included. From expression (13), $[n(z,p)/n_e n(z+1,1)]/(z+1)^3$ depends only on θ_e and the b factor. Therefore the statistical balance equations for hydrogenic ions expressed in terms of

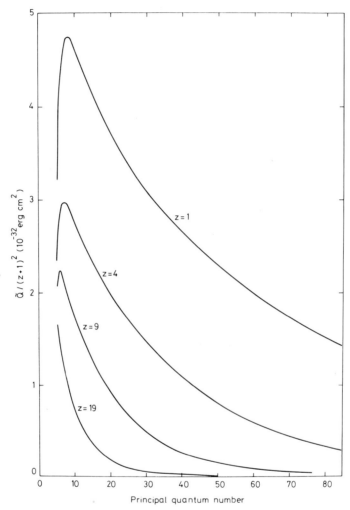

Fig. 4. Variations of scaled dielectronic cross section $\bar{Q}/(z + 1)^2$ with principal quantum number and ion charge; $z + 1$ is the charge on the recombining ion. The cross section is summed over angular states of the same principal quantum number. Representative core oscillator strength $f(i \to j) = 1.0$ and reduced core transition energy $\Delta\varepsilon(i, j) = 0.1$ are used.

the b's will be approximately independent of z for specified θ_e and η_e (see Bates et al., 1962a).

The dielectronic coefficient α_d has a more complicated z dependence owing to the last factor in formula (28). The scaling rules therefore have more limited value when dielectronic recombination is active and also when metastable populations are involved, but remain of use for rough extrapolation and as a basis for interpolation of populations along isoelectronic sequences.

3. Atomic Radiation from Low Density Plasma   71

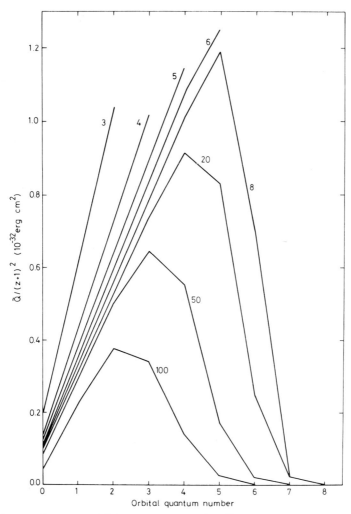

Fig. 5. Variation of scaled dielectronic cross section $\bar{Q}/(z + 1)^2$ with n and l where n is the principal quantum number and l the orbital quantum number of the outer electron in the recombined state. Representative values of $f(i \to j) = 1$ and $\Delta\varepsilon(i, j) = 0.1$ are used. The recombining ion charge, $z + 1 = 2$, and the curves are labeled with the principal quantum number.

IV. The Collective Viewpoint of Ionization and Recombination

A picture was presented in Section II of the rate of change of each level population density of an ion species determined by a very large number of coefficients and requiring the solution of the equations of statistical balance. It is difficult to express these equations explicitly and meaningfully in mathematical form in the general case because of the number and variety of terms involved. These terms represent the many individual collisional and radiative processes by which the ions of an element can change from one quantum level and charge to another. In practice, of course, finite selections of terms must be made and it is the solutions in these cases that will be discussed. Most of the considerations that guide the choice of the terms have already been mentioned. They may be summarized briefly thus:

(i) The most probable collisions are elastic collisions leading toward equipartition of kinetic energy and hence to Maxwellian velocity distributions.

(ii) The plasma composition ensures that electron and proton collisions far outnumber collisions with minor species—but beware of special cases such as charge exchange.

(iii) Electrons being faster are much more effective than protons in causing ionization and excitation—except for proton excitation of some very closely spaced levels.

(iv) Densities are such that binary collisions far outnumber multiple-particle collisions—but sometimes three-body recombination is important.

(v) The optically thin approximation is adopted, i.e., photoionization and photoexcitation are neglected.

These guidelines have been used to justify specific selections of terms for the equation of statistical balance, but they do not represent a completely logical argument for the exclusion of terms which have still to be studied in detail. However, using the guidelines the equation of statistical balance contains two kinds of terms:

(i) Radiation terms of the forms $n(z, p)A(p, q)$ and $n_e n(z + 1, p) \alpha(z + 1, p, z, q)$, where $A(p, q)$ and $\alpha(z + 1, p, z, q)$ are the relevant spontaneous transition probabilities and radiative recombination coefficients, respectively. Here the notation for the recombination coefficients has been changed to indicate explicitly the initial and final charge states of the ion.

(ii) Collisional terms which are products of the relevant particle population densities with a collisional coefficient, for example, $n_e n(z, p)X(z, p, q)$.

The individual coefficients have been discussed in Section III. In the present section it is shown how it is possible to define to a good degree of approximation composite coefficients of ionization and recombination. This leads to an important reduction in the number of differential equations of statistical balance that require to be solved. It will be seen that the relaxation time constants for the population densities to reach a steady state or quasi-steady state provide the key to these simplifications. Consideration of these time constants leads to the grouping of ions under the following three headings:

3. Atomic Radiation from Low Density Plasma

(i) Ions in excited levels (not metastable).
(ii) Ions in ground levels.
(iii) Ions in metastable levels.

The time constant to be associated with some quantum level of an excited ion is equal to (the sum of spontaneous transition probabilities plus the products of particle densities with collisional coefficients)$^{-1}$; thus these times are equal to or less than the radiative lifetimes. It will be seen that compared with the others such times are very short.

Ions in their ground level are stable against spontaneous radiative decay and their lifetimes are determined by collision processes which can lead to excitation, ionization, or recombination. The time constant associated with a ground-level ion is the reciprocal of the sum of products of particle densities with relevant collisional coefficients and is therefore inversely proportional to the plasma density. They are typically many factors of ten longer than radiative lifetimes of excited levels.

Metastable ion populations cannot be depleted rapidly by spontaneous emission processes since their only available radiative decay paths are via relatively slow non-electric-dipole transitions. At high electron densities collisional redistribution can take over from radiation so that at low densities metastable levels can have long lifetimes, but above some critical electron density have lifetimes dependent on density. They therefore form an intermediate type of population between ground and excited populations with important consequences.

Table I shows relaxation time constants for a number of principal quantum levels of hydrogenic ions. These are given for a range of electron temperatures and densities and demonstrate the varying influence of collisional and radiative factors. Following the scaling laws for hydrogenic ions already described the time constants scale as the fourth power of the nuclear charge.

It is concluded from data of this kind that when for some reason a particular distribution of population densities is disturbed the excited levels rapidly reestablish a distribution which is referred to as quasi-steady-state, while the ground and possibly the metastable levels take somewhat longer. This argument is used to justify putting the time derivative of the excited-level population $dn(z, p)/dt$ equal to zero for all levels but the ground and metastable. Thus an infinite number of differential equations (one for each quantum level, of which there are an infinite number) is reduced to an infinite set of ordinary simultaneous equations plus a small number of differential equations (one each for the ground and metastable levels). Effectively, these equations assume that the excited levels are in instantaneous equilibrium with the ground and metastable levels of the same and neighboring ions $n(z - 1, g)$, $n(z - 1, m)$, $n(z, g)$, $n(z, m)$, $n(z + 1, g)$, and $n(z + 1, m)$ at each moment of time. Here g and m indicate ground and metastable levels, respectively.

TABLE I

Relaxation Time Constants τ (in s) for Hydrogenic Ions[a]

θ_e (K)	η_e (cm^{-3})	$p = 1$	$p = 2$	$p = 3$	$p = 15$
4,000	10^8	3.1^{+11}	2.1^{-9}	1.0^{-8}	1.1^{-6}
4,000	10^{18}	3.1^{+1}	1.4^{-11}	2.8^{-13}	1.2^{-16}
16,000	10^8	1.2^{+2}	2.1^{-9}	1.0^{-8}	2.1^{-6}
16,000	10^{18}	1.2^{-8}	1.8^{-12}	1.7^{-13}	2.5^{-16}
64,000	10^8	5.4^{-1}	2.1^{-9}	1.0^{-8}	3.5^{-6}
64,000	10^{15}	5.4^{-8}	7.4^{-10}	2.3^{-10}	4.9^{-13}
64,000	10^{18}	5.4^{-11}	1.1^{-12}	2.3^{-13}	4.9^{-16}
256,000	10^8	1.3^{-1}	2.1^{-9}	1.0^{-8}	5.5^{-6}
256,000	10^{12}	1.3^{-5}	2.1^{-9}	9.8^{-9}	9.8^{-10}
256,000	10^{15}	1.3^{-8}	9.3^{-10}	4.1^{-10}	9.8^{-13}
256,000	10^{18}	1.3^{-11}	1.7^{-12}	4.3^{-13}	9.8^{-16}

[a] Scaled time constants $(z + 1)^4 \tau(p)$ are tabulated where z is the hydrogenic ion charge number and p the principal quantum number. Values are given at reduced temperatures θ_e and electron densities η_e [cf. Eq. (29)].

The next step of simplification reduces the infinite set of simultaneous equations to a finite set plus a set of specific equations for the populations of the most highly excited levels. Again it is best to use a physical rather than a mathematical argument to justify the step. This goes as follows. The more highly excited an ion is, the larger its structure and therefore the greater its probability of suffering a collisional transition to another level. At the same time the probability of radiative decay decreases with increasing principal quantum number. Thus there is always some level above which radiative processes may be neglected in comparison with collisions. For a situation where the colliding particles have a Maxwellian distribution of velocities, collisions in one direction are exactly balanced by collisions in the inverse direction. This leads to these levels having a Saha–Boltzmann distribution of population densities with the temperature defined by that of the Maxwellian. Thus above some level which has been called the collision limit it becomes a good approximation to describe the population densities of the excited levels by Saha–Boltzmann equations.

With three groupings of the levels (a: ground and metastable, b: other levels below the collision limit, and c: levels above the collision limit) and an appropriate selection of coefficients it becomes possible to solve the set of equations that now takes the place of the equations of statistical balance. The first solution of this kind was for hydrogen and hydrogenlike ions and is due to Bates et al. (1962a). These authors took account of radiative decay processes and only those collisions involving electrons with the other components. For the densities with which they were concerned the population density of the 2s metastable level exchanges rapidly with that of the 2p level

3. Atomic Radiation from Low Density Plasma

(see a recent paper by Ljepojevic et al., 1984), so there is no need to take separate account of the metastable time constant. Thus they were able to express the population densities of the quantum levels in the following forms:

$$n(z, p) = n(z, g)C_1(p) + n(z + 1, g)C_2(p), \qquad (30)$$

$$\frac{dn(z, g)}{dt} = -n_e n(z, g)S_{c-r} + n_e n(z + 1, g)\alpha_{c-r}; \qquad (31)$$

$C_1(p)$ and $C_2(p)$ are composite collisional coefficients which express the total range of ways in which level p may be populated from the ground level g in the case of $C_1(p)$ and from the fully stripped ion of charge $z + 1$ in the case of $C_2(p)$. Above the collision limit $C_1(p)$ tends to zero and the term $n(z + 1, g)C_2(p)$ becomes the appropriate expression in the Saha–Boltzmann equation. The coefficients S_{c-r} and α_{c-r} are called collisional–radiative ionization and recombination coefficients and are composite coefficients representing the multiplicity of routes by which the ground level may be depopulated and populated, respectively. The nature of the coefficients can be clarified from a mechanistic viewpoint. At finite densities ionization can occur both by direct collisional ionization from the ground level and also by multistep excitation and deexcitation followed finally by an ionizing collision. S_{c-r} takes proper account of these processes and discounts those electrons that eventually return to the ground level. In the limit of zero electron density S_{c-r} is simply the direct ionization rate out of the ground level. S_{c-r} increases with electron density, reaching a new limit determined by the sum of all excitation and direct ionization rates out of the ground level. Likewise at finite densities recombination of electrons into upper levels is dominated by three-body recombination. Thus α_{c-r} is greater than the summed radiative recombination rates into all levels but tends to that sum as the electron density tends to zero. All these coefficients are functions of temperature and electron density and have been tabulated for wide ranges of both in the following series of papers: Bates et al. (1962a,b); Bates and Kingston (1963); McWhirter and Hearn (1963); Hutcheon and McWhirter (1973); Ljepojevic et al. (1984). None of these includes helium or heliumlike ions in their calculations.

Some of the numerical values obtained by these authors are illustrated in Fig. 6, which shows collisional–radiative ionization and recombination rate coefficients in the limit of low electron densities ($n_e/Z^7 \lesssim 10^8$ cm^{-3} with Z the nuclear charge) for hydrogen and hydrogenlike ions. Figure 7 illustrates the way in which collisional processes take over from radiative processes with increasing principal quantum number of the level considered. The variation of the relative importance of these processes with density is shown in Fig. 8. It may be seen that above a reduced density of about 10^{16} cm^{-3} collisional processes dominate even for the ground level. There is a very much smaller variation with electron temperature.

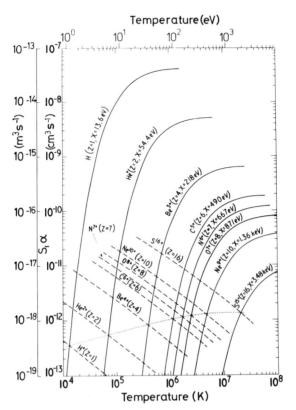

Fig. 6. Variation of collisional–radiative ionization rate coefficients $S_{c-r}(z)$ (continuous lines) and recombination rate coefficients $\alpha_{c-r}(z)$ (dashed line) for hydrogen and hydrogenlike ions with electron temperature T_e in the limit of low electron density n_e. [From Bates et al. (1962a).] Z denotes the nuclear charge. The intersection points indicated by the fine dotted line correspond to equal hydrogenic ion and bare-nucleus abundances in ionization equilibrium.

Before going on to describe calculations of composite coefficients for other types of ions there is another issue of wider implication which is better considered here. It is frequently necessary to add up the populations of all the ions of an element in all their quantum levels to discover the total. Since the population densities of the uppermost levels are proportional to the squares of the corresponding principal quantum numbers and would therefore appear to lead to infinite total populations the problem needs some consideration. In conditions of high electron density where collisions dominate the depopulating mechanisms for all levels including the ground level, the Saha–Boltzmann equation gives a good description of the population distribution among the levels of a particular ion. In practice, when this equation is applied to a specific situation the exponential factor has such values as ensure that highly excited level populations are small compared

3. Atomic Radiation from Low Density Plasma

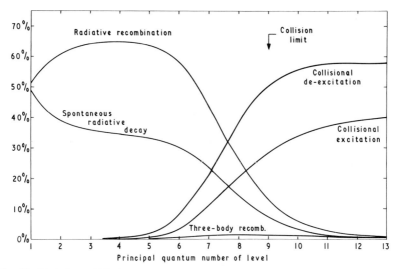

Fig. 7. Variation with principal quantum number of the relative magnitudes of the five processes populating the first thirteen principal quantum levels of hydrogenic ions in a plasma. The reduced electron density $n_e/Z^7 = 10^{10}$ cm^{-3} and reduced electron temperatures $T_e/Z^2 = 6.4 \times 10^4$ K. Z is the nuclear charge number.

with levels close to and including the ground level provided levels having very high principal quantum numbers (more than some hundreds) are excluded. Under conditions of lower density, where the Saha–Boltzmann equation may not be used down to the lower levels, the dominance of radiative processes ensures that the ground and low-lying metastable levels have even higher populations with respect to the upper levels than would be given by the Saha–Boltzmann equation. Thus again in calculating the total population only these low levels need be included provided the very highest levels

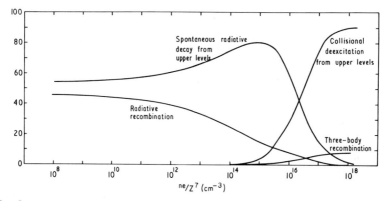

Fig. 8. Variation of the relative magnitudes of the four processes populating the ground level of hydrogenic ions with reduced electron density n_e/Z^7. The reduced electron temperature $T_e/Z^2 = 3.2 \times 10^4$ K. Z is the nuclear charge number.

are excluded. The physical justification for excluding these very highly excited levels is that in plasmas of finite density the orbits of the excited electrons are so large (radius proportional to the square of the principal quantum number) as to include many neighboring particles so that excited ions of charge z become indistinguishable from those of charge $z + 1$. This means in effect that the population of ions of charge $z + 1$ includes very highly excited ions of charge z. Despite its apparent arbitrariness this definition leads to a consistent and sufficiently precise determination of population densities for the tenuous plasmas with which this chapter is concerned. Thus in identifying the levels having a major part of the total population of a series of ions it is those near the ground that need to be taken into account—in particular the ground level itself and metastable levels close to the ground level. Metastable levels that are close to the ionization limit (as in hydrogenlike and heliumlike ions) have small populations with respect to the ground level in most circumstances of relevance here for the reasons given above.

For ions having one bound electron or more, dielectronic recombination can be an important process. Terms representing this process must therefore be included in the statistical balance equations for all ions except bare nuclei. Since the dielectronic process significantly populates very highly excited levels, dependent on the core excitation details, usually several hundred principal quantum shells have to be included. As another consequence of this, the collisional–dielectronic recombination coefficient α_{c-d} shows much greater sensitivity to electron density at low densities. α_{c-d} is illustrated for two contrasting cases in Figs. 9 and 10. Figure 9 shows $Ca^+ + e$ recombination, for which the core excitation energy is small. The very large rise in α_{c-d} at an electron temperature sufficient to excite the core is notable, as is the great sensitivity to increasing density. Figure 10, on the other hand, for $He^+ + e$ recombination, shows the smaller dielectronic rise typical of an ion with a large core excitation energy. In both these cases, as the free-electron density becomes very large, the α_{c-d} curves approach the collisional–radiative slope of $T_e^{-9/2}$. The large number of levels to be included in these calculations has necessitated the introduction of a numerical procedure called "matrix condensation" (Burgess and Summers, 1969, 1976), which renders the problem tractable. Figure 11 shows the variation of the high-level b factors [see Eq. (13)] for Fe^{14+}, indicating the role of such states in the overall recombination. Using the above methods Summers (1974, 1979) has calculated a large number of cases. Some correction of these calculations is now warranted in the light of the comments in Section III on the dielectronic process. S_{c-d} follows the same pattern as S_{c-r} for hydrogen. The onset of stepwise ionization shown here is found qualitatively in some laboratory measurements of ionization rates (Rowan and Roberts, 1979).

The specific calculations due to Summers (1974, 1979) mentioned above neglect the influence of metastable levels. The way in which these levels

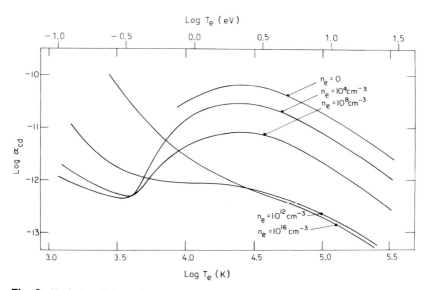

Fig. 9. Variation of the collisional–dielectronic recombination coefficient α_{c-d} for $Ca^+ + e$ with T_e and n_e. The large rise as T_e increases at low n_e is characteristic and due to the dielectronic process. The dielectronic recombination contribution is progressively suppressed as n_e increases. [Reprinted courtesy of A. Burgess and H. P. Summers and *The Astrophysical Journal*, published by the University of Chicago Press; © 1969 The American Astronomical Society.]

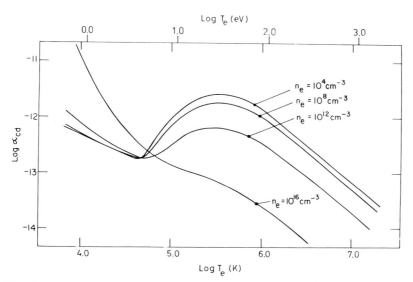

Fig. 10. Variation of the collisional–dielectronic recombination coefficient α_{c-d} for $He^+ + e$ with T_e and n_e. The behavior is broadly as in Fig. 9, but the overall dielectronic contribution is less due to the higher core excitation energy. Since the dielectronic recombination occurs mainly through lower levels, it becomes suppressed at higher electron densities than for $Ca^+ + e$. [Reprinted courtesy of A. Burgess and H. P. Summers and *The Astrophysical Journal*, published by the University of Chicago Press; © 1969 The American Astronomical Society.]

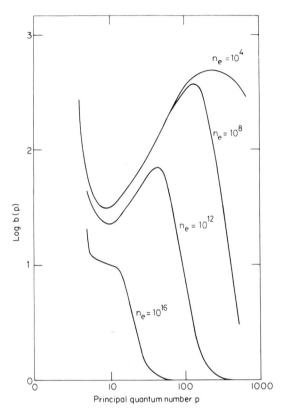

Fig. 11. Variation of the b factors for principal quantum shell populations of Fe^{14+} with principal quantum number p and electron density n_e for the fixed electron temperature $T_e = 2 \times 10^6$ K (170 eV). Very large deviations from Saha–Boltzmann values are evident. Coupling of the highly excited state populations to the free electrons by collisions causes the reduction to $b \sim 1$ as n_e (in cm^{-3}) increases. The large population inversions result from the dielectronic process. [Reprinted courtesy of A. Burgess and H. P. Summers and *The Astrophysical Journal*, published by the University of Chicago Press; © 1969 The American Astronomical Society.]

could be included has already been outlined above but no practical calculations have yet been done using these methods. It is evident that in this more complete model of ionization and recombination each ionization stage is represented by its group of ground and metastable populations, these populations being coupled to ground and metastable groups of neighboring charge stages by composite coefficients.

Some partial calculations have been done and are valuable in illustrating the nature of the effects to be expected. They disregard the time-constant differences (with some justification) and derive composite coefficients for the metastable levels. Thus Nussbaumer and Storey (1975) computed an ionization balance for a few stages of carbon in which they assumed that dielectronic recombination from metastable levels ought, in the first approximation, to be omitted. Vernazza and Raymond (1979) made similar allowances for a further range of ions. In a paper by Summers (1977) separate angular level populations were resolved and an elaboration of the "matrix condensation" procedures was developed. The method was applied to hydrogenlike and heliumlike ions with dielectronic recombination omitted. Blaha (1972) identified the particular problem of fine-structure levels belonging to the ground term of the recombining ion and their influence on dielectronic recombination. The recombining-ion ground-level population at finite densities is distributed amongst its fine-structure components and so disrupts the simple dielectronic process through the possibility of further autoionization. Blaha (1972) called this secondary autoionization. Summers (1974) made estimates of this correction.

The discussion so far has viewed ionization and recombination as taking place between adjacent ionization stages. While this will be true in recombining plasmas and generally in plasmas not too far removed from ionization balance, it will not be so in strongly ionizing plasmas. In this latter case, the free electrons may have energies substantially in excess of that required to remove a single outer electron from the typical ion present. Multiple ionization and ionizing events leaving the target in highly excited levels are energetically possible. Mewe and Schrijver (1978) have made rough allowance for the latter case in ionizing plasmas, although their work is directed at the variation of line intensities for special transient conditions rather than at the general picture of ionization.

V. The Distribution among the Stages of Ionization

It follows from the discussion of composite coefficients in the previous section that, in calculating spectral intensities, the treatments of (i) excitation and (ii) ionization and recombination may be handled separately. This section is concerned with the latter. In its simplest form the equation describing the rate of change of the population density of an ion of charge z [denoted now simply by $n(z)$ and nuclear charge z_0] may be written thus:

$$\frac{dn(z)}{dt} = n_e[n(z-1)S(z-1, z) - n(z)\alpha(z, z-1)$$
$$- n(z)S(z, z+1) + n(z+1)\alpha(z+1, z)] \quad (32)$$

where $S(z, z+1)$ is the composite ionization coefficient and $\alpha(z, z-1)$ is the composite recombination coefficient. The description of the ionization history of the element of nuclear charge z_0 then requires the simultaneous solution of a set of these equations for $z = 0$ to $z = z_0$ with suitable boundary conditions. In practice this solution must be done numerically using a computer and this is particularly true for plasmas where the electron temperature and density are changing with time.

Although most work directed to a time-dependent solution of the ionization equations is based on simple forms for the coefficients, it should be noted that

(1) Equation (32) ignores the possibility of multiple ionization where an electron impact is sufficiently energetic that it causes two or more bound electrons to be ejected. (Multiple recombination is less likely.)

(2) Also ignored is the effect of metastable levels.

The nature of the solution of the set of Eqs. (32) describing the ionization/recombination process depends on the particular circumstances in which they are applied. There is little that can be said of a general nature, so a specific solution will be presented by way of illustration. This formed part of an experimental study of which the object was to measure ionization rate coefficients in a laboratory plasma. Experiments of this kind have been reviewed by Burgess et al. (1977) and the specific example used here for illustration is due to Lang (1985).

The plasma used for the experiment had a small admixture of neon in hydrogen and was produced in a 40-kJ theta-pinch device. The time-dependent electron temperature and density were both measured by laser scattering and confirmatory values of the density were derived from the absolute intensity of continuum radiation. The results are illustrated in Fig. 12. Using these values of temperature and density the time-dependent ionization/recombination equations (32) were solved numerically to predict the time history of the various ionization stages of neon, which are also shown in Fig. 12. These results are typical of low-density pulsed plasmas and show how the ions of successively higher charges appear sequentially during the discharge lifetime until the plasma starts to cool about the time of theta-pinch current maximum. In order to make comparison with observed times of appearance of the corresponding spectral lines the excitation rate coefficients were introduced into the calculation in the manner described in the next section in order to predict the time history of the spectral intensities. The values of the ionization rate coefficients were deduced by adjusting their values so that the charge states were in agreement with spectroscopic observations.

The plasma parameter that has dominant influence on the time it takes for the ions to reach their steady state of ionization balance (where ionization is balanced by recombination) is the electron density. On the other hand, the charge state reached in steady-state ionization balance is predominantly determined by the electron temperature. An approximate estimate of the time to reach the steady state may be made by reference to Fig. 6, which illustrates the magnitudes of the ionization and recombination rate coefficients for H-like ions. It may be seen in this figure that the corresponding curves for these two coefficients cross over a band of values given by $S = \alpha \simeq 10^{-12}$ cm^3 s^{-1}. Similar plots for other ions yield a similar result to within a

3. Atomic Radiation from Low Density Plasma

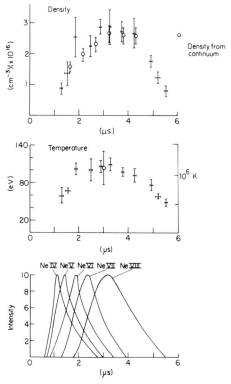

Fig. 12. Variation with time of electron density n_e and electron temperature T_e in a theta-pinch plasma composed of 5% neon and 95% helium excited by a 35-kV discharge. T_e and n_e are deduced by laser scattering. Also shown is the variation with time of resonance spectral line intensities from different neon ions. Intensities are normalized to a peak intensity of 10 and are indicative of the time histories of neon ionization stages during the discharge. [From Lang (1985).]

factor 10. Since in a steady-state ionization balance the ionization and recombination rate coefficients are about equal for the ions that have greatest populations it may be concluded that the time to reach the steady state is given approximately by

$$t_{\text{ion}} = 10^{12} n_e^{-1} \text{ s}. \tag{33}$$

This argument is presented in more detail by McWhirter (1960), who also points out that for the current range of fusion research plasma devices there is a need to take account of the time dependence of these atomic processes. Despite a significant increase in plasma containment times that has been achieved over the last two decades this last statement remains true.

In comparison with the time-dependent solution the steady-state solution ($dn(z)/dt = 0$ for all z) is much simpler and is more widely used. By applying the steady-state condition to the set of equations represented by (32) and after a little manipulation it may be shown that

$$n(z)/n(z-1) = S(z-1, z)/\alpha(z, z-1). \qquad (34)$$

The solution of this equation shows a strong dependence on electron temperature and a very much weaker dependence on electron density—through the dependence of the coefficients S and α on these parameters. There have been many calculations of ionization equilibria. Frequently used results include Jordan (1969), Summers (1979), and Jacobs et al. (1977a,b, 1979). The results of Summers's calculations for neon are illustrated in Fig. 13.

A major difference between these calculations, which essentially agree about which processes to include, is the choice of the expression for the ionization rate coefficient. In a review in which they compare ionization rate expressions with both theoretical and experimental data Burgess et al. (1977) show that there is little to choose between them and that disparities between theory and experiment make it difficult to predict ionization rates to an accuracy better than about a factor 2 by such simple methods. Further differences between the ionization balance calculations result from improvement and correction of details of the dielectronic process particularly (see

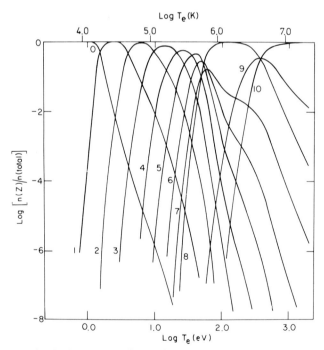

Fig. 13. Fractional abundances of neon ions in ionization equilibrium at the electron density $n_e = 10^{18}$ cm^{-3}. [From Summers (1974).] Z is the ion charge number.

discussion in Section III). For ions of small charge (low T_e) both these factors make a difference of about ±20% in the temperature of peak population whereas for ions of higher charge it increases by a factor of about 2 at temperatures of some millions of degrees.

The work by Nussbaumer and Storey (1975) and Vernazza and Raymond (1979) attempts to put into the calculation effects due to metastable levels and shows greater consistency in interpreting the solar spectrum than had been achieved previously. It has recently been pointed out by Baliunas and Butler (1980) that charge transfer can substantially modify the ionization balance of ions of small charge (such as Si^+ and Si^{2+}). A full set of calculations including all these processes and the best rate coefficient data has not yet been done.

VI. Spectral Line Intensities

In this section the preceding general theory for the calculation of spectral line intensities is applied to some specific ions. The theoretical behavior of the line intensities is fundamental to the practical application of spectroscopic diagnostic methods to measure such quantities as the electron temperature and density and to the calculation of the radiated power loss. Leaving aside practical questions (of the interpretation of the signals recorded using a spectrometer), attention here is concentrated on identifying the essential principles on which the methods depend. Some of the main ion types successfully exploited for spectroscopic diagnostic analysis are surveyed and the simplifications of the general theory appropriate to them discussed. The starting point is the expression for the intensity of a spectral line arising from a transition from an upper quantal level p to a lower level q of the ion $\mathcal{A}(z)$ in an optically thin plasma:

$$I(p, q) = n(z, p)A(p, q). \tag{35}$$

The radiative decay coefficients A have already been described in Section III, so the problem of calculating a spectral intensity reduces to the calculation of the population density $n(z, p)$ of the upper quantal level p. The multiplicity of atomic processes which can enter the differential equation in $dn(z, p)/dt$ and which may influence the population density has been discussed in Section III. For the fairly low-density, high temperature plasmas with which this chapter is concerned it has generally been judged that adequate simplification is attained by including only those populating (positive) terms describing excitation by electron collision with the ion $\mathcal{A}(z)$ in its ground or metastable level and a term to take account of the contribution due to recombination of the ion of the next charge up $\mathcal{A}(z + 1)$. In a few situations described later it is also necessary to include the effects of positive-ion collisions. The depopulation (negative terms in the equation) of these excited levels is generally taken to be dominated by spontaneous radiative decay.

In view of the time constants associated with the various rates in the equation for $dn(z, p)/dt$, the quasi-steady state approximation discussed in Section IV pertains except at very high densities. So setting $dn(z, p)/dt$ equal to zero, except when p represents the ground or metastable level, an explicit equation for $n(z, p)$ is obtained, namely,

$$0 = n_e[n(z, g)X'(g, p) + n(z, m)X'(m, p) + n(z + 1, g)\alpha'(p)]$$
$$- n(z, p) \sum_{r<p} A(p, r). \qquad (36)$$

Substituting for $n(z, p)$ in Eq. (35) gives

$$I(p, q) = n_e[n(z, g)X'(g, p) + n(z, m)X'(m, p) + n(z + 1, g)\alpha'(p)]$$
$$\times \frac{A(p, q)}{\sum_{r<p} A(p, r)} \qquad (37)$$

where X' and α' denote effective coefficients (for example, X' takes account of excitation to higher levels followed by cascade).

In the remainder of this section, Eq. (37) is applied to ions in the hydrogenlike to boronlike isoelectronic sequences. These are the simplest systems, yet draw out the main possibilities and illustrate the range of physical processes of which various authors have taken account.

A. Hydrogenlike Ions

Hydrogen and hydrogenlike ions, having only a single orbital electron, do not suffer autoionization, nor are they created by dielectronic recombination. Their emissions have been measured and calculated in more detail and over wider ranges of conditions than any other ion type. Two categories of line emission are widely studied. These are (a) the microwave lines corresponding to transitions where the principal quantum number changes by one and is greater than about 100 (often referred to as $n\alpha$ lines); and (b) the series of lines terminating on low levels and originating from upper levels with principal quantum number $\lesssim 15$.

The microwave emissions are of no relevance to fusion plasmas; however, the populations of the emitting levels are important to the overall recombination process. This has been described in Section IV. The simplifications entailed in Eq. (37) do not apply to the microwave emissions, the populations being strongly influenced by collisions with free particles. The works referenced in Section IV elaborate on their calculation. It may be noted that since the highly excited states of nonhydrogenic ions are similar to hydrogenic states, these hydrogenic calculations are generally valid for ions with only the further addition of the dielectronic recombination process.

The line emissions in the second category are taken to arise from levels depopulated by cascade. The boundary between the emitting levels of the two categories; that is, where collisions and radiation emission are equally effective in causing depopulation, is called the collision limit (McWhirter and Hearn, 1963). This limit depends on electron density and ion charge. Since hydrogenic states of the same principal quantum number but different angular quantum numbers are nearly degenerate, collisional rates for transitions between such states are very large (Pengelly and Seaton, 1964), with proton collisions more effective than electron collisions. Insofar as these rates are generally much larger than other rates which change the principal quantum number, it is usually judged satisfactory to apply Eq. (37) to bundled principal quantum shell populations. That is, populations are summed over degenerate angular states which are assumed relatively statistically populated. The collision limit then relates to these bundled populations. At very low densities or high z's, this approach loses validity and more elaborate angularly resolved treatments are required. Figure 14 illustrates the transition from bundled to unbundled populations in hydrogen using the angularly resolved code of Summers (1977).

Ljepojevic *et al.* (1984) have examined the influence of the 2s metastable level in hydrogenic ions on the other level populations and find that it may be neglected. The ratio of the metastable population to the ground-level population is always found to be less than 2×10^{-5}. Since the excitation coefficients from the 2s level are also small, the term describing excitation from the 2s level may be omitted. Specific calculations in the bundled approximation have been done by Bates and Kingston (1963) for atomic hydrogen and by McWhirter and Hearn (1963) for hydrogenic ions and have already been mentioned.

A number of computer codes have been written implementing general-purpose multilevel treatments (Brocklehurst, 1970, 1971; Burgess and Summers, 1976; Summers, 1977).

B. Heliumlike Ions

Helium atoms and heliumlike ions have two metastable terms, namely, 1s2s ^{1}S and 1s2s ^{3}S. These and the rest of the term scheme are illustrated in Fig. 15. The ^{1}S metastable level can decay to the ground $1s^2$ ^{1}S level by the emission of two photons, and the ^{3}S metastable level indirectly, by transfer first to the 1s2p $^3P_1$ level followed by an intercombination transition. Also important for the decay of the triplets to the ground level when the density is

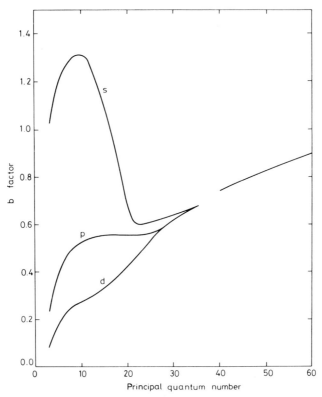

Fig. 14. Variation of *b* factors for hydrogen with quantum number for the electron temperature $T_e = 10^4$ K (0.86 eV) and electron density $n_e = 10^4$ cm^{-3}. This emphasizes the variation of the *b* factors for different angular levels (s, p, and d) of the same principal quantum number. For levels above 30 collisional transitions between degenerate levels are sufficiently rapid for the averaged principal quantum shell approximation to hold. Results are for a plasma optically thick in the Lyman lines. [From Summers (1977).]

low is a relativistic magnetic dipole transition from the ^{3}S metastable level. (There is also a magnetic quadrupole transition out of the 1s2p ^{3}P$_2$ level which has little practical effect.)

Consider the mechanisms of excitation of the heliumlike ions as the electron density decreases from a large value, paying particular attention to metastable levels. At electron densities so great that the rate of collisional excitation and deexcitation exceeds even the radiative rate of the dipole allowed transitions all level populations are close to their Saha–Boltzmann values. At somewhat lower densities when the radiative rates of the dipole allowed transitions take over from collisions the excited levels rapidly decay to the metastable (2^{1}S and 2^{3}S) and ground (1^{1}S) levels. The relative populations of these three levels are determined by the balance of collisional excitation into the metastable levels with collisions that depopulate them. Note

3. Atomic Radiation from Low Density Plasma

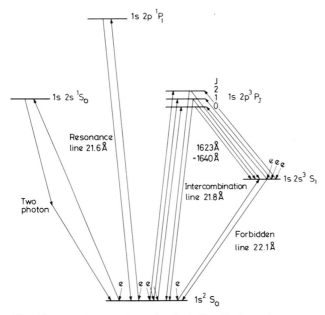

Fig. 15. Partial term scheme for the heliumlike ion O^{6+} (O VII).

that for the 2^3S level, excitation to any other triplet does little to depopulate it since all such levels decay back to the metastable level—except for the 2^3P_1 level which for ions heavier than carbon decays more readily to the 1^1S ground level than to 2^3S. As density get lower radiative decay of this 2^3P_1 level becomes the most probable route for the triplets to return to the 1^1S ground level. Down to this density range the 2^1S metastable level decays collisionally mainly through the 2^1P level. However, at lower densities the probability of two-photon decay takes over. Finally, at the lowest densities the forbidden $1^1S_0 - 2^3S_1$ transition provides a significant decay route for the triplets to return to the ground level. The calculation of these spectral intensities for O VII has been treated by a number of authors (see Doyle, 1980; Gabriel and Jordan, 1969). The interplay of these processes is illustrated for O VII in Fig. 16, where the regions of dominance of the various radiative decay ratios are marked. This calculation was done for an electron temperature of 3×10^6 K for a plasma close to its steady state without including any contribution due to the effects of recombination. Under the chosen conditions the effect of recombination on these line intensities is unlikely to be more than about 10%. However, departures from the steady state in the direction of a recombining plasma can lead to greater effects. This aspect of the calculation is discussed by Peacock and Summers (1978), who are interested in using the line ratios for diagnostics including an estimate of how far the ion ratios depart from their steady-state values.

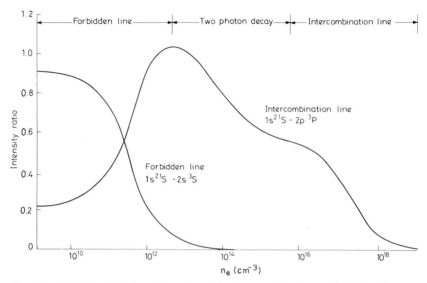

Fig. 16. Variation of the intensity ratios of the intercombination and forbidden lines to the resonance line of O VII with electron density n_e for the electron temperature $T_e = 3 \times 10^6$ K (260 eV). The mechanism controlling the ratios is identified in three density ranges. There is assumed to be no recombination contribution to the formation of these lines.

C. Satellite Lines

This section is introduced between the sections on He-like and Li-like ions because at this time it is the satellites associated with the 1^1S–2^1P resonance line of He-like ions, and due to the decay of doubly excited Li-like ions, that have been most widely studied. They arise primarily as a consequence of dielectronic recombination of He-like ions but can also be due to inner-shell excitation of the corresponding Li-like ions. The theory of these lines has been developed by Gabriel (1972), Bhalla *et al.* (1975), and Bely-Dubau *et al.* (1979a,b). The object of their work was to provide an explanation for the complicated pattern of these satellites and to develop a diagnostic method for the estimation of the electron temperature and extent of departure of the plasmas from steady-state ionization balance.

As described in Section III of this chapter the dielectronic recombination process is made up of three distinct stages, viz., capture, stabilization, and cascade. It is the stabilization process that can give rise to the satellites. The most prominent satellites arise when capture takes place into the $n = 2$ quantum shell of the ion to form a doubly excited Li-like ion having configuration $1s\,2l\,2l'$ where l and l' are angular quantum numbers. The decay of one of those outer electrons to the ground 1s configuration gives rise to a satellite of wavelength close to the wavelength of the $1s^2\ ^1S$–$1s2p\ ^1P$ line of the

3. Atomic Radiation from Low Density Plasma

corresponding He-like ion. Its intensity may be estimated using the concepts developed in Section III.E for dielectronic recombination, viz.,

$$I_s = n_e n(z + 1, g) \frac{h^3}{2(2\pi m T_e)^{3/2}} \frac{\omega(z, j, nl)}{\omega(z + 1, g)} \frac{A_r A_a}{A_r + A_a} \exp\left(-\frac{E_s}{kT_e}\right) \quad (38)$$

where E_s is the energy difference between the upper level of the satellite line denoted by (z, j, nl) and the heliumlike ground level. The heliumlike ion has charge $z + 1$. Thus the ratio of the intensity of a satellite line to the resonance line is independent of n_e and $n(\text{He})$ and depends only on the electron temperature T_e. Some satellite line emission may be generated by excitation of the Li-like ions, and it is this that leads to the possibility of using the ratio as a diagnostic to estimate departures from the steady-state ionization balance. Figure 17 shows the satellite lines of Ca XIX as observed in the sun.

D. Lithiumlike Ions

Lithiumlike ions have three bound electrons which in their ground level have the configuration $1s^2\,2s\,^2S$. These ions have no metastable level and for

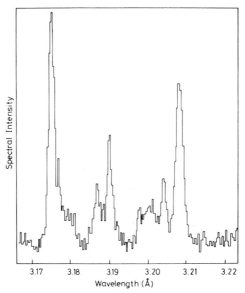

Fig. 17. Solar spectrum in the vicinity of the heliumlike resonance line $1s^2\,^1S–1s2p\,^1P$ of Ca XIX recorded on the bent-crystal spectrometer on the Solar Maximum Mission satellite, showing the forbidden and intersystem lines of Ca XIX together with the associated satellite lines of Ca XVIII. The prominent satellite lines arise from the spectator electron in the $n = 2$ shell, while those with spectator in the $n \geq 3$ shell contribute to and cause asymmetry of the resonance line.

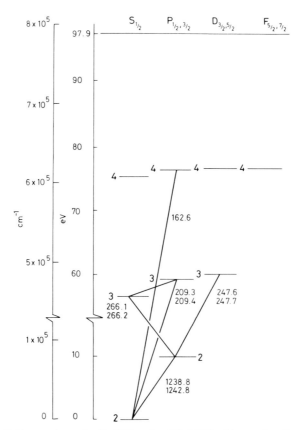

Fig. 18. Partial term scheme for the lithium ion N^{4+} (N V) with wavelengths in angstroms.

this reason their spectral intensities are the simplest of all to calculate. The term scheme of N V is typical of these ions and is illustrated in Fig. 18. It may be noted that the 2p ^{2}P$_{1/2,3/2}$ first excited level lies only 10 eV above the ground level and that this is a tenth of the ionization potential compared with hydrogenlike ions, for example, where the excitation potential of the first excited level is three-quarters of the ionization potential. From this arises the important property of lithiumlike and other similar ions of being efficient radiators of energy. It also makes lithiumlike ions useful for temperature diagnostics since the large differences in excitation potential of the 2p and, say, 3p levels provide line ratios such as 2s–2p to 2s–3p that are very sensitive to electron temperature. The lack of a metastable level means that except at very high densities there is no density dependence to confuse the situation.

However, the small value of the first excitation potential leads to a complication in calculating spectral intensities. When with increasing electron density the rate of collisional excitation of the 2p levels becomes about equal to the rate of their radiative decay the population of the 2p levels starts to build up to be comparable with that of the 2s level. Then the 2p levels act in a manner similar to a metastable level and stepwise excitation to higher levels can take place via the 2p level. This is particularly important for excitation to the 3d level and for N V may be shown to become an important effect at electron densities greater than about 10^{14} cm^{-3}. Lithiumlike ions have been well studied and sufficient atomic data is available to calculate the magnitude of such effects. Cochrane and McWhirter (1983) have published a comprehensive review of the available collisional excitation rates.

E. Berylliumlike Ions

Berylliumlike ions have a $1s^2 2s^2$ $^1S_0$ ground configuration and $1s^2 2s2p$ $^3P_{0,1,2}$ metastable levels. These and other levels are shown in Fig. 19, which illustrates the partial term scheme. Like lithiumlike ions the levels with principal quantum number 2 lie at relatively small excitation potentials above the ground level (compared with their ionization potentials) and this causes these transitions to be relatively strong. The metastable levels can decay to the ground level by an intercombination transition $2s^2$ ^1S–$2s2p$ $^3P_1$ due to a breakdown in LS coupling and by a magnetic quadrupole transition $2s^2$ ^1S–$2s2p$ $^3P_2$. The small values of these radiative decay probabilities compared with neighboring dipole rates coupled with the relatively small excitation potentials of the metastable levels means that these levels can acquire population densities comparable with the ground level. The population ratios between the metastable levels and the ground level varies with electron density in a way that depends on the balance between the collisional excitation rate and the combination of radiative and collisional deexcitation rates of the metastable levels. The data necessary for these calculations has been produced by Berrington et al. (1977). Three density ranges can be identified depending on which radiative process controls the relevant excited levels. Some line ratios are illustrated in Fig. 20 for O V where the ranges of

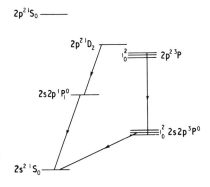

Fig. 19. Partial term scheme for berylliumlike ions.

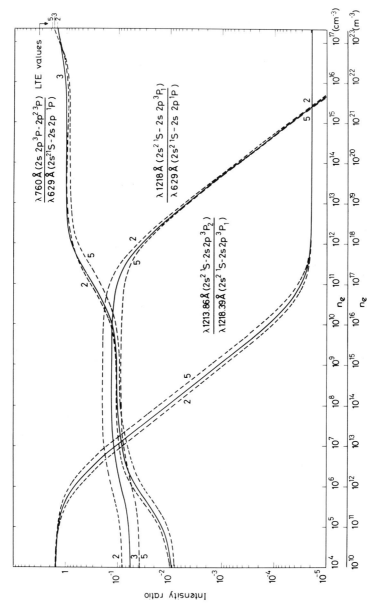

Fig. 20. Variation of the intensity ratios (with intensities in photons) of the spectrum lines of O V with electron density n_e for different electron temperatures T_e. The cross section data are taken from Berrington et al. (1977). Electron temperatures: --- 2, 2×10^5 K, 17 eV; —— 3, 3×10^5 K, 26 eV; --- 5, 5×10^5 K, 43 eV.

sensitivity to electron density are determined by the balance between collision rates and the following radiative rates:

(a) $10^5 < n_e < 10^6$ cm^{-3}, magnetic quadrupole 2s^2 ^{1}S$_0$–2s2p ^{3}P$_2$;
(b) $10^{11} < n_e < 10^{12}$ cm^{-3}, intercombination line 2s^2 ^{1}S$_0$–2s2p ^{3}P$_1$;
(c) $n_e \sim 10^{17}$ cm^{-3}, dipole transition 2s^2 ^{1}S$_0$–2s2p ^{1}P.

In calculating these intensity ratios account has been taken of the electron collision rates between all pairs of levels having principal quantum number 2 and the radiative rates between the same levels. In addition, for these ions, it is necessary to include effects due to proton collisions, with the 2s2p ^{3}P$_{0,1,2}$ metastable levels which are relatively efficient at interchanging populations among these sublevels. At electron densities just lower than that required to ensure LTE the metastable level has a relatively small probability of radiative decay. Its population density is determined by the balance between its collisional excitation from the ground level and collisional transitions into the singlet levels 2p^2 ^{1}D and 2s2p ^{1}P. Thus although collisions dominate the population departs from its LTE value.

The calculation of the population distribution among the $J = 0$, 1, and 2 sublevels of the 2s2p ^{3}P metastable level is critical for the determination of the intensities of all lines of these ions. One of the most important of the processes in this respect is the collisional excitation of the 2p^2 ^{3}P$_{0,1,2}$ levels from the metastable. It was first pointed out by Malinowski (1975) that the collisional rate coefficients between individual pairs of sublevels designated by their J quantum numbers could be very different from each other. The relative rates are approximately proportional to the f values connecting the sublevels. From this it follows that the relative intensities of the six components of the 2s2p ^{3}P–2p^2 ^{3}P multiplet depend on the electron density since the relative populations of the 2s2p ^{3}P $J = 0$, 1, and 2 sublevels depend on density. This is illustrated in Fig. 21 for O V and is also based on the calculations of Berrington et al. (1977). The same density regions identified in Fig. 20 are apparent. The figures also show that these line ratios are insensitive to the electron temperature. Others, involving levels having higher principal quantum numbers, show a strong temperature dependence.

F. Boronlike Ions

The ground configuration of boronlike ions is 1s^{2}2s^{2}2p ^{2}P$_{1/2,3/2}$ and there is a metastable level with the configuration 1s^{2}2s2p^2 ^{4}P$_{1/2,3/2,5/2}$. A partial term scheme for boronlike ions is illustrated in Fig. 22. A calculation of the spectral intensities of this ion has been done by Flower and Nussbaumer (1975). The new features introduced by these ions arise from the fact that the ground level has two fine-structure components having J values $\frac{1}{2}$ and $\frac{3}{2}$. For this particular ion the relative populations of these sublevels are adequately

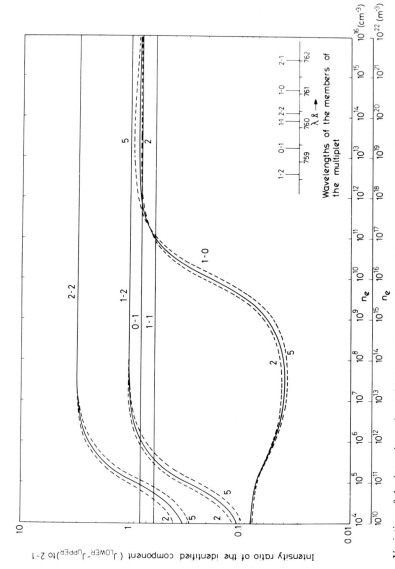

Fig. 21. Variation of the intensity ratios (with intensities in photons) of the members of the $2s2p\ ^3P_{0,1,2}-2p^2\ ^3P_{0,1,2}$ multiplet of O V with electron density n_e for different temperatures T_e. Cross-section data are taken from Berrington *et al.* (1977). Electron temperature values: --- 2, 2×10^5, 17 eV; ——— 3, 1×10^5 K, 26 eV; --- 5, 5×10^5 K, 43 eV.

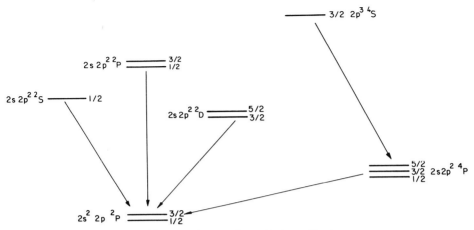

Fig. 22. Partial term scheme for boronlike ions.

described by the Boltzmann equation down to some minimum electron density where the radiative decay rate from the upper ($J = \frac{3}{2}$) to the lower ($J = \frac{1}{2}$) begins to have an influence. In any case it is necessary to take separate account of the sublevels in doing the collisional–radiative calculation since the rate coefficients are different. This leads to the possibility of a different kind of temperature and density diagnostic based on departures of the ground sublevels from their Boltzmann values. Otherwise the calculation of spectral intensities for this ion sequence is much the same as for Be-like ions, including the need to take account of positive-ion collisions in redistributing the populations of the sublevels of not only the metastable but also the ground level.

VII. Radiated Power Loss

In this section the theory for the calculation of the total power radiated by the kinds of plasmas met with in fusion research and in astrophysics is developed. It turns out that in both areas power loss by radiation can be the largest single mechanism of power loss. The radiation appears in three forms, viz., line radiation; recombination continuum; and bremsstrahlung arising from electrons making bound–bound, free–bound and free–free transitions in the fields of positive ions. In principle the calculation amounts to a prediction of the complete spectrum. However, in practice a whole range of simplifications are adopted to make the calculation tractable. These are discussed below.

One of the earliest calculations was by Pottasch (1965), who was able to confirm by detailed calculations that the small concentrations of elements heavier than hydrogen were responsible for the great majority of the energy radiated by the tenuous outer atmosphere of the sun and stars. Further calculations using more reliable atomic data were published in a series of papers by Cox and Tucker (1969) and their co-workers (Raymond *et al.*, 1976). An early estimate of the power loss due to iron impurity in a fusion reactor is due to McWhirter (1959). More recent calculations utilizing an "average-ion model" are due to Post *et al.* (1977a,b) and are done for a large number of elements with fusion experiments in mind. Accuracies are estimated by the authors to be in the range of "factors 2 or 4." These calculations are based on the assumption of steady-state ionization balance and data are presented in a form convenient for incorporation in a computer code. Another recent calculation is due to Summers and McWhirter (1979) and while concentrating on the steady-state ionization balance includes tables of coefficients that can be used in circumstances where there are departures from the steady state. They treat a range of elements up to molybdenum and use data selected from the atomic physics literature to give a claimed accuracy of ±50%. Despite the range in the quality of the atomic data used the agreement between all the calculated values is remarkably good. The greatest differences arise for elements of high atomic number and appear to be partly due to uncertainties in the values of the coefficients entering the ionization balance calculation.

In the discussion which now follows the results of Summers and McWhirter will be used to illustrate details of the radiated power loss calculation. Most of the assumptions considered and adopted in previous sections are again made. It is useful to summarize the key assumptions once more:

(a) The plasma is effectively optically thin to its own radiation. This means that, although there may be substantial reabsorption of resonance radiation, the probability of reemission is much greater than that for deexcitation by collisions. It appears that for magnetically confined laboratory plasmas and for the sun's atmosphere above the chromosphere this is generally a safe assumption.

(b) Again magnetic field effects are neglected. The most serious aspect of this assumption appears to be the possibility that synchrotron radiation can contribute to the radiated power loss. For the kinds of plasma of interest here the radiation lies in the infrared and may be optically thick.

(c) Densities are sufficiently low that stepwise ionization through the excited levels may be neglected. The consequence of this is that all the energy that goes to excitation by electron collision always appears ultimately as radiation.

(d) The presence of metastable ions is ignored. The problem of metastables has been explored in Section IV. It affects both the ionization balance and the spectral intensities, but there appears so far to have been no attempt to include these effects specifically in a calculation of radiated power loss. Although it is unlikely that metastables make very large differences to the results it will only be when these effects are included that a firm conclusion can be drawn. In the meantime, in the present treatment it will be assumed without justification, that all ions decay instantly to their ground levels and that their metastable levels may be ignored.

Other assumptions have been adopted by some authors, the most important of which is that steady-state ionization balance holds. This question was considered in Section V where it was shown that it is not an acceptable assumption for most plasmas being studied at the present time in connection with fusion. However, its adoption leads to substantial simplification and this has been taken as sufficient justification for doing it. The present treatment for the most part also assumes steady-state ionization balance.

As already mentioned three forms of radiation contribute to the radiated power, viz., lines, recombination continuum, and bremsstrahlung. It is convenient to retain this general classification for the purpose of the calculation, but included in the recombination radiation are those photons produced as a result of dielectronic recombination and cascade following radiative recombination. The remaining line radiation component (referred to as leading line radiation) is that line radiation emitted as a result of collisional excitation including cascade when the excitation is to a higher level. Since, for each of the three forms, radiation is emitted as a result of an electron having a collision with an ion the total intensity of all forms may be expressed in terms of products of the electron and ion densities:

$$\text{Total radiated power} = n_e n(\mathcal{A}) P(\mathcal{A}) \qquad (39)$$

where $n(\mathcal{A})$ is the total abundance of element $\mathcal{A}$ and $P(\mathcal{A})$ is called the "power loss function" for that element. Then

$$P(\mathcal{A}) = n_e n(\mathcal{A}) \sum_{z=0}^{z_0} \frac{n(\mathcal{A}, z, g)}{n(\mathcal{A})} P(\mathcal{A}, z), \qquad (40)$$

where $P(\mathcal{A}, z)$ is called the power loss coefficient for the ion $\mathcal{A}(z)$. Finally,

$$P(\mathcal{A}, z) = P_L(\mathcal{A}, z) + P_R(\mathcal{A}, z) + P_B(\mathcal{A}, z). \qquad (41)$$

The subscripts L, R, and B identify the components due to the leading lines, recombination radiation, and bremsstrahlung, respectively. It will be seen that the $P(\mathcal{A}, z)$ coefficients arise from prompt processes in the same way as collisional excitation coefficients so that in calculating them there is no need to take account of time dependence. They are almost independent of density but depend on the electron temperature. On the other hand the ionization balance ratio $n(\mathcal{A}, z, g)/n(\mathcal{A})$ can depend on the time history of the plasma in the manner discussed in Section V, i.e., it can be a function of time, density, and temperature. On the other hand, if the plasma parameters change sufficiently slowly that the ions have time to reach their steady state of ionization balance, the ratio $n(\mathcal{A}, z, g)/n(\mathcal{A})$ becomes independent of time and almost independent of density but still dependent on electron temperature. It has become usual in these circumstances (the steady-state ionization balance approximation) to carry out the summation over the ionic charges and present values of $P(\mathcal{A})$ as a function of electron temperature, ignoring the small dependence on density.

The methods of evaluating the three components of the radiated power loss coefficients for individual ions must now be considered in turn. That due to the leading lines, $P_L(\mathcal{A}, z)$, is the sum over all possible lines arising due to excitation of which any one line is derived from Eq. (37). The term arising from recombination is dropped and the line intensity expressed in energy units:

Power radiated by direct excitation in line p → q =

$$n_e[n(z, g)X(g, p) + n(z, m)X(m, p)] \frac{A(p, q)}{\sum_r A(p, r)} \Delta E(p, q). \qquad (42)$$

Note that in this expression it is the direct excitation rate $X(p, q)$ rather than the effective rate $X'(p, q)$ which enters.

Since it does not matter whether the decay is directly to the ground or metastable level or by cascade the ratio of the A's may be dropped and the Eq. (42) expressed thus:

Power radiated as a result of excitation to level p =

$$n_e[n(z, g)X(g, p) + n(z, m)X(m, p)]\Delta E(s, p) \qquad (43)$$

where s is either g or m depending on whether the decay or cascade of p ends up in the ground or metastable level. As pointed out above, metastable levels are ignored in these calculations in order to simplify them although there is no precise physical justification, and with this assumption the right-hand side of Eq. (43) is replaced by

$$n_e n(z, g)X(g, p)\Delta E(g, p). \qquad (44)$$

The summation of this expression over all levels p of the ion gives $P_L(\mathcal{A}, z)$:

$$P_L(\mathcal{A}, z) = \sum_p X(g, p)\Delta E(g, p). \qquad (45)$$

Since there are an infinite number of levels p it is necessary to adopt some procedure to make the calculation tractable. It is generally judged to be satisfactory to restrict the summation to the first few levels since it is these that give rise to the strongest lines. Thus, for example, Summers and McWhirter (1979) adopted the approximation that it was necessary to include only those levels whose principal quantum number was equal to or one greater than that of the ground level. The error introduced by such approxi-

3. Atomic Radiation from Low Density Plasma

mations is judged to be small compared with the ±50% uncertainty in the accuracy of the basic atomic data used.

The coefficient $P_R(\mathcal{A}, z)$ representing the recombination radiation component is made up of contributions arising from dielectronic recombination (both the stabilizing and cascading photons) and radiative recombination (including cascade). Ideally, as with line radiation, account should be taken of metastable levels both in the initial and recombined ion. Again no data including these effects is available. They can be expected to be less important for this component than for the leading lines. Thus ignoring metastables the value of $P_R(\mathcal{A}, z)$ is given by

$$P_R(\mathcal{A}, z) = \sum_p \Delta E(q, p) \sum_{nl} \alpha_d(q, p; nl) + \chi \alpha_{c-d} + \chi \beta \alpha_r \qquad (46)$$

where $\Delta E(q, p)$ is the energy of the stabilizing photon, $\alpha_d(q, p; nl)$ is the (zero-density) dielectronic recombination coefficient with capture into the level identified by quantum numbers nl, χ is the ionization potential of the recombined ion, α_{c-d} is the total collisional–dielectronic recombination coefficient, β is a factor lying between 0.2 and 1 to take account of the thermal energy of electrons taking part in radiative recombination, and α_r is the total radiative recombination rate coefficient. The same limits on the summation over p may be applied as in the case of the $P_L(\mathcal{A}, z)$ component.

The third component $P_B(\mathcal{A}, z)$ is due to bremsstrahlung and, because of its dependence on the square of the charge on the ion, it is the dominant component for ions of large charge even in circumstances where there are remaining bound electrons. It is important therefore to pay particular attention to its evaluation, especially for these highly charged ions. Since the free–free collisions giving rise to bremsstrahlung are dominated by the Coulomb field with little influence from the structure of bound electrons it is of little consequence if the ions are in a metastable or ground level. These problems are discussed in some detail by Summers and McWhirter, where an expression for $P_B(\mathcal{A}, z)$ may be found.

Figure 23 illustrates the nature of the radiated power loss function for various elements in steady-state ionization balance. It may be seen that the variation in the values of the functions for low-Z elements such as carbon span almost three decades whereas for molybdenum it is little more than one decade. This arises from the dominance of the leading-lines components for low-Z ions where the sharp peaks and troughs reflect the shell structure of the ions giving way for high-Z ions to the increasing influence of bremsstrahlung, which varies monotonically with electron temperature. The relative magnitudes of these components are illustrated for oxygen in Fig. 24.

As has been remarked already, the power loss calculated on the basis of the steady-state ionization balance assumption is relatively simple and is sometimes used in circumstances where it is not fully justified. In order to illustrate the difference that this can make, a full time-dependent calculation

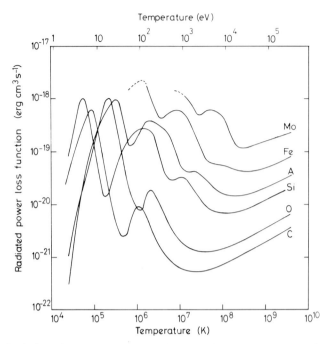

Fig. 23. Variation of radiated power loss functions for carbon, oxygen, silicon, argon, iron, and molybdenum with electron temperature T_e at low electron density. [From Summers and McWhirter (1979). Copyright 1979 The Institute of Physics.]

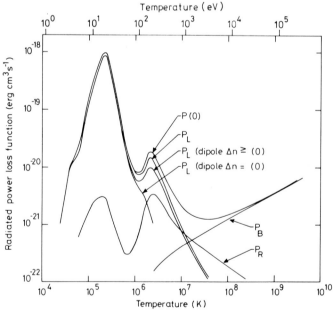

Fig. 24. Variation of the components of the radiated power loss function of oxygen with electron temperature T_e. Leading-line power loss P_L, recombination power loss P_R and bremsstrahlung power loss P_B are shown with the line power loss further resolved into the contribution from $\Delta n = 0$ dipole transitions and $\Delta n \geq 0$ dipole transitions. [From Summers and McWhirter (1979). Copyright 1979 The Institute of Physics.] (Δn is the change in the principal quantum number in the excitations.)

3. Atomic Radiation from Low Density Plasma 103

of the power loss due to the neon (radiating in the plasma discussed in Section V) is compared with the value that would be found with the assumption of steady state. Results are illustrated in Fig. 25 which show that large errors can arise due to the assumption of steady state.

It may be remarked that there are a number of other quantities derived from radiative power loss and ionization equilibria which are often useful. Thus the broad spectral distribution of the radiated power may be produced by appropriate grouping when assembling the components of Eq. (41). Also, a valuable power loss diagnostic may be obtained by relating the overall power loss to the intensities of a small number of readily observed spectrum lines. When time dependence is important, it can be useful to have a value

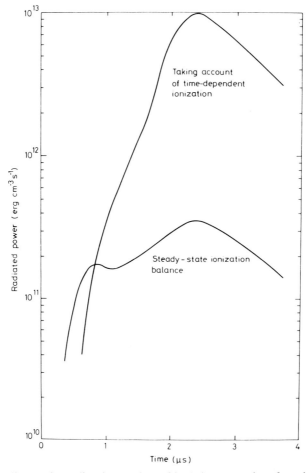

Fig. 25. Difference in predicted power loss with (a) the assumption of steady-state ionization balance and (b) taking account of time-dependent ionization. Both are for the plasma whose parameters are illustrated in Fig. 12.

for the total energy required to ionize an impurity atom to a given state of ionization, rather than the radiated power loss.

Finally, in this chapter the influence of impurity radiation on the power balance of a fusion reactor is discussed. This follows the treatment of Summers and McWhirter (1979) which is based on the ideas presented by Lawson (1957). The latter showed that a condition to be satisfied by a successful fusion reactor is that the product of hydrogen density and particle containment time must be greater than about 10^{14} cm^{-3} s. The physical basis of this argument is that more energy should be produced by the thermonuclear reactions than is radiated as hydrogen bremsstrahlung—assumed to be the dominant power loss mechanism. It is clear from the foregoing parts of this chapter that relatively small concentrations of impurities can account for more radiated energy than hydrogen bremsstrahlung and it is therefore important to attempt to predict the effect of impurities on the overall power balance of a fusion reactor.

For simplicity in the present treatment it is assumed that there is only one impurity element $\mathcal{A}$ present in a hydrogen plasma consisting otherwise of equal parts of deuterium and tritium. The power radiated by such a plasma is made up of that due to the hydrogen plus components from each of the ions $\mathcal{A}(z)$ that is present. The total may be expressed thus:

$$[n(\text{H})]^2 P_{\text{rad}} = n(\text{H}) n_e \left(P(\text{H}) + \sum_z \frac{n(\mathcal{A}, z)}{n(\text{H})} P(\mathcal{A}, z) \right)$$

$$= [n(\text{H})]^2 \left(1 + \frac{n(\mathcal{A})}{n(\text{H})} \sum_z \frac{n(\mathcal{A}, z)}{n(\mathcal{A})} z \right)$$

$$\times \left(P(\text{H}) + \frac{n(\mathcal{A})}{n(\text{H})} \sum_z \frac{n(\mathcal{A}, z)}{n(\mathcal{A})} P(\mathcal{A}, z) \right), \qquad (47)$$

where in calculating the electron density n_e account has been taken of the electrons released from the impurity ions together with the need to maintain plasma neutrality. The ratio $n(\mathcal{A})/n(\text{H})$ denotes the concentration with respect to hydrogen of the impurity $\mathcal{A}$ and $n(\mathcal{A}, z)/n(\mathcal{A})$ its state of ionization. The quantities $P(\text{H})$ and $P(\mathcal{A}, z)$ are the power loss functions defined earlier in this section. P_{rad} is therefore a composite power loss function.

At this stage steady-state ionization balance is assumed but this question is returned to later. Various hydrogen–impurity mixtures are taken for illustration and the resulting values of P_{rad} plotted in Fig. 26. On the same figure are plotted values of the deuterium–tritium power production coefficient $P(\text{D–T})$ taken from Glasstone and Lovberg (1960). In calculating the latter it was assumed that the ion and electron temperatures are equal and that the thermonuclear neutrons escape from the plasma without loss whereas the other reaction products are contained.

The lower temperature points at which the radiation curves cross the

3. Atomic Radiation from Low Density Plasma

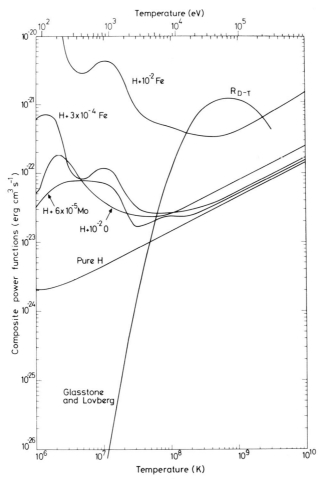

Fig. 26. Composite radiation power loss functions [see Eq. (47)] for various impurity mixtures compared with the deuterium–tritium fusion power generation coefficient R_{D-T}. R_{D-T} is taken from Glasstone and Lovberg (1960) and the radiative data from Summers and McWhirter (1979). R_{D-T} assumes that neutrons escape. Power = $n^2 R$, where R is the quantity plotted and n is total hydrogen density = $n_D + n_T$ and $n_D = n_T$.

fusion power curves mark the lowest temperature that the plasma must be raised to in order that the fusion power should overcome the radiation loss. It is called the ignition temperature and it may be seen that, due to the steep temperature dependence of the $P(D-T)$ curves, the presence of impurities causes only a modest increase in its value. Figure 26 also shows that the radiation loss curves in general have shallow minima in the region of the ignition temperature. This arises because of the dominance of line radiation at lower temperatures and means that during the initial plasma heating phase the radiation loss problem is likely to be more severe. It is during this phase also that the assumption of steady-state ionization balance is least likely to hold and this too probably means that more power is radiated than otherwise.

In order to quantify the level at which the radiation from a particular impurity element $\mathcal{A}$ becomes significant it is satisfactory to calculate the concentration $n(\mathcal{A})/n(H)$ at which the total radiated energy at the new ignition temperature is twice the value for pure hydrogen at its ignition temperature. These concentrations are plotted in Fig. 27 as a function of the nuclear charge of the impurity element and they indicate the concentration levels at

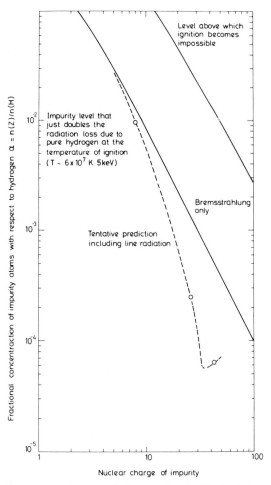

Fig. 27. Impurity concentrations required to produce double the hydrogen bremsstrahlung power loss and to quench the fusion reaction at the temperature of ignition. [From Summers and McWhirter (1979). Copyright 1979 The Institute of Physics.]

which each element's radiation takes over from hydrogen bremsstrahlung as the dominant component.

Because of the maximum in the fusion power curve, it is clear from Fig. 26 that at certain impurity levels the radiation loss curves fail to intersect the latter. This would result in the impurity radiation completely quenching the fusion reaction. The impurity concentration required for the radiation curves to just touch the fusion power curve in the region of its maximum is also plotted in Fig. 27 and therefore indicates the impurity level required to just quench fusion.

In addition to examining the power balance of the fusion plasma itself, as above, it is also useful (again following Lawson, 1957) to look at the energy balance of the total system. For this it is necessary to define a particle containment time t which for a pulsed reactor may be the duration of the pulse. It is assumed also that the thermonuclear neutrons are trapped and therefore are a positive contribution to the energy balance. Then the total fusion energy released in time t is

$$[n(\mathrm{H})]^2 5.03 P(\mathrm{D-T}) t, \tag{48}$$

where the factor 5.03 accounts for the neutron energy. The energy required to heat the plasma to ignition temperature and to account for radiation losses is given by

$$[n(\mathrm{H})]^2 P_{\mathrm{rad}} t + n(\mathrm{H}) K, \tag{49}$$

where K is a factor determined by the energy required to heat the plasma; K should include contributions due to the particle kinetic temperature, ionization energy, and additional radiation loss occurring during the transient heating phase. A complete estimate of this factor is not available and it is therefore assumed that K includes only the particle temperature; thus

$$K = \frac{3}{2} kT \left(2 + \sum_z \frac{n(\mathcal{A}, z)}{n(\mathrm{H})} (1 + z) \right). \tag{50}$$

The ratio

$$R = [n(\mathrm{H})]^2 5.03 P(\mathrm{D-T}) t / \{[n(\mathrm{H})]^2 P_{\mathrm{rad}} t + n(\mathrm{H}) K\} \tag{51}$$

is plotted against temperature for pure hydrogen and with 1% iron impurity and for various $n(\mathrm{H})t$ products in Fig. 28. The minimum value of R for a feasible reactor depends on the overall efficiency of the power-generating system and is taken to be 2 (Lawson, 1957). It may be seen that $n(\mathrm{H})t = 10^{14}$ cm^{-3} s only just satisfies this criterion for a plasma with 1% iron impurity.

It is now possible to return to the assumption of steady state ionization balance on which much of the foregoing discussion is based. For this assumption to be valid it is necessary that the time for impurities to reach steady-state ionization balance be significantly less than the required particle

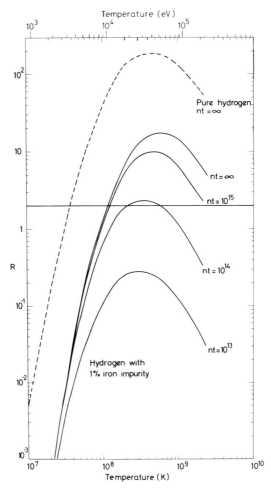

Fig. 28. Ratio R of the energy yield to that required to produce the plasma in a fusion reactor containing 1% of iron impurity. [From Summers and McWhirter (1979). Copyright 1979 The Institute of Physics.]

containment time. It has already been remarked in Section V that the time to reach the steady state is given to about a factor 10 by

$$n_e t_{\text{ion}} = 10^{12} \text{ cm}^{-3} \text{ s}. \tag{52}$$

This may be compared with the Lawson criterion $n(\text{H})t > 10^{14}$ cm^{-3} s to justify the assumption of steady-state ionization balance.

References

Alder, K., Bohr, R., Huus, T., Mottelson, B., and Winther, A. (1956). *Rev. Mod. Phys.* **28,** 432.
Bahcall, J. B., and Wolf, R. A. (1968). *Astrophys. J.* **152,** 701.
Baliunas, S. L., and Butler, S. E. (1980). *Astrophys. J.* **235,** L45.

Bates, D. R., and Damgaard, A. (1949). *Philos. Trans. R. Soc. London, Ser. A* **242**, 101.
Bates, D. R., and Kingston, A. E. (1963). *Planet. Space Sci.* **11**, 1.
Bates, D. R., Kingston, A. E., and McWhirter, R. W. P. (1962a). *Proc. R. Soc. London, Ser. A* **267**, 297.
Bates, D. R., Kingston, A. E., and McWhirter, R. W. P. (1962b). *Proc. R. Soc. London, Ser. A* **270**, 155.
Bely, O., and Faucher, P. (1970). *Astron. Astrophys.* **6**, 88.
Bely, O., and Van Regemorter, H. (1970). *Annu. Rev. Astron. Astrophys.* **8**, 329.
Bely-Dubau, F., Gabriel, A. H., and Volonté, S. (1979a). *Mon. Not. R. Astron. Soc.* **186**, 405.
Bely-Dubau, F., Gabriel, A. H., and Volonté, S. (1979b). *Mon. Not. R. Astron. Soc.* **189**, 801.
Berrington, K. A., Burke, P. G., Dufton, P. L., and Kingston, A. E. (1977). *J. Phys. B* **10**, 1465.
Bhalla, C. P., Gabriel, A. H., and Presynakov, L. P. (1975). *Mon. Not. R. Astron. Soc.* **172**, 359.
Blaha, M. (1972). *Astrophys. Lett.* **10**, 179.
Brocklehurst, M. (1970). *Mon. Not. R. Astron. Soc.* **148**, 417.
Brocklehurst, M. (1971). *Mon. Not. R. Astron. Soc.* **153**, 471.
Burgess, A. (1958). *Mon. Not. R. Astron. Soc.* **118**, 477.
Burgess, A. (1964a). *Proc. Symp. At. Collision Processes Plasmas, Culham Lab., UKAEA, Rep. No.* AERE-R4818.
Burgess, A. (1964b). *Mem. R. Astron. Soc.* **69**, 1.
Burgess, A. (1965a). *Ann. Astrophys.* **28**, 774.
Burgess, A. (1965b). *Astrophys. J.* **141**, 1588.
Burgess, A., and Chidichimo, M. C. (1983). *Mon. Not. R. Astron. Soc.* **203**, 1269.
Burgess, A., and Percival, I. C. (1968). *Adv. At. Mol. Phys.* **4**, 109.
Burgess, A., and Seaton, M. J. (1960). *Mon. Not. R. Astron. Soc.* **120**, 121.
Burgess, A., and Summers, H. P. (1969). *Astrophys. J.* **157**, 1007.
Burgess, A., and Summers, H. P. (1976). *Mon. Not. R. Astron. Soc.* **174**, 345.
Burgess, A., and Tworkowski, A. (1976). *Astrophys. J.* **205**, L105.
Burgess, A., Hummer, D. G., and Tully, J. A. (1970). *Philos. Trans. R. Soc. London, Ser. A* **266**, 225.
Burgess, A., Summers, H. P., Cochrane, D. M., and McWhirter, R. W. P. (1977). *Mon. Not. R. Astron. Soc.* **179**, 275.
Chapman, A., and Cowling, T. G. (1970). "The Mathematical Theory of Non-Uniform Gases," 3rd ed. Cambridge Univ. Press, London and New York.
Cochrane, D. M., and McWhirter, R. W. P. (1983). *Phys. Scripta* **28**, 25.
Cowan, R. D. (1980). "Theory of Atomic Structure and Spectra." Univ. of California Press, Berkeley.
Cox, D. P., and Tucker, W. H. (1969). *Astrophys. J.* **157**, 1157.
Crandall, D. H., Phaneuf, R. A., Hasselquist, B. E., and Gregory, D. C. (1979). *J. Phys. B* **12**, L249.
Doyle, J. G. (1980). *Astron. Astrophys.* **87**, 183.
Dubau, J., Loulergue, M., and Steenman-Clark, L. (1980). *Mon. Not. R. Astron. Soc.* **190**, 125.
Eissner, W., Jones, M., and Nussbaumer, H. (1974). *Comput. Phys. Commun.* **8**, 270.
Faucher, P. (1975). *J. Phys. B* **8**, 1886.
Flower, D. R., and Nussbaumer, H. (1975). *Astron. Astrophys.* **45**, 145.
Frank, M. (1980). Ph.D. Thesis, Cambridge Univ., Cambridge, England.
Froese-Fischer, C. (1977). "The Hartree–Fock Method for Atoms: A Numerical Approach." Wiley, New York.
Gabriel, A. H. (1972). *Mon. Not. R. Astron. Soc.* **160**, 99.
Gabriel, A. H., and Jordan, C. (1969). *Mon. Not. R. Astron. Soc.* **145**, 241.
Gabriel, A. H., and Jordan, C. (1972). "Case Studies in Atomic Collision Physics," Vol. 2. North-Holland Publ., Amsterdam.

Glasstone, S., and Lovberg, R. H. (1960). "Controlled Thermonuclear Reactions." Van Nostrand, New York.
Golden, L. B., and Sampson, D. H. (1977). *J. Phys. B.* **10**, 2229.
Golden, L. B., and Sampson, D. H. (1980). *J. Phys. B* **13**, 2645.
Golden, L. B., Sampson, D. H., and Omidvar, K. (1978). *J. Phys. B* **11**, 3235.
Henry, R. J. W. (1981). *Phys. Rep.* **68**, 3.
Hutcheon, R. J., and McWhirter, R. W. P. (1973). *J. Phys. B* **6**, 2668.
Jacobs, V. L., Davis, J., Kepple, P. C., and Blaha, M. (1977a). *Astrophys. J.* **215**, 690.
Jacobs, V. L., Davis, J., Kepple, P. C., and Blaha, M. (1977b). *Astrophys. J.* **211**, 605.
Jacobs, V. L., Davis, J., Rogerson, J. E., and Blaha, M. (1979). *Astrophys. J.* **230**, 627.
Jordan, C. (1969). *Mon. Not. R. Astron. Soc.* **142**, 501.
Kastner, S. O., and Bhatia, A. K. (1979). *Astron. Astrophys.* **71**, 211.
Kunze, H. J. (1972). *Space Sci. Rev.* **13**, 565.
Landman, D. A., and Brown, T. (1979). *Astrophys. J.* **232**, 636.
Lang, J. (1985). *J. Phys. B.* (to be published).
Lawson, J. D. (1957). *Proc. Phys. Soc. London, Sect. B* **70**, 6.
Ljepojevic, N. N., Hutcheon, R. J., and McWhirter, R. W. P. (1984). *J. Phys. B.* (to be published).
Lotz, W. (1967a). *Astrophys. J., Suppl. Ser.* **14**, 207.
Lotz, W. (1967b). *Z. Phys.* **206**, 205.
Lotz, W. (1968). *Z. Phys.* **216**, 241.
Lotz, W. (1970). *Z. Phys.* **232**, 101.
McWhirter, R. W. P. (1959). *U.K. At. Energy Res. Establ., Rep.* **AERE-R2980**.
McWhirter, R. W. P. (1960). *Proc. Phys. Soc. London* **75**, 520.
McWhirter, R. W. P., and Hearn, A. G. (1963). *Proc. Phys. Soc. London* **82**, 641.
Magee, N. H., Mann, J. B., Mertz, A. L., and Robb, W. D. (1977). *Los Alamos Sci. Lab.* [*Rep.*] *LA* **LA-6691MS**.
Magee, N. H., Mann, J. B., Mertz, A. L., and Robb, W. D. (1980). *Los Alamos Sci. Lab.* [*Rep.*] *LA* **LA-8267MS**.
Malinowski, M. (1975). *Astron. Astrophys.* **43**, 101.
Menzel, D. H., and Peckeris, C. L. (1935). *Mon. Not. R. Astron. Soc.* **96**, 77.
Mewe, R., and Schrijver, J. (1978). *Astron. Astrophys.* **65**, 115.
Moores, D. L. (1972). *J. Phys. B* **5**, 286.
Moores, D. L. (1978). *J. Phys. B* **11**, L403.
Moores, D. L. (1979). *J. Phys. B* **12**, 4171.
Moores, D. L., Golden, L. B., and Sampson, D. H. (1980). *J. Phys. B* **13**, 385.
Nussbaumer, H., and Storey, P. H. (1975). *Astron. Astrophys.* **44**, 321.
Peacock, N. J., and Summers, H. P. (1978). *J. Phys. B* **11**, 3757.
Peacock, N. J., Hughes, M. H., Summers, H. P., Hobby, M., Mansfield, M. W. D., and Fielding, S. J. (1979). *Plasma Phys. Controlled Nucl. Fusion Res., Proc. Int. Conf., 7th, Innsbruck, 1978* **1**, 303.
Pengelly, R. M., and Seaton, M. J. (1964). *Mon. Not. R. Astron. Soc.* **127**, 165.
Percival, I. C., and Richards, D. (1975). *Adv. At. Mol. Phys.* **11**, 2.
Post, D. E., Jensen, R. V., Tartar, C. B., Grasberger, W. H., and Lokke, W. A. (1977a). *Princeton Plasma Phys. Lab., Rep.* **PPPL-1352**.
Post, D. E., Jensen, R. V., Tarter, C. B., Grasberger, W. H., and Lokke, W. A. (1977b). *At. Data Nucl. Data Tables* **20**, 397.
Pottasch, S. R. (1965). *Bull. Astron. Inst. Neth.* **18**, 7.
Raymond, J. C., Cox, D. P., and Smith, B. W. (1976). *Astrophys. J.* **204**, 290.
Rowan, W. L., and Roberts, J. R. (1979). *Phys. Rev. A* **19**, 90.
Sahal-Brechot, S. (1974). *Astron. Astrophys.* **32**, 147.
Sampson, D. H. (1974). *Astrophys. J., Suppl. Ser.* **28**, 309.

Sampson, D. H. (1978). *J. Phys. B* **11,** 541.
Sampson, D. H., and Golden, L. B. (1978). *J. Phys. B* **11,** 541.
Sampson, D. H., and Golden, L. B. (1979). *J. Phys. B* **12,** L785.
Sampson, D. H., and Parks, A. D. (1974). *Astrophys. J., Suppl. Ser.* **28,** 323.
Seaton, M. J. (1958). *Mon. Not. R. Astron. Soc.* **118,** 504.
Seaton, M. J. (1959). *Mon. Not. R. Astron. Soc.* **119,** 81.
Seaton, M. J. (1962). *In* "Atomic and Molecular Processes" (D. R. Bates, ed.), Chap. 11. Academic Press, New York.
Seaton, M. J. (1962b). *Proc. Phys. Soc. London* **79,** 1105.
Seaton, M. J. (1964). *Planet. Space Sci.* **12,** 55.
Seaton, M. J. (1975). *Adv. At. Mol. Phys.* **11,** 83.
Seaton, M. J., and Storey, P. J. (1976). *In* "Atomic Processes and Applications" (P. G. Burke and B. L. Moisewitsch, eds.), Chapter 6, pp. 133–197. North-Holland Publ., Amsterdam.
Spitzer, L. (1962). "Physics of Fully Ionised Gases," 2nd ed. Wiley (Interscience), New York.
Summers, H. P. (1974). *Mon. Not. R. Astron. Soc.* **169,** 663.
Summers, H. P. (1979). Rep. AL-R-5. Rutherford Appleton Lab.
Summers, H. P. (1977). *Mon. Not. R. Astron. Soc.* **178,** 101.
Summers, H. P., and McWhirter, R. W. P. (1979). *J. Phys. B* **12,** 1979.
Thomson, J. J. (1912). *Philos. Mag.* **23,** 449.
Van Regemorter, H. (1962). *Astrophys. J.* **136,** 906.
Vernazza, I. E., and Raymond, J. C. (1979). *Astrophys. J.* **228,** L89.
Wiese, W. L., and Younger, S. M. (1976). *In* "Beam Foil Spectroscopy" (I. A. Sellin and D. J. Pegg, eds.), Vol. 2, pp. 951–960. Plenum, New York.
Wiese, W. L., Smith, M. W., and Glennon, B. M. (1966). *Natl. Stand. Ref. Data Ser. (U.S. Natl. Bur. Stand.)* **NSRDS-NBS** 4.
Wiese, W. L., Smith, M. W., and Miles, B. M. (1969). *Natl. Stand. Ref. Data Ser. (U.S. Natl. Bur. Stand.)* **NSRDS-NBS** 22.

4
Properties of Magnetically Confined Plasmas in Tokamaks

John T. Hogan

Fusion Energy Division
Oak Ridge National Laboratory
Oak Ridge, Tennessee

List of Symbols.	113
I. Introduction.	114
A. Purpose and Scope of Chapter	114
B. Organization	115
II. Magnetic Configuration.	116
A. Flux Surfaces.	116
B. Special Configurations and Maintenance.	119
III. Moment Equations	122
A. Particle Balance.	122
B. Energy Balance.	124
C. Adiabatic Compression	125
IV. Particle Balance	126
A. Model for Particle Balance	126
B. Neutral-Particle Distribution.	127
C. Global Confinement Time and Local Diffusivity	129
V. Energy Balance.	130
A. Electron Confinement: Conduction and Convection.	131
B. Electron Confinement: Impurity Effects	132
C. Ion Confinement.	135
VI. Impurity Transport	136
References.	138

List of Symbols

ϕ	Toroidal azimuthal angle
ψ, χ	Poloidal, toroidal magnetic fluxes
r	Minor radius of plasma
f	RB_ϕ
ρ, a	Minor radius of flux surface, maximum minor radius of plasma

R	Major radius of plasma in torus
$\mathbf{B}$	Magnetic field
B_p, B_ϕ	Poloidal, toroidal components of magnetic field
I_ϕ	Toroidal plasma current
τ	Characteristic thermalization time on a flux surface
$\tau_{P,E}$	Particle, energy lifetimes
t	Time
η	Plasma resistivity
$\boldsymbol{\varepsilon}$	Electric field vector, component parallel to $\mathbf{B}$
$\mathbf{v}_D$	Plasma diffusion velocity
Z, Z_k	Charge of an ion, charge of kth species of ion
K	Total number of ionic species
$n_{e,i,z}$	Electron, hydrogenic ion and impurity ion densities
n_0	Neutral atom density
$f_{e,i,0}$	Velocity space density of electrons, ions, neutral hydrogen
q	Safety factor
i	Rotational transform
$q_{e,i}$	Heat flux density carried by electrons, hydrogenic ions
$\Gamma_{e,p,z}$	Flux density of electrons, hydrogenic ions, impurity ions
$P_{e,i}$	Pressure of electrons, ions
$\chi_{e,i}$	Thermal diffusivity of electrons, ions
D	Particle diffusivity
$\langle\sigma v\rangle_{\text{ioniz}}$	Rate coefficient for hydrogen ionization by electron impact
σ_{cx}	Cross section for charge exchange between H^0, H^+
$m_{e,i,z}$	Mass of electron, hydrogenic ion, impurity ion
$g(T_e)$	Temperature dependence of radiative cooling rate in coronal equilibrium
$\nu_{i,\nu}$	Collision frequency of hydrogen ions with other hydrogen ions, impurity ions
$\nu^*_{i,e}$	Ratio of ion–ion (electron–ion) collision frequency to ion (electron) bounce frequency
$f(\nu^*, r/R, Z_{\text{eff}})$	Dependence of ion thermal diffusivity on ν^*, r/R, Z_{eff}
β, β_P, β_T	Ratio of plasma pressure to total, poloidal, toroidal magnetic field pressure
$Q_{\text{NB,RF}}$	Neutral-beam and electromagnetic (e.g., radio-frequency) power density sources

I. Introduction

A. Purpose and Scope of Chapter

The rise in applications of atomic physics studies to magnetic fusion research has led to an increasingly active collaboration among fusion and atomic physicists. The atomic physics data often make feasible the only measurement of a particular quantity of interest. It is often difficult for atomic physicists to judge the extent to which applications of their work actually affect the course of fusion research. This can have important ramifications, for often some choice must be made, in the light of potential fusion application, as to whether continued or more detailed elaboration of a theoretical technique or measurement is more valuable than emphasis on a new area of research.

This chapter is a description of some key physics areas in fusion research, with attention given to the role played by atomic physics processes. There are many steps to be taken between the initial establishment of feasibility of a fusion concept and the eventual production of economically competitive fusion-generated power. Major technological hurdles remain to be surmounted, and a substantial effort is yet required to understand fusion plasma behavior to the level needed for the production of high-power-density, long-pulse-length tokamak discharges that are foreseen as the embodiment of a working fusion reactor. With other concepts (tandem mirror, EBT, stellarators, fast pinches) there is a similar evolution in store. The requirements for models of plasma behavior undergo a qualitative change as each concept moves toward application. In the initial stages of tokamak research, for example, it was sufficient to know that the bulk of the plasma energy was radiated by *some* impurity; the need to improve control over vacuum procedures was thereby established. As plasmas came to be dominated by diffusion or convection losses, it was necessary to identify the major impurity species which played a role, and to consider sputter- and evaporation-resistant materials for wall and limiter design. As temperatures in the reactor regime are approached, serious attention is focused on the detailed dynamics of impurities which must inevitably be produced despite the introduction of plasma configurations which employ magnetic divertors or more complex limiter designs. Any diagnostic technique or model prediction which concerns impurities in the plasma must rely on the quality of the atomic data used.

B. Organization

The first and most important concern in magnetic fusion is to establish the proper magnetic confinement geometry. For a low-β plasma (β is the ratio of plasma pressure to magnetic field pressure) this requirement is met by proper location of coils and transformers external to the plasma. For the high-β plasmas needed to produce economically attractive fusion power output, the plasma evolution itself makes a major contribution to the configuration. This subject is discussed in Section II.

Although much progress has been made, it is still rare to have an accurate experimental accounting for the particle and energy budget in the plasma.

Modeling studies are often used to establish consistency (or lack of it) with a proposed theoretical or empirical model. There is, of course, a heavy reliance on atomic data in this area. The general framework of the particle and energy balance is discussed in Section III; the conceptual basis for establishing conservation equations for particles and energy lies in the velocity-space kinetic equations that describe the drifts and collisional deflection of individual particles. In Section IV the particle balance is described and, in Section V, the energy balance. Special attention is required for the study of impurity dynamics, since diagnostic techniques have been improved in recent years, and this is described in Section VI.

II. Magnetic Configuration

A discussion of the role of atomic processes in the plasma particle and energy balance must begin with consideration of the magnetic geometry which provides the essential confinement. Added to this pedagogical imperative, there is an important trend in present tokamak research involving the creation of specially shaped plasma cross sections in which atomic processes, in the form of impurity behavior, play an important role. The new configurations are intended to provide a better margin of stability (hence a more powerful fusion energy source) and also to provide manageable disposal of the plasma heat and particle efflux by means of a magnetic divertor.

A. *Flux Surfaces*

The main physical ingredients of the tokamak configuration were described in Barnett (Chapter 1, this volume). The toroidal field B_ϕ, the toroidal current I_ϕ, and its associated (poloidal) magnetic field B_p produce a set of helically spiraling field lines encircling the symmetry axis of the torus. If we were to follow a single field line our motion would (usually) map out a two-dimensional surface in three-space. This surface is also a toroidal figure, and a series of such surfaces can be generated starting at the magnetic axis (Fig. 1) and proceeding outward. The poloidal magnetic flux

$$\psi(\rho) = 2\pi R \int_{\text{magnetic axis}}^{\rho} d\rho' \, B_p(\rho') \tag{1}$$

increases monotonically for typical distributions of B_p, and so labels the succeeding surfaces on which the field lines lie. (Here R is the major plasma radius, ρ the minor radial coordinate, and ρ' the integration variable.)

This geometric concept of a set of nested, toroidal surfaces, formed by following magnetic field lines, is essential to the consideration of transport processes. For the simple geometry shown in Fig. 1, a charged particle with

4. Properties of Magnetically Confined Plasmas in Tokamaks

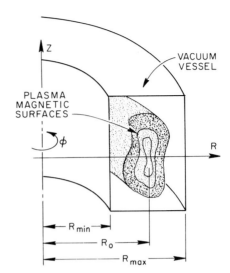

Fig. 1. Geometry of the tokamak. A vacuum vessel surrounds the plasma, which is axisymmetric with shaping in the minor cross section. Magnetic surfaces are the locus of magnetic field lines.

velocity nearly parallel to the magnetic field would circulate around the torus with an orbit lying very close to the flux surface. If its velocity were oriented perpendicular to the field, its orbit would be localized near this flux surface, but would have a recurrent drift away from it and toward it as the particle drifted in the toroidal direction. The flux surface retains its importance when nonideal confinement is considered. If collisions between single particles or motion induced by microscopic electric and magnetic turbulence are admitted, deviation from the set of flux surfaces will occur. These deviations will occur as discrete jumps from surface to surface, depending on the collision frequency or the strength of the turbulence. There will thus be a slower drift (or diffusion) from surface to surface combined with a rapid circulation around the torus on the surface. The virtually unimpeded circulation of particles around the torus on flux surfaces ensures that any variation of quantities on a flux surface will be obliterated rapidly (in a time $\tau \sim R/V_{\text{therm}}$) and that transport *across* flux surfaces is much slower than transport within the surfaces. Hence a three-dimensional spatial transport problem can be reduced to a one-dimensional diffusion process.

The definition of poloidal flux in Eq. (1) needs generalization if more complex, noncircular, cross section shapes are to be considered. In fact, the geometry is not specified by the major and minor radii of the torus (R and ρ), but rather by the magnetic flux itself. We retain Eq. (1), but use it to define a fictitious spatial coordinate ρ which labels surfaces of constant poloidal flux. The poloidal flux is defined by $\psi = \int \mathbf{B} \cdot d\mathbf{S}_\perp$, where $d\mathbf{S}_\perp$ is a surface element linking the torus (Fig. 2). The magnetic field can be described by two scalar functions of position:

$$\mathbf{B} = f\nabla\phi + \nabla\phi \times \nabla\psi, \qquad (2)$$

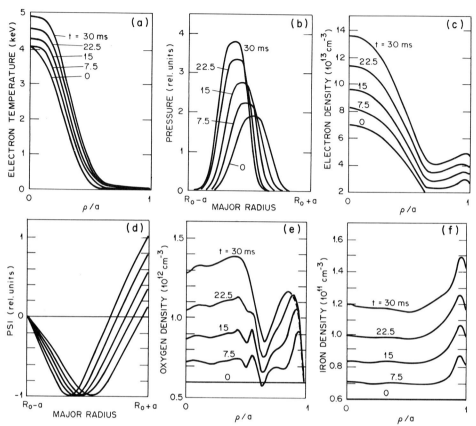

Fig. 2. Calculated adiabatic compression in the TFTR device; evolution of (a) temperature, (b) pressure, (c) electron density, (d) poloidal flux, (e) oxygen density, (f) iron density, as compression proceeds over an interval of 30 ms. Semiradial plots (magnetic axis at $\rho = 0$) are shown, except for (b) and (d). In the latter, the inward motion in major radius produced by compression is evident.

which automatically satisfies $\nabla \cdot \mathbf{B} = 0$: a three-component vector is reduced to two independent quantities f and ψ. In addition, for axisymmetry, it follows that $f(r, t) = f[\psi(\mathbf{r}, t), t]$; i.e., f can depend on spatial coordinates only through ψ.

The Maxwell field evolution equations, in axisymmetry, lead to an equation for ψ if we add Ohm's law (Grad and Hogan, 1970),

$$\frac{\partial}{\partial t}\psi + \mathbf{v}_D \cdot \nabla\psi = \eta_\parallel \Delta^* \psi - \frac{\varepsilon_\parallel}{2\pi R}. \tag{3}$$

v_D is the plasma diffusion velocity across the flux surface; $\Delta^* \equiv r^2 \, \mathrm{div}(r^{-2}\nabla)$ and $\varepsilon_\parallel$ is the parallel electric field. In a resistive plasma, the poloidal flux will diffuse in space at a rate determined by the plasma parallel electrical resistiv-

4. Properties of Magnetically Confined Plasmas in Tokamaks

ity. The exact calculation of $\eta_\parallel$ in toroidal geometry produces a complex expression (Hinton and Hazeltine, 1976; Birge et al., 1977), but the important factors are contained in Spitzer's original result for a simple spatially homogeneous plasma

$$\eta_\parallel \simeq 0.088 \frac{Z_{\text{eff}}}{T_e^{3/2}}, \quad Z_{\text{eff}} \equiv \frac{\sum_{k=1}^K n_k Z_k^2}{\sum_{k=1}^K n_k Z_k}, \tag{4}$$

where Z_K is the charge of the fully stripped ion. The scattering of electrons by impurities dominates the resistivity in many experiments, and is thus the controlling feature in the diffusion of ψ, and therefore of the magnetic geometry itself.

The picture of transport developed so far is that rapid circulation occurs *within* a toroidal flux surface, with slower diffusive or convective motion of the plasma occurring normal to these surfaces, toward the outer boundary and eventual loss. As described by Eq. (3) the set of flux surfaces itself also moves slowly (compared with the duration of the experiment), carrying the rapid internal motion with it. Depending as it does on Z_{eff}, the basic confinement geometry is strongly influenced by plasma impurity content. The diagnosis and control of impurity-related processes is, of course, an area where atomic physics considerations enter most decisively.

B. Special Configurations and Maintenance

Discussion of the motivation for choosing the cross section shapes under study would lead us too far afield (see Furth, 1975; Wesson, 1978). A brief description of the problems being studied will indicate the importance of the resistive diffusion process and, by implication, of the role of atomic physics in this area.

1. q-*profile*

A key physical parameter determining stability in the tokamak is the so-called safety factor

$$q \equiv \frac{\partial \chi}{\partial \psi} = \frac{2\pi}{i}. \tag{5}$$

The toroidal magnetic flux $\chi \equiv \int \mathbf{B} \cdot \mathbf{dS}$ is provided by the main field. The quantity i is the rotational transform; it is the poloidal angular displacement made in following a field line once around the torus (i.e., if $q = 3$, $i = \frac{2}{3}\pi$, and the field line rotates clockwise by 60° upon return to the starting plane).

Providing an optimal q-profile is of central importance to obtaining the plasma conditions of high pressure and good confinement necessary for reactors. Unfortunately, the spatial distribution of the rotational transform profile is only imperfectly known at present. Some techniques under study for measuring q are

(1) measuring the spatially distributed emission of neutral atoms produced by charge transfer between a low-current diagnostic neutral beam and the plasma ions (Goldston, 1977), and

(2) measurement of the Faraday rotation caused by the poloidal magnetic field of an FIR laser signal, as a function of position in the plasma (Hutchinson *et al.*, 1979).

These diagnostics are in a developmental stage, since their goal is to provide instantaneous, real-time information about the evolving q profile. In the absence of an ideal system, there is secondary information which allows an estimate of the safety factor:

(1) Observation of characteristic, repetitive "sawtooth" oscillations from the plasma core which are believed to be stimulated when $q \lesssim 1$ (von Goeler *et al.*, 1974; Equipe TFR, 1978a).

(2) Measurement of oscillations in the externally measured amplitude of fluctuations in the poloidal (i.e., plasma-current-produced) magnetic field (Mirnov and Semënov, 1971). The spatial structure $e^{im\theta + in\phi}$ can be used with theoretical saturated-instability models (Carreras *et al.*, 1979) to infer the location of the poloidal flux surface position ρ_S where $q(\rho_S) = m/n$. This surface is thought to be the source of such field oscillations.

2. *Concept of Flux-Conserving Tokamak*

While the q-profile measurement requires considerable development, the importance of the measurement is also very great. The prospects for attaining a high β require that an optimal q-profile be established in the plasma before heating. Then, if heating is applied rapidly, this safety factor profile will be "frozen" in the plasma (Mukhavatov and Shafranov, 1971; Clarke and Sigmar, 1977). This flux-conserving tokamak (FCT) concept has led to increased optimism over the chances for the tokamak configuration to achieve economically adequate values. The q-profile is frozen because it evolves in space and time according to

$$\left.\frac{\partial q}{\partial t}\right|_\psi = \frac{\partial}{\partial \psi} (\eta_\parallel \langle E \cdot B \rangle),$$

$$\langle \alpha \rangle \equiv \frac{\int \frac{dS}{|\nabla \psi|} \alpha}{\int \frac{dS}{|\nabla \psi|}}. \tag{6}$$

where α is any physical quantity and ε is the electric field. Hence, for a high-temperature plasma, the dependence of $\eta_\parallel$ on T_e [Eq. (4)] assures that $\partial q/\partial t \approx 0$, if the impurity level is not too high.

4. Properties of Magnetically Confined Plasmas in Tokamaks

At any rate, the diagnosis of q-profile evolution, and the spectroscopic analysis of impurity contributions to it, are areas in which atomic physics enters the analysis of the FCT process.

3. *Shaping the Plasma Cross Section*

The FCT evolution is important for attaining high-β states in all cross section shapes, but the stability of this plasma state is thought to depend sensitively on the form of the cross section. The vertical elongation of the circular cross section into an ellipse, and the deformation of the outer half of the ellipse away from the central torus symmetry axis to form a D-shape, are felt to allow a substantial increase in the stable β which can be maintained (Todd et al., 1979; Shafranov and Yurchenko, 1968). The D-shape is produced with the help of external coils that attract the currents flowing in the plasma. The plasma current distribution must be broad, i.e., not concentrated at the magnetic axis, in order to secure the benefits of shaping. As shown in Wesley et al. (1981), the formation of a doublet-shaped plasma cross section (Ohkawa, 1968) is made impossible if heavy recycling and impurity radiation at the plasma boundary[†] raise the resistivity at the plasma edge, depressing T_e (remember $\eta_\parallel \propto T_e^{-3/2}$). If the current is forced to the core, the flux surfaces there are mainly circular (a "droplet" configuration). If, however, the current is uniformly distributed, then the external coils are effective, and shaping to suit confinement requirements is successful. Again, atomic physics processes enter in establishing the control of recycling and impurity radiation which is needed to provide broad current profiles.

4. *Configuration Maintenance*

While the FCT process can allow immediate access to high-β states and the external coil configuration can provide an optimal shape if the current profile is broad enough, the eventual goal is to produce a plasma in steady state condition. (Failing that, the longest possible discharge time is desired.) The consequences of Eq. (6), which governs the evolution of the q-profile, must be considered. Even for a high-temperature plasma, with a very small $\eta_\parallel$, the long-time resistive diffusion process will eventually produce a change in the q-profile, perhaps away from the ideally desired configuration (Zakharov and Shafranov, 1980). Typical evolution times are 10–100 s. Calculations of resistive evolution with a simple computational model which, however, contains the essential physical ingredients of resistive diffusion and high-β show that states with $\bar{\beta} \approx 10\%$ (an adequate level for reactor economics) can be maintained in the presence of resistive diffusion (Charlton et al., 1980). Anomalously high resistivity, or steadily increasing resistivity due to increasing impurity contamination, would hasten the evolution process.

[†] See Harrison (Chapter 7, this volume).

III. Moment Equations

The essential features of the plasma confinement geometry have been dealt with in Chapter 1. The field configuration described in Section II is axisymmetric. Ideas are now being discussed which entail nonsymmetric modifications of the magnetic field to provide a bundle divertor (Stott et al., 1977) or the creation of an "ergodic" region in the plasma edge to decrease the temperature near the wall or limiter (Karger and Lackner, 1977). The motion of individual particles in the magnetic fields of modern fusion devices is exceedingly complex, and even the question of the eventual long-term confinement of a *single particle* in nonaxisymmetric fields is an open question (Weitzner, 1980). Nonetheless, the usual statistical mechanical notions allow us to consider the collective motion of a large number of particles, none of which we can characterize well, and to assemble a reasonably useful picture of the macroscopic behavior of the plasma. This reduction is discussed in many texts and only the results will be used here.

With a specified magnetic geometry, given by poloidal and toroidal magnetic flux distribution, and a resulting set of nested toroidal magnetic flux surfaces, it is possible to calculate the diffusion across this set of surfaces caused by particle collisions. This "classical" process gives the irreducible minimum loss rate for a confined plasma. When details of the specific toroidal geometry are included in the calculation, the term "neoclassical" is used to denote the results, but the essential features are quite simple: the rapid drift motion of particles within a magnetic surface is decoupled from the much slower diffusion of particle and energy across the surfaces, which is caused by interparticle collisions. These simple physical ingredients require rather complex calculations for the resulting fluxes, but the theory is in a relatively complete state and is described in an extensive review (Hinton and Hazeltine, 1976). We quote the main results here since this simplest of physical pictures for the plasma, while it is believed to be incomplete, serves as a useful starting point for a discussion of the plasma particle and energy balance.

A. Particle Balance

Starting with the calculated distribution of particles in velocity and physical space $f_{\text{neo}}(v, r)$, the diffusion time-scale evolution of the plasma spatial density

$$n_{e,i}(\mathbf{r}, t) = \int d^3v\, f_{e,i}(\mathbf{r}, \mathbf{v}, t) \tag{7}$$

can be followed; the flux-surface averaged density equation is

$$\frac{\partial}{\partial t}[n_e(V, t)V'(\psi, t)] = \frac{\partial}{\partial \psi}\left(\left\langle n_e \frac{\mathbf{v}_D \cdot \nabla \psi}{|\nabla \psi|}\right\rangle\right) + \Sigma_e. \tag{8}$$

4. Properties of Magnetically Confined Plasmas in Tokamaks

The evolution of the density of electrons (ions) on a flux surface ($n_{e,i}V'$) is determined both by the cross-surface diffusion velocity and by the local sources Σ_e of electrons (ions). The volume contained within a surface is defined by $V(\psi) = \int_{\text{magn axis}}^{\psi} d^3r$. The infinitesimal volume contained within surfaces labeled by ψ and $\psi + d\psi$ is

$$dV \equiv V'(\psi)\, d\psi = d\psi \frac{d}{d\psi} \int_{\text{magn axis}}^{\psi} d^3r = d\psi \int_{\psi} \frac{dS}{|\nabla\psi|}. \tag{9}$$

The general expression for evolution of the density in space and time in a geometry consisting of nested, axisymmetric magnetic flux surfaces is

$$\frac{\partial}{\partial t}(n_e V') = \frac{\partial}{\partial \psi}(\Gamma_e V') + \Sigma_e. \tag{10}$$

The magnitude of the cross-surface flux (Γ_e: no./cm² s⁻¹) will depend on the physical model chosen. The neoclassical expressions for Γ_e are given in Hinton and Hazeltine (1976), and alternatives describing other models will be discussed in Section IV.

The local source of electrons Σ_e consists of

(1) ionization of neutral hydrogenic atoms,
(2) ionization of impurity ions, and
(3) trapping of injected neutral beam atoms by charge transfer to background hydrogenic ions.

The proton and impurity fluxes are also calculated in neoclassical theory. [See Hirshman and Sigmar (1981) for a review of impurity-related neoclassical calculations.] The electron, hydrogenic ion, and impurity fluxes are related by the condition of ambipolarity so that

$$\Gamma_e = \Gamma_p + \sum_k Z_k \Gamma_k, \tag{11}$$

where the sum is taken over all charge states. A simplified model for the evolution of electron, hydrogen ion, and impurity densities is typically used. Rather than calculate solutions to the evolution equations for electrons (or hydrogen density), plus the k species of impurity charge-state densities (Amano and Crume, 1978), an *average-ion* impurity density is calculated by summing and averaging all the evolution equations of the impurity densities (Düchs *et al.*, 1977). The resulting impurity equation is

$$\frac{\partial n_z}{\partial t} = \frac{1}{V'} \frac{\partial}{\partial \psi} \left(D \frac{\partial n_z}{\partial \rho} + C_{pz} g_1(\langle z \rangle, n_p, n_z, n_z') + C_{zz} g_2(\langle z \rangle, n_p, n_z, n_z') \right). \tag{12}$$

Coupling of Eq. (11) with an average-ion quasineutrality condition forms a closed set of equations describing the plasma particle balance. The sources of electrons, hydrogen ions, and impurities, both distributed in space and entering at the boundary, are described in Section IV. Models for D, C_{pz}, C_{zz}, g, and g_2 are quite complex.

B. Energy Balance

Taking the energy moment of the kinetic transport equations yields the energy balance equations for electrons and ions

$$\frac{3}{2}\frac{\partial}{\partial t}(P_e V'^{5/3}) = \frac{\partial}{\partial \psi}(\langle q_e + \tfrac{3}{2}\Gamma_e T_e \rangle)$$
$$- Q_{LR,\text{ion brem}} - Q_{ei} + Q^e_{NB,RF} + Q^e_{D-T} + Q_{OH} \tag{13}$$

and

$$\frac{3}{2}\frac{\partial}{\partial t}(P_i V'^{5/3}) = \frac{\partial}{\partial \psi}(\langle q_i + \tfrac{3}{2}\Gamma_e T_i \rangle) + Q_{ei} - Q_{cx} + Q^i_{NB,RF} + Q^i_{D-T}. \tag{14}$$

In general, collisional coupling between hydrogenic and impurity ions is strong, so that only the electron and one ion (hydrogen and impurity) temperature need be identified.

The heat flux $q_{e,i}$ will again depend on the physical model being considered. For "neoclassical" (drift/collisions) processes, these fluxes are given in Hinton and Hazeltine (1976). The neoclassical $q_{e,i}$ depend on the local gradients in T_e, T_i, n_e, and n_i.

The source and loss terms for energy are processes not associated with particle motion. They are

Q_{cx}: Loss of energy from plasma ions by charge transfer to cooler hydrogen atoms.

$Q_{LR,\text{ion brem}}$: Losses of electron energy due to inelastic collisions of electrons with impurity ions and hydrogenic atoms, leading to excitation and ionization. The resulting radiation from the plasma is in the form of line and continuum radiation, with line radiation dominating the emission from present devices, and continuum radiation (bremsstrahlung) assumed to be important in clean future high-temperature reactors. Electron cyclotron bremsstrahlung is currently used as a diagnostic for T_e, and it may be an important bulk energy loss for reactors with $T_e = 20$ keV.

$Q^{e,i}_{D-T}$: Both D–D and D–T reactions produce energetic charged-particle reaction products (Glasstone and Lovberg, 1960). In present tokamak experiments energetic tritons have been detected in the PLT beam heating experiments (Colestock et al., 1979). In reactors, the α particles produced from D–T reactions will be a dominant electron heat source. Hence, D–T α production is an important plasma heating source to be considered in extrapolating present confinement models to future devices.

$Q^{e,i}_{NB,RF}$: The injection of intense, energetic neutral beams (see Cordey, Chapter 6B, this volume) provides the dominant heating source in the present generation of experiments. The injected neutrals are captured and converted to energetic contained ions by means of electron capture by a background ion. (In reactors which may use injection at energies of 200 keV/amu or above, the capture process is apt to be impact ionization by the background ions.) The energetic fast ions then thermalize with the background, sharing the initial injected energy between background electrons and ions. This thermalization process seems well described by classical collision theory, perhaps because the particle potential energy delivered by fluctuating fields due to plasma turbulence is small compared to the injected energy. The development of alternative heating sources has recently shown great progress (Hosea *et al.*, 1981; Hwang *et al.*, 1982). There are many similarities between ion cyclotron resonance heating (ICRH) and the fast-ion thermalization process. Atomic physics processes are more significant in the neutral-beam injection scheme.

Q_{OH}: The Joule heating of the electrons caused by ohmic dissipation. The applied toroidal electric field draws a toroidal current against the resistance of the medium: small-angle deflections of the electrons away from the ϕ direction caused by collisions with hydrogen ions and impurities.

Q_{ei}: The collisional coupling between electrons and ions. This classical equilibration process (Hinton and Hazeltine, 1976) varies as $n^2(T_e - T_i)/T_e^{3/2}$ and so forces T_e and T_i to be nearly equal at high density.

C. Adiabatic Compression

The properties of the energy balance equation in their simplest form, without cross-field transport, are most evident when we consider adiabatic compression of the plasma. In this scheme, proposed by Furth and Yoshikawa (1970), the plasma is moved inward in major radius during a time which is short compared with times required for appreciable transport of particles and energy across flux surfaces. Hence we can neglect Γ_e, q_e, and q_i in Eqs. (8), (13), and (14). What remains is the behavior of the plasma as the volume between neighboring magnetic surfaces is squeezed. The adiabatic form of Eqs. (8), (13), and (14) requires

$$n_e V' = \text{const}, \quad P_e V'^{5/3} = \text{const}, \quad P_i V'^{5/3} = \text{const}. \tag{15}$$

To assure stability, the total plasma current is changed so that the safety factor q is constant. As shown in Fig. 2, this inward motion of the plasma leads to a strong increase in the maximum local density n_e, although the total number of electrons in the system remains fixed. Also, the temperatures increase, although not as much as the density.

This conceptually simple test of the energy and particle balance equations has been successfully carried out experimentally, with reasonable agreement with theory (Bol et al., 1972; Daughney and Bol, 1977). (The compression in the experiment was comparable to the electron energy diffusion time, so that the ideal conditions were not precisely satisfied.)

In the next section we examine the particle and energy balance models in more detail.

IV. Particle Balance

A. Model for Particle Balance

The bookkeeping represented by the conservation equation for electron number density [Eq. (8)] becomes more complicated when we consider comparisons of theoretical expressions with experimental results. To simplify the discussion at the outset, it is best to choose a circular, low-β plasma case, in which the magnetic geometry reduces to the familiar cylindrical symmetry (assuming R/a is large). If, in addition, we forsake the complex neoclassical expressions for Γ_e in Eq. (10), with a simpler, *ad hoc*, Fick's law,

$$\Gamma_e = -D \frac{\partial n_e}{\partial r}, \tag{16}$$

then the essentials of the problem will be clearer. For a pure plasma (no impurities), without beam heating, the evolution of density can be described by

$$\frac{\partial}{\partial t} n_e(r, t) = \frac{1}{r} \frac{\partial}{\partial r} \left(rD \frac{\partial n_e}{\partial r} \right) + n_e n_0 \langle \sigma V \rangle_{\text{ioniz}}. \tag{17}$$

Many of the complexities of the particle balance are present even in this simple model.

The simplest quantity of interest is τ_p, the "particle confinement time." It should give a rough measure of the residence time in the plasma of an electron created by ionization of H^0. For comparisons of experiment with theory the diffusivity D should be measured directly as a function of position. Integrating (17) over the volume, we obtain the rate of change of the total number of electrons in the plasma:

$$\frac{dN_e}{dt} = 4\pi^2 RaD(a) \frac{\partial n_e(a, t)}{\partial r} + 4\pi^2 R \int_0^a dr \, r n_0 n_e \langle \sigma v \rangle_{\text{ioniz}}. \tag{18}$$

4. Properties of Magnetically Confined Plasmas in Tokamaks

In steady state, an often-used definition of the global particle replacement time is

$$\tau_p \equiv \frac{N_e}{4\pi^2 R \int_0^a dr\, r n_e n_0 \langle \sigma v \rangle_{\text{ioniz}}}. \tag{19}$$

The estimate of the total ionization rate is made by measuring H_α (or H_β) light emission using spectroscopic techniques described in Breton *et al.* (1980). Finding the neutral-gas distribution in a tokamak plasma is a complex question in its own right, and some discussion is warranted.

B. Neutral-Particle Distribution

The disposition of neutral atoms in the plasma has received a great deal of experimental and theoretical attention. The major physical processes are described by the neutral transport kinetic equation

$$\mathbf{v} \frac{\partial f_0(\mathbf{r}, \mathbf{v})}{\partial \mathbf{r}} + n_e \langle \sigma v \rangle_{\text{ioniz}} f_0 + \int d^3v'\, f_i(\mathbf{v}') f_0(\mathbf{v}) |\mathbf{v} - \mathbf{v}'| \sigma_{\text{cx}}(|\mathbf{v} - \mathbf{v}'|)$$

$$= \int d^3v'\, f_0(\mathbf{v}') f_i(\mathbf{v}) |\mathbf{v} - \mathbf{v}'| \sigma_{\text{cx}}(|\mathbf{v}' - \mathbf{v}|). \tag{20}$$

The spatial transport of neutrals with velocity $\mathbf{v}'$ is modified by annihilation of particles with velocity $\mathbf{v}$ due to electron impact ionization and by loss due to charge transfer with the plasma ions. These losses are compensated by gains of particles by charge exchange of neutrals with plasma background particles which have velocity $\mathbf{v}$.

Sources of neutral particles at the plasma edge are (Stott and McCracken, 1979)

(a) molecules desorbed on the wall from the plasma,
(b) atomic particles reflected from the wall or limiter,
(c) atoms and molecules deposited near the limiter,
(d) molecules which have diffused back to the surface of the vacuum chamber from the interior, and
(e) atoms or molecules deliberately injected through a gas valve to increase or maintain the density (gas puffing).

These incident neutrals travel freely toward the plasma core until they are ionized or until they encounter a plasma hydrogen ion and charge exchange. In the latter case, a more energetic neutral is produced (a plasma proton acquires an electron from the cold neutral, but keeps its thermal energy) and so this neutral can (if its velocity is so directed) penetrate even farther toward the plasma core. This cascade, upward in energy and inward toward the core, can lead to the existence of a considerable neutral density even in the central plasma.

The existence of neutrals in the center forms the basis for a useful diagnostic technique. Charge exchange between the central neutral population and the plasma ions yields an efflux of high-energy particles which can be detected and energy analyzed outside the plasma (Petrov, 1976). The central ion temperature can be deduced from this flux, if the particle flux is not too heavily attenuated as it leaves the plasma.

Typical neutral-density spectra are shown in Fig. 3 of Clarke and Hogan (1974). There is a shell of high neutral density at the plasma edge. This means that the radiance of H_α or H_β light, which is used to infer the total ionization rate, is heavily weighted toward the plasma periphery. Abel inversion can be used to estimate the central ionization rate, but this technique has been found to be inaccurate in some cases.

The neutral-particle balance becomes more complicated when extremes of the parameter ranges or nonideal conditions are considered.

The Alcator A device has produced electron densities in excess of 10^{15} cm^{-3} (Gaudreau et al., 1977), whereas a model based upon electron impact ionization coupled to $H^0 + H^+$ charge exchange predicted $n_0(0) \sim 10^3$ cm^{-3}. In order to explain the high level of neutral flux coming from the hot core the

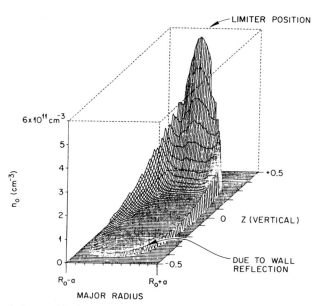

Fig. 3. Typical neutral hydrogen density spatial profile (averaged over toroidal angle). This calculated case assumes ISX-B device parameters and a neutral source from a top-rail limiter. $n_{H_0}(R, Z)$ is shown, as $n_{H_0}(r, \phi, Z)$ is averaged around the torus. Sharp localization near the (limiter) source is seen, as well as the somewhat lesser contribution away from the source due to wall reflection.

effects of three-body recombination were found to be important (Dnestrovskii *et al.*, 1979), and should be included even in the treatment of the low-temperature edge region.

When impurities are present, an additional annihilation term appears on the left-hand side of Eq. (20). The electron capture reaction

$$H^0 + A^{n+} \longrightarrow H^+ + A^{(n-1)+} \tag{21}$$

can have large cross sections, especially at the low energies characteristic of the plasma edge. Substantial theoretical and experimental work has been done recently, and charge exchange effects have been found to be important, especially for the case of carbon impurities (Phaneuf, 1981; Bienstock *et al.*, 1982; Hogan, 1982).

C. Global Confinement Time and Local Diffusivity

The result of the neutral-particle considerations is that there is at present no widely used and reliable technique to determine the local diffusivity. The global confinement time can be measured in the steady state [from Eq. (19)], but this τ_p is heavily weighted toward events in the periphery, and not in the plasma core.

In the absence of direct measurement, reliance has been placed on computer simulation to build a model for particle confinement. The first results of these calculations showed that the neoclassical rates for particle diffusion are much too slow to match experimental results (Hinton *et al.*, 1972). The particle confinement time predicted by the neoclassical model is too large by more than an order of magnitude.

A key experimental test for discriminating among models is provided by gas puffing. Starting with an established steady state discharge, cold gas is admitted through a pulsed valve and the plasma response in space and time is studied. The analysis of such experiments has led to formulation of several empirical models for particle confinement.

A semiempirical model has been used by Hughes (1978) to model gas puffing results in Alcator. In the model outward diffusivity due to current-driven drift wave turbulence (which is much larger than D^{neoclass}) balances against an inward neoclassical "pinch" flux originally proposed by Ware (1970) and incorporated within neoclassical theory. Thus the electron flux is

$$D^{\text{anom}} \simeq 10^{17}/n_e(r) \quad \text{cm}^2 \text{ s}^{-1}. \tag{22}$$

The difficulty with this and, indeed, all models to date for gas puffing, is that the experiment yields electron density profiles which decrease monotonically from the center to the edge, while the models typically predict the appearance of a transient "bump" in the profile, indicating that D in the model is too low (Düchs *et al.*, 1977).

In a recent study of gas puffing experiments on JIPPT-2 (Shimizu *et al.*,

1980), a model with an empirical D and the Ware pinch was shown to match the results under certain assumptions, i.e., by choosing

$$D^{\text{anom}} \simeq 2D_{\text{alcator}}^{\text{anom}} \simeq 2 \times 10^{17}/n_e \quad \text{cm}^2 \text{ s}^{-1} \tag{23}$$

and assuming that neutrals reflected from the wall enter the edge with an energy to tens of electron volts.

Düchs and Pfirsch (1975) also stressed the sensitivity of the particle confinement and models to the assumed edge neutral energy and to recombination effects.

Enhanced transport due to a postulated "ion mixing mode" has been found to lead to a diffusivity which could reproduce low-density Alcator gas puffing results by Antonsen et al. (1979).

The gas puffing paradox might be resolved by knowledge of cross sections for electron capture from H^0 by multiply charged (impurity) ions in the plasma. The effect of this process is to absorb electrons, which then alter the ionization stage of the impurities but are not available to produce unexplainable bumps on the density profile.

Pellet fueling serves as an independent test of particle confinement which avoids many of the uncertainties in atomic physics and plasma–wall interaction. Frozen hydrogen pellets containing from 1–300% of the total number of particles already in the discharge are accelerated to speeds of 1 km/s and injected into the plasma (Milora, 1980). There the pellet is ablatively ionized by the background plasma and a large, local source of particles is produced. The subsequent decay of this large density bump can be modeled and a local diffusivity proposed. Calculations of the results of ISX-B pellet injection experiments have shown that an empirical diffusivity

$$D \simeq 2 \times 10^{17}/n_e(r) \quad \text{cm}^2 \text{ s}^{-1} \tag{24}$$

gives a workable fit to the evolution date (Milora, 1980).

The pellet injection experiments serve as a further test of the adiabatic form of the particle and energy conservation equations [Eqs. (8), (13), and (14)]. The pellet ablation time (~ 100 s) is much shorter than the particle or energy confinement time; hence, the adiabatic response to the sudden increase in n_e takes the form of an instantaneous decrease in T_e which is needed to keep the total electron energy fixed. In cases in which a significant stored fast-ion component has been produced by beam injection, the pellet ablation process will consume much of the fast-ion energy and so the diffusivity which is measured by the pellet injection technique may not be representative of high-temperature plasma confinement processes.

V. Energy Balance

Atomic processes enter most decisively in the energy transport balance. Inelastic collisions of electrons with multiply charged (impurity) ions have

4. Properties of Magnetically Confined Plasmas in Tokamaks

received the bulk of the attention, since the stimulation of spectral line radiation has the largest effect on the plasma parameters. In this section we will consider the electron and ion energy balance separately, and describe the most widely used confinement models as well as noting the areas to which atomic processes contribute.

A. Electron Confinement: Conduction and Convection

The discussion of electron energy balance has been anticipated by the particle transport considerations of Section IV. The neoclassical (drift plus binary collisions) rates for electron fluxes Γ_e are calculated to be approximately m_e/m_p slower than the neoclassical ion rates. (This is a central fact in the neoclassical model, for it requires the creation of an electrostatic potential to retard the ion flux and maintain ambipolarity.) The neoclassical electron energy flux is similarly found to be much smaller than the observed electron losses. Comparisons of the neoclassical rates with experiment showed a large discrepancy (Hinton et al., 1972). As can be seen from the electron energy balance equations, however, there are source and loss terms which make the analysis complex.

The transport analysis uses measured T_e and n_e profiles, along with measured T_i on the magnetic axis, voltage, and current, to calculate as many of the energy balance variables as possible. Then the local D, χ_e, and χ_i are inferred by fitting the conduction and convection losses (q_e and $\frac{3}{2}T_e\Gamma_e$). Calculations show that there are significant contributions from each of the terms in Eq. (13), and none can be neglected everywhere in the plasma. There are, in addition, possible experimental uncertainties which would place significant error bars on the quantities in the analysis.

There are many figures of merit for energy confinement now in use. It is useful to consider a local $\tau_{E_e}(r)$ as

$$\tau_{E_e}(r) = \frac{\int d^3r\, \frac{3}{2} n_e T_e}{Q_{\rm OH} + Q_{\rm NB} - Q_{\rm ei} - Q_{\rm LR} - \dot{Q}_{n_e}}, \tag{25}$$

where $\dot{P}_{n_e} = \frac{3}{2} T_e\, \partial n_e/\partial t$. The global energy confinement time corresponding to $r = a$ would then be $\tau_{E_e}(a)$. The desired information, the value of the conduction and convection plasma transport losses (as distinguished from $Q_{\rm ei}$ and radiation losses) is defined as

$$\chi_e^{\rm exp}(r) \equiv \frac{Q_{\rm con}}{4\pi^2 R r n_e\, dT_e/dr}, \tag{26}$$

where $Q_{\rm con}$ is the local conduction and convection loss rate. The heat flux is expressed by a simple Fourier law

$$q_e = -n_e \chi_e \frac{dT_e}{dr}. \tag{27}$$

Neoclassical calculations would add dependence of q_e on dn_e/dr, $\varepsilon_\parallel$, etc. Of course, Q_{con} is determined experimentally as the difference between a calculated local power input rate $Q_{OH} + Q_{NB}^e$ and a loss due to transport radiation and electron–ion coupling. Hence the τ_e^{exp} so determined has the difference of large and nearly equal numbers in the numerator, and a term near zero (for the core plasma, $dT_e/dr \sim 0$) in the denominator. Hence high accuracy is not claimed for the results.

Nonetheless, the neoclassical rates appear to be too small to account for observations and so recourse has been made to empirical scaling for the global electron confinement time. The Alcator scaling law

$$\tau_E = 1.5 \times 10^{-15} \bar{n}_e a^2 \tag{28}$$

was deduced from high-density Alcator-A experiments (Cohn et al., 1976) and has been found to correlate well with a number of experiments (Hugill and Sheffield, 1978; Pfeiffer and Waltz, 1979). Lacking direct experimental data for χ_e, modeling is used to compare predictions. A local electron transport model, based on Alcator scaling, has found wide use:

$$\chi_e \simeq 5 \times 10^{17}/n_e \simeq 5 D_{Alcator}^{anom}. \tag{29}$$

This model, combined with a particle diffusivity $D = \frac{1}{5}\chi_e$, has been used as a basis for design studies of the INTOR (International Tokamak Reactor) (INTOR Zero Phase Report, 1980).

Global scaling does not discriminate between competing processes and it is not clear to what extent impurity radiation cooling and associated profile modification determine the empirical scaling.

In this regard, it is useful to mention another empirical scaling law which is widely used. The so-called Murakami limit for the maximum density attainable in an ohmically heated tokamak is (Murakami et al., 1976)

$$\bar{n}_e^{max} = 0.02 B_T \text{ (G)}/R \text{ (cm)}. \tag{30}$$

The quantity B_ϕ/R is also proportional to $j_\phi(0)$, and thus, perhaps, to the input power density in these ohmically heated discharges.

This empirical relationship has been analyzed experimentally by Mirnov and Semënov (1978), who find that the density limit is directly related to the stimulation of MHD instabilities in the initial phase of the discharge. An additional scaling ($\sim Z_{eff}^{-1}$) was also found to be valid.

B. Electron Confinement: Impurity Effects

The role of impurities in the electron energy balance is important and sometimes dominant. The models available for calculating radiative losses are relatively crude, and allow only qualitative analysis, especially for heavy-metal impurities (W, Mo). A widely used compilation of coronal equi-

4. Properties of Magnetically Confined Plasmas in Tokamaks

librium radiative cooling rates has been presented by Jensen *et al.* (1977) for the elements most commonly found in tokamak plasmas. Using the coronal equilibrium assumption (balancing electron impact ionization against radiative and dielectronic recombination) the radiative loss is

$$Q_{\text{rad}} = n_e n_z g(T_e), \tag{31}$$

where $g(T_e)$ depends on the element considered. The density dependences imply that, for a fixed impurity density and T_e, the radiative loss increases directly with n_e. Impurities are produced by charge exchange and by direct plasma bombardment of the wall and/or limiter, and since $Q_{\text{LR}} \propto n_e$, increasing density will result in an increased transfer of plasma thermal energy to the walls by photons, but the photon flux does not cause further impurity production. In this way, by shifting the form of the thermal load from particles to photons, impurity radiation can play a role in scaling the gross electron confinement time.

Other immediate consequences of radiative losses can be seen by reexamining the adiabatic compression process described earlier in Fig. 2. A calculation with W as an impurity in the plasma, rather than Fe, shows a markedly different evolution (Fig. 4). Rather than a rise in both n_e and T_e, as the plasma is compressed, we see that n_e, n_z, and hence the radiative cooling rate ($\sim n_e n_z$) sharply rise. This added cooling overwhelms the energy balance and the plasma temperature decays.

A similar effect has been observed experimentally. The presence of a sizable contamination of heavy metals (W, Mo) in the PLT and DITE experiments has led to the observation of hollow electron temperature profiles (Bol *et al.*, 1979; Hugill *et al.*, 1977). The electron temperature is lower on the magnetic axis than elsewhere in the plasma, because the electron density is highest there.

The effects of radiation cooling by heavy-metal impurities have, to date, been circumvented in experiments by using low-Z (e.g., carbon) limiters. As shown in Bol *et al.*, (1979), the carbon limiter produced a radiative emission profile which is peaked near the plasma edge in PLT beam heating experiments, while the steel limiters (perhaps with a small amount of W) gave rise to a centrally peaked radiative loss. In the steel-limiter case, the radiative loss occurs "in parallel" with the heat input, leading to degraded confinement. In the C-limiter case the radiative losses occur "in series" so that the rate of loss is not increased, but rather the nature of the losses of energy changes to photon emission from charged particle flux.

This latter "halo" effect has led to a proposal to shield the plasma core from impurities by producing a cool plasma mantle (Gibson and Watkins, 1977) of impurities at the plasma edge. These impurities would serve to radiate the thermal energy harmlessly to the walls. The implementation of such a scheme requires (a) greatly improved atomic data (to ensure that the measurement of impurity properties is adequately precise) and (b) a deep-

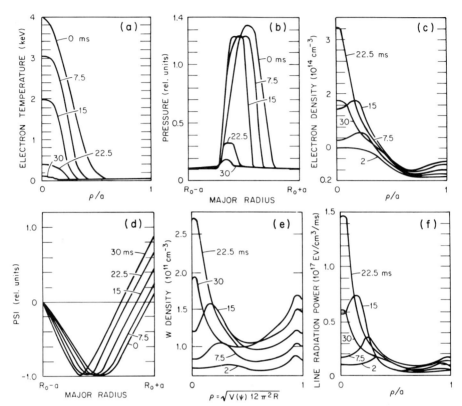

Fig. 4. Calculated adiabatic compression in the TFTR device. Parameters are as described in Fig. 2, *except:* (e) shows evolution of the tungsten density (in place of Fe in Fig. 2) and (f) follows the increase in calculated line radiation. The rapid increase in density (as the square of the compression factor) produces a sharp *drop* in T_e, and the loss of confinement.

ened understanding of the processes of impurity transport in the plasma in order to control the spatial position of the impurity. This subject will be discussed in Section VI.

The coronal equilibrium expressions of Jensen *et al.* (1977) are not adequate to describe the radiative cooling in present neutral-beam heating experiments. As shown by Abramov (1979) the neutral atoms injected for heating can suffer electron capture by plasma impurities. Beam-stimulated electron capture serves as the dominant recombination process in high-temperature plasmas and leads to a significant depression of the stage of impurity ionization. If the impurity content is large enough, the radiative cooling brought about by a lowered ionization state can destroy the effectiveness of auxiliary heating. This was observed in the DITE experiment, where 1-MW neutral-beam injection produced virtually no detectable electron heating (Axon *et al.*, 1981).

C. Ion Confinement

The ion dynamics appear to coincide with neoclassical expectations. The ion heat flux

$$q_i = -n_i \chi_i \frac{dT_i}{dr}, \qquad \chi_i = 0.68(r/R)^{1/2} \hat{\rho}_i^2 \nu_{ii} f(\nu_i^*, r/R, Z_{\text{eff}}) \tag{32}$$

is predicted to depend strongly on the ratio ν_i^* of the collision frequency ν_{ii} to the bounce frequency of ions trapped in local mirrors produced by the $1/R$ variation of the toroidal field. Experimental measurement of χ_i is less advanced than that of χ_e because T_i profiles are less readily available than Thomson scattering or electron cyclotron emission measurements of $T_e(r)$. Even so a variation by a factor 400 in ν_i^* (from $\nu_i^* \sim 8$ in high-density Alcator experiments to $\nu_i^* \sim 0.02$ with high-temperature PLT neutral-beam heating) results in values of χ_i that are within a factor of 1–5 of the neoclassical predictions. While this uncertainty is large, it lies within the experimental uncertainty, and it is much smaller than the range of variation in ν_i^* (Artsimovich, 1972; Brusati et al., 1978; Murakami et al., 1979a; Gondhalekhor et al., 1979; Equipe TFR, 1978b).

The implications of this scaling for fusion are quite significant because $\chi_i^{\text{neo}} \sim T_i^{-2}$; that is, the losses diminish as the temperature increases.

As mentioned earlier, the ion energy balance measurements are somewhat limited. The central ion temperature is inferred from the charge exchange neutral spectrum, and a profile can be constructed by injection of a diagnostic (low-current) neutral beam (Afrosimov et al., 1979). The estimate of the $T_i(r)$ profile is often made by numerical modeling of the $N_0(r)$ profile and then extracting $T_i(r)$ from the measured neutral efflux. The model for N_0 requires, as discussed earlier, an improved treatment of the effects of H^0 impurity charge exchange, in the 10 eV–2 keV energy range, before much reliance can be placed on it.

A possible remedy for this model dependence is afforded by the technique using resonance fluorescence stimulation of hydrogen light (Razdobarin et al., 1979). Measurements of $n_0(r)$ profiles in the FT-1 tokamak have been reported quantitative date on $n_0(r)$ profiles which would enhance the accuracy of the $T_i(r)$ profiles.

The most important effect of impurities on ion confinement arises through the dependence of q_i on Z_{eff}. The other area where atomic processes figure is in the charge exchange term in the ion energy balance [Eq. (14)]. The concentration of neutral density near the edge implies that the charge exchange loss is most effective in converting thermal energy in the ion component into a charge exchange neutral flux. This neutral efflux can then distribute the energy loss uniformly around the vacuum chamber, rather than causing local overheating (and possible evaporation) of an exposed limiter surface.

The fact that the hydrogen-ion heat flux obeys neoclassical scaling means that the neoclassical model should be given consideration when impurity dynamics are studied.

VI. Impurity Transport

Perfect magnetic confinement can, of course, not be realized. There must be some contact between the plasma and the external wall and, inevitably, some impurities must be introduced. The atomic physics of the ionization, excitation, and recombination processes which determine the effect of impurities on the magnetic configuration evolution, and on the particle and energy balance, is incomplete. The situation is especially bad for heavy-metal impurities ($Z \geq 26$). Concern over the deleterious effects of impurities is heightened by an observation arising from classical diffusion calculations. It has been shown that in a three-species plasma (electrons, protons, and charge-Z impurity ions) the eventual steady state distribution of impurity ions will be preferentially and sharply peaked in the core of the plasma, (Braginskii, 1965; Taylor, 1961):

$$n_z(0)/n_z(a) = [n_p(0)/n_p(a)]^z. \tag{33}$$

This result arises from the smallness of Γ_e compared with ion fluxes in classical theory, so that the ambipolar condition ($\Gamma_e = \Gamma_p + z\Gamma_z$) requires oppositely directed ion fluxes ($\Gamma_p \approx -z\Gamma_z$) when Γ_e is neglected.

The classical theory has been worked out in toroidal axisymmetric geometry for a multispecies plasma (Rutherford, 1974; Hinton and Moore, 1974; Hirshman and Sigmar (1981). The resulting neoclassical impurity theory is complex, in that separate diffusion equations are required for each of the impurity ionization stages, and the equations are nonlinear and strongly coupled. Hence to date there has not been a completely valid calculation of the expected impurity transport as predicted by a self-consistent neoclassical model (Crume et al., 1983).

It might be expected that the neoclassical impurity transport model could be judged without detailed calculation, simply by determining whether impurity concentrations are centrally peaked or not. Indeed, this criterion has often been cited in experimental papers. Unfortunately, the impurity neoclassical predictions are strongly influenced by the magnitude of the electron particle flux Γ_e. This flux determines, through the ambipolarity condition, the level of the self-consistent electrostatic field which controls the crossfield diffusion rates. As Stringer (1977) has shown, the existence of a nonnegligible Γ_e caused by anomalous electron transport, in the form

$$\Gamma_e^{\text{anom}} = -D_b \frac{\partial n_e}{\partial r}, \tag{34}$$

4. Properties of Magnetically Confined Plasmas in Tokamaks

results in a reduction and possible elimination of the peaking effect. The modified equilibrium distribution is

$$n_z(0)/n_z(a) = [n_p(0)/n_p(a)]^{z-g}, \tag{35}$$

where

$$g \simeq 0.7 \frac{ze^2 B_p^2}{(r/R)^{1/2} m_i T_i \nu_i} \left(1 - \frac{m_p}{m_z} \frac{\nu_i}{\nu_z} z\right) D_b.$$

As we have discussed in Section III, the experimental evidence favors the existence of an electron flux which is larger than the neoclassical value, and so the "peaking" criterion is not a reliable guide to impurity behavior. The anomaly in impurity dynamics is thus a consequence of a nonneoclassical electron flux ($\Gamma_e \neq 0$). In addition to the inherent complexity of the theory, the study of impurity dynamics has been hindered by the relative lack of knowledge about the impurity source. The natural impurity production mechanisms are strongly related to plasma variables which are poorly measured at present.

To circumvent this, a number of experiments have been carried out in which impurities are injected in a controlled manner. Small amounts of argon were injected into T-10 discharges and observed to diffuse into the center where the argon density became saturated rather than increasing indefinitely (Berlizov et al., 1981). The effect of periodic "sawtooth" oscillations in the plasma core was cited as a specific mechanism for providing the nonneoclassical (anomalous) Γ_e required to stop the central accumulation of impurities. Impurities have been injected into the plasma by means of a laser blowoff technique described in Cohen et al. (1975). A quartz plate coated with the impurity of interest is exposed to a laser light pulse and the impurities are driven into the plasma at a definite time. The subsequent evolution of the impurity density can be followed spectroscopically. The effects of heavy-metal impurity radiation have been shown by this technique. On the ISX-A device, tungsten was deliberately injected using the laser blowoff technique (Murakami et al., 1979b), and the electron temperature profile changed from peaked to hollow form and the impurity radiation was greatly enhanced.

A number of techniques are now being developed to make quantitative measurements of impurity densities. The diagnostic techniques in use have been reviewed (Equipe TFR, 1978a) and a review of spectroscopic results has been given by Suckewer (1980).

Laser-induced resonance fluorescence stimulation of impurity ions near the plasma edge has led to an absolute number density measurement on the ISX tokamak (Isler et al., 1981). A high-energy T1 beam technique has been used on the PLT tokamak (Cecchi et al., 1979); neutral atoms are injected into the edge plasma, are ionized there, and the secondary ions are velocity analyzed upon emerging from the plasma. The magnitude of the electrostatic potential is sought from experiments of this type, since it is the controlling variable for impurity diffusion rates.

The empirical study of impurity transport has been pursued as well. Recent measurements on the Alcator-A tokamak have employed the laser blowoff technique using Si as the trace impurity (Marmar *et al.*, 1980). It has been found that the Si does not continuously accumulate on axis, and that the (finite) impurity confinement time is proportional to the mass of the background plasma ions. While important in its own right, it is interesting to observe that a similar mass dependence has been observed on PDX (Hawryluk, 1982) for τ_E, and if τ_P and τ_E prove to be related, this variation may be describing the effect of Γ_e on impurity transport.

References

Abramov, V. (1979). *JETP Lett. (Engl. Transl.)* **29,** 501.
Afrosimov, V. A., Kislyakov, A. I., Khudoleev, A. V., and Shchemelinin, S. G. (1979). *Sov. J. Plasma Phys. (Engl. Transl.)* **29,** 501.
Amano, T., and Crume, E. (1978). *Oak Ridge Nat. Lab.* [*Rep.*] (*U.S.*) **ORNL-TM-6363.**
Antonsen, T., Coppi, B., and Englade, R. (1979). *Nucl. Fusion* **19,** 641.
Artsimovich, L. (1972). *Nucl. Fusion* **12,** 215.
Axon, K., *et al.* (1981). *Plasma Phys. Controlled Nucl. Fusion Res., Proc. Int. Conf., 8th, Brussels, 1980* **1,** 413.
Bienstock, S., Heil, T. G., Bottcher, C., and Dalgarno, A. (1982). *Phys. Rev. A* **25,** 2850.
Berlizov, A., *et al.* (1981). *Plasma Phys. Controlled Nucl. Fusion Res., Proc. Int. Conf., 8th, Brussels, 1980* **1,** 23.
Birge, B., Hirshman, S., and Hawryluk, R., (1977). *Nucl. Fusion* **17,** 611.
Bol, K., *et al.* (1972). *Phys. Rev. Lett.* **29,** 1495.
Bol, K., *et al.* (1979). *Plasma Phys. Controlled Nucl. Fusion Res., Proc. Int. Conf., 7th, Innsbruck, 1978* **1,** 11.
Braginskii, S. (1965). *In* "Reviews of Plasma Physics" (M. A. Leontovich, ed.), p. 251. Consultants Bureau, New York.
Breton, C., DeMichelis, C., and Mattioli, M. (1980). *Fontenay-aux-Roses Rep.* **EUR-CEA-FC-1060.**
Brusati, M., Davis, S., Hosea, J., Strachan, J., and Suckewer, S. (1978). *Nucl. Fusion* **18,** 1205.
Carreras, B., Waddell, B., and Hicks, R. (1979). *Nucl. Fusion* **19,** 1423.
Cecchi, J., Kozub, C., and Munson, C. (1979). *Bull. Am. Phys. Soc.* **24,** 986.
Charlton, L., Dory, R., and Nelson, D. (1980). *Phys. Rev. Lett.* **45,** 24.
Clarke, J., and Hogan, J. (1974). *J. Nucl. Mater.* **53,** 1.
Clarke, J., and Sigmar, D. (1977). *Phys. Rev. Lett.* **38,** 70.
Cohen, S., Cecchi, J., and Marmar, E. (1975). *Phys. Rev. Lett.* **35,** 1507.
Cohn, D., Parker, R., and Jassby, D. (1976). *Nucl. Fusion* **16,** 31.
Colestock, P., Strachan, J., Ulrickson, M., and Chrien, R. (1979). *Phys. Rev. Lett.* **43,** 768.
Daughney, C., and Bol, K. (1977). *Nucl. Fusion* **17,** 367.
Dnestrovskii, Y., Lysenko, S., and Kislyakov, A. (1979). *Nucl. Fusion* **19,** 293.
Düchs, D., and Pfirsch, D. (1975). *Bull. Am. Phys. Soc.* **21,** 1125.
Düchs, D., Post, D., and Rutherford, P. (1977). *Nucl. Fusion* **17,** 565.
Equipe TFR (1978a). *Nucl. Fusion* **18,** 647.
Equipe TFR (1978b). *Nucl. Fusion* **18,** 1271.
Furth, H. (1975). *Nucl. Fusion* **15,** 487.

4. Properties of Magnetically Confined Plasmas in Tokamaks

Furth, H., and Yoshikawa, S. (1970). *Phys. Fluids* **13,** 2539.
Gaudreau, M., *et al.* (1977). *Phys. Rev. Lett.* **39,** 1266.
Gibson, A., and Watkins, M. (1977). *Eur. Conf. Controlled Fusion Plasma Phys., Conf. Proc., 8th, Prague* **1,** 31.
Glasstone, S., and Lovberg, R. (1960). "Controlled Thermonuclear Reactions." Van Nostrand, Princeton, New Jersey.
Goldston, R. (1977). Ph.D. Thesis, Princeton Univ., Princeton, New Jersey.
Gondhalekhor, A., *et al.* (1979). *Plasma Phys. Controlled Nucl. Fusion Res., Proc. Int. Conf., 7th, Innsbruck, 1978.*
Grad, H., and Hogan, J. (1970). *Phys. Rev. Lett.* **24,** 1337.
Hawryluk, R. (1982). Personal communication.
Hinton, F., and Hazeltine, R. (1976). *Rev. Mod. Phys.* **48,** 239.
Hinton, F., and Moore, T. (1974). *Nucl. Fusion* **14,** 639.
Hinton, F. L., Wiley, J. C., Düchs, D. F., Furth, H. P., and Rutherford, P. H. (1972). *Phys. Rev. Lett.* **29,** 698.
Hirshman, S., and Sigmar, D. (1981). *Nucl. Fusion* **21,** 1079.
Hogan, J. (1982). In "Proceedings of the International Conference on the Physics of Electronic and Atomic Collisions" (S. Datz, ed.), p. 769. North-Holland Publ., Amsterdam.
Hosea, J., Boyd, D., *et al.* (1981). *Plasma Phys. Controlled Nucl. Fusion Res., Proc. Int. Conf., 8th, Brussels, 1980* **2,** 95.
Hughes, M. (1978). *Princeton Plasma Phys. Lab., Rep.* **PPPL-1411.**
Hugill, J., and Sheffield, J. (1978). *Nucl. Fusion* **18,** 15.
Hugill, J., *et al.* (1977). *Eur. Conf. Controlled Fusion Plasma Phys., Conf. Proc., 8th, Prague* **1,** 39.
Hutchinson, D. P., Staats, P. A., Vander Sluis, K. L., Wilgen, J. B., and Ma, C. H. (1979). *Bull. Am. Phys. Soc.* **24,** 984.
Hwang, D., *et al.* (1982). *Plasma Phys. Controlled Nucl. Fusion Res., 10th, Baltimore, 1982* (to be published by IAEA, Vienna).
INTOR Zero Phase Report (1980). IAEA, Vienna.
Isler, R., Milora, S., *et al.* (1981). *Plasma Phys. Controlled Nucl. Fusion Res., Proc. Int. Conf., 8th, Brussels, 1980* **1,** 53.
Jensen, R., Post, D., Grasberger, W., Lokke, W., and Tarter, B. (1977). *At. Nucl. Data Tables* **17,** 309.
Karger, F., and Lackner, K. (1977). *Bull. Am. Phys. Soc.* **22,** 1184.
Marmar, E., Rice, J., and Allen, S. (1980). *Phys. Rev. Lett.* **45,** 2025.
Milora, S. (1980). *J. Fusion Energy* **1,** 15.
Mirnov, S., and Semënov, I. (1971). *Sov. At. Energy (Engl. Trans.)* **30,** 22.
Mirnov, S., and Semënov, I. (1978). *Sov. J. Plasma Phys. (Engl. Transl.)* **4,** 27.
Mukhavatov, V., and Shafranov, V. (1971). *Nucl. Fusion* **11,** 605.
Murakami, M., Callen, J., and Berry, L. (1976). *Nucl. Fusion* **16,** 347.
Murakami, M., *et al.* (1979a). *Phys. Rev. Lett.* **42,** 655.
Murakami, M., *et al.* (1979b). *Plasma Phys. Controlled Nucl. Fusion Res., Proc. Int. Conf., 7th, Innsbruck, 1978* **1,** 269.
Ohkawa, T. (1978). *Kakuyugo-Kenkyu* **20,** 557.
Petrov, M. (1976). *Fiz. Plazmy (Moscow)* **2,** 371.
Pfeiffer, W., and Waltz, R. (1979). *Nucl. Fusion* **19,** 51.
Phaneuf, R. (1981). *Phys. Rev. A* **24,** 1138.
Razdobarin, G., *et al.* (1979). *Nucl. Fusion* **19,** 1439.
Rutherford, P. (1974). *Phys. Fluids* **17,** 1782.
Shafranov, V., and Yurchenko, E. (1968). *Sov. Phys.—JETP (Engl. Transl.)* **26,** 682.
Shimizu, K., Ikamotu, M., and Amano, T. (1980). *Res. Rep.—Nagoya Univ., Inst. Plasma Phys.* **IPPJ-483.**

Stott, P., and McCracken, G. (1979). *Nucl. Fusion* **19,** 889.
Stott, P., Wilson, C., and Gibson, A. (1977). *Nucl. Fusion* **17,** 481.
Stringer, T. (1977). *Proc. Int. Sch. Plasma Phys., Varenna, Italy.*
Suckewer, S. (1981). *Phys. Scripta* **23,** 72.
Taylor, J. (1961). *Phys. Fluids* **4,** 1142.
Todd, A., *et al.* (1979). *Nucl. Fusion* **19,** 743.
Von Goeler, S., Stodiek, W., and Sauthoff, W. (1974). *Phys. Rev. Lett.* **33,** 1201.
Ware, A. (1970). *Phys. Rev. Lett.* **25,** 916.
Weitzner, H. (1980). Rep. MF/97. Courant Inst. Math. Sci., New York.
Wesley, J., *et al.* (1981). *Plasma Phys. Controlled Nucl. Fusion Res., Proc. Int. Conf., 8th, Brussels, 1980* **1,** 35.
Wesson, J. (1978). *Nucl. Fusion* **18,** 87.
Zakharov, L., and Shafranov, V. (1980). "Problems in the Evolution of Toroidal Configurations," Rep. IAE-3075. Kurchatov Inst., Moscow; Engl. transl. available as *Oak Ridge Nat. Lab.* [*Rep.*] (*U.S.*) **ORNL-TR-4667** (1980).

5
Diagnostics

5A
Diagnostics Based on Emission Spectra

N. J. Peacock

Culham Laboratory
Abingdon, Oxfordshire
England

I.	Introduction	143
II.	Ionization Equilibrium	144
	A. Charge States in Diffusive Equilibrium	146
	B. Charge Transfer Recombination	155
III.	Atomic Level Populations	162
	A. Forbidden Lines	164
	B. Optical Opacity	168
IV.	Spectral Features and Their Diagnostic Application to Fusion Plasmas	169
	A. K-Shell Excitation	171
	B. L-Shell Excitation	176
	C. M-Shell Excitation	180
V.	Neutral-Beam Spectroscopy	181
VI.	Basic Atomic Physics	183
	References	186

I. Introduction

Emission spectroscopy has played an important role in the study of high-temperature laboratory plasmas since the earliest days of the controlled fusion research program; see for example the early experiments on Zeta (Wilson, 1962; Jones and Wilson, 1962). Significant early spectroscopic studies were also carried out in "θ-pinch" sources (Sawyer et al., 1963) and in the "plasma focus" (Peacock et al., 1969, 1971). Much of the emission from these fusion devices originated from low concentrations of highly stripped impurity ions either unavoidably present in the fuel ions of hydrogen and its isotopes or added deliberately as diagnostic indicators. In the early studies on these often short-lived plasmas, interest in the spectra was confined mainly to the diagnostic measurement of temperature, the importance of the contaminants on the overall plasma behavior never having been clearly established. The problem of contaminants persists even with present-day magnetic confinement systems. Impurities in tokamaks for example can govern the overall plasma behavior. Radiation loss, plasma resistivity, and impurity ion transport can all affect the overall energy balance and stability of the plasma. Spectroscopic techniques offer a unique method for identifying the impurity ion species and for measuring their concentrations. For this specific purpose no alternative diagnostic method has been successful.

Much of this subchapter refers to impurity ions in tokamaks since the most intensive effort in recent years has been directed towards these devices. In tokamaks, kilo-electron-volt temperatures are common in the core of the device with electron densities in the range 5×10^{12}–5×10^{14} cm^{-3}, the plasma lasting typically for many ion confinement times.

The discussion is limited to diagnostic methods based on emission spectra from mainly highly ionized atoms. Laser spectroscopic techniques such as resonance fluorescence using tunable lasers, an important and rapidly developing field (Burgess, 1978) are not included.

The text on *Plasma Spectroscopy* by Griem (1964) and review articles by Cooper (1966), McWhirter (1965), Bogen (1968), and Breton and Schwob (1974) form a useful background to this discussion. More recent and pertinent articles to tokamak spectroscopy are by De Michelis and Mattioli (1981), Hinnov (1979, 1980), Peacock (1980), and Suckewer (1981).

II. Ionization Equilibrium

The question of ionization–recombination balance and the equilibrium charge state of impurities in a fusion device is of intrinsic interest to the spectroscopist, depending as it does on the magnitude of the various atomic rate process (McWhirter and Summers, Chapter 3, this volume). In tokamaks the plasma is often sufficiently reproducible to contemplate the evaluation of atomic collision rates from programmed changes in the plasma parameters. Time-dependent line intensities from highly stripped ions during the marginally stable "sawtooth" mode of operation in the TFR tokamak (Breton *et al.*, 1978) have indeed been interpreted in terms of the magnitude of the ionization and recombination rates.

The problem is of more than academic interest to the plasma physicist since impurities contribute to and can even dominate the local effective ion charge seen by the electrons

$$Z_{\text{eff}} = \sum_{Z,z} \frac{N(Z^{z+})z^2}{n_e} \quad (1)$$

and since radiation loss from charge states which are not in coronal balance can exceed that calculated on a time-independent coronal model. In most plasma devices, though, the introduction of a finite confinement time, equivalent to lowering the confinement factor $n_e\tau$ (cm^{-3} s) below the coronal asymptotic value, has only a small effect on the average impurity ion charge $\langle z \rangle$ as indicated in the iron ion calculations in Fig. 1. The effective ion charge can be derived from $\langle z \rangle$ by using the approximation

$$Z_{\text{eff}} \simeq (1 + f_{\text{Fe}}\langle z \rangle_{\text{Fe}}^2)/(1 + f_{\text{Fe}}\langle z \rangle_{\text{Fe}}), \qquad (2)$$

where f_{Fe} is the fractional population of Fe impurity in the hydrogen plasma.

Radiation losses are not much affected either, as can be appreciated again from Fig. 1, except at high temperatures when the time-independent coronal conditions predict a minimum in the radiation from bare Fe nucleii, i.e., Fe^{26+}. The problem of noncoronal equilibrium radiation loss has been discussed by many authors (see, e.g., Jensen *et al.*, 1977; Uchikawa *et al.*, 1980). Radiation loss from transient, nonequilibrium, impurity ion populations is discussed by Hopkins and Rawls (1977) and by Summers and McWhirter (1979).

A more serious concern to the plasma physicist is the related problem of impurity ion diffusion. The stationary populations of the different ion charges depend not only on a correct treatment of the atomic processes but also on the motion of ions through the plasma.

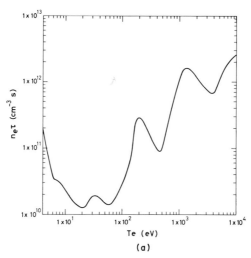

Fig. 1. (a) $n_e\tau$ (cm^{-3} s) values required to reach steady state iron ion populations as a function of electron temperature. (b) Average iron ion charge as a function of temperature for various values of $n_e\tau$ (cm^{-3} s); $\langle z \rangle_{\text{Fe}} = \sum_{z=1}^{26} f_z z$, where f_z is the fractional abundance of iron in charge state z. (c) Radiated power (W cm^{+3})/(Fe ion) × (electron density) from an iron plasma as a function of temperature for various values of $n_e\tau$ (cm^{-3} s). [From Carolan and Piotrowicz (1981)]. (*Continued on next page.*)

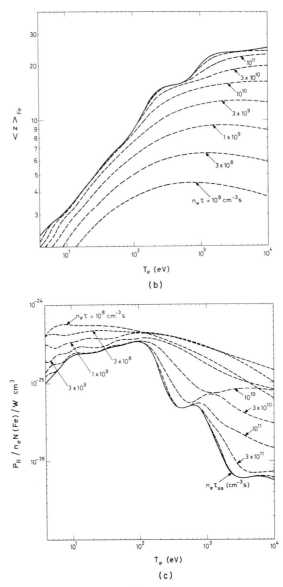

Fig. 1b and 1c.

A. Charge States in Diffusive Equilibrium

Typically in toroidal plasma devices, the electron temperature and density decrease monotonically from the core, their spatial functions often being approximately parabolic. Each impurity ion charge state is usually constrained to a fairly well-defined radial shell (Hinnov, 1980) which lies closer

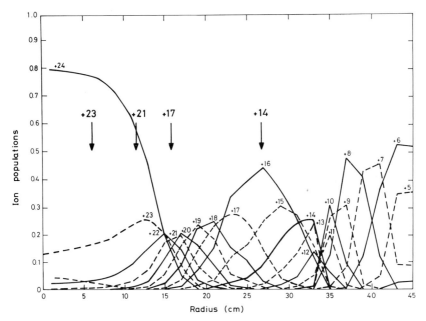

Fig. 2. Coronal equilibrium calculations of iron ion concentrations in PLT tokamak compared to radial positions at which the observed ion species indicated by arrows reach a peak abundance. [From Hulse et al. (1980a).]

to the high-temperature core, as illustrated in Fig. 2, than is predicted by coronal equilibrium.

Through spectroscopic studies of the stationary ion charge profiles (Hawryluk et al., 1979; TFR Group, 1978a,b, 1980) or, better still, through time-dependent charge-state profiles arising from test ion injection (see, e.g., Marmar et al., 1975; Cohen et al., 1975) the plasma physicist might hope to measure the sign and magnitude of impurity transport. At one extreme, low radial diffusion velocity might be expected to produce spatial ion profiles in coronal equilibrium, while at the other extreme, rapid transport of the ions through the plasma will lead to noncoronal ionization. It is important to realize that noncoronal ionization balance does not necessarily imply nonstationary plasma conditions; the former might easily be produced in a local plasma region through a balance between rapid influx and efflux of impurities.

The time dependence of the concentrations of an ion charge state z and atomic number Z can be written in the general form

$$\frac{\partial N}{\partial t}(Z^{z+}) + \mathbf{V} \cdot \nabla N(Z^{z+}) + \nabla \cdot [D(Z^{z+})\nabla N(Z^{z+})]$$

$$= n_e\{\alpha(Z^{(z-1)+})N(Z^{(z-1)+}) + \beta(Z^{(z+1)+})N(Z^{(z+1)+})$$

$$- [\alpha(Z^{z+}) + \beta(Z^{z+})]N(Z^{z+})\}. \tag{3}$$

The left-hand side of (3) describes the transport of impurity Z^{z+} due to plasma flow, e.g., rotation, etc., with mean velocity $\mathbf{V}$ plus the diffusive term $D(Z^{z+})$. The right-hand side of (3) describes the atomic physics processes, $n_e \alpha(Z^{z+}) = n_e \langle \sigma v \rangle_{\text{ion } z}$ being the ionization rate, while the recombination rate is

$$n_e \beta(Z^{z+}) = n_e \langle \sigma v \rangle_{\text{rad } z} + n_e \langle \sigma v \rangle_{\text{diel } z} + N(\text{H}^0) \langle \sigma v \rangle_{\text{c/x}},$$

where n_e is the electron density, $N(\text{H}^0)$ is the neutral density, and the σ's are the appropriate cross sections for ionization, radiative, dielectronic, and charge exchange recombination.

Impurity ion diffusion is a complex process in toroidal systems, depending as it does on time-varying parameters such as the temperature profile, the density gradients of the plasma particles, and their collisionality (Stacey, 1981). Experimentally (Terry et al., 1977b; Burrell et al., 1980), diffusion is found to be a multidimensional problem. The detailed physics has been discussed widely in the literature (see, e.g., TFR Group, 1978a,b, 1980; Hulse, 1983; Stringer, 1979; Hawryluk et al., 1979; Coppi and Sharky, 1979). Conceptually, particle transport is often separated into classical collisional processes and anomalous terms, viz.,

$$\frac{\partial N(Z^{z+})}{\partial t} = \frac{1}{r} \frac{\partial}{\partial r} r \left(D_A \frac{\partial N(Z^{z+})}{\partial r} - \Gamma(Z^{z+}) \right)$$

$$+ \text{(ionization and recombination terms)}, \qquad (4)$$

where D_A is the anomalous diffusion coefficient and $\Gamma(Z^{z+})$ is an expression for classical diffusion with toroidal geometry superimposed (Connor, 1973). This "neoclassical" particle flux can be expressed as

$$\Gamma(Z^{z+}) = K[zN(Z)\nabla N_i - N_i \nabla N(Z)], \qquad (5)$$

where K is a toroidal geometric factor. The steady state solution to Eq. (5) in the absence of temperature gradients indicates a concentration of the highest charge states in the high-temperature core with the partial exclusion of the hydrogenic fuel ions N_i, viz.,

$$\frac{N(Z^{z+}, r = 0)}{N(Z^{z+}, r = a)} = \left(\frac{N_i, r = 0}{N_i, r = a} \right)^z. \qquad (6)$$

This neoclassical model would lead to catastrophic radiation losses from the hot core of long-lived thermonuclear plasmas, hence the concern to measure the location of the impurity ions.

In some tokamak conditions (see, e.g., Burrell et al., 1981) neoclassical transport is observed, while in other tokamaks anomalous transport is often indicated. Anomalous transport of particles due to particle–wave collisions and bulk plasma motion is thought, in most cases, to dominate the diffusive process and to smear out the impurity concentration gradients (TFR Group,

5A. Diagnostics Based on Emission Spectra

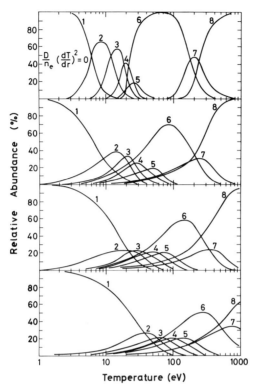

Fig. 3. Diffusive equilibrium concentrations of oxygen ion charge states for a hydrogen plasma in which the total elemental concentrations of hydrogen and oxygen contaminant do not vary with plasma radius r, or with $T_e(r)$. The oxygen concentration is sufficient to balance the power conducted into this plasma region by ionization and radiation loss. Reading from top to bottom the diffusion parameter $(D/n_e)(dT_e/dr_1)^2$ assumes the following values: 0, 5×10^{-8}, 5×10^{-7}, and 5×10^{-6} cm^3 s^{-1} (eV)^2 at $T_e(r_1) = 1$ keV. At $T_e(r) = 1$ eV the temperature gradient is assumed to be zero. For each value of the diffusion parameter, the diffusivity D is assumed constant and independent of charge state. [From Ashby and Hughes (1981).]

1978a). The smearing effect of anomalous diffusion on the stationary-state concentrations of oxygen ions is well illustrated in Fig. 3 by the model calculations of Ashby and Hughes (1981). In an analysis of the magnitude of the anomalous diffusion term D_A, which is required to account for the observed $T_e(r)$, $n_e(r)$ profiles in the PLT tokamak, Hinnov et al. (1978) derived values at the core $D_A(r = 0) \simeq 5 \times 10^3$ cm^2 s^{-1}, somewhat lower values at the $q = 1$ magnetic surface [for a definition of the "safety factor" q, see Stacey (1981)] and increasing to $D_A(r = a) \simeq 2 \times 10^4$ cm^2 s^{-1} at the walls. Application of Eq. (4) has yielded particle fluxes of the order $\Gamma \sim 10^{14}$ cm^{-2} s^{-1} for lighter impurities such as oxygen and $\Gamma \sim 10^{11}$–10^{12} cm^{-2} s^{-1} for metallic impurities in the Alcator tokamak (Terry et al., 1977a). The fully detailed expansion of Eq. (4) contains terms which will drive the impurities either inwards or outwards but an overall diffusive particle flux of the form

$$\Gamma(Z^{z+}) = -D(Z^{z+}) \left(\frac{\partial N(Z^{z+})}{\partial r} + \frac{2N(Z^{z+})r}{a^2} \right), \quad (7)$$

where a is the plasma minor radius, has been found to give more realistic agreement with observed particle transport in tokamaks (Behringer et al., 1981; TFR Group, 1978a; Coppi and Sharky, 1979).

It may be helpful to introduce the concept of an effective ion confinement time through the relation

$$\Gamma \simeq N(Z^{z+})V(Z^{z+}), \quad (8)$$

where $V(Z^{z+})$ is an effective ion diffusion velocity. τ_c, the effective confinement time, is then

$$\tau_c = \Delta r/V(Z^{z+}), \quad (9)$$

where Δr is a characteristic shell thickness for the charge state z with ionization potential ψ_z; $\Delta r \simeq [(1/\psi_z)dT_e/dr]^{-1}$. In order that there be coronal equilibrium, the effective confinement in the local temperature region must exceed the ionization time; i.e.,

$$n_e\tau_c \geq n_e\tau_{ion}, \quad (10)$$

i.e.,

$$n_e \left(\frac{1}{\psi_z} \frac{dT_e}{dr} \right)^{-1} \frac{1}{V} (Z^{z+}) \geq \frac{1}{\alpha_z}.$$

Thus for coronal ionization balance

$$V(Z^{z+}) \leq n_e\psi_z\alpha_z \left(\frac{dT}{dr} \right)^{-1}. \quad (11)$$

Inserting values for $\alpha_z = (n_e\tau_{ion})^{-1}$, 10^{-10}–10^{-12} cm^3 s^{-1} for Fe in Fig. 1, and taking a value for ψ_z of between 1 and 2 keV, typical of the neon-shell Fe ions, and with $dT/dr \sim 4 \times 10^1$ eV cm^{-1} and $n_e \sim 1 \times 10^{13}$ cm^{-3}, typical of existing tokamaks, then $V(Z^{z+})$ lies in the range of 10^3–10^5 cm s^{-1}, the lower value of 1 cm ms^{-1} corresponding to the higher temperatures in Fig. 1.

Following Griem (1979) we can express the coronal criterion analytically using his expressions for the ionization rate coefficients, viz.,

$$\alpha = \frac{10\xi}{\alpha'c} \left(\frac{\hbar}{m} \right)^2 \left(\frac{E_H}{kT_e} \right)^{1/2} \frac{E_H}{\psi_z} \left(\int_x^\alpha \exp(-x) \frac{dx}{x} \right); \quad (12)$$

ξ is the number of valency electrons outside of the closed shell, α' is the fine-structure factor, $x = E/kT_e$, and the other symbols are conventional or previously defined. Thus, from Eq. (12),

5A. Diagnostics Based on Emission Spectra

$$\frac{1}{\alpha} (m^{-3} s) = 1.6 \times 10^{13} \left(\frac{kT_e}{E_H}\right)^{1/2} \frac{\psi_z}{E_H} [\quad]^{-1}$$

and we can take the exponential integral in square brackets ~ 1.7. The criterion for coronal equilibrium is then

$$V(Z^{z+}) \lesssim \xi \times 10^{-13} n_e E_H \left(\frac{\partial T}{\partial r}\right)^{-1} \tag{13}$$

(with energy units in keV, n_e in m^{-3}, and V in m s^{-1}).

The maximum impurity flux required to preserve coronal balance from Eq. (8) is then given by

$$\Gamma \lesssim \xi \times 1 \times 10^{-13} n_e^2 f_z \frac{E_H}{kT_e} E_H \left(\frac{\partial T}{\partial r}\right)^{-1}. \tag{14}$$

Again, inserting typical metal concentrations $f_z \sim 10^{-3}$, we derive at a temperature of 1 keV a limiting flux of $\Gamma \lesssim \xi \times 1.6 \times 10^{15}$ ions m^{-2} s^{-1}, in order to ensure coronal equilibrium. The experimental values derived for the parameters Γ, f_z, and $V(Z)$ (Hinnov *et al.*, 1978; Breton *et al.*, 1976; Hawryluk *et al.*, 1979) can be of the same order as those predicted by the analytic expressions, Eqs. (11), (13), and (14). In the present generation of tokamaks, as distinct from thermonuclear reactor conditions, coronal equilibrium cannot therefore be taken for granted. This particularly applies to the cooler peripheral plasma where the temperature gradient is steepest, ψ_z is relatively small, and the ionization shells are relatively narrow.

In most tokamak experiments nearly steady-state concentrations are reached in the hot core after the initial transient diffusion of ions from the walls or from the current limiter. In the core plasma of the PLT tokamak (see, e.g., Hinnov, 1980), the apparent deviations of the stationary radial locations of the impurity ions from their coronal equilibrium positions can be accounted for by a slow diffusion $V(Fe^{z+}) \lesssim 0.1$ m s^{-1}, while in the TFR tokamak (TFR Group, 1980; Breton *et al.*, 1982), the ionized Ni ion profiles can be explained by uncertainties in the atomic rate coefficients. It is worth emphasizing that in tokamak experiments the stationary ion locations are in general less sensitive to the magnitude of the diffusive processes than to the atomic processes, and the possibility of enhanced recombination through charge exchange with low concentrations of neutrals cannot be discounted.

The sensitive temperature dependence of the ionization rates at the peak abundance temperature of the ion ensures that in the case of a steeply varying temperature profile diffusion has little affect on the mean radial location of the ions, even though as illustrated in Fig. 3 an appreciable radial smearing of each ion does occur. For example, using the data in Fig. 2, Hulse *et al.* (1980a) show that introduction of an anomalous diffusion coefficient $D_A \sim 4000$ cm^2 s^{-1} or, alternatively, modifications to the atomic rates by only a factor of 2, are required to bring the theoretical ion concentration profiles into near coincidence with the experimental observations. Uncertainties of a factor of 2 in the magnitude of the atomic rates are not uncommon.

The relative sensitivity of the impurity ion profiles to atomic rates rather than diffusion rates can be understood simply in terms of the respective ionization and diffusion times. At an electron density of 10^{13} cm^{-3} an ionization time for a typical impurity ion is $\tau_{ion} \sim 10 \rightarrow 1$ ms. Taking a radial scale length for the ion concentration as ~ 10 cm then [Eq. (9)], the radial diffusion velocity has to be $V(Z^+) \gtrsim 10^3 - 10^4$ cm s^{-1} in order materially to distort the ionization balance. This value of $V(Z^{z+})$ is already a rather high value for tokamaks (TFR Group, 1976, 1982), where $V(Z^{z+}) \lesssim 10^3$ cm s^{-1} is more usual in the high-temperature core region.

Derivation of ion transport models from observed stationary impurity ion profiles in a near equilibrium plasma is therefore a tedious task in terms of the detailed profile data required, viz., $T_e(r)$, $n_e(r)$, $N(Z^{z+})(r)$, $N(H^0)(r)$. The interpretation finally is not unambiguous. A more sensible approach is to monitor the space and time variation of injected bursts of foreign impurity ions (see, e.g., Marmar et al., 1980; Burrell et al., 1981).

In most tokamak experiments accumulation of impurities is not observed after the initial transient influx. A radially homogeneous impurity concentration, $N(Z)(r) = $ const, is commonly found (TFR Group, 1978a,b; Hinnov, 1980), though a scaling $N(Z)(r) \propto n_e(r)$ is not unusual (Clark et al., 1982). "Burst injection" of impurity ions into tokamak devices like Alcator (Marmar et al., 1975, 1980; Cohen et al., 1975) indicate that the impurities are progressively ionized as they diffuse towards the core. Some of these highly ionized species, however, find their way back out to the peripheral plasma region.

As diffusion indicators one- and two-electron ions have a special appeal. The total recombination rate of free electrons with impurity ions which are fully stripped or have one- or two-electron configurations is small relative to other ion species. Impurity nuclei are special cases since the dielectronic recombination process for these ions is not possible. Even for one- and two-electron systems, the dielectronic rate (Burgess, 1965; Merts et al., 1976) is still small on account of the large excitation energy of the K-electrons. Ion species with L- and M-shell configurations, on the other hand, have large dielectronic rates which, for intermediate values of Z at least, much exceed the collisional radiative recombination rates, as illustrated in Fig. 4. The ionization rates (Summers, 1974) normalized to the temperature at which a particular species is in the majority is also lower for the more highly stripped ions. An immediate consequence of the relative magnitude of these rates is the propensity of nuclei and one- and two-electron ions to diffuse further through the plasma without changing their charge state. The H- and He-like ions therefore occupy a wider plasma radius than do the less highly charged ions.

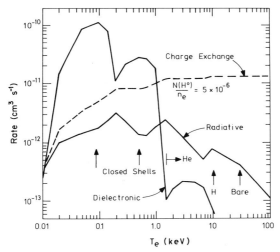

Fig. 4. Recombination rates (cm^3 s^{-1}) of the most abundant ion species of iron in coronal equilibrium. For the H- and He-like species the dielectronic rates are relatively low, while charge transfer recombination, from neutral hydrogen, whose concentration relative to the electron density is taken as $N(H^0)/n_e = 5 \times 10^{-6}$, is relatively large. [From R. A. Hulse (personal communication, Princeton Plasma Physics Laboratory).]

The outwards diffusion of the one- and two-electron systems to the cooler peripheral plasma, balanced by the inward ion flux, causes several spectral features which are important indicators of the magnitude of the diffusion rates. First, the recombination continuum intensity per ion, with $\langle \sigma v \rangle_{\text{rec}} \propto T_e^{-3/2}$, shows an enhancement at larger plasma radii. Radiation capture of the free electrons by H-like argon seeded into the PLT tokamak (Brau *et al.*, 1980) gives rise to steps in the x-ray recombination as illustrated in Fig. 5. These steps, occurring at 4.12 keV, the ionization potential of Ar XVII, are most pronounced at the larger radii where the temperature $T_e \simeq 0.5$ keV, indicating departures from ionization balance for the Ar^{17+} ion in this plasma region. The extent of this departure is interpreted by Brau *et al.* (1980) in terms of an outwards diffusion rate, which from Eqs. (4) and (8) can be expressed as

$$\frac{1}{r}\frac{\partial}{\partial r}[r^2 V(\text{Ar}^{17+})]\frac{\partial N(\text{Ar}^{17+})}{\partial r} = -n_e N(\text{Ar}^{17+})\beta(\text{Ar}^{17+}), \quad (15)$$

where all the other coefficients for Ar^{17+}, apart from the radiative recombination terms, are negligible at $T_e \simeq 0.5$ keV and $n_e \simeq 2 \times 10^{13}$ cm^{-3}. Diffusion velocities $V(\text{Ar}^{17+}) \sim 10^3$ cm s^{-1} are derived for the PLT tokamak from Eq. (15) (Brau *et al.*, 1980).

Another manifestation of the outwards diffusion of one-electron systems in tokamaks is the appearance of allowed and intercombination lines of He-like ions in the outer regions of the plasma (Peacock *et al.*, 1979). The

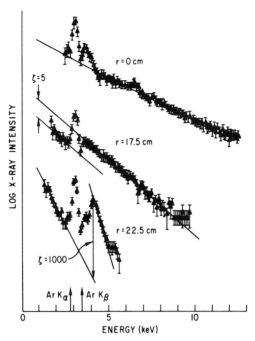

Fig. 5. Continua with superimposed K_α ($n = 1-2$) and K_β ($n = 1-3$) Ar lines and Ar free-bound continuum "steps" in the spectrum from the PLT tokamak seeded with argon. The central electron temperature $T_e(r = 0) \sim 1.5$ keV, while $T_e(r = 22.5$ cm$) \sim 0.5$ keV; $n_e \sim 2 \times 10^{13}$ cm^{-3}. ζ is the enhancement factor of the continuum intensity due to recombination (Ar^{17+} + e → Ar^{16+}) over the bremsstrahlung intensity. In coronal equilibrium ζ is much less than that shown for plasma radii $r \gtrsim 20$ cm. [From Brau et al. (1980).]

intensities of the 1s^2–1s2p, $^3P_1$, $^1P_1$ resonance lines in the cooler outer region of the plasma in the DITE tokamak indicate a much higher occupancy for these levels than would be predicted on a coronal balance model, owing to the effect of the recombination process O^{7+} + e → O^{6+}. Using theoretical indicator plots of the singlet and triplet level populations for O^{6+} as a function of temperature and density, Peacock and Summers (1978) interpreted the departure of the 1s^2–1s2p, $^3P_1$, $^1P_1$ resonance lines from their stationary intensities in terms of an outward diffusion of O^{7+} with a velocity $V_r \sim 10^2$ cm s^{-1} (Peacock et al., 1979).

Both of the spectral features, the x-ray recombination edge and the triplet/singlet ratio, are very attractive from the point of view of diagnostic measurements of the one-electron ion diffusion velocities. Essentially, the techniques depend on radial measurements of only one spectral feature (in the case of the triplet/singlet, this feature is two neighboring lines) and no absolute instrument calibration is involved. Ordinarily in order to measure ion diffusion rates, the radial distributions of the absolute concentrations of several neighboring ion species is required. A caveat to the use of these

B. Charge Transfer Recombination

The problem of energetic particle loss from magnetically confined plasmas through resonance charge exchange

$$H^+ + H^0 \to H^0 + H^+ \tag{16}$$

between the hot confined H^+ ions and lower-temperature neutrals at the plasma periphery has been appreciated for many years in mirror machine studies. The relatively recent use of energetic (>1 MW) H^0, H_2^0, D^0, etc., beams for heating tokamaks has highlighted the role of the charge transfer process in these devices. The process is important, first because the absolute magnitude of the electron exchange cross section is very high, as much as 10^{-14} cm^{-2}, and second because the cross section for charge exchange on impurities

$$Z^{z+} + H^0 \to Z^{(z-1)+} + H^+ \tag{17}$$

scale approximately as z, the ion charge state. Charge transfer rates even with relatively low concentrations of neutrals and impurities can be comparable to or even exceed other recombination rates, as indicated for Fe ions in Fig. 4. Some of the effects of charge exchange reactions between atomic hydrogen and impurities on parameters like the ionization balance and radiation loss in tokamaks have been discussed by Hulse *et al.* (1980b) and by Puiatti *et al.* (1981). However, other parameters such as beam penetration into the plasma and energy loss through reactions as in Eq. (16) are affected.

In calculating the charge exchange cross section, only the relative interaction velocity V_r of the particle is important. For beam velocities $V_n(H^0) \gg V_{th}(Z^{z+})$, the thermal velocity of the impurity ions, we need make no distinction between V_r and V_n for beam injection experiments into plasmas of fusion interest.

The dynamics of the three-body charge transfer interaction involves quite different physical processes as the incident particle energy is varied. It is not surprising therefore that no single model can encompass the wide interaction energy range (1 eV–100 keV) of interest. However, various theoretical models have been used, their validity depending on the atomic structure of the particles and on the magnitude of V_n relative to v_0, the hydrogen ground state orbital velocity, ($v_0 = e^2/\hbar = 2.188 \times 10^8$ cm s^{-1}, which is equivalent to a relative interaction energy of 23.3 keV/amu for the impacting particles). These models include the distorted-wave approximation (Ryufuku and Watanabe, 1978, 1979a,b), the absorbing-sphere model valid for $V_n < 2 \times 10^8$ cm s^{-1} (Olson and Salop, 1976), the tunneling model (Grozdanov and Janev, 1978), and the classical trajectory approximation (Olson and Salop, 1977), the latter being valid for $V_n > v_0$. Surveys of these models and their applicability to the separate energy ranges are given by Janev and Grozdanov (1980) and by Janev and Presnyakov (1981).

A first approximation to the charge exchange cross section can be derived, following Knudsen et al. (1981), from the classical treatment of Bohr and Lindhard (1954). The classical approach is reasonable since the de Broglie wavelength $\lambda = \hbar/MV_n$ of the projectile is much smaller than the collision diameter for the exchange interaction. Using the Bohr–Lindhard model, the captive cross section at low impacting velocity V and for a highly charged state z is given by

$$\sigma_1 = \pi R_r^2 = \pi a_0^2 z \frac{a}{a_0} \left(\frac{V}{v_0}\right)^{-2} \simeq 10^{-16} z \quad \text{cm}^2 \tag{18}$$

in the case where the orbital velocity v of the electron equals v_0. In Eq. (18), $a_0 = \hbar^2/me^2$ is the Bohr radius and the "release radius" for the transfer electron is given by

$$ze^2/R_r^2 = mv^2/a. \tag{19}$$

For slow particles and high z therefore the cross section is independent of the relative interaction velocity V_r of the particle and is linearly proportional to the charge state of the ion.

At high velocities when $V_r > v_0$, the probability of electron capture is $(v/a)(R_c/V_r)$, where the "capture radius" R_c is given by

$$ze^2/R_c = \tfrac{1}{2}mV_r^2, \tag{20}$$

the cross section for fast particles is then

$$\sigma_2 = \pi R_c^2 \frac{v}{a} \frac{R_c}{V_r} = 8\pi a_0^2 z^3 \left(\frac{a}{a_0}\right)^{-1} \frac{v}{v_0} \left(\frac{V_r}{v_0}\right)^{-7} \tag{21}$$

and the capture cross section falls off rapidly with beam velocity, scaling as $z^3 V_r^{-7}$. An implicit assumption in this model is that quantum effects can be ignored; i.e., that there is a quasicontinuum of electron states to which the electron can be attached. This is generally true for $z \geq 4$.

In practice the theoretical and experimental data appear to fit reasonably well along universal curves of σ_c/z^k when plotted against E (keV/amu)/z^l, with $k \simeq 1.0$ and $l \simeq 0.5$.

Figure 6 illustrates the magnitude and shape of the capture cross section for a typical reaction, $O^{6+} + H^0 \rightarrow O^{5+} + H^+$, as a function of the relative velocity of the particles (Crandall, 1979). The main results of the Bohr–Lindhard model i.e., constant value of σ_{cx} for $V_r < 2 \times 10^8$ cm s^{-1} and its rapid decrease for $V_r > 2 \times 10^8$ cm s^{-1}, are to be noted. At very low velocities, $V_r \sim 10^7$ cm s^{-1}, theoretical models require consideration of quasimolecular formation and collision-induced transitions, while at very high impacting velocities, $V_r \gtrsim 10^9$ cm s^{-1}, inner quantum shell interactions

5A. Diagnostics Based on Emission Spectra

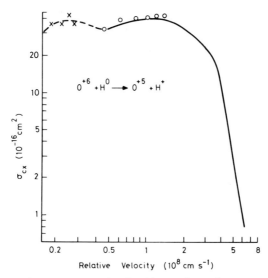

Fig. 6. Charge exchange cross section for $O^{6+} + H^0 \rightarrow O^{5+} + H^+$ as a function of the relative interaction velocity of the particles [From Crandall (1979); unpublished data: (×) R. A. Phaneuf and I. Alvarez (ORNL, 1981), (○) D. H. Crandall, F. W. Meyer, and R. A. Phaneuf (ORNL, 1979).]

need to be considered. At sufficiently high V_r ion impact ionization eventually becomes the dominant process (Freeman and Jones, 1974). The ion particle ionization rates are compared with charge exchange rates in a series of recent experiments on carbon and oxygen by Shah and Gilbody (1981).

Topical charge exchange cross section measurements (see, e.g., Crandall et al., 1979; Meyer et al., 1979), are discussed in the proceedings of the "IAEA Technical Committee Meetings on Atomic and Molecular Data for Fusion," where Gilbody (1981), for example, summarizes the data on charge exchange reactions between H^0 and multiply charged ions.

The magnitude of the charge transfer rate relative to recombination resulting from the capture of free electrons, i.e., collisional dielectronic recombination, is illustrated in Fig. 7. The charge exchange coefficient at $V_r \sim v_0 \sim \bar{v}_e$ (the thermal velocity of the plasma electrons at $T_e = 27.25$ eV) is four orders of magnitude higher than the free-electron recombination coefficient. At the higher electron temperatures typical of tokamak plasmas and for ion charge states greater than $z = 6$, the difference in the coefficients can easily be an order of magnitude greater, $\sim 10^5/1$. The charge exchange rate $N(H^0)\langle\sigma v\rangle_{cx}$ is, of course, independent of electron density whereas the total free-electron recombination rate $n_e(\langle\sigma v\rangle_{rad} + \langle\sigma v\rangle_{diel})$, which is the sum of the collisional–radiative and –dielectronic rates, increases linearly with electron density. Charge transfer is therefore only likely to be a competitive or dominant recombination process when the neutral-atom concentration is $N(H^0) \gtrsim 10^{-5} n_e$ (see also Fig. 4).

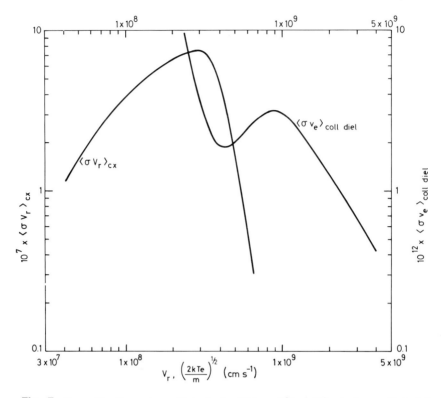

Fig. 7. Recombination rate coefficients for $O^{6+} \rightarrow O^{5+}$; $\langle\sigma V_r\rangle_{cx}$ is the coefficient for charge exchange $O^{6+} + H^0 \rightarrow O^{5+} + H^+$ as a function of the interaction velocity V_r; $\langle\sigma v_e\rangle_{\text{coll diel}}$ is the collisional dielectronic rate (Summers, 1974) for $n_e = 10^{12}$ cm^{-3} as a function of the most probable electron velocity.

When neutral atoms are present, charge transfer to the impurities represents an additional recombination process the inclusion of which leads to a revised ionization balance equation, viz.,

$$\frac{N(Z^{(z-1)+})}{N(Z^{z+})} = \frac{\beta_r(Z^{z+}) + \beta_{\text{diel}}(Z^{z+}) + \zeta\beta_{cx}(Z^{z+})}{\alpha(Z^{(z-1)+})}, \qquad (22)$$

where $\zeta = N(H^0)/n_e$, and the right-hand side is the ratio of the total, collisional–dielectronic plus charge transfer recombination rate to the total electron impact ionization rate, including inner-shell effects.

A result of the charge exchange term is to decrease the effective charge state on the ions at a given plasma temperature, as illustrated for ionized iron in Fig. 8. In some circumstances the radiation loss is considerably increased and the overall power balance altered by the inclusion of charge transfer recombination (Krupin *et al.*, 1979; Hulse *et al.*, 1980b). The effect

5A. Diagnostics Based on Emission Spectra

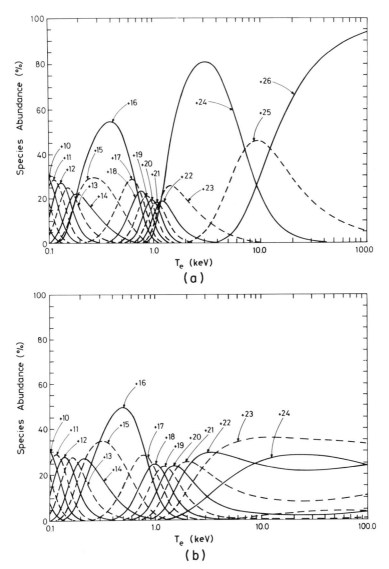

Fig. 8. Iron ion abundances (a) in coronal equilibrium and (b) with additional charge transfer recombination from 20-keV/amu H^0 neutrals with a concentration $N(H^0)/n_e = 10^{-5}$. [From Hulse *et al.* (1980b).]

on the radiation loss is greatest when the plasma temperature has a value such that the dominant steady state, impurity ion species has one or two electrons. For iron this temperature lies between about 2 and 10 keV as illustrated in Fig. 9. One- or two-electron configurations persist over a wide range of temperatures (Fig. 8), and are typical of the impurity ion species to

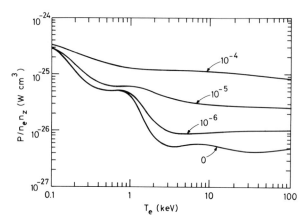

Fig. 9. Radiation loss per Fe impurity ion from a hydrogen plasma with $n_e = 2 \times 10^{13}$ cm^{-3}. The curves are parameterized to take into account the fraction $N(H^0)/n_e$ of beam neutrals present. H^0 beam energy is 20 keV/amu. [From Hulse *et al.* (1980b).]

be found in the well-confined core plasma in tokamaks. The *fractional* increase in radiation due to charge transfer can be most dramatic therefore in localized plasma regions.

Experimental evidence for charge transfer recombination in tokamaks has come recently from spectroscopic studies of the time evolution of the emissivities of different ion species during H^0 beam injection into PLT (Suckewer *et al.*, 1980c). A decrease in the emission from the dominant ionization species, e.g., Fe^{24+} in the core of the PLT plasma, with a concomitant rise in the emissivity of Fe^{23+} during neutral injection, is ascribed to charge transfer recombination. It is to be noted that, though each recombination event will result in spontaneous decay of the excited level into which the electron has been captured, the main cause of the change in the ion emissivities, which is also observed in regions remote from the local beam deposition, is electron excitation of the altered ion concentrations resulting from Eq. (22). In similar pulsed-beam experiments on the DITE tokamak (Clark *et al.*, 1982), the temporal behavior of the impurity line emission, the dependence of the impurity radiation on beam injection current and on plasma density, and the spatial distribution of the impurity charge state can all be accounted for by the inclusion of charge transfer recombination during the H^0 beam injection period. The rather dramatic variation in the radial distribution of the impurity ions shown in Fig. 10 is due to the imprint, via charge exchange recombination, of the H^0 beam profile as it penetrates the plasma in the DITE tokamak.

Temporal variation of the *level* populations as distinct from the ion populations of the target ions during charge transfer constitutes a more direct piece of evidence for the charge transfer recombination process. Theoreti-

5A. Diagnostics Based on Emission Spectra

cally (Janev and Grozdanov, 1980), capture into the separate quantum levels is estimated to maximize at a principal quantum level of

$$\bar{n} = z\{1 + [(z - 1)/(2z - 1)^{1/2}]\}^{-1/2}. \tag{23}$$

The distorted-wave model (Ryufuku and Watanabe, 1978, 1979a,b) predicts, in approximate agreement with the above,

$$\bar{n} = z^{0.774}. \tag{24}$$

For O^{8+} ions, for example, recombination takes place preferentially into the $\bar{n} = 5$ quantum level. Clearly for recombination to be "stabilized" by radiative decay $\bar{n}$ must be much less than the so-called collision limit n_c, at which collisional deexcitation is ten times the radiative decay, a criterion that for hydrogenic ions can be expressed (Griem, 1964) as

$$n_c \simeq 1.26 \times 10^2 Z^{14/17} n_e^{-2/17} \left(\frac{kT_e}{\psi_z}\right)^{1/17}. \tag{25}$$

In tokamak plasmas $\bar{n} < n_c$ is generally the case. The first study of level population changes during beam injection into tokamaks was made by Isler (1977). The anomalously rapid increase in O^{7+} Balmer-α at 102.36, 0.51 Å

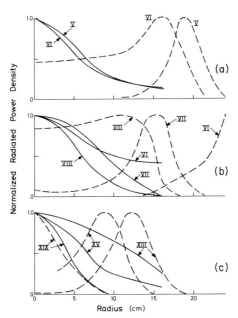

Fig. 10. Computed profiles (broken lines) of the equilibrium power density radiated from (a) carbon, (b) oxygen, and (c) titanium impurity ions with no H^0 beam injection into the DITE tokamak. The full curves are the experimental profiles during a 1-MW, 25-kV, H^0 beam injection pulse. The difference between these two sets of profiles can be accounted for by charge exchange between H^0 and the impurity ions. [From Clark et al. (1981).]

during the beam heating pulse in the ORMAK tokamak has been interpreted by Isler (1977) as due to $H^0 + O^{8+} \rightarrow (O^{7+})^*$. On the other hand, the relatively slow increase exhibited by O^{7+} Lyman series could be accounted for by electronic excitation, which dominates the emissivities of these resonance lines. It is more difficult to account for the equally slow increase in O^{7+}, Balmer-β, which should, like its Balmer-α counterpart, also show the "instantaneous" effect of charge exchange. However, the detailed balance of populating and depopulating processes varies with the angular momentum quantum states. Charge exchange is thought to occur preferentially into high l states at low energy and low l (spherically symmetric) states at high interaction energies (Janev and Grozdanov, 1980). The temporal emission of the Lyman and Balmer lines would be expected to vary during beam injection and these features can be respectively interpreted (see Section V) in terms of changes in the ionization balance and in the absolute number of charge exchange events. Clearly, the populations of the different excited n, l quantum levels resulting from charge exchange (see, e.g., Salop, 1979) and their redistribution due to radiative decay and to collisions need to be computed in order to fully understand the detailed line emission features in a charge exchange experiment. Abramov et al. (1978) have attempted this type of analysis with some success for the Isler (1977) experiments.

The importance of charge transfer processes involving background thermal (1–100 eV) neutrals at the plasma edge should not be neglected. Apart from inhibiting beam transport, charge transfer recombination may need to be included in the analysis of spectroscopic features, particularly those discussed by Peacock and Summers (1978) and Brau et al. (1980), which are aimed at deriving diffusion rates.

III. Atomic Level Populations

In order to relate the volume emissivity of line ij

$$I(i, j) = [n(j)/4\pi]A_{ij} \qquad (26)$$

to the ground state population of the ion $n(g)$, it is necessary to have a model describing the instantaneous occupancy of the levels in the ion. $I(i, j)$ itself is derived from an observable quantity, the line radiance (Section IV), while the radiative probability A_{ij} (s^{-1}) can be calculated. For allowed dipole transitions the transition probability is

$$A_{ij} = \frac{6.67 \times 10^{15}}{\lambda_{ij}^2} \frac{\omega_i}{\omega_j} f_{ij}, \qquad (27)$$

where the ω's are the statistical weights of the levels, f_{ij} the absorption oscillator strength, and λ_{ij} is in angstroms. In practice all but the ground state or quasiground state metastable levels will reach a stationary occupancy on a

5A. Diagnostics Based on Emission Spectra

time scale $\Delta \tau$ (s) $\sim 10^{-7}/Z^4$ of the order of or less than the radiation time (McWhirter and Hearn, 1963). This is considerably shorter than the effective ion confinement time in a local plasma region of a tokamak.

In the simplest case of an excited level which is coupled by strong dipole radiation to the ground state and has negligible competing coupling processes with other levels, the coronal condition applies:

$$n(g)\langle\sigma v\rangle_{\text{ex}}^{gj} \rightleftharpoons n(j)A_{gj}. \tag{28}$$

The excitation rates $\langle\sigma v\rangle$ and the models used in their calculation have been discussed widely in the literature, the article by De Michelis and Mattioli (1981) citing many of the references. It is evident that the rates are specific to the atomic configuration, ion species, and impacting particle energy range.

For an approximate calculation of dipole excitation, however, we can use the expression involving the collision strength Ω_{ij}:

$$\langle\sigma v\rangle_{ij} \text{ cm}^3 \text{ s}^{-1} = \frac{8.63 \times 10^{-6}}{\omega_i k T_e^{3/2}} \int_{E_{ij}}^{\infty} \Omega_{ij}(E) \exp(-E/kT_e) \, dE,$$

$$= \frac{8.63 \times 10^{-6}}{T_e^{1/2}} \frac{\Omega_{ij}}{\omega_i} \exp - \frac{E_{ij}}{kT_e}, \tag{29}$$

with E_{ij} in rydbergs and T_e in degrees kelvin.

The basic atomic parameters, Ω_{ij}, A_{ij}, f_{ij}, and E_{ij} are often derived from sophisticated calculations which require computer codes such as those generated at Los Alamos (Cowan, 1977) or the University College London (see, e.g., Eissner and Seaton, 1972; Saraph, 1972). A compilation of electron impact excitation of currently available theoretical and experimental data for highly stripped ions is given by Merts *et al.* (1980), while data on some neon-shell ions of Ti, Cr, Fe, and Ni are given by Mason and Storey (1980), Bhatia and Mason (1980), Bhatia *et al.* (1980), and Feldman *et al.* (1980). The transition energy E_{ij} if not known can again be derived using atomic structure computer codes (see, e.g., Eissner *et al.*, 1974; Cowan, 1981). An overview of the availability of atomic data for fusion plasma analyses is given by Wiese (1978).

In the case of high-energy distant encounters the Coulomb–Bethe approximation for optically allowed transitions represents a further simplification and leads to the expression

$$\langle\sigma v\rangle_{ij} \text{ cm}^3 \text{ s}^{-1} = 1.6 \times 10^{-5} \frac{f_{ij}\bar{g} \exp(-E_{ij}/kT)}{E_{ij}(kT_e)^{1/2}}, \tag{30}$$

with E_{ij}, kT_e in electron volts. The relation between Ω and $\bar{g}$ is

$$\Omega_{ij} = (1.97 \times 10^2/E_{ij})\omega_{ij} f_{ij} \bar{g}, \tag{31}$$

with E_{ij} in electron volts.

An approximate formula for the Gaunt factor $\bar{g}(kT_e/E_{ij})$ has been calculated by Van Regemorter (1962), and in the case of electron excitation of optically allowed transitions in positive ions has the value $\bar{g} = \text{const}$ [≈ 0.2 for $(kT_e/E_{ij})^{1/2} < 1$]. This value is widely used by plasma spectroscopists for $\Delta n \neq 0$ transitions in highly stripped ions. Interpolation formulas for $\bar{g}$ which can also be used for optically forbidden or quadrupole transitions have been proposed by Mewe (1972) and Mewe et al. (1980). A comparison of the $\bar{g}$ approximation with available theoretical data has been discussed by Younger and Wiese (1979).

In the general case, the individual level populations have to be derived from a solution of the rate equations linking the level of interest with other levels in the ion, z, and with cascade processes from the next ion stage, $z + 1$:

$$N(j) = U_j n_e N(z, g) + V_j n_e N(z + 1, g), \tag{32}$$

where U and V are coefficients depending only on the temperature and density of the plasma. In practice the amount of atomic data required and the procedures for calculating U and V are formidable, but appropriate computer codes have been constructed by several groups (see, e.g., Summers, 1977), and have been modified for use in particular cases (see, e.g., Peacock and Summers, 1978). Often it is sufficient to consider the ground population and that of neighboring levels only, neglecting recombination and the effects of weakly coupled upper levels. The result of such a simplified code (Gordon et al., 1982) is shown in Fig. 11 for the levels within the ground configurations $2s^2 2p^2$, $2s2p^3$ of carbonlike Ti whose level structure is illustrated in Fig. 12. Rather similar calculations of the level populations in L-shell ions of the common metals have been made by Feldman et al. (1980) and by Bhatia et al. (1980).

A. Forbidden Lines

Magnetic-dipole and spin-forbidden transitions, between levels within the ground configuration and between those and other $\Delta n = 0$ excited configurations of $3s^k 3p^l$ and $2s^k 2p^l$, are of particular interest since they provide convenient visible and near-UV diagnostic lines for astrophysical plasmas and tokamaks.

The spontaneous decay rates of various allowed and forbidden transitions are compared in Fig. 13 with the collisional rates typical of tokamak plasma conditions. The dominance of the collisional rates over the A_{ij} values of forbidden lines at low transition energies would lead us to expect, on simple-minded considerations, that only these forbidden lines with $A_{ij} > 10^4 \text{ s}^{-1}$ and transition energies $E > 10$ eV would be present in the tokamak spectrum. However, even lower-energy transitions, particularly those within the $2s^k 2p^l$ ground configurations of ionized metals, have been observed recently in

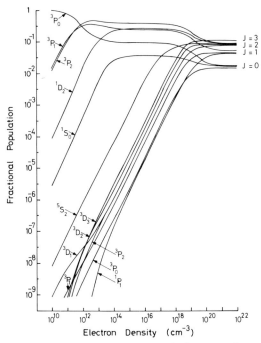

Fig. 11. Fractional populations of some of the $2s^22p^2$ and $2s2p^3$ levels of Ti XVII at a temperature corresponding to the peak abundance of the ion in corona equilibrium.

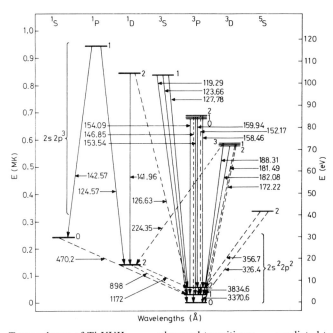

Fig. 12. Term scheme of Ti XVII. ———, observed transitions; ---, predicted transitions.

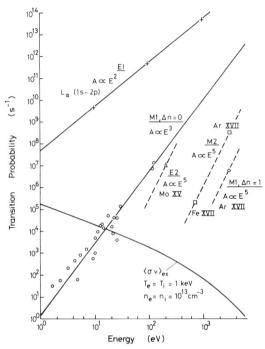

Fig. 13. Spontaneous decay rates as a function of level separation E (eV) for various types of transitions: -+-, electric dipole (1s–2p); -○-, magnetic dipole; -□-, magnetic quadrupole; -△-, electric quadrupole. Collisional excitation rates $\langle \sigma v \rangle n_e$ for typical tokamak conditions are plotted for comparison.

tokamaks and studied by Suckewer and Hinnov (1978), Suckewer et al. (1980a,b), and Lawson et al. (1981). The emission intensities of lines with A_{ij} values as low as 10^2–10^4 s^{-1}, can be seen by inspection of the Ti XIV, Ti XVIII emission from the DITE tokamak in Fig. 14. The almost constant populations of the $2s^2 2p^l$ ground configuration levels at $n_e > 10^{13}$ cm^{-3}, relative to the levels in the excited configurations $2s2p^{l+1}$, which in general increase linearly with n_e (Fig. 11), indicate why the long-wavelength (M1) forbidden lines are prominent only in the emission from *low*-density high-temperature laboratory plasmas such as tokamaks. The use of these lines for diagnostic purposes is discussed in Section IV. Suffice it to say at this stage that both the intensity profiles and the absolute intensities embody information of interest to the plasma diagnostician. Their absolute intensities can be used, through level population calculations such as that illustrated in Fig. 11 to deduce ionic concentrations.

The plots in Fig. 13 suggest that forbidden lines which lie at wavelengths of less than a few hundred angstroms should readily be observed in tokamaks. Apart from intersystem lines of the configurations $2s^2 2p^k \rightarrow 2s2p^{k+1}$ which for the common metals lie typically in the region 80–200 angstroms,

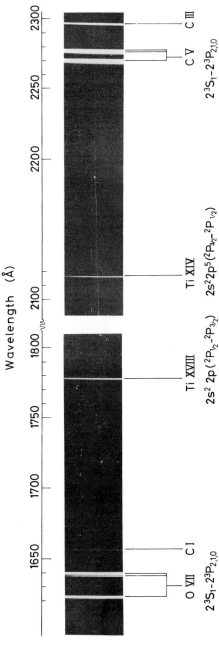

Fig. 14. Forbidden transitions (spontaneous decay rate $A \sim 10^3$ s^{-1}) within the $1s^2sp^n$ ground configuration of ionized titanium impurities in the DITE tokamak ($n_e \sim 2 \times 10^{13}$ cm^{-3}, $T_e \sim 800$ eV) are seen to have intensities comparable to the allowed ($A \sim 10^8$ s^{-1}) lines from the $1s2s\,^3S_1$–$1s2p\,^3P_{2,1,0}$ triplets of oxygen and carbon impurities: $N(O) \sim 10 \times N(\text{Ti})$.

these shorter-wavelength forbidden lines involve changes in total quantum number, i.e., $\Delta n > 0$. An example is the (E2) resonance decay of the $3d4s(J = 2)$ levels in the first excited configuration of Ni-like ions. In Mo XV, for example, these transitions, lying at 58.833 and 57.920 angstroms and with A_{ij} value $\sim 10^7$ s^{-1}, appear as the strongest line features in the XUV spectrum of the DITE tokamak when operated with a current limiter of molybdenum (Mansfield et al., 1978).

The first excited configuration of neonlike ions also gives rise to the forbidden (M2) $3p^53s$ $^3P_2$ transition to the ground state which occurs for example at 17.08 angstroms in Fe XVII. It is of interest that the relative intensities of *all* the Fe multiplet components, whether allowed or forbidden and involving resonance decay from $2p^5(^2P)$ 3d, 3s levels, are markedly similar in the sun and tokamak plasmas (Peacock, 1980). In contrast, only the allowed components such as $3p^53s$ $^3P_1$, $^1P_1$ are strong features of high density, $n_e \sim 10^{21}$ cm^{-3}, laser-produced plasmas at temperatures of the order of $T_e \sim 0.5$ keV.

In the case of two-electron ions the first excited configurations 1s2l give rise to important intersystem resonance lines which for low atomic numbers are forbidden on L–S coupling considerations as in the case of 1s2p $^3P_1$, or are highly forbidden as in the case of the M(2) and M(1) radiative decays from the 1s2p $^3P_2$ and 1s2s ^3S levels. The intersystem A values, however, scale rapidly with atomic number as indicated by the lines through Ar XVII in Fig. 13, where A is approximately proportional to $(\Delta E)^5$. In the x-ray line spectra of Fe and Ti (see, e.g., Bitter et al., 1979a, 1981) and of elements such as Cl (see, e.g., Källne et al., 1982) these intersystem lines feature prominently and are distributed among the satellite transitions

$$1s^k nl - 1s^{k-1} nl, n'l', \qquad k = 1 \text{ or } 2, n' = 2, \tag{33}$$

which lie to the long-wavelength side of the allowed $1s^2$ $^1S_0 - 1s2p$ $^1P_1$ resonance line. The diagnostic potential of the intersystem and satellite lines have been investigated by Freeman et al. (1971), Gabriel and Jordan (1972), Gabriel (1972), and Bhalla et al. (1975), and examples of the relevant x-ray line spectra are discussed in Section IV. A summary comment pertaining to this section is that fusion devices such as tokamaks, with relatively low collisionality and with the capability of generating highly stripped ions under near-equilibrium conditions, offer close to ideal plasma conditions for studying metastable levels and forbidden transitions.

B. Optical Opacity

Up until now we have assumed that the level populations [Eq. (32)] do not involve radiation absorption and that the emergent line intensity is dependent only on the line-of-sight integral of the volume emissivity. In practice this is only true for small optical depths where the optically thin approximation

$$I(\nu, x = D) = S(\nu)\tau(x) = \varepsilon(\nu)D \tag{34}$$

is a particular case of the equation of radiation transfer (see, e.g., Mihalas, 1978), viz.,

$$\frac{dI(\nu, x)}{d\tau(\nu)} = I(\nu, x) - S(\nu, x), \tag{35}$$

where D is the depth of the emitting region, $S(\nu)$ is the plasma source function, and $\varepsilon(\nu)$ is the emissivity.

The optical depth is given by

$$\tau(\nu) = \int_0^\nu N(i)L(\nu) \frac{h\nu_0}{4\pi} B_{ij} \, dx, \tag{36}$$

where B_{ij} is the Einstein absorption coefficient and $L(\nu)$ is the absorption line-shape factor which when integrated over the profile gives unit probability for absorption. $L(\nu)$ (Hz^{-1}) is a function of the intrinsic plasma broadening, which may be a single or convoluted function depending on the broadening mechanisms—Doppler, Zeeman, Stark, mass motion, etc.

The optically thin approximation (Cooper, 1966) demands that $\tau(\nu_0) \leq 0.2$, and for the most likely case of resonance absorption by the ground state, $i = 1$, and for Doppler broadening, the optically thin criterion is

$$DN(z) \, (\text{cm}^{-2}) \leq 16\Delta\nu_\text{D}/f_{12}, \tag{37}$$

where $\Delta\nu_\text{D}$ (Hz) is half of the half-width of the line.

Typical impurity densities in tokamaks are $\leq 10^{-3}n_\text{e}$ for metals (Ti, Fe, Ni, etc.), and $\leq 10^{-2}n_\text{e}$ for light elements (e.g., oxygen or carbon). Ion temperatures for the highly ionized ions are typically ~1 keV. With these parameters, it is evident that photon absorption even into the ground state can be ignored in the calculations of impurity level populations in tokamaks.

In the event that opacity cannot be ignored a full solution of the equations describing the level populations must include radiative transfer for the optically thick lines. A reasonable simplification (see Drawin and Emard, 1976) can be made to the equations by multiplying the Einstein A_{ij} values by the optical escape factor Λ_{ij} introduced by Holstein (1951).

IV. Spectral Features and Their Diagnostic Application to Fusion Plasmas

For a plasma temperature in the range $0.1 < kT_\text{e} < 10$ keV we expect most of the radiated power to lie in the soft x-ray region (100–10 angstroms) and in the x-ray region proper ($\lambda < 10$ angstrom). In the case of optically thin bremsstrahlung the spectral intensity will peak at a photon energy corresponding to $\sim 2kT_e$. This radiation will be enhanced by free–bound recombination into incompletely stripped ions as illustrated for example in Fig. 5. In practice the free–bound radiation

$$P_{\text{f-b}} = CZ^4/T_e^{3/2} \exp(\psi_{n,\infty} - h\nu)/kT_e \tag{38}$$

is a factor of $\psi_{n,\infty}/kT_e$ higher than the bremsstrahlung for the same ion (whose ionization potential from level n is $\psi_{n,\infty}$). With a fractional impurity concentration of f_z the enhancement over the hydrogen bremsstrahlung is $f_z Z^4 \psi_{n,\infty}/kT_e$, and this can easily be a factor of 10^3 or more for recombination into the ground state of H- or He-like ions (large $\psi_{n,\infty}$) and in the peripheral region of the plasma (low T_e, see Fig. 5). The regions of the x-ray spectrum which are uncluttered by line emission and free–bound continua steps are widely used to measure the high-energy tail of the free-electron energy spectrum, thus giving the parameter T_e on a routine basis.

The most prominent features of the spectrum are emission lines extending from the visible into the x-ray region with photon energies as high as $h\nu \lesssim 10kT_e$. The spectral distribution of the prominent lines for a representative impurity, iron, is shown schematically in Fig. 15. Pertinent to each ion species there are two important sets of allowed transitions, those at longer wavelengths involving no change in principal quantum number $\Delta n = 0$ and the $\Delta n \gtrsim 1$ transitions at shorter wavelengths. In H- and He-like ions of the metals these latter K-shell lines lie in the x-ray region, where high-resolution crystal dispersion techniques find a ready application (Bitter et al., 1979a, 1981; Källne et al., 1982; Platz et al., 1980; TRF Group, 1981).

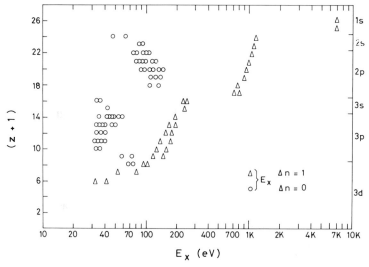

Fig. 15. Excitation energies E_x of $\Delta n = 1$, $\Delta n = 0$ transitions in iron ions of charge state z.

5A. Diagnostics Based on Emission Spectra

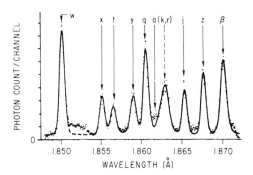

Fig. 16. Spectrum of Fe XXV $1s^2$ $^1S_0$–$1s2p$ $^1P_1$ allowed line w and its associated satellites and forbidden lines emitted from the PLT tokamak. [From Bitter *et al.* (1979a).] The line annotation is that ascribed by Gabriel (1972).

A. K-Shell Excitation

A section of the x-ray line spectrum from the PLT tokamak (Bitter *et al.*, 1979a) is shown in Fig. 16. The resolution of the crystal dispersion instrument used by these authors, $\lambda/\delta\lambda \gtrsim 1.5 \times 10^4$, was sufficiently high not only to separate out the main allowed and forbidden components of the $1s^2$–$1s2l$ transitions of Fe XXV from the satellites of Fe XXIV, Fe XXIII, etc., but also to permit detailed comparison of the line shapes and intensities with theory.

The features x, y, and z of Fig. 16 are the intersystem lines arising from the resonance decay of the levels $1s2p$ $^3P_{2,1}$ and $1s2s$ $^3S_1$, respectively, the feature β is the Fe XXIII Be-like satellite $1s^22s^2$ $^1S_0$–$1s2s^22p$ $^1P_1$, while the other annotated lines correspond to the Fe XXIV satellites. These Li-like ion transitions are of the type $1s^2nl$–$1s2pnl$ and their intensities have been modeled extensively by Gabriel (1972) and by Bhalla *et al.* (1975). The contribution of higher series numbers with $n \geq 3$ has also been considered by Bely-Dubau *et al.* (1979a,b) and is required to account for the energy in the long-wavelength wing of the allowed line, w in Fig. 16. Finally the effect of non-Maxwellian distributions of the electrons on the overall intensity pattern of these Li-like ion satellites has been considered by Gabriel and Phillips (1979). While the impetus for these model calculations has been the interpretation of solar emission, the calculations are also applicable to tokamak plasmas. Indeed, because the laboratory plasmas are well diagnosed and controllable, a comparison between the tokamak emission and the model calculations provides a sensitive test to the theory (Bitter *et al.*, 1979a,b).

Confining our attention to the Li ion satellites, these multiply excited levels can be populated by direct inner-shell excitation of the Li ion and/or by dielectronic recombination from the He-like ion,

$$e + X^{(Z-2)+} \Leftrightarrow X^{(Z-3)+}[n'l + 1 : n''l'']$$

$$\Rightarrow X^{(Z-3)+}[nl : n''l''] + h\nu,$$

dielectronic capture of the electron being followed by a stabilizing radiative decay with emission of a photon rather than the more likely reverse process of autoionization. The main contributions come from $n' = 2, 3$ and $6 > n'' > 3$, so that satellites tend to cluster around the first few members of the allowed resonance series. Provided the doubly excited level is an autoionizing level with the $1s^2$ + e continuum, the recombination process and the dielectronic satellite intensity is then temperature dependent. The ratio of the dielectronic satellite to w the allowed line (Bely-Dubau et al., 1979a,b), is given by

$$\frac{I_s}{I} = \frac{(3)^{1/2}}{2} \frac{m\pi a_0^2}{h} \frac{1}{\bar{f}\bar{g}} \frac{E_0}{kT_e} \frac{g_s}{g_1} \frac{A_R A_a \exp[(E_0 - E_s)/kT_e]}{A_a + \Sigma A_r} = F_1(T_e)F_2(S), \quad (39)$$

where $F_1(T_e)$ is a function only of T_e and E_0 and E_s, the energies of the He-like resonance and Li-like satellite levels above the He-like ion ground state, with statistical weights g_1 and g_s; $F_2(S) = g_s A_r A_a/(A_a + \Sigma A_r)$ is a function only of the autoionizing A_a and radiative A_r decay rates; $\bar{g}$ is a Gaunt factor ~ 0.2; and $\bar{f} \sim 0.6$ is the effective oscillator strength of the allowed line. Such a line is j, i.e., $1s^22p\ ^2P^0_{3/2}-1s2p^2\ ^2D_{5/2}$, in Fig. 16 and has been used by Bitter et al. (1979a) to derive T_e (Fig. 17). Other Li-like ion satellites, e.g., $1s2p^2$ $^2P^e$, i.e., a of Fig. 16, which, neglecting configuration interaction effects, cannot couple by autoionizing transitions with the continuum for reasons of parity and angular momentum, can only be present by inner-shell excitation. The most intense of these inner-shell excited satellites of Fe XXIV is $1s^22s$ $^2S_{1/2}-1s2p2s(^1P)^2P^0_{3/2}$, i.e., line q in Fig. 16. The relative intensities of lines w, q, and β are proportional mainly to the respective ion populations N(Fe XXV), N(Fe XXIV), and N(Fe XXIII), since their upper excited levels have almost the same energy above their ground states. These electron impact excited lines define the equilibrium state of ionization with a characteristic temperature $T_z = f[N(z)/N(z - 1)]$. Comparison of T_z with T_e gives the thermodynamic state of the plasma. Sufficient discrepancy between T_z and T_e has been noted by Bitter et al. (1979a) (Fig. 17) to throw some doubt either on the assumption of coronal ionization balance or on the rate coefficients used in the model calculations.

During high-power neutral (H_0 and D_0)-beam heating of the PLT tokamak (Suckewer et al., 1980c) a substantial lowering of the ionization state is observed without an accompanying lowering of T_e. This is attributed to an additional charge transfer recombination term, Eq. (22), from the beam neutrals. The effect on the x-ray spectrum, Fig. 16, i.e., an increase in I_q/I_w with an almost constant ratio of I_j/I_w, is consistent with the calculated change in the equilibrium charge state during beam injection.

The diagnostic potential of the group of lines shown in Fig. 16 is not yet exhausted. The broadening of the allowed line w has been interpreted in terms of the ion thermal motion after due care is taken to subtract the energy in the long-wavelength wing (Fig. 16) due to higher series member satellites

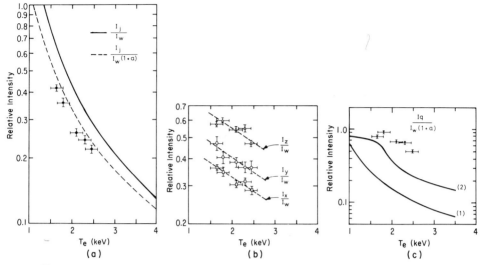

Fig. 17. Analysis of the line intensities of the Fe XXV $1s^2$ $^1S_0$–$1s2p$ $^1P_1$ line w and its associated $n = 1$–2 lines shown in Fig. 16. The ratio of w to other line features is fitted to model calculations (see text). The intensity I_j (a) is a dielectronic satellite from the $1s2p^2$ $^2D_{5/2}$ level from which T_e may be derived; model calculations are from Bhalla *et al.* (1975) and Bely-Dubau *et al.* (1979a,b). $I_{x,y,z}$ are $1s2p$ ^3P, ^3S level decays (b) and I_q (c) is the decay from the inner-shell excited level $1s2p2s$ $^2P^0_{3/2}$, its ratio with w being mainly a function of $N(Fe^{23+})/N(Fe^{24+})$. The full lines (c) are based on coronal balance calculations: (1) Jordan (1969), (2) Summers (1974), Bitter *et al.* (1979a).

(Bitter *et al.*, 1979b). With a somewhat improved resolving power $\lambda/\Delta\lambda = 2.3 \times 10^4$ the width of the allowed line w at 2.6099 angstroms in the x-ray line spectrum of Ti-XXI has been used by Bitter *et al.* (1981) to monitor the changes in ion temperature $T_i(\max) = 5.5$ keV, during beam injection into the PDX tokamak. The relative intensities of the intercombination lines x, y, and z are mainly density dependent beyond a threshold density, and in the case of the iron lines, from the PLT tokamak (Bitter *et al.*, 1979a), satisfactory agreement is found (Fig. 17) with the theory of Freeman *et al.* (1971).

The K-shell spectra of the other important class of impurity, namely ions of the light elements, e.g., C, N, O, etc., lie in the soft x-ray region of the spectrum where crystal dispersion and grating dispersion instruments do not achieve as high a resolving power as is common in the x-ray region. The scaling of the dielectronic recombination coefficient with Z (McWhirter and Summers, Chapter 3, this volume) means that for the lighter ions the satellites to the parent lines are typically weak and poorly resolved as in the O VII spectra of Fig. 18. Diagnostic interest in the He-like ions is mainly confined in this case to allowed and intercombination lines and to the spatial variation of their intensities, which can indicate a departure from stationary ionization balance (Peacock and Summers, 1978).

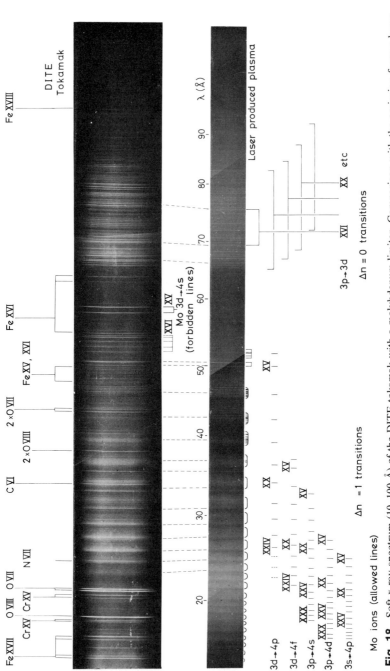

Fig. 18. Soft x-ray spectrum (10–100 Å) of the DITE tokamak with a molybdenum limiter. Comparison with the emission from a laser-irradiated solid Mo target plasma ($n_e \sim 10^{21}$ cm^{-3}) facilitates identification of the DITE forbidden lines. Note that the 1s–2p emission from oxygen in the tokamak spectrum is accounted for by the allowed and intercombination lines, other satellite transitions being weak or absent.

5A. Diagnostics Based on Emission Spectra

The light elements have commonly the highest impurity concentrations $\Sigma_z N(Z^{z+}) \gtrsim 0.01 n_e$, and can make the dominant contribution to the effective ion charge Z_{eff} in the plasma. In order to calculate $Z_{\text{eff}}(r)$, the local concentrations need to be known on an absolute basis.

Provided that the observable data, namely the line-of-sight radiances $E(h, t)$ (photons cm^{-2} sr^{-1} s^{-1}), are available for a number of chordal heights h from the axis of symmetry, at least in the case of cylindrical symmetry, the volume emissivity $I(r, t)$ (photons cm^{-3} s^{-1}) can be recovered by the Abel inversion procedure (see, e.g., Bockasten, 1961). Numerical techniques for reconstructing asymmetric emissions have been discussed, among others, by Meyers and Levine (1978) and by Sauthoff and Von Goeler (1979). The ion concentrations are then derived from $I(r, t)$ as described in Section III. Spatial scans of the plasma discharge are taken typically on a shot-to-shot basis in the x-ray and soft x-ray regions since imaging optics and reflecting surfaces are inefficient at wavelengths between 1 and 100 angstroms.

Calibration of the incident photon flux against detector response on an absolute basis is also something of a problem in the soft x-ray region. The absolute response of an instrument at characteristic K_α wavelengths can be measured with a gas-flow proportional counter and an x-ray diode source (Morgan *et al.*, 1966) or from separate measurements of the diffraction element efficiency and the detector response (Hobby and Peacock, 1973). Synchroton sources and even K-shell excitation of targets irradiated by nuclear disintegration products also have their uses. More conveniently in fusion experiments, the plasma itself often provides pairs of emission lines, one at long wavelength in the visible or VUV, the other in the XUV or soft x-ray region but both originating from the same upper level. The emissivity of the longer wavelength of the pair is readily measured using a separate VUV spectrometer which itself has been calibrated against a radiation transfer standard. A MgF$_2$ window deuterium lamp is a particularly suitable radiation standard for $\lambda \gtrsim 1150$ angstroms (Key and Preston, 1980). The emissivity of the XUV or soft-ray line of the branching pair is then only a function of the longer-wavelength emissivity and the appropriate A values (Hinnov, 1979; Irons and Peacock, 1973). Convenient branching-ratio pairs belonging to the He-like ions of light elements in tokamaks are (1s^2 ^{1}S$_0$–1s2p^{3}P$_1$)/(1s2s ^{3}S$_1$–1s2p ^{3}P$_1$) with wavelengths at 21.8 angstroms/1638.4 angstroms in O VII and 40.73 angstroms/2277.3 angstroms in C-V (Figs. 18 and 14). Theoretical emissivities for these and other O VII lines from the same principal quantum levels are shown in Fig. 19 as a function of T_e. Their dependence on electron density is discussed by McWhirter and Summers (Chapter 3, this volume).

Since the temperature in the core of tokamaks can reach well above 1 keV the lighter ions in these circumstances will exist as bare nucleii whose presence will only be registered through their contribution to the x-ray continuum. Active, particle beam probe techniques, Section V, where electrons are transferred to bound levels by charge exchange is an attractive method for measuring the concentrations of these bare nucleii.

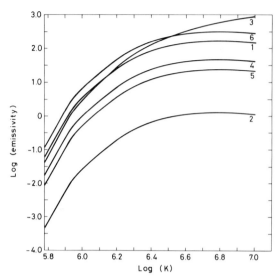

Fig. 19. Model calculations of the emissivity (photons s^{-1} per ion) of the O VII transitions 1s^2–1s2p, 1s2s–1s2p as a function of temperature at a density $n_e = 1 \times 10^{13}$ cm^{-3}: (1) $^1S_0$–$^3P_1$, (2) $^1S_0$–$^3P_2$, (3) $^1S_0$–$^1P_1$, (4) $^3S_1$–$^3P_0$, (5) $^3S_1$–$^3P_1$, (6) $^3S_1$–$^3P_2$. [From Gordon et al. (1982).]

B. L-Shell Excitation

As can be seen on inspection of Fig. 15, lines from metal ions which are isoelectronic with neon-shell elements lie in two fairly well-defined spectral bands, with the allowed $\Delta n > 0$ transitions in the region 10 angstroms–20 angstroms and the $\Delta n = 0$ transitions lying approximately between 100 and 200 angstroms.

Appropriate dispersion systems for the wavelength region $\lambda \lesssim 20$ angstroms have been discussed by Peacock (1981). For the $\Delta n > 0$ transitions a broadband crystal spectrometer with a shot-by-shot spatial scan capability is an effective impurity concentration monitor. An extended wavelength coverage, equivalent to a finite range of Bragg angles at the crystal surface, can be achieved with a curved crystal bent convex in the de Broglie mode (Peacock et al., 1969; Bromage et al., 1977), or by continuous rotation of a flat crystal in time (Von Goeler et al., 1981). Only sufficient wavelength resolution to identify the separate lines is necessary, as is illustrated in Fig. 20 by the L-shell spectrum of iron taken with a flat rotating crystal viewing the PLT tokamak (Von Goeler et al., 1983).

The ease with which the $\Delta n = 0$ lines can be excited coupled with their relatively long wavelengths ensures that these particular transitions (Hinnov, 1976) are important diagnostic features in plasmas with $T_e \lesssim 1$ keV. For this reason the energy levels of the $2s^k2p^l$ configurations have been critically compiled in a series of publications by Edlén (see, e.g., Edlén, 1981).

5A. Diagnostics Based on Emission Spectra

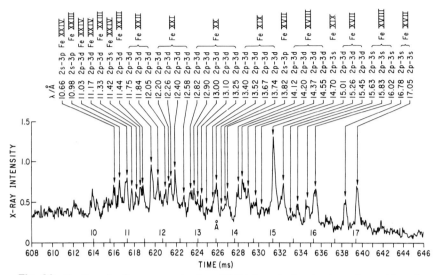

Fig. 20. Section of soft x-ray spectra from PLT tokamak showing $\Delta n > 0$ transitions from L-shell iron ions. [From Von Goeler et al. (1983).]

The long-wavelength $\Delta n = 0$ transitions at $\lambda \gtrsim 100$ angstroms are appropriately dispersed using grazing-incidence grating techniques with rotating mirror optics placed between the spectrometer and the plasma to effect a continuous chordal scan of the plasma cross section (Breton et al., 1979). Characteristically, the wavelengths of the allowed $\Delta n = 0$ transitions in a given element tend to *increase* with ion charge (Fig. 15), in contrast to the $\Delta n > 0$ transitions. Figure 21 illustrates a typical spectrum of the L-shell $\Delta n = 0$ metal ions between 100 and 200 Å, mostly from ionized titanium interspersed with $\Delta n > 0$, L-shell transitions from oxygen in the DITE tokamak. The characteristic lines from each ion species are well separated relative to those in the soft x-ray region. The 2s–2p doublet of Li-like Ti XX, for example, appears at 259.30, 309.09 Å, while the equivalent F-like Ti XIV transitions are at 121.98 and 129.44 Å (Lawson et al., 1981). In respect of bandwidth grazing-incidence grating spectroscopy generally outperforms crystal dispersion so that there are considerable advantages to be gained from monitoring these $\Delta n = 0$ transitions in metal ions. The volume emissivity of the most intense $\Delta n = 0$ lines (expressed in photons cm^{-3} sr^{-1} s^{-1}) exceeds the $\Delta n > 0$ x-ray line emissivities by well over an order of magnitude. Photoelectric detector efficiencies moreover are relatively high in the few 100-angstrom region.

At longer wavelengths still, in the VUV and visible typically, the $\Delta n = 0$ forbidden (M1) transitions between levels within the ground configurations also provide useful diagnostic information. The low A values somewhat restrict their use to plasma densities $n_e \lesssim 10^{14}$ cm^{-3}. However, both the ion

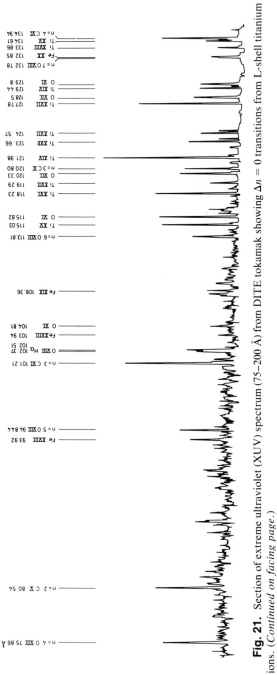

Fig. 21. Section of extreme ultraviolet (XUV) spectrum (75–200 Å) from DITE tokamak showing $\Delta n = 0$ transitions from L-shell titanium ions. (*Continued on facing page.*)

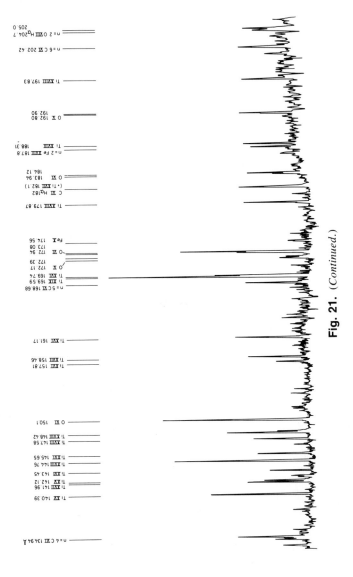

Fig. 21. (*Continued.*)

temperature and plasma rotation in the PLT tokamak, for example, have been measured from the intensity profiles of the $2s^22p^3$ $^2D_{5/2-3/2}$ transition in Fe XX at 2665.1 angstroms (Eubank *et al.*, 1979; Suckewer *et al.*, 1979). Feldman *et al.* (1980), Bhatia and Mason (1980), Mason *et al.* (1979), and Doschek and Feldman (1976) have outlined additional diagnostic applications of the forbidden and allowed $n = 2$–2 transitions for measuring T_e and n_e in tokamak plasmas. The intensity ratio of the Fe XX 2665-angstroms forbidden line to the $2s^22p^3$ $^4S_{3/2}$–$2s2p^4$ $^4P_{5/2}$ resonance line at 132.85 angstroms, for example, would be a reasonable monitor of electron density. Perhaps the most useful information from such lines, however, are impurity ion concentrations which may readily be derived from their absolute intensities using level population codes (Gordon *et al.*, 1982). L-shell excitation of the light elements, oxygen, carbon, etc., near the cool ($\lesssim 100$ eV) plasma boundary, can evidently play an important role in controlling the level of metal contaminants in a tokamak (Suckewer and Hawryluk, 1978). Radiation loss, a relatively benign process at the plasma edge, screens out the influx of wall impurities by depressing the edge temperature.

C. M-Shell Excitation

The above comments on the importance of the L-shell shell ions, as ion concentration indicators and as a source of radiation loss, apply with even more force to the less highly stripped $3s^k3p^l3d^m$ configurations. Both the $\Delta n = 0$ and $\Delta n > 0$ transitions in the metals lie at grazing-incidence grating wavelengths or longer, as illustrated in Fig. 18 for Mo. The $\Delta n = 0$, $3p^l3d^m$–$3p^{l-1}3d^{m+1}$ transitions in iron are notable and for a few years remained unidentified features of the spectrum of the sun and of the Zeta toroidal pinch (Fawcett *et al.*, 1963; Gabriel *et al.*, 1966). In the Zeta fusion device, modest electron temperatures $T_e \lesssim 100$ eV were typically achieved. The $\Delta n = 0$ forbidden lines between levels within the ground configurations provided some of the earliest diagnostic evidence in the visible and VUV for the sun's high-temperature corona.

The 3p → 3d spectral lines in elements such as Ti through Fe are well separated, identifiable, and useful individually for diagnostic purposes. In an element with as high an atomic number as Mo, however, the spectrum (Fig. 18) already consists of overlapping bands due to inner subshell excitation and open subshell excited configurations (Mansfield *et al.*, 1978).

The M-shell ions of the iron period are found typically in the outer peripheral regions of tokamaks. Because of the relatively large volume they occupy and their high efficiency as radiators (Fig. 1) the radiation loss from metal-contaminated high-temperature tokamaks ($\gtrsim 1$ keV) tend to be dominated by the plasma boundary (Harrison, 1980).

V. Neutral-Beam Spectroscopy

We have noted in Section II that by virtue of the high value of the charge transfer cross section, neutral atoms, e.g., H^0, even at low concentrations can alter the spatial distribution of charge states, ionization balance, and radiation loss. A corollary is the diagnostic use of neutral-particle beam probes for measuring the concentration of impurity ions such as bare nuclei which normally have a weak radiation signature. Electron capture from the H^0 beam increases the luminosity of the impurity, while crude spatial resolution can be achieved in the case where a well-collimated beam propagates radially through the separate ionization shells. Provided that the beam propagation period exceeds an ionization time a new balance will be set up between electron impact ionization and charge exchange recombination viz.,

$$N(Z^{z+})\zeta\beta_{cx} = N(Z^{(z-1)+})\alpha_{el,\,ion}, \qquad (40)$$

i.e., for each electron impact ionization event we have a charge transfer recombination event. In a tokamak fusion device we typically find $\tau_{circ} \lesssim \tau_{el,\,ion} < \tau_{diff}$, where τ_{circ} is the toroidal period for the freely circulating ions, $\tau_{el,\,ion}$ is the ionization time, and τ_{diff} is the radial diffusion time from the beam interaction region; then Eq. (40) represents the charge state averaged around the minor axis of the toroid.

In order to derive $N(Z^{z+})$, say, we need to measure radiation from the bound level populations of $Z^{(z-1)+}$ given by

$$n(j)\sum_l A_{lj} = \left[n_e N(Z^{(z-1)+}, i)\sum_i^{k>j}\alpha_{el,\,x}(Z^{(z-1)+}, i, k)\right.$$

$$\left. + N(H^0)N(Z^{z+})\beta_{cx}(Z^{z+}, 1, k)\right]$$

$$\times \left(\frac{A_{jk}}{\sum_{l=1}^{k-1} A_{lk}} + \frac{A_{k-1,k}}{\sum_{l=1}^{k-1} A_{lk}}\frac{A_{j,k-1}}{\sum_{l=1}^{k-2} A_{lk}} + \cdots\right); \qquad (41)$$

here we have assumed a sufficiently low electron density that collisional deexcitation is ignored. The terms within the second set of large parentheses represent the probability P_{jk} of filling the j level by spontaneous decay per occupancy event into an upper k level. The terms within the square brackets describe the occupancy rate of the k level either by electron excitation of $N(Z^{(z-1)+})$, $\alpha_{el,\,x}$, or by charge transfer, β_{cx}, from $N(Z^{z+})$.

Electron capture selectively populates the relatively high n quantum state, Eq. (24), with a distribution among the l sublevels, as calculated by Salop (1979). Typically $n = 5$ for $O^{8+} + H^0 \rightarrow O^{7+*} + H^+$ and for $V_r \simeq 2 \times 10^8$ cm s^{-1} with an occupancy of the highest allowed azimuthal quantum states $l = n - 1$ being preferred. At higher collision velocities captures are smeared out over adjacent n values with a slight increase in the mean principal quantum number. Redistribution among the sublevels by collisions (Pengelly and Seaton, 1964) may need to be taken into account. The probability of a Balmer-α or Lyman-α photon being emitted from a one-electron impurity ion, per charge transfer event involving H^0 and a bare nucleus, is somewhat less than unity, but not very much less, since radiative decay to the $n = 3, 2$ levels preferentially takes place via $\Delta l = 1$ quantum jumps between *maximum l* values for a given n.

The probability of a photon ij being emitted per charge exchange event is

$$P_{ij} = \left(A_{ij} \Big/ \sum_{l=1}^{j-1} A_{lj}\right) P_{jk}, \tag{42}$$

and for the above charge transfer reaction in a plasma with a sub-kilo-electron-volt temperature and density $n_e \sim 10^{13}$ cm^{-3}, P_{cx} (Lyman-α, O VIII) $\simeq 0.8$–0.9, while P_{cx} (Balmer-α, O VIII) $\simeq 0.7$. These photons will appear as prompt radiation from the beam propagation region.

In contrast to the recombination photons, electron impact excitation, proportional to n^{-3} for one-electron ions, is most likely to populate the lower quantum levels. The probability of emission of a photon per ionization event, (i.e., per recombination event in the steady state) is of the order of unity for Lyman-α photons, $P_{el, x}$ (Lyman-α, O VIII) ~ 1, and for Balmer-α photons, $P_{el, x}$ (Balmer-α, O VIII) $< 10^{-1}$, under the above plasma conditions. These photons, moreover, will be delayed, relative to a charge transfer event, typically by an electron impact excitation time (Fig. 13), and thus may appear in plasma regions remote from the beam injection locality. Since the ionization and excitation times are closely related, pronounced asymmetries in the Lyman-α O VIII emission around the azimuth of a tokamak with local beam injection might be expected. Asymmetries, arising from a variation in the charge state of the freely circulating ions in the toroidal direction, are indeed observed in the DITE beam injection experiment (Clark *et al.*, 1982).

An immediate consequence of the large increase in the radiative decay from the high relative to the low quantum states during beam injection (the intensity ratio of the Balmer-α/Lyman-α of O VIII, for example might increase tenfold) suggests a diagnostic technique for mapping out the beam profile as it propagates through the plasma. The time dependence of the prompt recombination lines allows one to discriminate between background radiation at the same wavelength when using beam modulation techniques.

Application of charge transfer recombination to the measurement of impurity concentrations has been discussed by Gordeev and Zinov'ev (1980) and demonstrated by Afrosimov *et al.* (1978). Using a short-period, 0.2-ms, modulated H^0 beam whose current is sufficiently low not to alter the electron parameters in the plasma, Afrosimov *et al.* (1978) were able to interpret the absolute intensity of C^{5+} Lyman-α in terms of the concentration $N(C^{6+}) \simeq 10^{11}$ cm^{-3} of carbon nuclei in the core of the plasma. Somewhat similar H^0

5A. Diagnostics Based on Emission Spectra

beam probe experiments, by Zinov'ev et al. (1980) using Lyman-α emission from O^{7+} and even more effectively by Fonck et al. (1982) who have interpreted the spatially resolved emission from Balmer-α and higher quantum states of O^{7+} and C^{5+} in terms of the diffusion lifetime of ionized light atom impurities. These "remote sounding" techniques using particle beams are particularly valuable for estimating the temporal and spatial concentrations of those ions, such as bare nuclei, which do not register their presence by line emission. The study of the excited states of impurity ions that are populated through charge exchange recombination from modulated neutral beams in tokamaks has also been applied to the measurement of impurity ion temperature (Fonck et al., 1983); plasma rotation (Isler and Murray, 1983); and collisional-rate coefficients (Isler et al., 1982).

The importance of charge transfer processes involving thermal (1–100 eV) neutrals at the plasma edge should not be neglected. Isler and Crume (1978) suggest that intensity "anomalies" in oxygen lines at 115.8 and 81.9 angstroms in the ISX-A tokamak are due to preferential population at the $n = 4\,^2P$ levels of O^{5+} and the $n = 5$, ^3P levels of O^{6+} by charge exchange from the background thermal H^0 atoms. All other oxygen lines observed in these experiments can be accounted for adequately by straightforward electron excitation.

VI. Basic Atomic Physics

The construction of quasi-dc toroidal fusion devices, such as tokamaks with wave or particle beam, auxiliary heating methods, and with improved thermal insulation and particle confinement, offers the spectroscopist an opportunity to study collision and radiation processes involving highly stripped ions under plasma conditions hitherto inaccessible in the laboratory. Multi-kilo-electron-volt temperatures and a confinement parameter $n_e\tau_{conf} \sim 10^{12}$ cm^{-3} s in the core of the device dictates that highly ionized species such as H- and He-line ions of the common metals are generated. The low collisionality of these plasmas ensures that forbidden lines with A values $\gtrsim 10^2$ s^{-1} are noteworthy features of the spectra.

The near corona equilibrium conditions in the core plasma of a tokamak can be perturbed in a controlled manner, the time constant for a new equilibrium or for a relaxation to the original state being a function of the relevant collision rates, Eq. (3). In the case when there are "sawtooth" excursions in T_e and n_e arising from marginal instability at the q = 1 magnetic surface [$T_e(r = 0)$ typically drops by ~15% on a time scale of <1 ms, then recovers over a somewhat longer period], the time variation in the ion species can be related to the magnitude of the ionization and recombination rates. Relaxation to a lower ionization balance, for example, will take place in a time

$$\tau_{rel} \simeq [\beta(T_e)n_e]^{-1}.$$

During sawtooth operation of the TFR tokamak (Breton et al., 1978), the time variation of the $\Delta n = 0$, M-shell lines of Mo-XXXI and Mo-XXXII could be accounted for by rate coefficients which were within a factor of 2 of the Lotz (1967) values for electron impact ionization and the Burgess (1965) values for dielectronic recombination. An advantage of choosing the $\Delta n = 0$ transitions from the lowest excited configurations is the close coupling between these level populations and the ground state concentrations.

The equilibrium charge state could equally well be perturbed by injection of a time-modulated neutral (H^0) beam, in which case

$$\tau_{rel} \simeq [N(H^0)\beta_{cx}]^{-1}.$$

Rate coefficients relevant to species other than the contaminants which are unavoidably present in tokamaks can be studied by neutral-gas injection or ablation of laser-irradiated targets placed near the chamber walls.

Tokamak plasmas are very attractive also for atomic structure studies. The various ion species are conveniently located in distinct plasma regions. Even with mass motion effects induced by auxiliary beam heating, the quality of the soft x-ray and XUV line spectra, Figs. 16 and 21, are superior to that of most other laboratory sources in terms of their relative freedom of most other laboratory sources in terms of their relative freedom from source broadening, and from overlapping spectra from many ion stages of the same element. The L-shell ions (Edlén, 1981) and their forbidden lines (Lawson et al., 1981), are of particular interest, as are their collisionally excited and dielectronic satellites to the K-shell ion transitions (Merts et al., 1976; Bhalla et al., 1975). The H-like ion lines from the light elements provide, in high orders (Fig. 21), convenient calibration standards for the $n = 2-2$ transitions.

The energy levels of highly ionized one- and two-electron ions are of fundamental interest in the theory of atomic structure. High-Z ions for example are in some ways a better test of quantum electrodynamic theory (QED) than is the H atom, since the higher-order radiative terms scale as Z^n, $n \gtrsim 4$. The non-QED component of the energy levels 2p $^2P_{3/2,1/2}$ and 2s $^2S_{1/2}$ can in principle be calculated with a precision of the order of 1 in 10^5 so that subtraction of the non-QED component from precision experimental values of the energy levels should provide a good test of QED theory. Tests of the $2S_{1/2}-2P_{1/2}$ Lamb-shift measurements in one-electron ions have been reviewed by Kugel and Murnick (1977) and by Mohr (1976). For $Z > 9$, Lamb shifts have not been tested to an accuracy of better than 1%.

The use of a tokamak plasma as a spectroscopic source for measuring the term structure of one- and two-electron ions is an obvious possibility. As illustrated in Fig. 16 the appropriate highly charged ions are readily generated and, using high-dispersion crystal diffraction, precision wavelengths of the $n = 1-2$ lines can be measured (Bitter et al., 1979a). In order to make some impact on QED theory, however, a wavelength measurement of the centroid of the Lyman line to nearly one part in 10^6 is necessary, the Lamb shift contribution being one part in 2×10^4 for $z \simeq 20$. Such high-resolution spectroscopy on lines which suffer from plasma broadening (due to thermal and bulk rotational motion) and from radiation damping ($\Delta E = hA_{ij}$, proportional to Z^4), while not impossible, is a very formidable task.

It has been pointed out by several authors (see, e.g., Ermolaev, 1973; Berry et al., 1978) that Lamb-shift measurements need not be confined to one-electron ions and that two-electron ions may indeed offer higher precision. This is especially the case with the transitions 1s2s $^3S_1$–1s2p $^3P_{2,1,0}$. These $\Delta n = 0$ lines have energies scaling as Z in contrast to the Lamb-shift scaling as ($\propto Z^4$). The progressive increase in the energy of the Lamb shift relative to the $n = 2$, $^3S_1$–$^3P_{2,1,0}$ transition energy is such that at $Z = 20$, the QED contribution is ~1% (Ermolaev and Jones, 1974). The wavelength of the 1s2s–1s2p triplet lies typically in the region $100 \leq \lambda \leq 1000$ Å for $Z > 10$, and precision wavelengths for most of these transitions have still to be measured.

In two-electron systems the ionization potential is $I_{tot} = I_{rel} + E_1$, where I_{rel} is the electrostatic energy with relativistic corrections and $E_{1'}$ proportional to $\alpha^4 Z^4/n^3$, contains both the singlet–triplet fine-structure terms and the Lamb shift, i.e., $E_1 = E_{st} + E_{QED}$. Ermolaev and Jones (1972) indicate that configuration mixing corrections to the fine-structure splitting, $^3P_{2,1,0}$, is not important for $Z < 10$, where $E_{st} \simeq E_{QED}$, but must be included for $Z > 10$, where E_{st} exceeds E_{QED}.

As Z increases, spin–orbit coupling breaks down and progressive mixing of the $^1P_1$ and $^3P_1$ levels causes (E1) resonance decay of the intercombination line 1s^2 $^1S_0$–1s2p $^3P_1$. The probability of the 2^3P_2 level (M2) radiative decay to ground also increases rapidly along the isoelectronic sequence (proportional to Z^8) and is more probable than the branching 2^3P_2–2^3S_1 (E1) transition for $Z > 17$. The only member of the 1s2s–1s2p triplet likely to remain observable in the two-electron spectra from highly stripped ions is therefore $2s^2S_1$–$2p^3P_0$.

In tokamaks, the O VII triplet, 1s2s $^3S_1$–1s2p $^3P_{2,1,0}$ intensities can be intense as illustrated in Fig. 14. For the O-VII $n = 2$ triplet levels, ~0.1% is due to the Lamb shifts. In order to measure the latter, to one part in a hundred, say, a wavelength precision $\delta\lambda \simeq 0.02$ Å is required. In the normal incidence region of the spectrum from the DITE tokamak the wavelengths of the O VII and F VIII ions have been measured to about this accuracy (Stamp et al., 1981), and already indicate discrepancies between the experimental data and the best available theoretical values (Ermolaev and Jones, 1974; DeSerio et al., 1981. For higher-Z ions, say $Z = 20$, the situation is even more promising and a wavelength accuracy of 0.01 Å would allow extraction of the two-electron Lamb shift with a precision of one part in 10^3.

While the $n = 2$ triplet transitions for $Z > 10$ have yet to be observed in tokamak plasmas, there is no doubting the capability of these sources of producing two-electron spectra with high emissivity. Precision wavelength measurements from these plasmas will constitute a most interesting test of QED for electrons in an intense Coulomb field.

References

Abramov, V. A., Baryshnikov, E. F., and Lisitsa, V. S. (1978). *JETP Lett. (Engl. Transl.)* **27**, 464–467.
Afrosimov, V. V., Gordeev, Y. S., Zinov'ev, A. N., and Korotkov, A. A. (1979). *JETP Lett. (Engl. Transl.)* **28**, 500–502.
Ashby, D. E. T. F., and Hughes, M. H. (1981). *Eur. Conf. Controlled Fusion Plasma Phys., Proc., 10th, Moscow* **1**, paper J10.
Behringer, K., Engelhardt, W., and Fussmann, G. (1981). *IAEA Tech. Meet. Divertors, Impurity Control Garching,* FDR Germany. IPP Rep. III-73, pp. 25–28.
Bely-Dubau, F., Gabriel, A. H., and Volonté, S. (1979a). *Mon. Not. R. Astron. Soc.* **186**, 405–419.
Bely-Dubau, F., Gabriel, A. H., and Volonté, S. (1979b). *Mon. Not. R. Astron. Soc.* **189**, 801–816.
Berry, H. G., Deserio, R., and Livingston, A. F. (1978). *Phys. Rev. Lett.* **41**, 1652–1655.
Bhalla, C. P., Gabriel, A. H., and Presnyakov, L. P. (1975). *Mon. Not. R. Astron. Soc.* **172**, 359–375.
Bhatia, A. K., and Mason, H. E. (1980). *Astron. Astrophys.* **83**, 380–382.
Bhatia, A. K., Feldman, U., and Doschek, G. A. (1980). *J. Appl. Phys.* **51**, 1464–80.
Bitter, M., Hill, K. W., Sautoff, N. R., Efthimion, P. C., Meservey, E., Roney, W., Von Goeler, S., Horton, R., Goldman, M., and Stodiek, W. (1979a). *Phys. Rev. Lett.* **43**, 129–132.
Bitter, M., Von Goeler, S., Horton, R., Goldman, M., Hill, K. W., Sauthoff, N. R., and Stodiek, W. (1979b). *Phys. Rev. Lett.* **42**, 304–307.
Bitter, M., Von Goeler, S., Hill, K. W., Horton, R., Johnson, D., Roney, W., Sauthoff, N., Silver, E., and Stodiek, W. (1981). *Princeton Plasma Phys. Lab., Rep.* **PPPL-1751**.
Bockasten, K. (1961). *J. Opt. Soc. Am.* **51**, 943–947.
Bogen, P. (1968). *In* "Plasma Diagnostics" (W. Lochte-Holtgreven, ed.), pp. 424–477. North-Holland Publ., Amsterdam.
Bohr, N., and Lindhard, J. (1954). *Mat.-Fys. Medd.—K. Dan. Vidensk. Selsk.* **28**, No. 7.
Brau, K., Von Goeler, S., Bitter, M., Cowan, R. D., Eames, D., Hill, K., Sauthoff, N., Silver, E., and Stodiek, W. (1980). *Princeton Plasma Phys. Lab., Rep.* **PPPL-1644**.
Breton, C., and Schwob, J. L. (1974). *In* "Some Aspects of Vacuum Ultraviolet Radiation Physics" (N. Damany, B. Vodar, and J. Romand, eds.), pp. 241–284. Pergamon, Oxford.
Breton, C., De Michelis, C., and Mattioli, M. (1976). *Nucl. Fusion* **16**, 891–899.
Breton, C., De Michelis, C., Finkenthal, M., and Mattioli, M. (1978). *Phys. Rev. Lett.* **41**, 110–113.
Breton, C., De Michelis, C., Finkenthal, M., and Mattioli, M. (1979). *J. Phys. E* **12**, 894–898.
Breton, C., *et al.* (1982). *Assoc. Euratom-CEA Rep., Cent. Etud. Nucl., Fontenay aux Roses* **EUR-CEA-FC-1159**.
Bromage, G. E., Cowan, R. D., Fawcett, B. C., Gordon, H., Hobby, M. G., Peacock, N. J., and Ridgeley, A. (1977). *Culham Lab. Rep.* **CLM-R170**.
Burgess, A. (1965). *Astrophys. J.* **141**, 1588–1590.
Burgess, D. D. (1978). *In* "The Physics of Ionized Gases" (R. K. Janev, ed.), pp. 543–577. Inst. Phys., Belgrade.
Burrell, K. H., Wong, S. K., and Amano, T. (1980). *Nucl. Fusion* **20**, 1021–1036.
Burrell, K. H., Wong, S. K., Muller, C. H., III, Hacker, M. P., Ketterer, H. E., Isler, R. C., and Lazarus, E. A. (1981). *Nucl. Fusion* **21**, 1009–1014.

5A. Diagnostics Based on Emission Spectra

Carolan, P. G., and Piotrowicz, V. A. (1982). *Culham Lab. Rep.* **CLM-P672**; (1983) *Plasma Phys.* **25**, No. 10, 1065–1086.
Clark, W. H. M., Cordey, J. G., Cox, M., et al. (1981). *Nucl. Fusion* **2**, No. 3, 333–345.
Cohen, S., Cecchi, J. L., and Marmar, E. S. (1975). *Phys. Rev. Lett.* **35**, 1507–1510.
Connor, J. W. (1973). *Plasma Phys.* **15**, 765–782.
Cooper, J. (1966). *Rep. Prog. Phys.* **29**, 35–130.
Coppi, B., and Sharky, N. (1979). *Phys. Plasmas Close Thermonucl. Conditions, Proc. Course, Varenna, Italy* **1**, 47. *Comm. Eur. Communities* **EUR FU BRU/XII/476/80**.
Cowan, R. D. (1981). "The Theory of Atomic Structure and Spectra." Univ. of California Press, Berkeley.
Cowan, R. S. (1977). *Los Alamos Sci. Lab.* [*Rep.*] *LA* **LA-6679MS**.
Crandall, D. H. (1979). *Int. Semin. Ion–Atom Collisions (ISIAC VI), Tokai-mura, Jpn.*
Crandall, D. H., Phanenf, R. A., and Meyer, F. W. (1979). *Phys. Rev. A* **19**, 504–514.
De Michelis, C., and Mattioli, M. (1981). *Nucl. Fusion* **21**, 677–754.
DeSerio, R., Berry, H. G., Brooks, R. L., Hardis, J., Livingston, A. E., and Hinterlong, S. J. (1981). *Phys. Rev. A* **24**, 1872–1888.
Doschek, G. A., and Feldman, U. (1976). *J. Appl. Phys.* **47**, 3083–3087.
Drawin, H. W., and Emard, F. (1976). *Physica B+C (Amsterdam)* **85B+C**, 333–356.
Edlén, B. (1981). *Phys. Scr.* **22**, 593–602.
Eissner, W., and Seaton, M. J. (1972). *J. Phys. B* **5**, 2187–2198.
Eissner, W., Jones, M., and Nussbaumer, H. (1974). *Comput. Phys. Commun.* **8**, 270–306.
Ermolaev, A. M. (1973). *Phys. Rev. A* **8**, 1651–1657.
Ermolaev, A. M., and Jones, M. (1972). *J. Phys. B* **5**, L225–227.
Ermolaev, A. M., and Jones, M. (1974). *J. Phys. B* **7**, 199–207.
Eubank, H., Goldston, R. J., et al. (1979). *Plasma Phys. Controlled Nucl. Fusion Res., Proc. Int. Conf., 7th, Innsbruck, 1978* **1**, 167–198.
Fawcett, B. C., Gabriel, A. H., Griffin, W. G., Jones, B. B., and Wilson, R. (1963). *Nature (London)* **200**, 1303–1304.
Feldman, U., Doschek, G. A., Cheng, C.-C., and Bhatia, A. K. (1980). *J. Appl. Phys.* **51**, 190–201.
Fonck, R. J., et al. (1982). *Phys. Rev. Lett.* 49, No. 10, 737–740.
Fonck, R. J., Goldston, R. J., Kaita, R., and Post, D. E. (1983). *Appl. Phys. Lett.* 42, No. 3, 239–241.
Freeman, F. F., Gabriel, A. H., Jones, B. B., and Jordan, C. (1971). *Philos. Trans. R. Soc. London, Ser. A* **270**, 127–133.
Freeman, R. L., and Jones, E. M. (1974). *Culham Lab. Rep.* **CLM-R137**.
Gabriel, A. H. (1972). *Mon. Not. R. Astron. Soc.* **160**, 99–119.
Gabriel, A. H., and Jordan, C. (1972). *In* "Case Studies of Atomic Collision Physics" (E. W. McDaniel and M. R. C. McDowell, eds.), Vol. 2, pp. 209–291. North-Holland Publ., Amsterdam.
Gabriel, A. H., and Phillips, K. J. H. (1979). *Mon. Not. R. Astron. Soc.* **189**, 319–327.
Gabriel, A. H., Fawcett, B. C., and Jordan, C. (1966). *Proc. Phys. Soc., London* **87**, 825–839.
Gilbody, H. B. (1981). *Phys. Scr.* **23**, 143–152.
Gordeev, Y. S., and Zinov'ev, A. N. (1980). *In* "The Physics of Ionised Gases" (M. Matic, ed.), pp. 215–250. Boris Kidric Inst. Nucl. Sci., Belgrade.
Gordon, H., Tully, Summers, H. P., and Tully, J. A. (1982). *Culham Lab. Rep.* **CLM-R229**.
Griem, H. R. (1964). "Plasma Spectroscopy." McGraw-Hill, New York.
Griem, H. R. (1979). *Phys. Plasmas Close Thermonucl. Conditions, Proc. Course, Vareına, Italy* **1**, 395–418. *Comm. Eur. Communities* **EUR FU BRU/XXI/476/80**.
Grozdanov, T. P., and Janev, R. K. (1978). *Phys. Rev. A* **17**, 880.
Harrison, M. F. A. (1980). *In* "Atomic and Molecular Processes in Controlled Thermonuclear Fusion" (M. R. C. McDowell and A. M. Ferendeci, eds.), pp. 15–70. Plenum, New York.

Hawryluk, R. J., Suckewer, S., and Hirshman, S. P. (1979). *Nucl. Fusion* **19**, 607–632.
Hinnov, E. (1976). *Phys. Rev. A* **14**, 1533–1541.
Hinnov, E. (1979). In "Diagnostics for Fusion Experiments," Proceedings of the International School of Plasma Physics, Varenna, Italy, 1978 (E. Sindoni and C. Wharton, eds.), pp. 139–148. Pergamon, Oxford.
Hinnov, E. (1980). In "Atomic and Molecular Processes in Controlled Thermonuclear Fusion" (M. R. C. McDowell and A. M. Ferendeci, eds.), pp. 449–470. Plenum, New York.
Hinnov, E., Suckewer, S., Bol, K., Hawryluk, R. J., Hosea, J., and Meservey, E. (1978). *Plasma Phys.* **20**, 723–734.
Hobby, M. G., and Peacock, N. J. (1973). *J. Phys. E* **6**, 854–857.
Holstein, T. (1951). *Phys. Rev.* **83**, 1159–1168.
Hopkins, G. R., and Rawls, J. M. (1977). *Nucl. Technol.* **36**, 171–186.
Hulse, R. A. (1983). *Nucl. Technol. Fusion* **3**, 259–272.
Hulse, R. A., Post, D. E., Brau, K., Von Goeler, S., and Hinnov, E. (1980a). *Bull. Am. Phys. Soc.* **25**, 868.
Hulse, R. A., Post, D. E., and Mikkelsen, D. R. (1980b). *J. Phys. B* **13**, 3895–3907.
Irons, F. E., and Peacock, N. J. (1973). *J. Phys. E* **6**, 857–862.
Isler, R. C. (1977). *Phys. Rev. Lett.* **38**, 1359–1362.
Isler, R. C., and Crume, E. C. (1978). *Phys. Rev. Lett.* **41**, 1296–1300.
Isler, R. C., and Murray, L. E. (1983). *Appl. Phys. Lett.* **42** (4), 355–357.
Isler, R. C., Crume, E. C., and Arnuirus, D. E. (1982). *Phys. Rev. A* **26** (4), 2105–2116.
Janev, R. K., and Grozdanov. T. P. (1980). In "The Physics of Ionised Gases" (M. Matic, ed.), pp. 181–213. Boris Kidric Inst. Nucl. Sci., Belgrade.
Janev, R. K., and Presnyakov, L. P. (1981). *Phys. Rep.* **70**, 1–107.
Jensen, R. V., Post, D. E., Grasberger, W. H., Tarter, C. B., and Lokke, W. A. (1977). *Nucl. Fusion* **17**, 1187–1196.
Jones, B. B., and Wilson, R. (1962). *Plasma Phys. Controlled Nucl. Fusion, Proc. Int. Conf., 1st, Salzberg, 1961* 889.
Jordan, C. (1969). *Mon. Not. R. Astron. Soc.* **142**, 501–521.
Källne, E., Källne, J., and Pradhan, A. K. (1982). *Phys. Rev. A* **27**, No. 14, 1476–1486.
Key, P. J., and Preston, R. C. (1980). *J. Phys. E* **13**, 886–870.
Knudsen, H., Haugen, H. K., and Hvelplund, P. (1981). *Phys. Rev. A* **23**, 597–610.
Krupin, V. A., Marchenko, V. S., and Yakovlenko, S. I. (1979). *JETP Lett. (Engl. Transl.)* **29**, 318–321.
Kugel, H. W., and Murnick, D. E. (1977). *Rep. Prog. Phys.* **40**, 297–343.
Lawson, K. D., Peacock, N. J., and Stamp, M. F. (1981). *J. Phys. B* **14**, 1929–1952.
Lotz, W. (1967). *Astrophys. J., Suppl. Ser.* **14**, 207.
McWhirter, R. W. P. (1965). In "Plasma Diagnostic Techniques" (R. H. Huddlestone and S. L. Leonard, eds.), pp. 201–264. Academic Press, New York.
McWhirter, R. W. P., and Hearn, A. G. (1963). *Proc. Phys. Soc.* **82**, 641–654.
Mansfield, H. E., Peacock, N. J., Smith, C. C., Hobby, M. G., and Cowan, R. D. (1978). *J. Phys. B* **11**, 1521–1544.
Marmar, E. S., Cecchi, J. L., and Cohen, S. A. (1975). *Rev. Sci. Instrum.* **46**, 1149–1154.
Marmar, E. S., Rice, J. E., and Allen, S. L. (1980). *Phys. Rev. Lett.* **45**, 2025–2028.
Mason, H. E., and Storey, P. J. (1980). *Mon. Not. R. Astron. Soc.* **191**, 631–639.
Mason, H. E., Doschek, G. A., Feldman, U., and Bhatia, A. K. (1979). *Astron. Astrophys.* **73**, 74–81.
Merts, A. L., Cowan, R. D., and Magee, N. H. (1976). *Los Alamos Sci. Lab.* [*Rep.*] *LA* **LA-6220MS**.
Merts, A. L., Mann, J. B., Robb, W. D., and Magee, N. H. (1980). *Los Alamos Sci. Lab.* [*Rep.*] *LA* **LA-8267MS**.
Mewe, R. (1972). *Astron. Astrophys.* **20**, 215–222.

5A. Diagnostics Based on Emission Spectra

Mewe, R., Schrijver, J., and Sylwester, J. (1980). *Astron. Astrophys.* **87,** 55–57.
Meyer, F. W., *et al.* (1979). *Phys. Rev. A* **19,** 515–525.
Meyers, B. R., and Levine, M. A. (1978). *Rev. Sci. Instrum.* **49,** 610–616.
Mihalas, D. (1978). "Stellar Atmospheres," Freeman, San Francisco, California.
Mohr, P. (1976). *In* "Beam Foil Spectroscopy" (I. A. Sellin and D. J. Pegg, eds.), Vol. 1, pp. 89–96. Plenum, New York.
Morgan, F. J., Gabriel, A. H., and Barton, M. J. (1968). *J. Sci. Instrum.* **1,** 998–1002.
Olson, R. E., and Salop, A. (1976). *Phys. Rev. A* **14,** 579–585.
Olson, R. W., and Salop, A. (1977). *Phys. Rev. A* **16,** 531–541.
Peacock, N. J. (1980). *In* "Physics of Ionised Gases" (M. Matic, ed.), pp. 687–714. Boris Kidric Inst. Nucl. Sci., Belgrade.
Peacock, N. J. (1981). *In* "Low Energy X-Ray Diagnostics" Attwood and Henke, eds.), AIP Conference Proceedings, No. 75, pp. 101–114. Am. Inst. Phys., New York.
Peacock, N. J., and Summers, H. P. (1978). *J. Phys. B* **11,** 3757–3774.
Peacock, N. J., Speer, R. J., and Hobby, M. G. (1969). *J. Phys. B* **2,** 798–810.
Peacock, N. J., Hobby, M. G., and Morgan, P. D. (1971). *Plasma Phys. Controlled Nucl. Fusion Res., Proc. Int. Conf., 4th, Madison, Wis.* **1,** 537–551.
Peacock, N. J., Hughes, M. H., Summers, H. P., Hobby, M. G., Mansfield, M. W. D., and Fielding, S. J. (1979). *Plasma Phys. Controlled Nucl. Fusion Res., Proc. Int. Conf., 7th, Innsbruck, 1978* **1,** 303–314.
Pengelly, R. M., and Seaton, M. J. (1964). *Mon. Not. R. Astron. Soc.* **127,** 165–175.
Platz, P., Ramette, J., Belin, E., Bonnelle, C., and Gabriel, A. (1980). *Assoc. Euratom-CEA Rep., Cent. Etud. Nucl., Fontenay aux Roses* **EUR-CEA-FC-1057.**
Puiatti, M. E., Breton, C., De Michelis, C., and Mattioli, M. (1981). *Assoc. Euratom-CEA Rep., Cent. Etud. Nucl., Fontenay aux Roses* **EUR-CEA-FC-1085.**
Ryufuku, H., and Watanabe, T. (1978). *Phys. Rev. A* **18,** 2005–2015.
Ryufuku, H., and Watanabe, T. (1979a). *Phys. Rev. A* **19,** 1538–1549.
Ryufuku, H., and Watanabe, T. (1979b). *Phys. Rev. A* **20,** 1828–1840.
Salop, A. (1979). *J. Phys. B* **12,** 919–928.
Saraph, H. E. (1972). *Comput. Phys. Commun.* **3,** 256–268.
Sauthoff, N. R., and Von Goeler, S. (1979). *IEEE Trans. Plasma Sci.* **PS-7**(3), 141–147.
Sawyer, G. A., Bearden, A. J., Henins, I., Jahoda, F. C., and Ribe, F. L. (1963). *Phys. Rev.* **131,** 1891–1987.
Shah, M. B., and Gilbody, H. B. (1981). *J. Phys. B* **14,** 2831–2841.
Stacey, W. M., Jr. (1981). "Fusion Plasma Analysis." Wiley, New York.
Stamp, M. F., Armour, I. A., Peacock, N. J., and Silver, J. D. (1981). *J. Phys. B* **14,** 3551–3561.
Stringer, T. E. (1979). *Phys. Plasmas Close Thermonucl. Conditions, Proc. Course, Varenna, Italy* **1,** 3–17. *Comm. Eur. Communities* **EUR FU BRU/XII/476/80.**
Suckewer, S. (1981). *Phys. Scr.* **23,** 72–86.
Suckewer, S., and Hawryluk, R. J. (1978). *Phys. Rev. Lett.* **40,** 1649–1651.
Suckewer, S., and Hinnov, E. (1977). *Nucl. Fusion* **17,** 945–953.
Suckewer, S., and Hinnov, E. (1978). *Phys. Rev. Lett.* **41,** 756–759.
Suckewer, S., Eubank, H. P., Goldston, R. J., Hinnov, E., and Sauthoff, N. R. (1979). *Phys. Rev. Lett.* **43,** 207–210.
Suckewer, S., Cecchi, J., Cohen, S., Fonck, R., and Hinnov, E. (1980a). *Princeton Plasma Phys. Lab., Rep.* **PPPL-1712.**
Suckewer, S., Fonck, R., and Hinnov, E. (1980b). *Phys. Rev. A* **21,** 924–927.
Suckewer, S., Hinnov, E., Bitter, M., Hulse, R., and Post, D. (1980c). *Phys. Rev. A* **22,** 725–731.
Summers, H. P. (1974). *Appleton Lab. Rep.* **JM-367**; reissued as **AL-R15** (1979).
Summers, H. P. (1977). *Mon. Not. R. Astron. Soc.* **178,** 101–122.
Summers, H. P., and McWhirter, R. W. P. (1979). *J. Phys. B* **12,** 2387–2412.

Terry, J. L., Chen, K. I., Moos, H. W., and Marmar, E. S. (1977a). *Johns Hopkins Univ., Dep. Phys., Rep.* **COO-2711-3**.
Terry, J. L., Marmar, E. S., Chen, K. L., and Moos, H. W. (1977b). *Phys. Rev. Lett.* **39**, 1615–1618.
TFR Group (1976). *Phys. Rev. Lett.* **36**, 1306.
TFR Group (1978a). *Assoc. Euratom-CEA Cent. Etud. Nucl., Fontenay aux Roses* **EUR-CEA-FC-947**.
TFR Group (1978b). *Plasma Phys.* **20**, 207–223.
TFR Group (1980). *Assoc. Euratom-CEA Cent. Etud. Nucl., Fontenay aux Roses* **EUR-CEA-FC-1033**; *Plasma Phys.* **22**, No. 8, 851–860.
TFR Group (1982). *Assoc. Euratom-CEA Cent. Etud. Nucl., Fontenay aux Roses* **EUR-CEA-FC-1160**; *Nucl. Fusion* **22**, No. 9, 1173–1189.
TFR Group, Dubau, J., and Loulergue, M. (1981). *J. Phys. B* **15**, 1007–1019.
Uchikawa, S., Griem, H. R., and Duchs, D. (1980). *Max-Planck-Inst. Plasmaphys. [Ber.] IPP* **IPP 6/199**.
Van Regemorter, H. (1962). *Astrophys. J.* **136**, 906–915.
von Goeler, S., et al. (1983). *Princeton Plasma Phys. Lab. Rep* **PPPL-1968**.
Wiese, W. L. (1978). *In* "Physics of Ionized Gases" (R. K. Janev, ed.), pp. 661–696. Inst. Phys., Belgrade.
Wilson, R. (1962). *J. Quant. Spectrosc. Radiat. Transfer* **2**, 467–475.
Younger, S. M., and Wiese, W. L. (1979). *J. Quant. Spectrosc. Radiat. Transfer* **22**, 161–170.
Zinov'ev, A. N., Korotkov, A. A., Krzhizhanovskii, E. F., Afrosimov, V. V., and Gordeev, Y. S. (1980). *JETP Lett. (Engl. Transl.)* **32**, 539–542.

5B
Laser Diagnostics

D. E. Evans

Euratom/UKAEA Fusion Association
Culham Laboratory
Abingdon, Oxfordshire
England

I.	Introduction	192
II.	Faraday Rotation	192
	A. Refraction and the Poincaré Sphere	192
	B. Characteristic Waves for Magnetized Plasma	194
	C. Faraday Effect as Rotation of Polarization	195
	D. Amplitude Ratio Method: Half-Shade Angle	196
	E. Polarization Modulation Method	197
	F. Amplitude-Independent Polarization Modulation	198
III.	Interferometry	199
	A. Magnetic-Field-Independent Phase Shift	199
	B. Phase Modulation: Zebra-Stripe Interferometer	200
	C. Phase Quadrature: Optical Method	202
	D. Phase Quadrature: Electrical Method	204
	E. Vibration-Compensated Quadrature Interferometer	204
	F. Phase-Comparison Double Interferometer	208
	G. Multiple-Beam Interferometry and Data Interpretation	210
IV.	Thomson Scattering	211
	A. Scattering by an Electron Gas: Scattering Form Factor	211
	B. Thomson Scattering as a Consequence of Refractivity	212
	C. Form Factor for a Field-Free, Warm, Thermal Plasma	214
	D. Impurities	216
	E. Ion Temperature in Fusion Research Tokamaks	217
	F. Hot Plasma Electron Temperature: Relativistic Treatment	221
	References	224

I. Introduction

All diagnostic methods based on transmitting electromagnetic radiation through plasma depend on a degree of coherence in the probing beam. The early availability of quasicoherent microwave sources made possible the extensive development of interferometry, Faraday rotation measurements, and turbulence scattering in the centimeter and millimeter wavebands. The coherence requirement is likewise responsible for the failure to extend these techniques to shorter wavelengths by the use of thermal sources. Indeed, the need to increase the range of plasmas accessible to investigation by reducing the probing wavelength could be met only when lasers became obtainable after 1960.

The appropriate way to describe the propagation of a beam of coherent radiation is by Gaussian optics (Siegman, 1971), and a complete account of its interaction with plasma, including scattering by the electrons, is given in terms of phase shifts by refraction theory.

Plasma with magnetic field behaves as a birefringent, optically active medium, propagation perpendicular to the field being purely birefringent, propagation parallel to it, purely optically active. It can be shown (Ramachandran and Ramaseshan, 1961) that for any direction of propagation there exist two characteristic waves with different phase velocities whose state of polarization remains unchanged during propagation, and the characteristic polarizations are orthogonal. A beam propagating in arbitrary direction can be resolved into two characteristic components, which will travel at different phase velocities determined by the corresponding characteristic refractive indexes, then recombine to produce an altered polarization and phase relative to a second beam that traversed the same distance of empty space. This is the basis of the Faraday rotation method for measuring magnetic field, and of interferometry designed to measure plasma electron density. It is also, though less obviously, the process underlying Thomson scattering. The present subchapter is devoted to a short account of these three topics.

II. Faraday Rotation

A. *Refraction and the Poincaré Sphere*

Faraday rotation can conveniently be discussed with reference to the Poincaré unit sphere (Ramachandran and Ramaseshan, 1961; Born and Wolf, 1964), points on whose surface can be made to correspond to all possible states of optical polarization in the following way. If the polarization of light propagating in the $+z$ direction is specified by its ellipticity $\chi = \tan^{-1} b/a$, and by the angle ψ between the ellipse major axis and the x-

5B. Laser Diagnostics

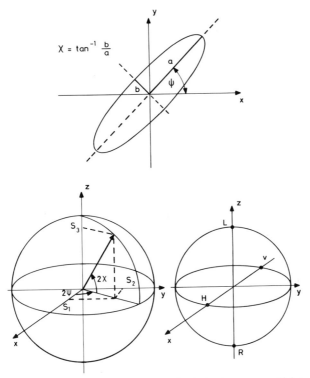

Fig. 1. Vibrational ellipse for electric vector of polarized wave, specifying ψ and χ, from which the Stokes parameters S_1, S_2, and S_3 are calculated, defining a point on the Poincaré sphere. H, horizontal; V, vertical; L and R, left- and right-hand circularly polarized.

coordinate axis, as shown in Fig. 1, then (χ, ψ) defines a unique point having Cartesian coordinates (Stokes parameters)

$$S_x = \cos 2\chi \cos 2\psi,$$
$$S_y = \cos 2\chi \sin 2\psi, \qquad (1)$$
$$S_z = \sin 2\chi$$

on the surface of the unit sphere. It will be seen that orthogonal polarizations occupy diametrically opposite points on the sphere, with plane-polarized light at the equator in the x–y plane, and left- and right-handed circularly polarized light at the north ($+z$) and south ($-z$) poles, respectively. A line of longitude is the locus of all ellipses at the angle ψ, and a line of latitude that of all ellipses having a fixed shape (χ) but all orientations.

Being orthogonal, the two characteristic waves referred to in the Introduction define an axis through the sphere's center, and it can be shown that evolution of polarization during propagation can be represented on the Poincaré sphere by a rotation about this axis equal to a phase difference $\varphi = (\omega/c)L(\mu_1 - \mu_2)$, where ω is the optical angular frequency, c the velocity of light, L the geometric distance of propagation, and μ_1 and μ_2 are the characteristic refractive indexes. For a homogeneous medium, if **n** is unit vector in the direction of this axis of rotation, and **u**, also a unit vector, defines the polarization of the incident light, the emergent polarization will be given by the vector

$$\mathbf{u}' = \mathbf{u} \cos \varphi + \mathbf{u} \cdot \mathbf{n}(1 - \cos \varphi)\mathbf{n} + (\mathbf{n} \times \mathbf{u}) \sin \varphi. \qquad (2)$$

B. Characteristic Waves for Magnetized Plasma

Assuming that the magnetic field **B** lies in the y–z plane at an angle θ to the direction of radiation propagation ($+z$), it can be shown that the characteristic waves are elliptically polarized with axes along the x and y coordinate axes, as shown in Fig. 2.

DeMarco and Segre (1972) introduce the parameter F, where

$$\frac{1}{F} = \frac{\Omega_c}{2(1 - \Omega_p^2)} \frac{\sin^2 \theta}{\cos \theta}, \qquad (3)$$

allowing one to write the Appleton–Hartree equation (Heald and Wharton, 1965) for the characteristic refractive indexes μ as

$$1 - \mu^2 = \frac{\Omega_p^2}{1 \pm (\Omega_{c\parallel}/F)[(1 + F^2)^{1/2} \mp 1]}. \qquad (4)$$

Here $\Omega_p \equiv \omega_{pe}/\omega$ and $\Omega_{c\parallel} \equiv (\omega_{ce}/\omega) \cos \theta$, ω_{pe} and ω_{ce} being the electron plasma frequency and the electron gyrofrequency, respectively.

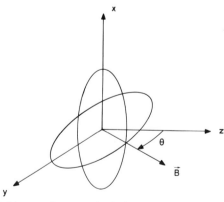

Fig. 2. Characteristic waves in magnetized plasma. Radiation propagates in $+z$ direction. Magnetic field **B** lies in y–z plane at angle θ to $+z$ axis. Vector indicated by bold character in text and by overarrow in figure.

The wave polarization is (Heald and Wharton, 1965)

$$\tan \chi \equiv \frac{E_x}{E_y} = i \frac{\Omega_p^2 \mp (1 - \mu^2)}{(1 - \mu^2)\Omega_{c\|}} = i \frac{(1 + F^2)^{1/2} \mp 1}{F}. \quad (5)$$

That this turns out to be pure imaginary means that the phase difference between the x and y components is $\frac{1}{2}\pi$, and the major axis of the ellipse lies in the x direction (Born and Wolf, 1964, p. 29). The coordinates of the characteristic polarizations on the Poincaré sphere are then determined by

$$\sin 2\chi = \mp \frac{F}{(1 + F^2)^{1/2}}, \quad \cos 2\chi = \frac{1}{(1 + F^2)^{1/2}}, \quad (6)$$

and $\psi = \frac{1}{2}\pi$ or 0.

This has the general meaning that radiation passing through homogeneous magnetized plasma will alter its polarization both in ellipticity and in angular attitude of the plane of polarization. However, it can be seen that if $F \gg 1$, the characteristic waves are almost right- and left-hand circularly polarized, and then the effect of the transit on polarization, represented by rotation about the z axis, is to leave the ellipticity unchanged while rotating the angle of the ellipse. In particular, radiation initially plane polarized experiences only a rotation of the plane of polarization, and this is popularly called Faraday rotation.

C. Faraday Effect as Rotation of Polarization

For $F \gg 1$ Eq. (4) gives the indexes of refraction of the characteristic waves as

$$1 - \mu^2 \approx \Omega_p^2 (1 \pm \Omega_{c\|})^{-1},$$

and the angle ψ/L through which the plane of polarization rotates per unit path length through the plasma becomes

$$\frac{\psi}{L} = \frac{\varphi}{2L} = \frac{1}{2}\frac{\omega}{c}(\mu_1 - \mu_2) = \frac{1}{2}\frac{\omega}{c}\frac{\Omega_p^2 \Omega_{c\|}}{1 - \Omega_{c\|}}$$

$$\approx \frac{1}{2}\frac{\omega}{c}\Omega_p^2 \Omega_{c\|} = 2.64 \times 10^{-13} \lambda^2 n_e B,$$

provided $\Omega_{c\|}^2 \ll 1$, λ is in centimeters, n_e is per cubic centimeter, and B is in tesla.

It was pointed out by Wood (1967) that in the $F \gg 1$ approximation the effect of magnetic field on the dispersion curve, μ as a function of ω, is merely to shift its origin by an amount $\pm \frac{1}{2}\omega_{ce\|}$, the Larmor frequency, but to leave its shape unchanged.

It cannot be taken for granted that the $F \gg 1$ condition, necessary for the foregoing simplification, will inevitably be met. For example, if a CD_3F laser

(λ = 1.22 mm) is being used to measure Faraday rotation in a plasma with B = 3 T, F is already smaller than unity when θ is greater than 80° and an incident plane-polarized wave will emerge with an elliptical component perpendicular to the incident plane of polarization which will be confused with the pure angular rotation of this plane.

To ensure the validity of the plane polarization approximation, the wavelength must be kept as small as possible, even though the Faraday rotation angle one seeks to measure is proportional to λ^2. Suppressing ellipticity has generally been held to be more important than a large rotation angle, however, and the measurement of small rotations of the plane of polarization has accordingly been the subject of considerable experimental ingenuity. Two distinct approaches are in use. One relies on ratios of amplitudes, the other on phase shifts.

D. Amplitude Ratio Method: Half-Shade Angle

A general description of the amplitude ratio method has been given by Falconer and Ramsden (1968). After traversing the plasma, the plane-polarized probe beam of intensity I_0 is divided into two equal parts by a polarization-sensitive beam splitter. Each part meets a polarizer oriented $\frac{1}{2}\varepsilon$ rad away from perpendicular to the initial polarization, as shown in Fig. 3. From polarimetry (Heller, 1960) $\frac{1}{2}\varepsilon$ is termed the half-shade angle.

If the plane of polarization has experienced no rotation, the signal in each of the two channels will be $\frac{1}{2}I_0 \sin^2 \frac{1}{2}\varepsilon$, but a rotation θ will give rise to signals $I_\pm(\theta) = \frac{1}{2}I_0 \sin^2 (\frac{1}{2}\varepsilon \pm \theta)$, and the ratio

$$R = \frac{I_+ - I_-}{I_+ + I_-} = \frac{\sin \varepsilon \sin 2\theta}{1 - \cos \varepsilon \cos 2\theta}$$

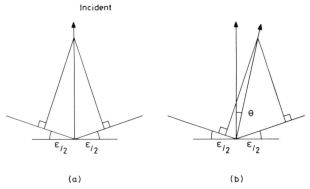

Fig. 3. Half-shade method for Faraday rotation measurement. Analyzers are set $\frac{1}{2}\varepsilon$ rad away from perpendicular (a) no rotation, (b) rotation through small angle θ.

5B. Laser Diagnostics

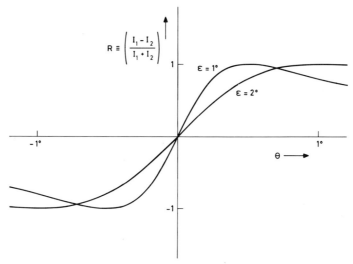

Fig. 4. Analysis of half-shade angle method. I_1 and I_2 are intensities transmitted through the two analyzers, set at angle $\frac{1}{2}\varepsilon$. Angle θ is rotation undergone by probe beam polarization.

which is independent of the incident power I_0, will have a value between ± 1. The sensitivity of this arrangement to small rotations is determined by the slope of the curve R as a function of θ near $\theta = 0$, viz.,

$$\left.\frac{dR}{d\theta}\right|_{\theta=0} = \frac{2 \sin \varepsilon}{1 - \cos \varepsilon}.$$

One configuration sometimes adopted (Dougal *et al.*, 1964; Brown *et al.*, 1977) is to set the analyzers at $\pm 45°$ to the incident plane of polarization, in which case R becomes simply $\sin 2\theta$ and $dR/d\theta|_{\theta=0} = 2$.

A more sensitive configuration is one in which $\frac{1}{2}\varepsilon$ is made less than $45°$. When ε is small, the sensitivity of the ratio R to small changes in rotation becomes $dR/d\theta|_{\theta=0} = 4/\varepsilon$. Figure 4, displaying R as a function of rotation angle θ, shows that a choice of $\frac{1}{2}\varepsilon$ smaller than the expected value of θ would be foolish, and that the best choice of $\frac{1}{2}\varepsilon$ would be about twice the maximum rotation to be measured.

The minimum rotation θ that can be measured by this method is set ultimately by detector noise, but window birefringence, imperfections in the analyzers, and simply the range of angles at which radiation impinges on the analyzers may all be important limitations.

E. Polarization Modulation Method

A quite different approach to small-angle Faraday rotation measurement, involving modulating the polarization angle of the incident radiation, was suggested by Kunz and Dodel (1978). The question of measuring poloidal field and its radial distribution in tokamak plasma by Faraday rotation has been extensively discussed (DeMarco and Segre, 1972; Craig, 1976; Segre, 1978) and was successfully performed for the first time using this technique in the TFR plasma by Kunz and Equipe TFR (1978).

A disk of material transparent at the appropriate probe wavelength and with a reasonable Verdet constant is placed within a coil and located so that the plane-polarized probe beam passes through it prior to encountering the plasma. For 337-μm radiation from an HCN laser, ferrite (Frayne, 1968; Birch and Jones, 1970) has been used, though a disk of suitable thickness absorbs about 80% of the incident power. A radio-frequency current in the coil causes the plane of polarization to oscillate with an angular amplitude α of a few degrees. A detector, viewing the result through an analyzer crossed with the initial polarization, produces a signal proportional to $I_0 \sin^2 (\alpha \sin \omega t) \approx \frac{1}{2} I_0 \alpha^2 (1 - \cos 2\omega t)$ for small α. Magnetized plasma in the path of the probe beam induces an additional rotation θ, resulting in a detector signal $I_0 \sin^2 (\theta + \alpha \sin \omega t)$, which is approximately $I_0[\theta^2 + \frac{1}{2}\alpha^2(1 - \cos 2\omega t) + 2\alpha\theta \sin \omega t]$ for small α and θ. Comparison of this with the plasma-free expression above shows that the plasma contributes a term at the modulation frequency ω: $I_\omega = I_0 2\alpha\theta \sin \omega t$, and this can be efficiently extracted from the detector output by the use of a lock-in amplifier referred to the radio-frequency oscillator.

Assuming their experimental performance on TFR to be limited by detector NEP, Kunz and Equipe TFR (1978) expected a minimum measurable rotation of 0.6 mrad for a modulation frequency of 100 Hz and a GaAs detector. In practice they realized a minimum of about 1 mrad and ascribed the discrepancy to amplifier noise.

F. Amplitude-Independent Polarization Modulation

Though very different from the half-shade angle technique, the method just described remains amplitude dependent, and its success depends on the constancy of the probe beam intensity. Any component of probe beam vibration at the modulation frequency will be wrongly attributed to plasma Faraday effect. Dodel and Kunz (1978) have suggested a modification that altogether removes the dependence on intensity, and converts the measurement to one of phase shift. The significant step is the replacement of the small amplitude oscillation imposed on the plane of polarization by a complete rotation through 2π. This is readily produced by superimposing right-hand and left-hand circularly polarized beams having a frequency difference $\Delta\omega$. The result is plane-polarized radiation whose plane of polarization rotates at the frequency $\frac{1}{2}\Delta\omega$. Such a beam, viewed through an analyzer by a square-law detector, induces a signal component $\cos \Delta\omega t$. Plasma Faraday rotation appears as a phase shift resulting in cos ($\Delta\omega t + \theta$). Provided the frequency of rotation $\Delta\omega$ is chosen so that the condition $|(2\pi/\Delta\omega) d\theta/dt| < \pi$, θ can be deduced from the time shift between zero crossings of the signal with plasma relative to the signal without, in strict analogy with, and retaining all the advantages of the procedure introduced by Veron (1974) for interferometry.

Frequency shifts can be generated by the rotating grating method used by Veron, or by the detuning of a pair of optically pumped lasers, as demonstrated by Wolfe et al. (1976).

III. Interferometry

A. Magnetic-Field-Independent Phase Shift

Interferometry is performed in laboratory plasmas to measure the electron density distribution, and is accordingly concerned with plasma refraction relative to free space, rather than the difference between characteristic wave refractive indexes as in Faraday rotation. Once again the choice of probe wavelength involves conflicting requirements. An interferometer measures phase difference $\Delta\varphi = 2\pi\lambda^{-1}L(1 - \mu)$, and the use of short wavelength means μ will be well approximated by $1 - \frac{1}{2}\Omega_p^2$, leading to the simple dependence $\Delta\varphi = r_e\lambda L n_e$, r_e being the classical electron radius, n_e the electron density, and L the path length. But short wavelength also implies few fringes to measure.

On the other hand, while generating relatively many fringes, long wavelength entails the danger of gross refraction by density gradients, culminating in total opacity as the plasma frequency is approached. The choice of long wavelength can also invalidate convenient approximations for μ. Inspection of the Appleton–Hartree formula [Eq. (4)] shows, however, that this can be avoided even at long wavelength provided the probe beam can be propagated as an ordinary mode ($\mathbf{E}\|\mathbf{B}$) perpendicular to the magnetic field $\mathbf{B}$, for then $\mu^2 = \mu_0^2 = 1 - \Omega_p^2$ without approximation. In a tokamak where $\mathbf{B}$ is nearly toroidal and interferometer beams are confined to a minor cross section, these conditions can readily be met.

Interferometers for studying magnetically confined plasmas have usually been assembled in the Mach–Zehnder or the Michelson configuration (Alpher and White, 1965; Jahoda and Sawyer, 1971). A typical primitive layout would consist of a pair of beam splitters and a pair of mirrors as in Fig. 5. The first beam splitter divides a beam of coherent radiation from a laser into two channels, one of which passes through the plasma under investigation, acquiring a phase shift $\varphi(t)$ relative to the other in the process. They then recombine on the second beam splitter and are finally directed onto a square-law detector giving rise to detection current $i \propto A \cos \varphi(t)$.

Extracting $\varphi(t)$ and hence the time history of the density from this detec-

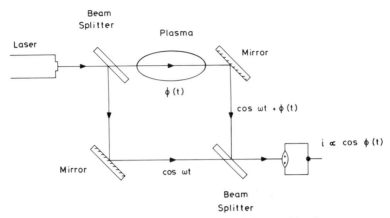

Fig. 5. Primitive Mach–Zehnder interferometer. Probe beats with reference on square-law detector to produce current proportional to cosine of phase difference introduced in probe by plasma.

tor current is complicated by the unwanted dependence on the amplitude A, which may itself be varying with time. It is further complicated by the ambiguity of the cosine function regarding the sign of its argument, which makes it impossible to distinguish increasing from decreasing φ, or to determine when a change in the sign of φ has occurred. Such an instrument will also be subject to spurious phase shifts due to vibration-induced variation in the optical path length. Remedies for the former failing have been sought in different forms of phase modulation technique, while vibration compensation arrangements have been invented to overcome the latter.

B. *Phase Modulation: Zebra-Stripe Interferometer*

An elegant solution to these problems was demonstrated by Gibson and Reid (1964), who referred to their phase modulation technique as "zebra striping" because of the appearance of the presentation on an oscilloscope. The scale of the problem of interpreting raw interference fringes can be judged from their confused appearance during a plasma discharge, shown in Fig. 6. By contrast phase shift in units of $\frac{1}{2}\lambda$ can be read directly from the accompanying zebra-stripe display. This gives an unambiguous indication of reversals in the changing optical path length, is entirely independent of fringe amplitude, and even lends itself to correction for vibration.

As applied to the three-mirror interferometer of Ashby and Jephcott (1963), the method works by imposing controlled changes in the optical path length in the arm of the interferometer that spans the plasma. The mirror on that arm is caused to oscillate and the wave form driving the oscillator is shown on an oscilloscope. The interference fringes generated by the moving mirror are not themselves displayed, but instead are used to modulate the oscilloscope brightness, one brightup of the beam corresponding to each fringe maximum. The result is that the waveform appears discontinuous as though traced by a line of bright dashes, and because the fringes are caused by the waveform, the mth fringe always produces its brightup at the same value of y displacement. When the time base is run very slowly so as to show many cycles of the waveform on the screen at once, the bright dashes belonging to the mth fringe coalesce to form a horizontal stripe. A set of successive fringes thus appears on the oscilloscope screen as a pattern of horizontal stripes whose spacing corresponds to the change in optical path length, $\frac{1}{2}\lambda$ in the Ashby–Jephcott arrangement, needed to go from one interference maximum to the next. This can be seen in Fig. 6.

Put quantitatively, the condition for an interference maximum or oscilloscope brightup is that the total optical path, consisting of the geometrical part L, the driven oscillation part $l_o(t)$, and the plasma contribution $l_p(t)$, be an integer multiple of $\frac{1}{2}\lambda$:

$$L + l_o(t) + l_p(t) = m(\tfrac{1}{2}\lambda).$$

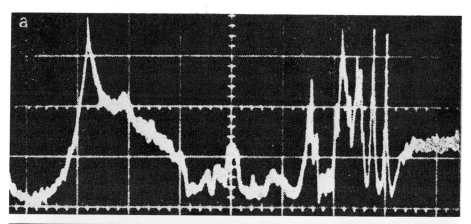

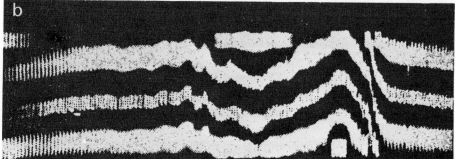

Fig. 6. Raw interferometer output (a) and zebra-stripe display (b) of a similar plasma discharge. [From Gibson and Reid (1964).]

Now the oscilloscope displays the oscillation waveform

$$y = \text{const} \times l_0(t),$$

which becomes $y = \text{const} \times [m(\frac{1}{2}\lambda) - l_p(t) - L]$ for the locus of a fringe or a brightup stripe. Here L is simply a constant displacement and may be tuned out and ignored.

Without plasma, the brightup locus is $y = \text{const} \times m(\frac{1}{2}\lambda)$, a set of horizontal straight stripes whose spacing is proportional to $\frac{1}{2}\lambda$. With plasma, the displacement in the y direction of each stripe follows faithfully the phase development of the plasma in time, and the stripe spacing constitutes a calibrated scale against which the displacement can be measured in units of $\frac{1}{2}\lambda$. The maximum rate of change of phase that can be followed is one fringe per oscillation cycle of the mirror.

Sensitivity is determined by the smallest fraction of a fringe that can be detected. It could be increased by the ratio of the wavelengths in a two-laser modification of the system. A short wavelength would generate the calibration scale, and a long wavelength, traversing the same arrangement of mirrors, would produce the stripe whose displacement measures plasma phase shift.

Furthermore, a two-wavelength scheme permits spurious vibration effects to be compensated. The expressions for the λ_1 stripe and the λ_2 stripe have as a common additive component the vibration-induced change in path length. The difference between such a pair of stripes is therefore vibration independent.

C. Phase Quadrature: Optical Method

Another scheme that approaches the ambiguity problem in a different way is the quadrature interferometer, employing two signals out of phase with each other by 90°. Just as in the simple instrument the detector signal is $S = A \cos \varphi$, in the quadrature device, the two signals are $S_1 = A \cos \varphi$ and $S_2 = A \sin \varphi$, whose ratio $S_2/S_1 = \tan \varphi$ is independent of the amplitude A. Unlike the simple system, therefore, this one is free from spurious influences such as refraction-induced varying fringe contrast. Versions of the quadrature interferometer described by Heym (1968) and by Berger and Lovberg (1970) derived their two signals from separate portions of the beam aperture and could accordingly have been subject to misalignment uncertainty. Buchenauer and Jacobson (1977) avoid this danger by separating polarization components of the beam to obtain their signal pair.

Figure 7a diagrams the modified Mach–Zehnder arrangement proposed by Buchenauer and Jacobson (1977). The laser beam is initially plane polarized, but a quarter-wave plate in the reference arm converts this to circular polarization. Following the second beam splitter, a Wollaston prism ana-

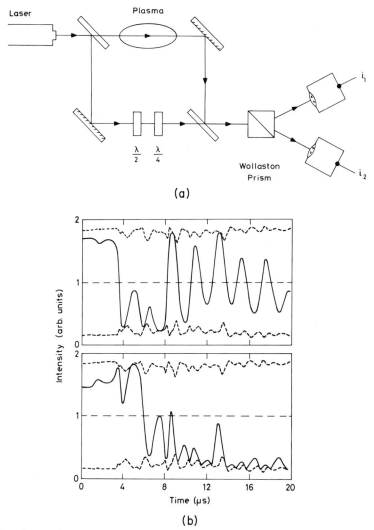

Fig. 7. (a) Phase quadrature arrangement, after Buchenauer and Jacobson (1977). Retardation plate ($\frac{1}{4}\lambda$) circularly polarizes reference beam as it enters the Wollaston prism. Probe remains plane polarized. The $\frac{1}{2}\lambda$ plate compensates residual ellipticity of reference beam induced by beam splitter and mirror. (b) Signals in quadrature produced by the pair of detectors above. Short dashes are calculated fringe envelope extrema. [From Buchenauer and Jacobson (1977).]

lyzes the circularly polarized reference into two mutually perpendicular plane-polarized parts 90° out of phase at the same time as it divides the plane-polarized probe beam into two plane-polarized but in-phase components. The four resulting beams are mixed in reference-plus-probe pairs on independent detectors, producing the signals illustrated in Figure 7b. It can be seen that whenever one signal crosses its zero line the other is at an extremum, and in general the amplitude $A = (S_1^2 + S_2^2)^{1/2}$ can be measured continuously throughout the plasma discharge, in contrast to what is possible in a simple interferometer. Moreover, the sign of $d\varphi/dt$ is at no point uncertain.

Finally it may be remarked that in the conventional instrument the uncertainty $\Delta\varphi$ in the determination of the phase shift is itself a function of φ, viz.,

$$\Delta\varphi = \Delta S/(A \sin \varphi),$$

and even diverges when φ is an integer multiple of π. In quadrature, by contrast, $\Delta\varphi_{\text{rms}} = \Delta S/A$, and is accordingly the same for all values of phase shift.

D. Phase Quadrature: Electrical Method

The quadrature technique is a powerful one and its extension to all probe beam wavelengths is desirable. But division of a beam of radiation into a pair of beams mutually orthogonal in phase can be accomplished optically only at wavelengths at which appropriate birefringent elements such as $\frac{1}{4}\lambda$ plates and polarizing prisms are readily available. Universal applicability can be approached if the quadrature is performed electrically rather than optically, and this can be done provided an intermediate frequency is introduced at which electronics can operate with comfort.

Baker and Lee (1978) obtained their intermediate frequency by Doppler-shifting their probe beam frequency with a translating mirror on the Michelson arrangement on Doublet III. Another more versatile technique is to use Bragg diffraction by sound waves driven in some suitably transparent medium to induce a convenient frequency shift (Jacobson and Call, 1978; Hugenholtz and Meddens, 1979). The probe beam, carrying the plasma-induced phase shift $\cos(\omega t + \varphi)$, is mixed on the detector in the usual way, with the frequency-shifted reference beam $\cos(\omega + \omega_B)t$ to generate a detector current only one component of which, $i_d \propto \cos(\omega_B t + \varphi)$ is at a frequency low enough to cause the electronics to respond. A commercially available quadrature phase comparator can then be used to divide this current into two channels, one of which is subject to a $\frac{1}{2}\pi$ phase shift relative to the other, before each is mixed with the Bragg cell driver waveform $\cos \varphi_B t$. This yields currents in quadrature $i_1 = A \cos \varphi$ and $i_2 = A \sin \varphi$, in direct analogy with the optical quadrature signals of the Buchenauer and Jacobson apparatus.

E. Vibration-Compensated Quadrature Interferometer

Gowers and Lamb (1982) have used the principle just discussed to build a CO_2 laser quadrature interferometer for use in reverse field pinch plasmas, in which the intermediate frequency is 40 MHz, and incorporated a subsidiary common-path HeNe laser interferometer for vibration compensation.

The experimental layout is essentially a Michelson configuration, as shown in Fig. 8. The 5-W 10.6-μm Gaussian laser beam is divided in two by the Ge acousto-optic modulator (Bragg cell), and the undeviated zeroth order traverses the plasma twice before combining with the first-order frequency-shifted reference beam on a liquid-nitrogen-cooled 100-MHz-bandwidth CdHgTe infrared detector.

By coaxially aligning the zeroth- and first-order HeNe laser beams diffracted in a dense flint glass Bragg cell with the two 10.6-μm beams, and by using optical components compatible with both visible and IR wavelengths, i.e., ZnSe for beam splitters and Au-coated mirrors, a second simultaneously operating interferometer at 0.63 μm wavelength was formed. Using almost the same optical path for both visible and CO_2 interferometers allowed vibration occurring almost anywhere in the interferometer assembly to be compensated, since the shorter wavelength is relatively much less sensitive to plasma phase shifts, and much more sensitive to vibration than the longer wavelength. Figure 9 shows the quadrature waveforms obtained from the 10.6-μm system and also those from the visible, and Fig. 10a shows

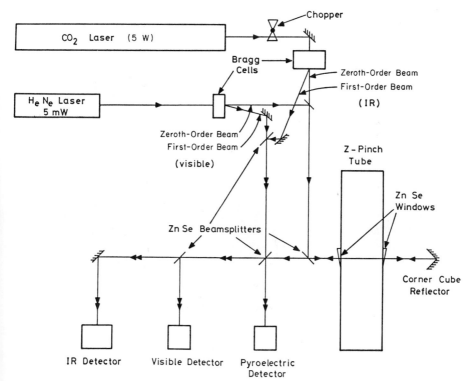

Fig. 8. Vibration-compensated quadrature interferometer of Gowers and Lamb (1982).

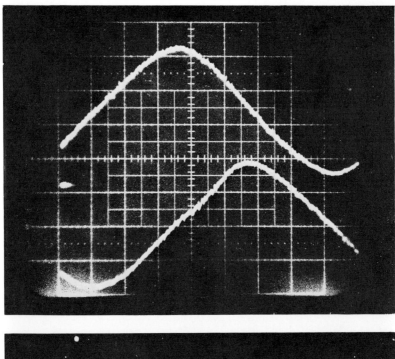

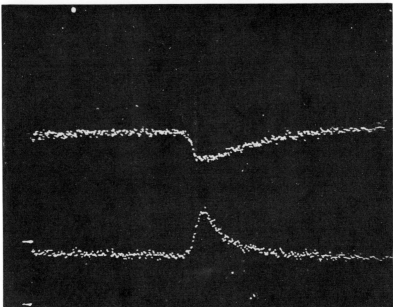

Fig. 9. Quadrature waveforms $\cos(\varphi_{pl} + \varphi_{vib})$ and $\sin(\varphi_{pl} + \varphi_{vib})$ for 10.6- and 0.63-μm interferometers. (a) HeNe laser interferometer, (b) CO_2 laser interferometer; both: 5 μs div^{-1}, 0.2 V div^{-1}. [From Gowers and Lamb (1982).]

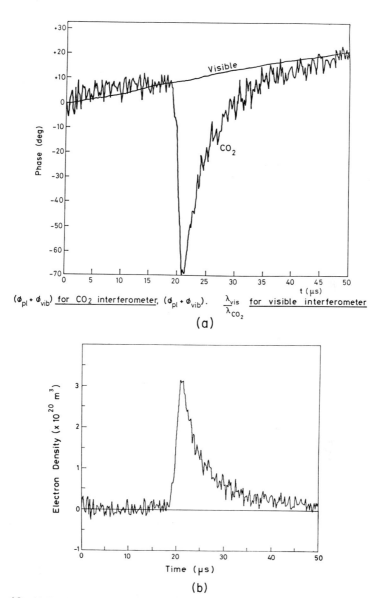

Fig. 10. (a) Total phase change ($\varphi_{pl} + \varphi_{vib}$) in degrees for 10.6 μm and 0.63/10.6 ($\varphi_{pl} + \varphi_{vib}$) for 0.63 μm. [From Gowers and Lamb (1982). Copyright 1982, The Institute of Physics.] (b) Phase change due to plasma alone given by the difference between the phase changes above. The result is converted to line electron density $\bar{n}_e = 1/L \int n_e(x)\,dx$.

the total phase change computed from both systems. Finally, Fig. 10b shows the result of subtracting the corrected visible phase change attributed to vibrations from the total 10.6-μm phase change, leaving the plasma phase change part alone. This has been converted, in the figure, to electron density integrated along the line of sight.

F. Phase-Comparison Double Interferometer

Double interferometers utilizing an intermediate frequency but replacing quadrature by a phase comparison technique have been used on TFR by Veron (1974, 1979) and Veron et al. (1977), and on Alcator by Wolfe et al. (1976). These systems, shown schematically in Fig. 11, differ principally in the means adopted to generate the shifted frequency.

Veron utilizes Doppler shift obtained by reflecting a beam in the subsidiary interferometer from a rotating cylindrical blazed grating. The change in frequency is $\Delta\omega = kv$, and for reflection at nearly 180° at an angle β to the radius r of the cylinder (Fig. 11a)

$$\Delta\omega = (4\pi/\lambda)v \sin \beta = 8\pi^2 \lambda^{-1} rN \sin \beta,$$

where N is in revolutions per second. With $\beta = 54°$, $\lambda = 337$ μm (HCN

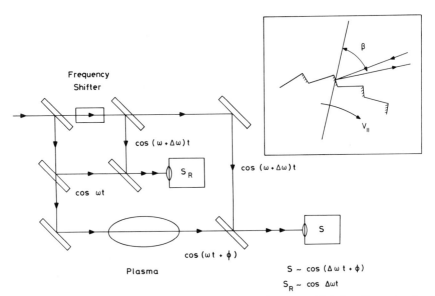

Fig. 11. Schematic of phase-comparison double interferometer. Main interferometer signal S is compared in phase with reference signal S_R. (a) Frequency shifter in the Veron (1974) version is a rotating cylindrical grating giving shift $\Delta\omega = (4\pi/\lambda)v \sin \beta$.

5B. Laser Diagnostics

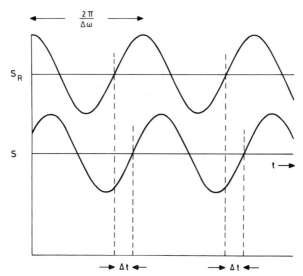

Fig. 12. Analysis of the phase-comparison interferometer signals $S \sim \cos(\Delta\omega t + \varphi)$ and $S_R \sim \cos \Delta\omega t$, where $\Delta\omega$ is the frequency shift. Time elapsed between a positive-slope zero crossing of S_R and the next positive-slope zero crossing of S is Δt. The plasma-induced phase shift $\varphi = \Delta t \Delta\omega$. [From Veron (1979).]

laser), $r = 6$ cm, and $N = 5$ rps (Veron *et al.*, 1977) this gives shifts of about 10 kHz.

The two detectors produce signals $S_R \approx \cos \Delta\omega t$ and $S \sim \cos(\Delta\omega t + \varphi)$, and the elapsed time Δt between a positive-slope zero crossing of S_R and the next positive-slope zero crossing of S determines the plasma-induced phase difference through $\varphi = \Delta t \Delta\omega$, as shown in Fig. 12. When the phase difference between S and S_R passes 2π, the elapsed time reaches a maximum equal to the period $2\pi(\Delta\omega)^{-1}$ of the intermediate frequency oscillation, then drops to zero and starts again. If one plots elapsed time as a function of real time during which the excursion of φ exceeds 2π, the display looks like the sketch in Fig. 13. Maximum sensitivity of Veron's instrument is limited by detector noise to about 0.01 fringe.

If plasma conditions are to remain comparatively static during one period of the intermediate frequency, the latter must satisfy $\Delta\omega \gg (2\pi/n_e) \, dn_e/dt$. As indicated above, the rotating grating offers typical frequency shifts of about 10 kHz. To procure the substantially larger frequencies near 1 MHz needed to measure much more rapidly changing plasma, Wolfe and his colleagues at MIT used a pair of 1-m optically pumped 119-μm methyl alcohol lasers. Controlling their cavity length difference to better than 1 μm by a feedback mechanism, this group succeeded in stabilizing their difference frequency to about ±2%.

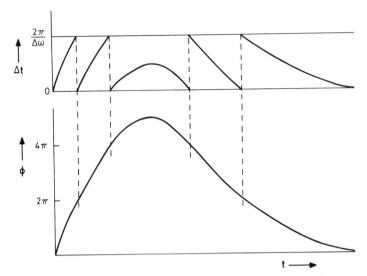

Fig. 13. Display from phase-comparison interferometer. Lower curve shows actual phase shift $\varphi(t)$ and upper, the interferometer display. [From Magyar (1981).]

G. Multiple-Beam Interferometry and Data Interpretation

Multiple-beam interferometry such as that performed by Veron (1979) on TFR produces a set of line integrals from which one can recover the two-dimensional distribution whose projections along the probe beams are the data. For an array of parallel beams and a circularly symmetric density distribution, the required inversion is the Abel transform (Bockasten, 1961; Bracewell, 1965). Distributions having higher-order symmetry or no symmetry at all, such as the quasi-D shapes predicted for JET, call for more advanced methods (Sauthoff and von Goeler, 1979).

Williamson and Evans (1982) have discussed this question in terms of conventional matrix inversion. They consider optimizing the location of the probe beams by maximizing their individual freedom from redundancy, and show that an asymmetric arrangement always leads to better reconstruction accuracy than a regular array. They introduce smoothing, and deal with its effect on resolution, channel redundancy, and reconstruction accuracy, and they argue that smoothing should be increased until the natural negative statistical correlations between adjacent picture elements of the two-dimensional source just vanish. They also demonstrate that source function shape can be identified rather sensitively with as few as ten channels provided it conforms to a set which is a member of an already known family.

IV. Thomson Scattering

A. *Scattering by an Electron Gas: Scattering Form Factor*

Laser radiation scattered by the free electrons in the plasma conveys spatially resolved information about electron and ion temperatures, impurities, magnetic field, and microturbulence. Thomson scattering has been the subject of numerous reviews (see, e.g., Evans and Katzenstein, 1969; DeSilva and Goldenbaum, 1971; Evans, 1974, 1976; Segre, 1975; Sheffield, 1975; Luhmann, 1979; Magyar, 1981). Bernstein *et al.* (1964) gives a thorough account of the plasma theory involved.

Every charged particle in the plasma scatters radiation, but because the scattering cross section is inversely proportional to particle mass squared, the contribution from the ions is negligible except insofar as they influence the motion of the electrons. Radiation scattering by individual electrons is dipole, and where φ is the angle between the incident radiation electric vector and the direction of the scattered ray, the differential cross section is

$$\frac{d\sigma}{d\Omega} = r_e^2 \sin^2 \varphi.$$

Integration over solid angle $d\Omega = 2\pi \sin \varphi \, d\varphi$ results in $\sigma = \frac{8}{3}\pi r_e^2 = 6.65 \times 10^{-25}$ cm^2.

The electric field radiated to a detector by an electron undergoing acceleration in an incident laser beam is calculated using the Liénard–Wiechert potentials of classical electromagnetic theory. Calculation of the phase factor of the field using the retarded time and the approximation of rectilinear motion over very short time periods leads to the emergence of the differential scattering vector $\mathbf{k} = \mathbf{k}_s - \mathbf{k}_0$, and the Doppler-shifted frequency $\Delta\omega = \mathbf{k} \cdot \mathbf{v}$, where $\mathbf{k}_0$ and $\mathbf{k}_s$ are the wave vectors of the incident and the scattered radiation, and $\mathbf{v}$ is the instantaneous electron velocity. The power spectrum of the scattered Poynting flux is found by taking the Fourier transform of the autocorrelation of the field at the detector (Wiener–Khinchine theorem), which results in

$$I = I_0 r_e^2 \sin^2 \varphi \, n_e l S(\mathbf{k}, \omega). \tag{7}$$

Here, I_0 is the incident laser power (watts), n_e the electron density, and l the length in the plasma over which scattering occurs.

The factor $S(\mathbf{k}, \omega)$ is given by

$$S(\mathbf{k}, \omega) = \frac{1}{2\pi n_e V} \int_{-\infty}^{+\infty} \exp[i(\omega - \omega_0)\tau] \langle N^*(\mathbf{k}, t) N(\mathbf{k}, t + \tau) \rangle \, d\tau, \tag{8}$$

where V is the scattering volume and $N(\mathbf{k}, t)$ is the Fourier transform of the point density of the electrons in that volume,

$$N(\mathbf{u}, t) = \sum_{j=1}^{N} \delta[\mathbf{u} - \mathbf{u}_j(t)].$$

Note that the integral of $S(\mathbf{k}, \omega)$ over the frequency spectrum is

$$S(\mathbf{k}) = \frac{1}{V} \frac{\langle |N(k, t)|^2 \rangle}{n_e}. \tag{9}$$

The simplest case for which $S(\mathbf{k})$ can be evaluated is that of Raman–Nath scattering, i.e., scattering by a monochromatic density perturbation represented by

$$n(\mathbf{r}) = n_e + \tilde{n}_e \cos \mathbf{k} \cdot \mathbf{r}.$$

The Fourier transform of this is

$$n(\mathbf{k}) = \tilde{n}_e V,$$

giving

$$S(\mathbf{k}) = \frac{1}{V} \frac{\tilde{n}_e^2 V^2}{n_e} = \frac{\tilde{n}_e^2}{n_e} V.$$

Substituting the above into Eq. (7) and assuming scattering in the dipole equatorial plane ($\sin \varphi = 1$) results in

$$I = I_0 r_e^2 \tilde{n}_e^2 l V \Omega,$$

where Ω, the collection solid angle, has been introduced. The scattering volume V and the solid angle Ω are evaluated from the scattering geometry. Assuming Gaussian optics and scattering taking place at a beam waist, halfwidth W_0, the scattering volume becomes $V = \pi W_0^2 l$, and the solid angle $\Omega = (1/\pi)(\lambda/2W_0)^2$. The scattering intensity therefore becomes

$$I = \tfrac{1}{4} I_0 r_e^2 \lambda^2 l^2 \tilde{n}_e^2, \tag{10}$$

which is the well-known result given by Brillouin's theory for scattering on ultrasonic waves.

B. Thomson Scattering as a Consequence of Refractivity

It was remarked in the Introduction that Thomson scattering, like interferometry and Faraday rotation, is an effect of refractivity. As a striking demonstration that this is so, it will now be shown, following Evans et al. (1982), that the Raman–Nath result derived above from scattering theory could equally well be obtained from considerations of refraction and diffraction alone.

5B. Laser Diagnostics

A basic theorem of Fourier optics (Goodman, 1968), stemming from the fact that the propagation of radiation can be completely described in terms of diffraction integrals, is that the amplitude distributions in focal planes on either side of a lens are Fourier transforms of each other, viz.,

$$U(\xi, \eta) \propto \int_{-\infty}^{+\infty} U(x, y) \exp\left(-i\frac{2\pi}{\lambda f}(x\xi + y\eta)\right) dx\, dy.$$

Consider a Gaussian beam with a beam waist at each focal plane. Let one waist lie in the plasma where a monochromatic wave or phase grating of the form $\Delta\varphi \sin(Kx - \Omega t)$ is traveling perpendicular to the optical axis and so imposes phase shifts on the radiation passing through it. The form of the phase shift depends on the varying refractive index, and is the same as that used in discussing the interferometer, namely $\Delta\varphi = 2\pi\lambda^{-1}L\Delta\mu = r_e\lambda L\bar{n}_e$ for plasma. Thus the Gaussian beam having undergone refraction by the phase grating, can be represented by

$$U(x, y) \exp[i\Delta\varphi \sin(Kx - \Omega t)]$$
$$= \frac{U_0}{\pi^{1/2}W_0} \exp\left(-\frac{x^2 + y^2}{2W_0^2}\right) \sum_{l=-\infty}^{\infty} J_l(\Delta\varphi) \exp[il(Kx - \Omega t)],$$

where the phase factor has been expanded into a sum of Bessel functions. The Fourier transform of this expression is performed to obtain $U(\xi, \eta)$ and the latter is multiplied by its complex conjugate to produce an intensity. Finally, under the assumption that phase shift $\Delta\varphi$ is small, the Bessel functions are replaced by their small-argument approximations, and the outcome of this whole procedure is the following expression for the intensity distribution of the beam in the front focal plane of the lens:

$$I(u) = (I_0/\pi^{1/2}W_f)\{\exp(-u^2)[1 - \tfrac{1}{2}(\Delta\varphi)^2]$$
$$+ \Delta\varphi \exp(-\tfrac{1}{2}v^2)\{\exp[-(u - \tfrac{1}{2}v)^2] - \exp[-(u + \tfrac{1}{2}v)^2]\} \cos \Omega t$$
$$+ (\tfrac{1}{2}\Delta\varphi)^2\{\exp[-(u - v)^2] + \exp[-(u + v)^2]\}$$
$$+ \cdots\}. \tag{11}$$

Here, W_f is the beam waist in the front focal plane, $u \equiv \xi/W_f$ is the front focal plane coordinate normalized to W_f, and the dimensionless parameter v relating spot size at the beam waist in the plasma, W_0, to the wave number K of the phase modulation, $v \equiv KW_0$, has been introduced.

The first term above is time independent and describes the undeviated but slightly attenuated transmitted beam. The small attenuation can be understood as the loss to the main beam of the radiation that appears in the third of the three main terms above. The latter is proportional to $(\tfrac{1}{2}\Delta\varphi)^2 = \tfrac{1}{4}r_e^2\lambda^2L^2\bar{n}_e^2$, and can be recognized as the Thomson scattering of a Gaussian beam by a monochromatic electron density wave. The spatial profile of this term, given by the expression in curly braces multiplying it, consists of two Gaussian maxima disposed symmetrically on either side of the main beam, and centered at $u = \pm v$, respectively. The equation $u = \pm v$ can readily be shown to be equivalent to the Bragg relation $K = 4\pi\lambda^{-1} \sin \frac{1}{2}\theta$ for small scattering angle θ. We emphasize again that this term has appeared spontaneously in a treatment based exclusively on refraction and diffraction.

C. Form Factor for a Field-Free, Warm, Thermal Plasma

A general expression for the form factor $S(\mathbf{k}, \omega)$ for a magnetic-field-free thermal plasma was calculated using the Nyquist dissipation theorem by Dougherty and Farley (1960), and using methods of plasma kinetic theory by Akhiezer et al. (1957), Salpeter (1960), Fejer (1960), and Hagfors (1961). They showed that in terms of the dielectric susceptibilities of the electrons and the ions, G_e and G_i, respectively, the form factor can be written

$$S(k, \omega) = (|1 - G_i|^2 F_e + Z|G_e|^2 F_i)/|1 - G_e - G_i|^2, \quad (12)$$

where F_e and F_i are the Maxwellian velocity distributions of the electrons and the ions. Moreover, G_e and G_i are both proportional to the Fried and Conte (1961) plasma dispersion function (Fig. 14)

$$W(x) = 1 - 2x \exp(-x^2) \int_0^x \exp(t^2)\, dt - i\pi^{1/2} x \exp(-x^2),$$

with $G_e = -\alpha^2 W$ and $G_i = -Z(T_e/T_i)\alpha^2 W$. Here Ze is the ion charge, T_e and T_i the electron and ion temperatures, and $\alpha \equiv (k\lambda_D)^{-1}$ is the ratio of the scattering scale length k^{-1} to the plasma Debye length λ_D. The variable $x \equiv \omega/kv$, v being the thermal velocity.

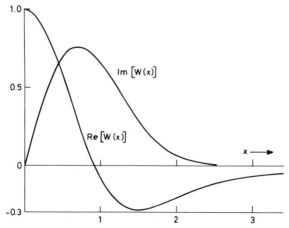

Fig. 14. The real and imaginary parts of the plasma dispersion function $W(x)$.

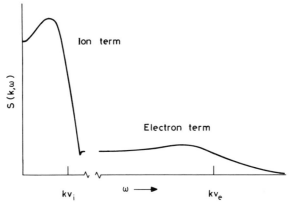

Fig. 15. Frequency spectrum of radiation scattered from a thermal plasma.

The form-factor frequency distribution has the general character illustrated in Fig. 15, the wide low-intensity part coming mainly from the first term in Eq. (12) and the narrow high-intensity central maximum coming from the second term. These have been called the electron and ion terms, respectively. When the parameter $\alpha \ll 1$ the ion term is virtually absent and $S(\mathbf{k}, \omega) \approx F_e$. When $\alpha \gtrsim 1$ the susceptibilities no longer vanish and the ion term dominates the spectrum.

Provided

$$\beta^2 \equiv \frac{\alpha^2}{1+\alpha^2} Z\left(\frac{T_e}{T_i}\right) < \frac{-1}{\mathrm{Re}[W(x)]_{\min}} \approx 3.5,$$

the form factor can be approximated by

$$S(k, \omega) = a_e \Gamma_\alpha(a_e x) + Z[\alpha^2/(1+\alpha^2)]^2 \Gamma_\beta(x),$$

which is the approximation due to Salpeter (1960). Here $a_e^2 \equiv (m_e/M)(T_i/T_e)$ and $\Gamma_{\alpha,\beta}(u) \equiv |1 + (\alpha, \beta)^2 W(u)|^{-2} \exp(-u^2)$ with $x \equiv \omega/kv_i$. Figure 16 shows how the spectral distribution of the ion term varies with the parameter β. With β small it is almost Gaussian, with width kv_i being a measure of ion temperature T_i. If β lies in the range $1 < \beta < \sqrt{3.5}$ the ion feature has maxima at the ion plasma frequency $\pm \omega_{pi}$ for $\alpha \sim 1$, and goes over into ion acoustic waves at frequency $k(KT_e Z/M)^{1/2}$ as α increases.

For more than a decade, experiments using giant pulse ruby lasers have confirmed the details of the thermal theory, in particular with respect to the collective ion feature. Most are summarized in Sheffield (1975). Recent advances have entailed the use of pulsed CO_2 lasers (see, e.g., Pasternak and Offenberger, 1977; Peebles and Herbst, 1978). Massig (1978a) even succeeded in measuring the thermal ion spectrum of a hydrogen arc using a 20-W cw CO_2 system.

Representative single discharge ion features measured in the Culham

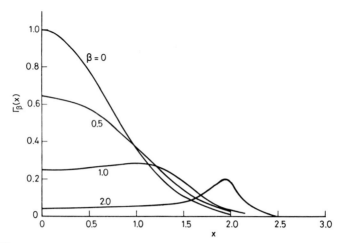

Fig. 16. Collective scattering ion term illustrating its dependence on α and temperature ratio T_e/T_i. $\Gamma_\beta = \exp(-x^2)/|1 + \beta^2 W(x)|^2$, $x \equiv \omega/kv_i$, $\beta^2 \equiv \alpha^2/(1 + \alpha^2)\bar{Z}T_e/T_i$; $W(x)$ is the plasma dispersion function.

plasma focus with a pulsed ruby and a gated optical multichannel analyzer (OMA) are shown in Fig. 17 (Kirk, 1982).

D. Impurities

The theory is readily extended to describe plasmas containing more than a single ion species, i.e., impurities, and the form factor for scattering then becomes (Evans, 1970)

$$S(k, \omega) = \frac{|1 - \Sigma\, G_j|^2 F_e + |G_e|^2(1/n_e) \Sigma\, Z_j^2 N_j F_j}{|1 - G_e - \Sigma\, G_j|^2}, \tag{13}$$

where now G_j is the susceptibility for ions of type j, N_j is the number of j-type ions, and F_j is their velocity distribution. Modest impurity levels can produce striking changes in the frequency spectrum, for example that of a hydrogen plasma contaminated by 5% fully stripped oxygen will be dominated by the oxygen. Stamatakis (1981) has used the above expression to calculate spectra for a hydrogen–deuterium plasma in which the relative abundance of the two components varies from pure hydrogen to pure deuterium. The result is shown in Fig. 18. Because the effective charge $\bar{Z} \equiv \Sigma\, Z_j^2 N_j/n_e$ appears explicitly, collective scattering may offer an attractive alternative to spectroscopy as a means of making a localized measurement of this important quantity (Evans and Yeoman, 1974; Bretz, 1977).

Kasparek et al. (1980) and Kasparek and Holzhauer (1981; 1983) have measured ion features in a hydrogen arc with admixtures of helium, nitro-

5B. Laser Diagnostics

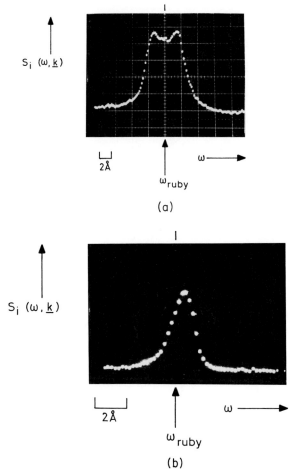

Fig. 17. Single discharge ion features measured in the Culham plasma focus with a pulsed ruby laser and a multichannel gated optical analyzer (OMA). (a) $\mathbf{k} = 6.9 \times 10^4$ cm^{-1} perpendicular to $\mathbf{j}$. $\theta = 45°$. Intensity approximately thermal. Scale is 2.4 Å per division. [From Kirk (1982).] (b) $\mathbf{k} = 1.26 \times 10^4$ cm^{-1} parallel to $\mathbf{j}$. Scattering angle $\theta = 8°$. About 30× thermal intensity. Scale is 1 Å unit per division.

gen, and argon. Their results (Fig. 19) are the first unambiguous experimental confirmation of the theoretical model just outlined for a multiconstituent plasma.

E. Ion Temperature in Fusion Research Tokamaks

Determination of ion temperatures near the center of fusion research plasmas using the conventional charge exchange neutrals technique will meet

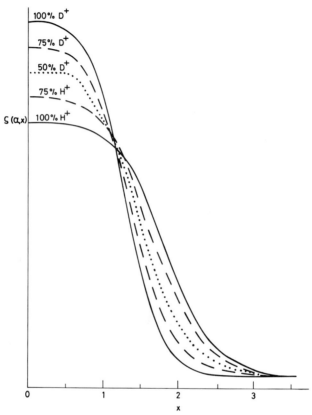

Fig. 18. Ion features for plasmas composed of hydrogen plus deuterium. Relative abundance of two components varies as shown. $T_e = T_i$, $\alpha = 1$, $x \equiv \omega/kv_i$. [From Stamatakis (1981).]

difficulties in the new generation of large tokamaks whose dimensions (typically ≳1 m minor diameter) will inhibit the escape of neutrals from the central regions. Collective Thomson scattering offers an attractive alternative, and pilot experiments at 10.6 μm on PDX (Taylor and Bretz, 1981) and at 385 μm on Alcator C (Woskoboinikow et al., 1981) are currently being conducted. Detection in both cases exploits the advantages of heterodyne receivers, including discrimination against stray radiation at small scattering angle and the capacity to extract weak signal from noise.

The theory of heterodyne detection (Cummins and Swinney, 1970) shows that when a weak optical field having its intensity distributed over frequency according to some function $I_s(\nu)$ and a strong local oscillator at a fixed unique frequency ω_{LO} are superimposed on a square-law detector under conditions that meet the requirements of the van Cittert and Zernike coherence theorem (Born and Wolf, 1964), the resulting detector current i has a power spectrum $P_i(\nu)$ proportional to i_{LO}, the detector current due to the

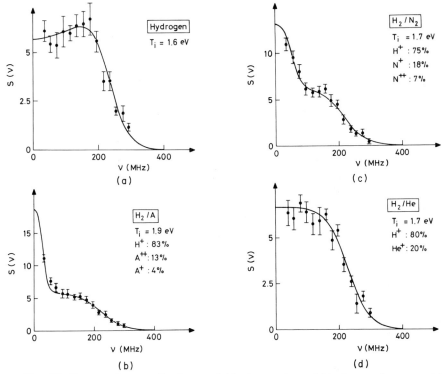

Fig. 19. Spectra of composite plasmas: (a) hydrogen alone, (b) hydrogen plus argon, (c) hydrogen plus nitrogen, (d) hydrogen plus helium. [From Kasparek and Holzhauer (1981).]

local oscillator alone, and to $I_s(\nu - \nu_{LO})$, the intensity spectrum of the weak radiation transposed to the beat frequency $\nu - \nu_{LO}$.

That this is plausible can be seen by taking the beat frequency component of the detector current to be proportional to the square of the sum of the electric fields associated with the signal and with the local oscillator, i.e.,

$$i \sim |E_s + E_{LO}|^2 = I_s + I_{LO} + 2(I_s I_{LO})^{1/2} \cos 2\pi(\nu - \nu_{LO})t.$$

The frequency-analyzing network in turn generates an output which is proportional to the square of the beat frequency term:

$$i_{out} \sim i^2 \sim I_{LO} I_s \sim i_{LO} I_s.$$

The power spectrum is given by [Cummins and Swinney (1970), Eq. 2.19]

$$P_i(\nu) = ei_{LO} + 2\pi i_{LO}^2 \delta(\nu)$$
$$+ i_{LO}\langle i_s\rangle \int_{-\infty}^{\infty} \{\exp[2\pi i(\nu + \nu_{LO})\tau]g_s^l(\tau)$$
$$+ \exp[2\pi i(\nu - \nu_{LO})\tau]g_s^l(\tau)^*\} \, d\tau. \quad (14)$$

In this expression $g_s^1(\tau)$ is the normalized first-order autocorrelation function of the weak radiation field, defined by

$$g_s^1(\tau) \equiv \frac{\langle E_s^*(t)E_s(t+\tau)\rangle}{\langle E_s(t)^2\rangle} = \frac{e\eta}{h\nu}\varepsilon_0 c \frac{\langle E_s^*(t)E_s(t+\tau)\rangle}{\langle i_s\rangle}$$

since

$$\langle i(t)\rangle = \frac{e\eta}{h\nu}\langle I(t)\rangle = \varepsilon_0 c\langle E(t)^2\rangle.$$

Here i_s is the detector current due to the weak field alone, $E_s(t)$ is its time-dependent electric field, η is the detector quantum efficiency, and the other symbols have their usual meanings. Substituting $g_s^1(\tau)$ into the integral term of $P_i(\nu)$ and dropping the first part of the integrand, which is at the sum rather than the difference frequency and is generally physically unobservable for that reason, yields

$$i_{LO}\frac{e\eta}{h\nu}\varepsilon_0 c \int_{-\infty}^{\infty} \exp[2\pi i(\nu - \nu_{LO})\tau]\langle E_s^*(t)E_s(t+\tau)\rangle^* d\tau,$$

which equals $i_{LO}(e\eta/h\nu)I_s(\nu - \nu_{LO})$ by the Wiener–Khinchine theorem.

The first term in Eq. (14) represents shot noise, the second, of which a factor is the Dirac delta function, is absent except at zero frequency, and the third is the heterodyne spectrum. The latter can now be seen to be an exact replica of the optical spectrum, but centered at a frequency equal to the difference between the weak field and the local oscillator frequencies. It is this power spectrum that one seeks to measure.

Using Eq. (14), the signal-to-noise ratio s in the detector photocurrent can be expressed as

$$s = \frac{\text{signal}}{\text{noise}} = \frac{i_{LO}(e\eta/h\nu)I_s(\nu - \nu_{LO})}{ei_{LO} + \text{electronics noise}}.$$

The foregoing expression shows that if the local oscillator is so strong that the first term in the denominator ei_{LO}, the shot noise, dominates other noise contributions, then s becomes independent of the local oscillator current i_{LO} and is given simply by

$$s = [I_s(\nu - \nu_{LO})]/(h\nu/\eta).$$

In these circumstances, the detector noise equivalent power (NEP), that is, the value the signal must take to make $s = 1$, is precisely $h\nu/\eta$.

The detector current i is either processed by an electronic spectrum analyzer (for details, see Sharp et al., 1980, Appendix) or digitized and processed numerically (Green et al., 1980). In either case an output proportional to $P_i(\nu)$, the detector current power spectrum, is generated. It is this output current that is actually measured, and it has both signal and noise components. The signal-to-noise ratio S in this output current is related to that of the detector current, s, through

$$S = [s/(1 + s)](1 + \Delta\nu T)^{1/2},$$

where $\Delta\nu$ is the bandwidth of the detector channel resolution interval and T is the integration time, or pulse length in the case of a pulsed laser experiment. Clearly output signal-to-noise S depends strongly on s only if the latter is less than one. If $s \gg 1$, $S \sim (1 + \Delta\nu T)^{1/2}$ and is consequently independent of input signal-to-noise. It is then determined exclusively by the parameters of the frequency-analyzing circuit and the pulse length, and no significant improvement can be effected by increasing the input signal-to-noise.

The value of S required to enable ion temperature to be determined to any desired accuracy has been investigated by the Lausanne group (Green et al., 1980) and by Sharp et al. (1981). In both studies, Monte Carlo techniques were used to generate numerical simulations of noisy spectra, and the best fits between simulated and theoretical distributions were identified by χ^2 testing. It was concluded that ion temperatures could be measured to an accuracy of a few tens of percent if $S \gtrsim 2$, and if $S = 10$, the accuracy of T_i could approach 10%. Estimates of impurity contamination were very much less accurate, but the presence of up to 2% fully stripped oxygen had no adverse influence on the temperature estimate. Perhaps the most important conclusion was that the required values of S could be achieved in the new hot tokamak plasmas only if laser pulses could be sustained for up to a few microseconds.

F. Hot Plasma Electron Temperature: Relativistic Treatment

Incoherent scattered radiation for which the form factor is simply the one-dimensional Maxwell velocity distribution of the electrons has served for many years as a routine and reliable means for measuring plasma electron temperature. While temperatures remained below a few hundred electron volts only imperceptible errors resulted from ignoring relativistic effects. But even by 1978 the PLT plasma temperature had reached 5 keV (Eubank et al., 1979), and much higher temperatures are anticipated in the new generation of fusion research assemblies now under construction or being contemplated. Revision of elementary theory to take relativity into account has been necessary to predict correctly the scattered radiation frequency distributions expected in these plasmas.

The relativistic expression for the frequency spectrum of radiation incoherently scattered by a hot plasma is found by integrating the contribution from a single electron over the relativistic Maxwell velocity distribution. For a single electron, the scattered electric field at a detector remote from the plasma can be shown to be

$$\mathbf{E} = r_e \frac{(1-\beta^2)^{1/2}}{(1-\beta_s)^3 u} \mathbf{E}_i [\beta_E^2(1-\cos\theta) - (1-\beta_i)(1-\beta_s)],$$

where the electron velocity $\mathbf{v} \equiv \boldsymbol{\beta}c$, and β_i, β_s, and β_E are components of $\boldsymbol{\beta}$ in the directions of the incident beam, the scattered beam, and the incident-beam electric field vector, respectively. The symbol θ represents the scattering angle between the incident and scattered beam directions, and incident and scattered beams are polarized perpendicular to the scattering plane. The u in the denominator is the distance from the electron to the detector.

We define the scattering cross section per unit solid angle $d\sigma/d\Omega$ as the rate at which energy reaches the detector during a finite period of electron acceleration, divided by the incident Poynting flux $\varepsilon_0 c E_i^2$. Following standard treatments [see, e.g., Jackson (1962), p. 472; Panofsky and Phillips (1955), p. 302; Landau and Lifshitz (1975), p. 194], we note that the energy reaching a detector from an electron undergoing acceleration from retarded time T_1 to T_2 is

$$\int_{t=T_1+[u(T_1)/c]}^{T_2+u(T_2)/c} \varepsilon_0 c E^2 u^2 \, d\Omega \, dt = \int_{t'=T_1}^{T_2} \varepsilon_0 c E^2 u^2 \, d\Omega \, \frac{dt}{dt'} \, dt',$$

where $t' = t - u(t')/c$ is the retarded time and $dt/dt' = 1 - \beta_s$. Accordingly,

$$\frac{d\sigma}{d\Omega} = r_e^2 \frac{1-\beta^2}{(1-\beta_s)^5} [\beta_E^2(1-\cos\theta) - (1-\beta_i)(1-\beta_s)]^2. \tag{15}$$

The foregoing expression differs from what one would calculate using simply the Poynting flux associated with the field $\mathbf{E}$ at the detector, by an extra factor $1 - \beta_s$ in the numerator. Pechacek and Trivelpiece (1967) drew attention to the need to include this factor and subsequently offered experimental evidence in support of their argument (Ward et al., 1971; Ward and Pechacek, 1972).

The relativistic Maxwell distribution, normalized to unity, is (Watson et al., 1960)

$$f(\beta) = \left[2\pi K_2\left(\frac{2c^2}{v_e^2}\right)\right]^{-1} \frac{c^2}{v_e^2} (1-\beta^2)^{-5/2} \exp\left[-2\frac{c^2}{v_e^2}(1-\beta^2)^{-1}\right],$$

where $v_e \equiv (2KT_e/m)^{1/2}$ and $K_2(\ )$ is a modified Bessel function of the second kind, of order 2 [tabulated in, for example, Abramowitz and Stegun (1965)]. It is easy to verify that the Doppler shift experienced by the radiation can be expressed as

$$\omega_s/\omega_i = (1-\beta_i)/(1-\beta_s),$$

where ω_s and ω_i are frequencies of the scattered and incident radiation,

5B. Laser Diagnostics

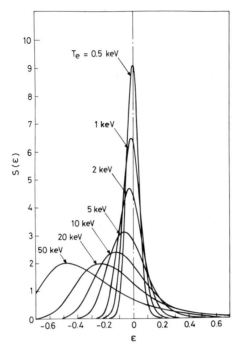

Fig. 20. Spectrum of radiation incoherently scattered by relativistic plasmas; $\theta = 90°$. [From Matoba et. al. (1979).]

respectively, so the scattered radiation frequency spectrum should be

$$r_e^2 S(\omega_s) = \int \int \int \frac{d\sigma}{d\Omega} f(\beta) \, \delta \left(\frac{\omega_s}{\omega_i} - \frac{1 - \beta_i}{1 - \beta_s} \right) d^3 \boldsymbol{\beta}.$$

A first-order approximation to this integral was obtained by Sheffield (1972), and Matoba et al. (1979) have published a second-order formula. By taking a mean value for the polarization term

$$q = \{1 - [(1 - \cos\theta)/(1 - \beta_i)(1 - \beta_s)]\beta_E^2\}^2 \leq 1$$

[see Eq. (15) above] Zhuralev and Petrov (1979) have succeeded in expressing the integral in closed analytic form, and Selden (1980) has reduced this to a result suitable for routine experimental analysis, viz.,

$$S(\varepsilon, \theta) = q[2K_2(2c^2/v_e^2)A(\varepsilon, \theta)]^{-1} \exp[-(2c^2/v_e^2)B(\varepsilon, \theta)],$$

where $\varepsilon \equiv \omega_i/\omega_s - 1$

$$A(\varepsilon, \theta) \equiv (1 + \varepsilon)^2 [2(1 + \varepsilon)(1 - \cos\theta) + \varepsilon^2]^{1/2},$$

$$B(\varepsilon, \theta) \equiv [1 + \varepsilon^2 [2(1 + \varepsilon)(1 - \cos\theta)]^{-1}]^{1/2}.$$

Figure 20 shows frequency distributions calculated with this expression for plasmas with temperatures up to 50 keV. They display the characteristic blue shift of the maximum associated with the forward bias in the radiation pattern of the relativistic electrons.

Hot plasma incoherent scattering has received so much attention because of its practical importance for measuring electron temperature. By contrast, the role of relativity in collective scattering has suffered neglect on the grounds, presumably, that the small frequency shifts involved correspond to subrelativistic phase velocities, ignoring the fact that the electrons actually doing the scattering may be very relativistic indeed. This is an area that would repay attention at the present time.

Acknowledgments

My thanks are due to all my colleagues who assisted in the preparation of this article. I am particularly indebted to David Muir, who advised me on Faraday rotation, and to Adrian Selden, who gave me the benefit of his understanding of the role of relativity in plasma scattering.

References

Abramowitz, M., and Stegun, I. A. (1965). "Handbook of Mathematical Functions." Dover, New York.
Akhiezer, A. I., Prokgoda, I. G., and Sitenko, A. G. (1957). *Zh. Eksp. Teor. Fiz.* **33**, 750.
Alpher, R. A., and White, D. R. (1965). *In* "Plasma Diagnostic Techniques" (R. H. Huddleston and S. L. Leonard, eds.), Chap. 10. Academic Press, New York.
Ashby, D. E. T. F., and Jephcott, D. F. (1963). *Appl. Phys. Lett.* **3**, 13.
Baker, D. R., and Lee, S.-T. (1978). *Rev. Sci. Instrum.* **49**, 919.
Berger, J., and Lovberg, R. H. (1970). *Science (Washington, D.C.)* **170**, 296.
Bernstein, I. B., Trehan, S. K., and Weenink, M. P. (1964). *Nucl. Fusion* **4**, 61.
Birch, J. R., and Jones, R. G. (1970). *Infrared Phys.* **10**, 217.
Bockasten, K. (1961). *J. Opt. Soc. Am.* **51**, 943.
Born, M., and Wolf, E. (1964). "Principles of Optics." Pergamon, Oxford.
Bracewell, R. (1965). "The Fourier Transform and Its Applications," p. 262. McGraw-Hill, New York.
Bretz, N. L. (1977). *Appl. Phys. Lett.* **31**, 372.
Brown, R., Deuchars, W. M., Illingworth, R., and Irving, J. (1977). *J. Phys. D* **10**, 1575.
Buchenauer, C. J., and Jacobson, A. R. (1977). *Rev. Sci. Instrum.* **48**, 769.
Craig, A. D. (1976). *Plasma Phys.* **18**, 777.
Cummins, H. Z., and Swinney, H. L. (1970). *Prog. Opt.* **8**, 135.
DeMarco, F., and Segre, S. E. (1972). *Plasma Phys.* **14**, 245.
DeSilva, A. W., and Goldenbaum, G. (1970). *In* "Plasma Physics" (H. R. Griem, ed.), Methods of Experimental Physics, Vol. 9, Part A, Chap. 3. Academic Press, New York.
Dodel, G., and Kunz, W. (1978). *Infrared Phys.* **18**, 773.
Dougal, A. A., Craig, J. P., and Gribble, R. F. (1964). *Phys. Rev. Lett.* **13**, 156.
Dougherty, J. P., and Farley, D. T. (1960). *Proc. R. Soc. London, Ser. A* **259**, 79.
Eubank, H., Goldston, R., and Arunasalam, V. (1979). *Plasma Phys. Controlled Nucl. Fusion Res., Proc. Int. Conf., 7th, Innsbruck, 1978*.
Evans, D. E. (1970). *Plasma Phys.* **12**, 573.
Evans, D. E. (1974). *In* "Plasma Physics" (B. E. Keen, ed.), Conference Series, No. 20, Chap. 6. Inst. Phys., London.
Evans, D. E. (1976). *Physica B+C (Amsterdam)* **82B+C**, 27.

Evans, D. E., and Katzenstein, J. (1969). *Rep. Prog. Phys.* **32**, 207.
Evans, D. E., and Yeoman, M. L. (1974). *Phys. Rev. Lett.* **33**, 76.
Evans, D. E., von Hellermann, M., and Holzhauer, E. (1982). *Plasma Phys.* **24**(7), 819–834.
Falconer, I., and Ramsden, S. A. (1968). *J. Appl. Phys.* **39**, 3449.
Fejer, J. A. (1960). *Can. J. Phys.* **38**, 1114.
Frayne, P. G. (1968). *J. Phys. D* **1**, 741.
Fried, B. D., and Conte, S. D. (1961). "The Plasma Dispersion Function." Academic Press, New York.
Gibson, A., and Reid, G. W. (1964). *Appl. Phys. Lett.* **5**, 195.
Goodman, J. W. (1968). "Introduction to Fourier Optics." McGraw-Hill, New York.
Gowers, C. W., and Lamb, C. (1982). *J. Phys. E* **15**, 343–346.
Green, M. R., Morgan, P. D., Siegrist, M. R., and Watterson, R. L. (1980). *CRPP (Centre de Recherches en Physique des Plasmas) Ec. Polytech. Fed. Lausanne, Rep.* **LRP 168/80.**
Hagfors, T. (1961). *J. Geophys. Res.* **66**, 1699.
Heald, M. A., and Wharton, C. B. (1965). "Plasma Diagnostics with Microwaves." Wiley, New York.
Heller, W. (1960). *In* "Physical Methods of Organic Chemistry" (Weissberger, ed.), Vol. I, Part 3, pp. 2180–2332. Wiley (Interscience), New York.
Heym, A. (1968). *Plasma Phys.* **9**, 1069.
Hugenholtz, C. A. J., and Meddens, B. J. H. (1979). *Rev. Sci. Instrum.* **50**, 1123–1124.
Jackson, J. D. (1962). "Classical Electrodynamics," Chap. 14. Wiley, New York.
Jacobson, A. R., and Call, D. L. (1978). *Rev. Sci. Instrum.* **49**, 318–320.
Jahoda, F. C., and Sawyer, G. A. (1971). *In* "Plasma Physics" (R. H. Lovberg, ed.), Methods of Experimental Physics, Vol. 9, Part B. Academic Press, New York.
Kasparek, W., and Holzhauer, E. (1981). *Univ. Stuttgart, Inst. Plasma Sci., Rep.* **IPF-81-5.**
Kasparek, W., Hirsch, K., and Holzhauer, E. (1980). *Plasma Phys.* **22**, 555–558.
Kirk, R. E. (1982). Ph.D. Thesis, Royal Holloway College, Univ. of London.
Kunz, W., and Dodel, G. (1978). *Plasma Phys.* **20**, 171–174.
Kunz, W., and Equipe TFR (1978). *Nucl. Fusion* **18**, 1729–1732.
Landau, L. D., and Lifshitz, E. M. (1975). "The Classical Theory of Fields" (M. Hamermesh, transl.). Pergamon, Oxford.
Luhmann, N. C., Jr. (1979). *In* "Infrared and Millimeter Waves, Vol. 2, Instrumentation" (K. J. Button, ed.), Chap. 1. Academic Press, New York.
Magyar, G. (1981). *In* "Plasma Physics and Nuclear Fusion Research" (R. D. Gill, ed.), Chap. 23. Academic Press, New York.
Massig, J. H. (1978a). *Phys. Lett. A* **66A**, 207–209.
Massig, J. H. (1978b). *Inst. Plasmaforsch., Univ. Stuttgart* **IPF-78-2.** (Prepr.)
Matoba, T., Itagaki, T., Yamauchi, T., and Funahashi, A. (1979). *Jpn. J. Appl. Phys.* **18**, 1127–1133.
Panofsky, W. K. H., and Phillips, M. (1955). "Classical Electricity and Magnetism." Addison-Wesley, Reading, Massachusetts.
Pasternak, A. W., and Offenberger, A. A. (1977). *Can. J. Phys.* **55**, 419–427.
Pechacek, R. E., and Trivelpiece, A. W. (1967). *Phys. Fluids* **10**, 1688–1696.
Peebles, W. A., and Herbst, M. J. (1978). *IEEE Trans. Plasma Sci.* **PS-6**, 564–567.
Ramachandran, G. N., and Ramaseshan, S. (1961). "Crystal Optics Encyclopedia of Physics," Vol. 25/1. Springer-Verlag, Berlin and New York.
Salpeter, E. E. (1960). *Phys. Rev.* **120**, 1528–1535.
Sauthoff, N. R., and von Goeler, S. (1979). *IEEE Trans. Plasma Sci.* **PS-7**, 141–147.
Segre, S. E. (1975). *In* "Plasma Diagnostics and Data Acquisition" (H. Eubank and E. Sindoni, eds.), pp. 265–301. Editrice Compositori, Bologna.
Segre, S. E. (1978). *Plasma Phys.* **20**, 295–307.
Selden, A. C. (1980). *Phys. Lett. A* **79A**, 405–406.
Selden, A. C. (1982). *Culham Lab. Rep.* **CLM-R220.**

Sharp, L. E., Sanderson, A. D., and Evans, D. E. (1980). *Culham Lab. Rep.* **CLM-P548**. (Prepr.)
Sharp, L. E., Sanderson, A. D., and Evans, D. E. (1981). *Plasma Phys.* **23**, 357.
Sheffield, J. (1972). *Plasma Phys.* **14**, 783.
Sheffield, J. (1975). "Plasma Scattering of Electromagnetic Radiation." Academic Press, New York.
Siegman, A. E. (1971). "An Introduction to Lasers and Masers." McGraw-Hill, New York.
Stamatakis, T. (1981). Personal communication (Appendix to JET report).
Taylor, G., and Bretz, N. L. (1981). *Bull. Am. Phys. Soc.* **26**, 991.
Veron, D. (1974). *Opt. Commun.* **10**, 95.
Veron, D. (1979). *In* "Infrared and Submillimeter Waves, Vol. 2, Instrumentation" (K. J. Button, ed.), Chap. 2. Academic Press, New York.
Veron, D., Certain, J., and Crenn, J. P. (1977). *J. Opt. Soc. Am.* **67**, 964–967.
Ward, G., and Pechacek, R. E. (1972). *Phys. Fluids* **15**, 2202–2210.
Ward, G., Pechacek, R. E., and Trivelpiece, A. W. (1971). *Phys. Rev. A* **3**, 1721–1723.
Watson, K. M., Bludman, S. A., and Rosenbluth, M. N. (1960). *Phys. Fluids* **3**, 741.
Williamson, J. H., and Evans, D. E. (1982). *IEEE Trans. Plasma Sci.* **PS-10**(2), 82–93.
Wolfe, S. M., Button, K. J., Waldman, J., and Cohn, D. R. (1976). *Appl. Opt.* **15**, 2645–2648.
Wood, R. W. (1967). "Physical Optics." Dover, New York.
Woskoboinikow, P., Mulligan, W. J., Cohn, D. R., Fetterman, H., Praddaude, H. C., and Lax, B. (1981). *Bull. Am. Phys. Soc.* **26**, 922.
Zhuralev, V. A., and Petrov, G. D. (1979). *Sov. J. Plasma Phys.* (*Engl. Transl.*) **5**, 3–5.

5C
Plasma Diagnostics Using Electron Cyclotron Emission

D. A. Boyd

Laboratory for Plasma and Fusion Energy Studies
University of Maryland
College Park, Maryland

I. Introduction	227
II. The Theory of Electron Cyclotron Emission . . .	229
A. Statement of the Problem	229
B. Calculation of the Absorption Coefficient α . .	230
C. Non-Maxwellian Distributions	232
III. Instrumentation.	233
A. Swept Heterodyne Receivers	233
B. Fourier-Transform Spectroscopy	234
C. Grating Spectroscopy	236
D. Fabry–Perot Interferometer	238
IV. Applications	240
A. Tokamaks	240
B. Mirrors.	244
C. Bumpy Tori	244
V. Concluding Remarks.	244
References	246

I. Introduction

If a plasma is threaded by a magnetic field, the plasma electrons move on helical trajectories about the magnetic field lines. The angular frequency with which they gyrate about the field lines is given by

$$\omega = eB/\gamma m, \tag{1}$$

where B is the local magnetic field strength, e and m are the electron charge and mass, respectively, and $\gamma = (1 - v^2/c^2)^{-1}$, where v is the electron velocity and c the velocity of light. This frequency is called the electron cyclotron frequency, and because of the acceleration involved in this circular motion the electron radiates electromagnetic radiation at the electron cyclotron frequency and multiples of that frequency. The radiation is called electron cyclotron emission, synchrotron radiation, and magnetic bremsstrahlung. All three terms will be encountered in the literature.

This radiation could be a serious energy loss channel for fusion reactors which confine plasmas with magnetic fields, and so early (before 1958) interest in electron cyclotron radiation was concerned with calculating the power radiated from a plasma with a temperature of several tens of kilovolts or several hundred million degrees kelvin (Trubnikov and Kudryavtsev, 1958; Hayakawa et al., 1958; Beard, 1959). A serious notion of using this radiation as a means of diagnosing plasma parameters had to wait until the papers of Engelmann and Curatolo (1973).

Tokamak devices at that time were able to produce plasmas with electron temperatures of about one kilovolt and contained these plasmas in magnetic fields with a strength of a few tesla. Since for these low-temperature plasmas $\gamma \sim 1$, if the magnetic field strength is known as a function of position, then so is the electron cyclotron frequency. Thus measurements at a particular frequency can be associated with a particular spatial location. At low harmonics $\omega = \omega_c$ or $2\omega_c$ it is possible that the plasma radiates like a blackbody and so the intensity of the radiation is directly proportional to the electron temperature. Consequently, a measurement of the intensity of the radiation as a function of the frequency could be transformed into data depicting the electron temperature as a function of spatial position.

At higher harmonics, usually $\omega > 2\omega_c$, the intensity of the radiation can be proportional to the product of the electron density and the electron temperature raised to the power of the harmonic number. Hence, if the temperature distribution in space is known from lower-frequency measurements, the electron density distribution in space can be deduced from an intensity versus frequency measurement.

Lastly, the emitted radiation at higher harmonics is linearly polarized with its electric vector perpendicular to the magnetic field direction at the place where the radiation is generated. If the radiation propagates out of the plasma with its original direction of polarization preserved, then a measurement of the plane of polarization can be used to deduce the orientation of the magnetic field lines within the plasma.

These are the three diagnostic possibilities outlined in the Engelmann–Curatolo paper and it is the success with the first of these that has generated the contemporary interest in the use of cyclotron radiation measurements as a diagnostic tool.

In Section II a brief description of the theory of electron cyclotron emission from plasmas will be given; in Section III, a description of the instruments used to make the measurements and, Section IV, examples of their application on various plasma devices for diagnostic purposes. Lastly, in Section V, some concluding remarks are made in an attempt to assess the present situation.

II. The Theory of Electron Cyclotron Emission

A. Statement of the Problem

In a fully ionized plasma at a temperature below 10 keV, the calculation of the electron cyclotron emission can be divided into a number of separate categories. The first division that can be made is on the basis of the kind of broadening that dominates the emission linewidth. When the emission is nearly perpendicular to the magnetic field lines, the width of the line is determined by the variation in γ of the electrons in the electron distribution. This is the relativistic regime. For emission at angles away from the perpendicular, the width of a line is dominated by Doppler broadening and obviously this is called the Doppler regime. Another division can be made on the grounds of plasma density. If the density is low, the electrons radiate as if they were in free space, where the dispersion relation is $\omega = kc$. If the density is higher, the electrons radiate into a dielectric whose properties are generated by the other electrons in the plasma. Hence the dispersion relation is altered and the contribution of the background electrons has to be taken into account when one is calculating the radiated power.

For diagnostic purposes, the quantity of importance is the specific intensity I. It is defined by

$$I \equiv dP/d\omega \, d\Omega \, dA, \tag{2}$$

where dP is the power radiated in an angular frequency interval $d\omega$, into a solid angle $d\Omega$ through a surface of area dA. If the transport of the radiation through the plasma is considered, then the specific intensity observed exterior to the plasma is given by

$$I = I_{BB}[1 - \exp(-\tau)], \tag{3}$$

where $I_{BB} = \omega^2 kT_e/8\pi^3 c^2$ and τ is the optical depth, which is related to the absorption coefficient α by the equation

$$\tau = \int \alpha \, ds, \tag{4}$$

where the integration is carried out along the line of observation, k is Boltzmann's constant, and T_e is the electron temperature in the region where the absorption coefficient has a significant value. In an inhomogeneous magnetic field, the width of the resonant region, that is, the region where α is significant, is given by

Relativistic regime:

$$\Delta S_R = \frac{B}{dB/ds} (2\pi l)^{1/2} \frac{kT_e}{mc^2}. \tag{5a}$$

Doppler regime:

$$\Delta S_D = \frac{B}{dB/ds}(2\pi)^{1/2}l\left(\frac{kT_e}{mc^2}\right)^{1/2}\cos\phi, \tag{5b}$$

where l is the harmonic number and ϕ is the angle between the observation direction and the direction of the magnetic field.

B. Calculation of the Absorption Coefficient α

At low densities, the absorption coefficient for $l \lesssim 5$ is given by

$$\alpha_l(\omega, \phi) = \frac{\pi}{2^l}\frac{\omega_p^2}{c}\frac{l^{2l-1}}{(l-1)!}\left(\frac{kT_e}{mc^2}\right)^{l-1}(\sin\phi)^{2l-2}(1+\cos^2\phi)\phi_l(\omega,\phi)$$

$$\times\left(\frac{1}{2}\pm\frac{(\sin^4\phi/4l)+\cos^2\phi}{[(\sin^2\phi)/4l^2+\cos^2\phi]^{1/2}(1+\cos^2\phi)}\right) \tag{6}$$

where, for $l = 1$, we must satisfy $(\omega_p^2/\omega^2)(mc^2/kT_e)^{1/2} \ll 1$, and for $l > 2$, we must satisfy $\omega_p^2/\omega^2 \ll 1$; Φ is an emission line profile function normalized so that $\int_{-\infty}^{\infty}\Phi\,d\omega = 1$. Plus (+) signifies extraordinary mode at $\phi = \frac{1}{2}\pi$, right-hand circularly polarized mode at $\phi = 0$. Minus (−) signifies ordinary mode at $\phi = \frac{1}{2}\pi$, left-hand circularly polarized mode at $\phi = 0$. At low temperatures, the line profile does not matter, so we set $\Phi = \delta(\omega - l\omega_c)$, and the optical depth is

$$\tau = \frac{\pi}{2^l}\frac{\omega_p^2}{\omega_c c}\frac{l^{2l-2}}{(l-1)!}\left(\frac{kT_e}{mc^2}\right)^{l-1}\frac{B}{dB/ds}(\sin\phi)^{2l-2}(1+\cos^2\phi)$$

$$\times\left(\frac{1}{2}\pm\frac{(\sin^4\phi)/4l+\cos^2\phi}{[(\sin^2\phi)/4l^2+\cos^2\phi]^{1/2}(1+\cos^2\phi)}\right). \tag{7}$$

This formula gives zero emission at $\phi = \frac{1}{2}\pi$ for the ordinary (−) mode; this is only correct to order v_{th}/c. To the next order, we get

$$\alpha_- = (kT_e/mc^2)\alpha_+ \quad \text{at} \quad \phi = \frac{1}{2}\pi. \tag{8}$$

At higher densities, things are more complicated. In the Doppler regime for quasiperpendicular propagation, the absorption coefficients for $l = 1$ are (Bornatici, 1981, 1982)

$$\alpha_+ = \frac{1}{(2\pi)^{1/2}}\left(2-\frac{\omega_p^2}{\omega_c^2}\right)^2\left(1+\frac{\omega_p^2}{\omega_c^2}\right)^2\left(\frac{\omega_c}{\omega_p}\right)^2\frac{\omega}{c}\left(\frac{kT_e}{mc^2}\right)^{1/2}\cos(\phi)\Phi(\xi), \tag{9}$$

$$\alpha_- = \frac{1}{(2\pi)^{1/2}}\frac{\omega_p^2}{\omega_c^2}\frac{\omega}{c}\left(\frac{kT_e}{mc^2}\right)^{1/2}\sec(\phi)\Phi(\xi), \tag{10}$$

where $\Phi(\xi)$ is a profile function given by

5C. Plasma Diagnostics Using Electron Cyclotron Emission

$$\Phi(\xi) = \pi \frac{\exp(-\xi)^2}{|Z(\xi)|^2} \quad \text{and} \quad \xi = \frac{\omega - l\omega_c}{\sqrt{2}(kT_e/mc^2)^{1/2}\omega N \cos\phi}; \quad (11)$$

$Z(\xi)$ is the plasma dispersion function and $N \equiv ck/\omega$ the refractive index.

Beyond the quasiperpendicular region shown here, a formulation valid at smaller angles is given by Stepanov and Pakhomov (1960), but it fails as $\theta \to 0$. At $\theta \simeq 0$ a formulation due to Silin (1955) is valid, but unfortunately there is an intermediate region where no analytic expression is accurate. For $l > 1$,

$$\alpha_\pm = \alpha_l \Phi_l \eta_\pm, \quad (12)$$

$$\alpha_l \equiv \frac{\pi}{2^l} \frac{l^{2l-1}}{(l-1)!} \frac{\omega_p^2}{c} \left(\frac{kT_e}{mc^2}\right)^{l-1} (\sin\phi)^{2l-2}(1 + \cos^2\phi), \quad (13)$$

$$\Phi_l \equiv \frac{\exp(-\xi^2)}{\sqrt{2\pi}(kT_e/mc^2)^{1/2}\omega N \cos\phi}, \quad (14)$$

$$\eta_\pm = \frac{N^{2l-3}\{1 - l[1 - (l^2-1)/l^2]f_\pm\}^2}{(1 + \cos^2\phi)(a_l^2 + b_l^2)^{1/2}}, \quad (15)$$

where

$$f_\pm = \frac{2[l^2 - (\omega_p/\omega_c)^2]}{2(l^2 - \omega_p^2/\omega_c^2) - \sin^2\phi \pm \rho_l}, \quad (16)$$

$$\rho_l \equiv \left[\sin^4\phi + \frac{4}{l^2}\left(l^2 - \frac{\omega_p^2}{\omega_c^2}\right)^2 \cos^2\phi\right]^{1/2}, \quad (17)$$

and

$$a_l^2 = \left[1 + \frac{[1 - (\omega_p/l\omega_c)^2]N^2\cos^2\phi}{[1 - (\omega_p/l\omega_c)^2 - N^2\sin^2\phi]^2} l^2\left(1 - \frac{l^2-1}{l^2}f_\pm\right)^2\right]^2 \sin^2\phi, \quad (18)$$

$$b_l^2 = \left[1 + \frac{1 - (\omega_p/l\omega_c)^2}{1 - (\omega_p/l\omega_c)^2 - N^2\sin^2\phi} l^2\left(1 - \frac{l^2-1}{l^2}f_\pm\right)^2\right]^2 \cos^2\phi, \quad (19)$$

$$N^2 = 1 - (\omega_p/l\omega_c)^2 f_+. \quad (20)$$

In the relativistic regime for $l = 1$,

$$\alpha_+ = \frac{1}{2}\left(2 - \frac{\omega_p^2}{\omega_c^2}\right)^{3/2}\left(\frac{\omega_c}{\omega_p}\right)^2 \frac{\omega_c}{c} \frac{kT_e}{mc^2} \frac{-\text{Im}[F_{5/2}(z)]}{|F_{5/2}(z)|^2} L\left(z, \frac{\omega_p^2}{\omega_c^2}\right), \quad (21)$$

$$\alpha_- = \frac{1}{\sqrt{2}}\left(1 - \frac{\omega_p^2}{\omega_c^2}\right)^{1/2} \frac{\omega_p^2}{\omega_c^2} \frac{\omega_c}{c} \frac{-\text{Im}[F_{7/2}(z)]}{|G_{7/2}|[\text{Re}(G_{7/2}) + |G_{7/2}|]^{1/2}}, \quad (22)$$

where

$$G_{7/2} \equiv 1 + \tfrac{1}{2}(\omega_p^2/\omega_c^2)^2 F_{7/2}(z), \quad (23)$$

$$F_q = \text{Re}(F_q) + i\,\text{Im}(F_q) \equiv -i\int_0^\infty \frac{\exp(izt)}{(1-it)^q} dt, \quad (24)$$

$$\text{Im}(F_q) = -[\pi/\Gamma(q)]|z_l|^{q-1}\exp(-|z_l|), \qquad (25)$$

$$z_l = \frac{mc^2}{kT_e}\frac{\omega - l\omega_c}{\omega}. \qquad (26)$$

Γ is the gamma function, and finite Larmor radius corrections give

$$L\left(z, \frac{\omega_p^2}{\omega_c^2}\right) = \left(1 - \frac{\omega_p^2}{\omega_c^2}F_{7/2}(z)\right)^2 + \frac{4}{5}\left(\frac{\omega_p^2}{\omega_c^2}\right)|z|\left(\text{Re}(F_{5/2}(z))\right.$$
$$\left. - \frac{\omega_p^2}{\omega_c^2}\text{Re}[F^*_{5/2}(z)]F_{7/2}(z)\right) + \frac{6}{35}\left(\frac{\omega_p}{\omega_c}\right)^4 [z|F_{5/2}(z)|]^2. \qquad (27)$$

At the higher harmonics $l > 1$,

$$\alpha_+ = N^{(2l-3)}(1 + p)^2 \frac{l^{2l-1}}{2^l l!}\left(\frac{\omega_p}{\omega_c}\right)^2 \left(\frac{kT_e}{mc^2}\right)^{l-2}\frac{\omega_c}{c}\{-\text{Im}[F_{l+3/2}(z)]\}, \qquad (28)$$

where $p = (\omega_p/\omega_c)^2/l(l^2 - 1 - \omega_p^2/\omega_c^2)$, and

$$\alpha_- = \frac{l^{2l-1}}{2^l l!}\left[1 - \left(\frac{\omega_p}{l\omega_c}\right)^2\right]^{l-1/2}\frac{\omega_p^2}{\omega_c^2}\left(\frac{kT_e}{mc^2}\right)^{l-1}\frac{\omega_c}{c}\{-\text{Im}[F_{l+5/2}(z)]\}. \qquad (29)$$

We are in the Doppler regime when

$$N^2\cos^2\phi > kT_e/mc^2, \qquad (30)$$

where $N \equiv ck/\omega$, the refractive index, and in the relativistic regime if

$$N^2\cos^2\phi < kT_e/mc^2. \qquad (31)$$

Quasiperpendicular means

$$\sin^4\phi \gg 4[1 - (\omega_p^2/\omega_c^2)]^2\cos^2\phi. \qquad (32)$$

C. Non-Maxwellian Distributions

Radiation from non-Maxwellian electron distributions containing relativistic electrons is easily detected as they radiate copiously. Unfortunately, the analysis of the data is difficult because the topic has been neglected theoretically. There is a beginning of an analysis in the paper by Celata and Boyd (1977) for runaway electrons in toroidal machines.

The relativistic electron rings in Bumpy Tori are candidates for diagnostic studies. The approach of Tsakiris and Davidson (1977) has some potential for development in this direction, and perhaps from it a theory of EBT ring diagnostics could grow. Recently Winske and Boyd (1983) and Tamor have made some progress on this problem. Relativistic and nonrelativistic, non-Maxwellian distributions which occur in magnetic-mirror-confined plasmas have been almost entirely neglected. A contribution has been made by Tsakiris and Ellis (1982). Diagnostic studies of the electrons in mirror machines will be handicapped until this theoretical discrepancy is replaced by the kind of understanding we have for the plasmas in toroidal devices.

III. Instrumentation

A. Swept Heterodyne Receivers

Heterodyne receivers have been used as cyclotron radiation detectors for a number of years. A very successful system has been built by a group at the Princeton Plasma Physics Laboratory (Efthimion et al., 1979). Up to about 90 GHz, the technology is straightforward with higher-frequency systems on the near horizon. Before describing the apparatus used in these measurements and the relative advantages and disadvantages of this approach, let us deal with the theory of the measurement.

We know that

$$I = \frac{dP}{d\omega \, d\Omega \, dA} = \frac{\omega^2 k T_e}{8\pi^3 c^2} [1 - \exp(-\tau)] \tag{33}$$

for a source at uniform temperature which fills the antenna, and

$$\int d\Omega \, dA = \lambda^2 = \frac{4\pi^2 c^2}{\omega^2}; \tag{34}$$

therefore

$$P = kT_e[1 - \exp(-\tau)] \, \Delta f. \tag{35}$$

If $\tau \gg 1$,

$$P \to kT_e \, \Delta f, \tag{36}$$

where Δf is the bandwidth of the receiver. Equation (36) indicates that the receiver power is directly proportional to the plasma temperature.

The receiver of Efthimion et al. (1979) is shown in Fig. 1. Basically, a backward wave oscillator (output frequency 60–90 GHz) is swept (10 ms) by an amplified ramp voltage signal; the output power is leveled by a ferrite modulator regulated by a feedback circuit. In the mixer, the signal from the plasma and the local oscillator power are mixed and the resulting output signal is amplified by a 400-MHz bandwidth amplifier. This amplified signal is then detected.

First, let me stress the advantages of this kind of system. Below 100 GHz, the equipment is relatively standard and of reasonable cost. It is rugged and, once set up, is easy to operate. With some effort, the sweep time which determines the temporal resolution could be reduced to 100 μs or less. The

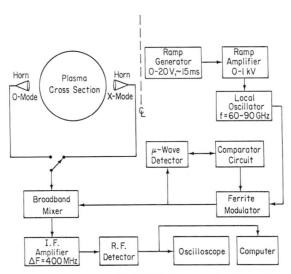

Fig. 1. Detailed block diagram of a 60–90 GHz fast-scanning heterodyne receiver built by Efthimion et al. (1979).

frequency resolution is excellent and since that determines the spatial resolution along the magnetic field gradient, it too is excellent. The sensitivity of the heterodyne receiver makes calibration a reasonable procedure.

Of course, such systems have some disadvantages. The backward wave oscillators can sweep about ±20% around their center frequency. This usually limits the spatial extent of the plasma which can be examined. Broadband mixers have a strong frequency dependence in their response and so calibration of these systems is problematical as the output signal for constant input power is a strong function of frequency. In addition, one is tempted to work at the lowest possible frequency since everything is easier there. But this can force one to work where $\tau \lesssim 1$, where one must rely on reflections within the plasma chamber to approach the blackbody level and accept the loss of spatial resolution. The low frequencies also mean that, for a limited window or horn size, the antenna pattern is broader than it would be at higher frequencies because of diffraction and so the transverse spatial resolution suffers.

B. Fourier-Transform Spectroscopy

Fourier-transform spectroscopy was introduced into plasma diagnostics by Costley et al. (1974). Since then it has been very successfully used on a number of large tokamak devices. The type of interferometer which has been used is the Martin–Puplett form (Martin and Puplett, 1970) of the Michelson interferometer. The instrument is operated in a rapid scanning mode and uses a polarizing wire grid as a beam splitter.

A sketch of such an interferometer appears in Fig. 2. Radiation enters from the right and encounters grid G_1. The parallel wires of G_1 are usually aligned horizontally. Hence radiation with its electric vector in a vertical direction passes through the grid and the radiation with the orthogonal polarization is reflected, in this illustration, to a monitor detector. The radiation transmitted by G_1 next encounters grid G_2, which has its wires oriented so that they are at 45° to the electric vector of the incoming radiation. Hence half the radiation is transmitted and half reflected by G_2. The reflected radiation is again reflected by the fixed mirror and grid G_2. The transmitted radiation reflects from the scanning mirror and is again transmitted by G_2. The recombined beams interfere and are reflected by G_1 toward the detector.

If the interferometer views a source with a continuous spectrum the output signal, assuming a detector with an output proportional to the radiation intensity and the scanning mirror displaced a distance x from the position at which both beams cover an equal optical path length, is given by

$$V(x) = 2 \int_0^\infty S(k)[1 + \cos(x)] \, dk, \tag{37}$$

where S is the spectral intensity at wave number k and $k = 1/\lambda$:

$$V(0) = 2 \int_0^\infty 2S(k) \, dk. \tag{38}$$

Hence

$$V'(x) \equiv V(x) - \tfrac{1}{2}V(0) = \int_0^\infty 2S(k) \cos(\pi kx) \, dk. \tag{39}$$

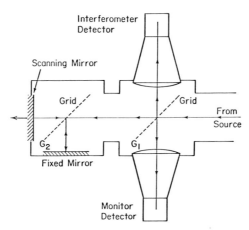

Fig. 2. Schematic diagram of a rapidly scanning, polarizing, Fourier-transform spectrometer of the Martin–Puplett type, developed by Costley *et al.* (1974).

$V'(x)$ is called the interferogram function. Taking the Fourier transform of $V'(x)$ enables us to recover the input spectrum.

Usually the postdetector amplifier rejects the continuous contribution $\frac{1}{2}V(0)$ by having an imposed low-frequency rolloff. The detectors are either indium antimomide crystals or germanium bolometers operating at liquid helium temperatures. Typically $1 < \Delta x < 2$ cm and the scanning time τ lies between 10 and 20 ms.

There are some considerable advantages to such a system. The instrument essentially measures the entire cyclotron emission spectrum in a single polarization state in each scan. The information content of the spectra is very large. Detector signals are large and hence the signal-to-noise ratios are large because of the large optical throughput. It is also true that such an interferometer is the easiest of all instruments to calibrate although no instrument in this spectral range is easy.

The disadvantages include modest spectral resolution (~ 5 GHz), modest temporal resolution (~ 15 ms), and the necessity of having substantial computing power available to do the Fourier transforms required to produce the spectra.

For devices with magnetic field above two tesla and electron temperatures above 100 eV, if one were going to have only a single cyclotron radiation diagnostic this system would certainly be the best choice.

C. Grating Spectroscopy

A grating spectrometer was one of the first instruments used as an electron cyclotron emission diagnostic (Lichtenberg *et al.*, 1964). Almost without exception these instruments are not used in a scanning mode; that is, the grating is not rotated during the plasma discharge. The most successful method of operation has been to use a system with multiple output channels with separate detectors for each channel. Thus several signals continuous in time are obtained and the system is particularly suited to studying fluctuations with the plasma. Between plasma discharges the grating may be rotated so that the output channels can be associated with different spatial locations. The multiple-output grating was introduced as a diagnostic by Rutgers and Boyd (1977).

An example of such an instrument assembled by G. D. Tait is shown in Fig. 3. Radiation enters the instrument through the input light pipe, is collimated by the spherical mirror M1, is diffracted by the eschellette grating, and is focused on the output exit slits by mirror M2. In this instrument there are five output channels and the wavelength range that can be covered by the exit slits is limited by the width of M2.

For the Czerny–Turner mount used in the above, the condition for constructive interference is given by

$$m\lambda = 2d \cos \varepsilon \sin(\theta + \alpha - \varepsilon), \tag{40}$$

5C. Plasma Diagnostics Using Electron Cyclotron Emission

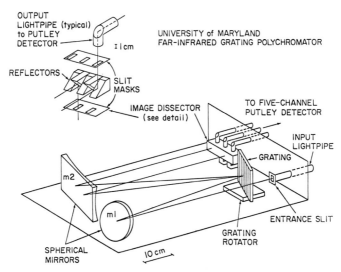

Fig. 3. Diagram of the five-channel echellete grating spectrometer assembled by G. D. Tait.

where 2ε is the angle between the incident and diffracted beams and θ is the angle through which the grating has been rotated from its zeroth-order position. The system is operated with $\theta + \alpha$ and ε varying from exit slit to exit slit. α is the angle of incidence on the grating when $\theta = 0$, and usually the exit slits are arranged so that for one slit $\alpha = \varepsilon$. d is the grating spacing; m is an integer, in this case $m = -1$.

The output intensity distribution can be written, when $\alpha = \varepsilon$, as

$$I = S^2(\theta) f^2(\theta), \tag{41}$$

where

$$S^2(\theta) = \frac{\sin^2 [N(\Delta/2)]}{\sin^2 (\Delta/2)} \quad \text{and} \quad \Delta = (2\pi/\lambda)2d \cos \varepsilon \sin \theta,$$

with N the total number of grooves in the grating. It can be shown that

$$f^2(\theta) = \frac{\sin^2[kd \sec \zeta \cos \varepsilon \sin(\theta - \zeta)]}{[kd \sec \zeta \cos \varepsilon \sin(\theta - \zeta)]^2}, \tag{42}$$

and hence

$$I = \frac{\sin^2(Nkd \sin \theta \cos \varepsilon) \sin^2[kd \sec \zeta \cos \varepsilon \sin(\theta - \zeta)]}{\sin^2(kd \sin \theta \cos \varepsilon)[kd \sec \zeta \cos \varepsilon \sin(\theta - \zeta)]^2}, \tag{43}$$

where ζ is the angle of inclination of the grating grooves.

Usually the instruments are operated with $|m| = 1$ and with $\theta \sim \zeta$, the "blaze" angle.

If w is the exit slit width and f_2 the focal length of mirror M2 the resolution is

$$\Delta\lambda = d(w/f_2)\cos\varepsilon\cos\theta, \tag{44}$$

and the resolving power

$$R = (2f_2/w)\tan\theta. \tag{45}$$

In practice the resolution is limited by throughput and signal-to-noise constraints.

The advantages of such a spectrometer include continuous temporal coverage at a number of spatial locations in the plasma, fairly good spatial resolution, and a simple procedure for data reduction. In addition, the device is simple and can be made quite cheaply. The disadvantages are that one requires multidetector calibration with a device that is intrinsically difficult to calibrate. Both for plasma measurements and particularly for calibration one needs very effective low-pass filters to remove the higher orders of diffraction. And the low throughput means that almost invariably signal-to-noise problems limit the performance of the system. However, for the multipoint, continuous temporal coverage required for plasma fluctuation studies, this system seems to have the most attractive properties.

D. Fabry–Perot Interferometer

The Fabry–Perot interferometer has been developed to an advanced level by Walker and collaborators (1981) and Baker (1982), although work began at MIT with Komm *et al.* (1975) and was continued by Hutchinson and Komm (1977). A number of groups have used the Fabry–Perot interferometer in the static mode where the device is tuned to a single frequency that can be changed between plasma discharges, but the real strengths of the instrument appear when it is rapidly scanned, in times approaching one millisecond.

In a Fabry–Perot with two identical parallel reflecting plates with an intensity transmission coefficient T and reflection coefficient R, the transmission through the interferometer at normal incidence is

$$\tau(x) = \left(1 - \frac{A}{T}\right)^2 \left[1 + \frac{4R}{(1-R)^2}\sin^2\left(\phi + \frac{2\pi x}{\lambda}N\right)\right]^{-1}, \tag{46}$$

where $A = 1 - R - T$ is the absorption coefficient, ϕ is the average of the phase angles of the complex reflection coefficients, and x is the distance between the reflecting surfaces; N is the refractive index of the material between the reflecting surfaces. If $R > 0.6$, the finesse is given by

$$F_R = \pi R^{1/2}/(1 - R). \tag{47}$$

The resolving power is $\rho = mF_R$ where m is the order of the interference.

In conditions where the absorption coefficient is negligible, the reflection coefficient and the phase angle for metal meshes, which are usually used as the reflecting plates, are given by (Ulrich, 1967)

$$R = [(1 + Z)^2 + 4Y^2]^{-1}, \tag{48}$$

$$\phi = \pi + \arctan[Y/(1 + Z)], \tag{49}$$

where

$$Y = \ln \cosec\left(\frac{a\pi}{g}\right) \bigg/ \left(\frac{g}{\lambda(1 - 0.27a/g)} - \frac{\lambda(1 - 0.27a/g)}{g}\right), \tag{50}$$

$$Z = \frac{g}{4a}\left(\frac{4\pi\varepsilon_0 c}{\lambda\sigma}\right)^{1/2}, \tag{51}$$

and λ is the wavelength of radiation incident on the mesh, c the velocity of light, σ the conductivity of the metal, ε_0 the permittivity of free space, a the strip half-width, and g the center-to-center distance between the strips of the mesh.

The actual finesse achieved in practice is limited by various "imperfections." It can be written

$$\frac{1}{F^2} = \frac{1}{F_R^2} + \frac{1}{F_\parallel^2} + \frac{1}{F_f^2} + \frac{1}{F_\theta^2}, \tag{52}$$

where F_R is reflection finesse defined previously. If one reflector is rotated so that its edge is displaced by a distance Δ from the plane parallel to the other reflector, then

$$F_\parallel = \lambda/2\Delta_\parallel. \tag{53}$$

If the surface of a reflector is not flat, but has a Gaussian distribution of surface defects with a root-mean-square amplitude Δ_{f1}, then

$$F_f = \lambda/4.7\Delta_{f1}. \tag{54}$$

However, if the surface is bowed in a spherical distortion, as is likely in a fast-scanning instrument, where the center of the reflector is displaced by a distance Δ_{f2}, then

$$F_f = \lambda/2\Delta_{f2}. \tag{55}$$

If the input beam of radiation is not a parallel beam, but is conically shaped with cone angle θ, then

$$F_\theta = 2/m\theta^2. \tag{56}$$

Figure 4 shows a diagram of the Costley rapid-scanning interferometer. Radiation enters through the side of an aluminum cylinder which is being

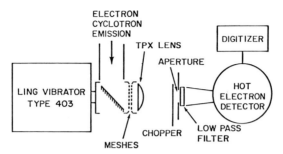

Fig. 4. Schematic diagram of the rapidly scanning Fabry–Perot interferometer developed by Walker and Costley.

vibrated rapidly. The radiation then reflects off a mirror inside the aluminum cylinder and is transmitted through an electroformed metal mesh which is attached to the end of the cylinder. Opposite this mesh is another mesh fixed to an adjustable mount attached to a micrometer-driven translation stage. These two meshes form the reflecting plates of the Fabry–Perot interferometer. Subsequently, the radiation is guided by a lens through a chopper, and a low-pass filter to the detector.

There are several advantages to the Fabry–Perot system when operated as a scanning instrument. It can have a high resolving power, and fast scan time, although to combine these requires good engineering. Inherently, the device is simple. However, there are disadvantages. The system is difficult to calibrate, and it can only scan a limited spectral range. Still, if its potential can be realized, it can be the simplest system for measuring temperature profiles.

IV. Applications

A. Tokamaks

The most successful application of electron cyclotron radiation diagnostics has been on tokamak devices. Each of the systems discussed in the previous section has been applied to a tokamak, and in this section examples of the kind of data acquired by these systems will be presented.

First, it will be useful to take a brief excursion into the theory of cyclotron emission as it applies to tokamaks in particular. If we keep to the simplest formulation, the main principles can be clearly expressed. With $l\omega_c \gg \omega_p$ where $\omega_p^2 = e^2 n / \varepsilon_0 m$ and R is the major radius coordinate, the optical depth is

$$\tau = \frac{e^2}{2\varepsilon mc} \frac{l^{2l-2}}{(l-1)!} n \left(\frac{kT_e}{2mc^2}\right)^{l-1} \frac{R}{\omega_c}. \tag{57}$$

5C. Plasma Diagnostics Using Electron Cyclotron Emission

Usually, the vacuum chamber of a tokamak has a high reflection coefficient at the frequencies associated with electron cyclotron emission. If we adopt a simple model for these reflections (Costley et al., 1974), the observed specific intensity is

$$I(\omega) = I_{BB}\{[1 - \exp(-\tau)]/[1 - \rho \exp(-\tau)]\}, \qquad (58)$$

where ρ is the reflection coefficient of the vacuum chamber walls. With $\tau \gg 1$

$$I(\omega) \to I_{BB} \propto (\omega^2 T_e), \qquad (59)$$

and since we have $\omega = l(e/m)(B_0 R_0/R)$, where B_0 is the magnetic field strength at the major radius $R = R_0$, a measurement of $I(\omega)$ can be transformed into one of $T_e(R)$. Defining $I/I_{BB} = I'$ and solving (58) for τ, we get

$$-\tau = \ln[(1 - I')/(1 - \rho I')]. \qquad (60)$$

If $\rho = 0$, and $T_e(R)$ is known, Eqs. (57) and (60) can be used to recover $n_e(R)$ from a measurement of $I(\omega)$. Accuracy decreases as $I' \to 1$. With $\rho \neq 0$ and unknown, it still may be possible to recover $n_e(R)$ (Boyd, 1980).

Figure 5 shows results obtained with the swept heterodyne receiver described in Section III. Only four scans are shown covering the 210-ms interval although about 15 scans were made during this time. The scalloped structure in the profiles is an artifact of the calibration procedure. The times associated with each scan are nominal times chosen by the diagnostician. The data reduction program then processes the data from the scan during which the nominal time occurred. Only half the electron temperature profile can be measured because of the limited bandwidth of the system.

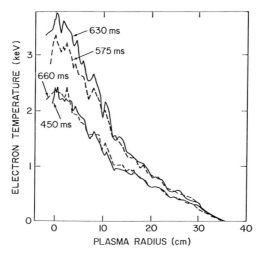

Fig. 5. Data from the scanning heterodyne receiver shown in Fig. 1. A set of fou temperature profiles measured before, during, and after injection of deuterium neutral beams into the plasma in the PLT tokamak.

The measurements were made before, during, and after injection of 2.1 MW with deuterium neutral beams between 450 and 600 ms into a low-density hydrogen plasma. The peaking of the electron temperature 30 ms after the beams switches off is produced by the still-circulating energetic ions.

Figure 6 illustrates the results from a Fourier-transform spectrometer system (Stauffer and Boyd, 1978) similar to the one discussed in Section III. In this case, the current in the discharge was ramped down to about one-third of its value at 200 ms. The temperature profile deduced from the $2\omega_c$ emission was measured from $r = -20$ to 40 cm every 25 ms, but only every second scan and radii less than $r = 15$ cm are plotted for the sake of clarity. For radii less than -20 cm, data reduction is confused by the overlap of the second and third cyclotron harmonics.

From this set of profiles it is possible to notice that the temperature falls rapidly at the plasma edge, but more slowly in the center. In addition, the equilibrium is disturbed and the center of the plasma shifts outwards from $r = 4$ to $r > 13$ cm.

The output of the multichannel grating spectrometer is different from that of the other systems described here in that it does not scan but follows the radiation intensity at fixed frequencies, and hence the electron temperature at several fixed spatial locations. Figure 7 shows some data taken with the system described in the previous section. In this case, the plasma was heated by the injection of 1.2 MW of neutral-beam power. Time histories of the electron temperature at three radial positions are shown. The signals were

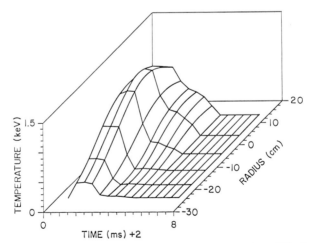

Fig. 6. Data from a Fourier-transform spectrometer similar to that of Fig. 2, developed by Stauffer and Boyd (1978). A set of temperature profiles is presented which were taken during an experiment on PLT in which the plasma current was ramped down to one-third of its peak value. Less than half the data collected are shown, to keep the figure reasonably clear.

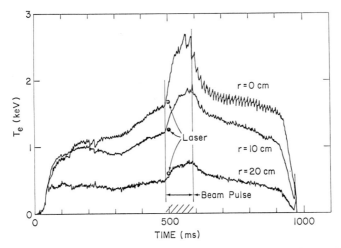

Fig. 7. Data from the multichannel grating spectrometer shown in Fig. 3. The electron temperature is shown as a function of time at three radial positions during a discharge within which hydrogen neutral beams were injected into the PLT tokamak. 1.2 MW $H^0 \rightarrow D^+$. $\bar{n}_e = 4.3 \times 10^{13}$.

sampled every millisecond and many temperature fluctuations with a longer period can be discerned in the data. This time resolution, simultaneously obtained at a number of points, is the greatest virtue of this device.

Figure 8 concludes the examples of tokamak data. It shows five successive temperature profiles obtained with the fast-scanning Fabry–Perot interferometer system described previously. The second-harmonic ($2\omega_c$) radiation from the DITE tokamak was scanned during the period from 9 to 22 ms after the start of the discharge. The fast scan time is particularly attractive, as is the finesse of about 30. Unfortunately, the entire temperature profile is

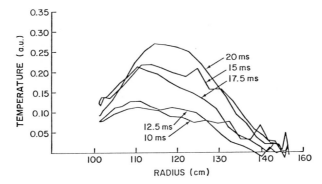

Fig. 8. Data from the Fabry–Perot interferometer shown in Fig. 4. Five successive electron temperature profiles obtained on the DITE tokamak are shown. The times shown refer to a zero time coincident with the initiation of the discharge.

not scanned because of the limitation on the maximum displacement of the moving reflector in the instrument.

B. Mirrors

The first experiments on mirror machines were done by Lichtenberg *et al.* (1964). They used a grating monochromator with a cryogenic InSb photodetector. These pioneering experiments gave the first data which could be compared with the theory of Trubnikov and Kudryavtsev (1958). Using several gratings, they measured the spectrum out to the tenth harmonic. An absolute calibration of their instrument made it possible to estimate the temperature and density of the hot electron plasma. The electron temperature approached 80 keV and the density about 10^{12} cm^{-3}.

Subsequently, a development of this work was pursued by Trivelpiece and collaborators at the University of Maryland. This effort reached its peak with the experiments of Tsakiris *et al.* (1978). In this case, a swept heterodyne receiver was used to measure the synchrotron radiation from a relativistic electron plasma. In the experiment, the measurement of the line shape of the radiation at the fundamental frequency was used to deduce the energy distribution function of the trapped electron ring and to study the evolution of the radial density profile. Figure 9 gives an example of their results. The evolution of the electron ring in the decaying magnetic mirror field is clearly demonstrated.

C. Bumpy Tori

Recently, two heterodyne systems have been applied to the study of the radiation from the relativistic electrons in the Oak Ridge and Nagoya Bumpy Tori. The Japanese measurements (Efthimion, 1980; Tanaka *et al.*, 1982) covered the frequency range 1–10 GHz, and produced line spectra in the vicinity of the fundamental and second-harmonic frequencies. An inaccurate analysis was used to estimate the "temperature" of the ring electrons.

The Oak Ridge experiments (Uckan *et al.*, 1980; Wilgen and Uckan, 1982) involved in the 80–110 GHz range. These frequencies correspond to the 10th to 15th harmonics of the fundamental cyclotron frequency. A featureless spectrum was observed. Good agreement with theory was claimed.

These experiments are the first to have been done and are, of necessity, very crude. However, they are certainly steps along an interesting road.

V. Concluding Remarks

As can be seen from the preceding discussion a number of different experimental devices have been used to measure the electron cyclotron emission

5C. Plasma Diagnostics Using Electron Cyclotron Emission 245

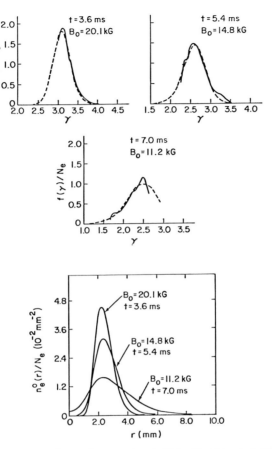

Fig. 9. Data from a swept heterodyne receiver built by Tsakiris *et al.* (1978). Three electron energy distribution functions and radial density profiles showing the evolution of a ring of relativistic electrons confined in a mirror magnetic field are shown.

spectrum emitted by plasmas in a variety of plasma confinement devices. Each approach has its own advantages and disadvantages. The choice of the apparatus for a particular measurement should be arrived at by a comparison of the available systems. There is no best system *a priori*.

A gratifying success has been achieved in measuring the electron temperature profile in tokamak and stellarator devices. Nonetheless, very little progress has been made outside of this specific measurement. There are as yet no density profile measurements, no magnetic field component measurements, and the extension of even the temperature profile techniques to other plasma confinement devices such as mirrors, bumpy tori, and pinches is in its infancy. At this stage it seems that the slow progress is a result of a lack of effort by experimentalists rather than any obvious insurmountable intrinsic difficulty.

On the theoretical side, a disturbing difficulty is associated with the significant number of erroneous results in the literature. In addition, there are a fairly large number of typographical errors in important expressions in a number of papers. Let the reader beware. Fundamentally good progress has been made in the theory over the last twenty years. The complications encountered when $\omega_p^2/\omega_c^2 \to 1$ and even 2 have been or are being worked out. The theory of emission from plasma with a Maxwellian velocity distribution is in a mature state, but non-Maxwellian distributions are poorly covered. Diagnostic techniques which may be possible for relativistic electrons have been barely touched by theoretical analysis.

Three extensions of current theory would be useful. It would be helpful to have an analytic formula for the absorption coefficient for waves propagating at small angles with respect to the magnetic field in high-density plasmas. Diagnostic applications on mirror machines would be the obvious benefactors. Second, it would be desirable to have the theory of the propagation of the state of polarization of the emitted radiation developed so that the external measurements of the polarization state could be related to the magnetic field orientation at the source. Last, our diagnostic capability would be increased if the theory of emission for non-Maxwellian relativistic distributions imbedded in a denser but low-temperature background plasma were written down in explicit detail, emphasizing the diagnostic possibilities.

Finally, two remarks are in order. In some sense in this subject the ultimate question was asked at the beginning. That is, what is the power lost through cyclotron radiation from a plasma at thermonuclear temperatures? Currently our diagnostic capability is such that we can tackle this problem experimentally in existing plasma confinement devices. Our ability to accurately predict these losses will be necessary for the design of economic fusion reactors, particularly those which use advanced fuels such as D–D and ^{3}He–D. And last, because of developments over the last ten years relatively simple systems have revolutionized the measurement of electron temperature in plasma confinement devices.

Acknowledgment

This work was supported by the U.S. Department of Energy and the National Science Foundation.

References

Baker, E. A. M., and Walker, B. (1982). *J. Phys. E.* **15**, 25.
Beard, D. B. (1959). *Phys. Fluids* **2**, 379.
Bornatici, M., Englemann, F., Novak, S., and Petrillo, V. (1981). *Plasma Phys.* **23**, 1127.
Bornatici, M. (1982). *Plasma Phys.* **24**, 629.
Boyd, D. A. (1980). *Int. J. Infrared Millimeter Waves* **1**, 45.
Celata, C. M., and Boyd, D. A. (1977). *Nucl. Fusion* **17**, 735.

5C. Plasma Diagnostics Using Electron Cyclotron Emission 247

Costley, A. E., Hastie, R. J., Paul, J. W. M., and Chamberlain, J. (1974). *Phys. Rev. Lett.* **33,** 758.
Efthimion, P. C. (1980). Personal communication.
Efthimion, P. C., Arunasalam, V., Bitzer, R., Campbell, L., and Hosea, J. C. (1979). *Rev. Sci. Instrum.* **50,** 949.
Englemann, F., and Curatolo, M. (1973). *Nucl. Fusion* **13,** 497.
Hayakawa, S., Hokkyo, N., Terashima, Y., and Tsuneto, T. (1958). *Proc. U.N. Int. Conf. Peaceful Uses At. Energy, 2nd, Geneva* A/conf. 15/P/1330.
Hutchinson, I. H., and Komm, D. S. (1977). *Nucl. Fusion* **17,** 1077.
Komm, D. S., Blanken, R. A., and Brossier, P. (1975). *Appl. Opt.* **14,** 460.
Lichtenberg, A. J., Sesnic, S., Trivelpiece, A. W., and Colgate, S. A. (1964). *Phys. Fluids* **7,** 1549.
Martin, D. H., and Puplett, E. (1970). *Infrared Phys.* **10,** 105.
Rutgers, W. R., and Boyd, D. A. (1977). *Phys. Lett. A* **62A,** 498.
Silin, V. P. (1955). *Tr. Fiz. Inst. im. P. N. Lebedeva, Akad. Nauk SSSR* **6,** 200.
Stauffer, F. J., and Boyd, D. A. (1978). *Infrared Phys.* **18,** 755.
Stepanov, K. N., and Pakhomov, V. I. (1960). *Sov. Phys.—JETP (Engl. Transl.)* **11,** 1126.
Tamor, S. (1982). Hot electron physics. *Proc. Workshop Hot Electron Rings, 2nd, ORNL, Oak Ridge, Tennessee* **2,** 689.
Tanaka, M., Hosokawa, M., Fujiwara, M., and Ikegami, H. (1982). *Proc. Workshop Hot Electron Rings, 2nd, ORNL, Oak Ridge, Tennessee* **1,** 339.
Trubnikov, B. A., and Kudryavtsev, V. S. (1958). *Proc. U.N. Int. Conf. Peaceful Uses At. Energy, 2nd, Geneva* A/conf. 15/P/2213.
Tsakiris, G. D., and Davidson, R. C., (1977). *Phys. Fluids* **20,** 436.
Tsakiris, G. D., Boyd, D. A., Hammer, D. A., and Trivelpiece, A. W. (1978). *Phys. Fluids* **21,** 2050.
Tsakiris, G. D., and Ellis, R. F. (1982). Phys. Publ. No. 83-048, Univ. of Maryland, Dept. of Physics and Astronomy.
Uckan, T., Wilgen, J., and Bighel, L. (1980). *Bull. Am. Phys. Soc.* **25,** 832.
Ulrich, R. (1967). *Infrared Phys.* **7,** 37.
Walker, B., Baker, E. A. M., and Costley, A. E. (1981). *Phys. E.* **14,** 832.
Wilgen, J., and Uckam, T. (1982). *Proc. Workshop Hot Electron Rings, 2nd, ORNL, Oak Ridge, Tennessee* **2,** 635.
Winske, D., and Boyd, D. A. (1983). *Phys. Fluids* **26,** 755.

5D
Particle Plasma Diagnostics[†]

C. F. Barnett

Physics Division
Oak Ridge National Laboratory
Oak Ridge, Tennessee

I.	Introduction	249
II.	Particle Diagnostic Atomic Physics	251
III.	Neutral-Particle Spectrometers Used in Determining Ion Temperatures	257
IV.	Plasma Ion Density and Effective Charge by Neutral-Beam Attenuation	278
V.	Beam Scattering Diagnostics	282
VI.	Impurity Ion Density	286
VII.	Magnetic Field Measurements	290
VIII.	Heavy-Ion Beam Probe	296
	References	303

I. Introduction

Since the late 1950s when high temperature plasmas became available for study in the laboratory, atomic particle diagnostics have been widely used throughout the world to determine plasma parameters. Historically, these diagnostics were first used in measuring the ion temperature on the Alpha plasma (Afrosimov *et al.*, 1960) in the Soviet Union and on the DCX-1 magnetically mirror-confined plasma in the United States (Barnett *et al.*, 1961). Intensive efforts have been expended throughout the intervening years to develop these techniques to their present state of sophistication. The terminology has evoled to describe the diagnostic methods as either passive or active. In the passive description atomic interactions occur between the plasma particles and produce neutral particles which escape the plasma periphery or boundary. Placing suitable detectors external to the plasma boundary provides data which can be related to the plasma parameters. An example of the passive technique is the determination of plasma ion

---

[†] Research sponsored by the Office of Fusion Energy, U.S. Department of Energy under Contract W-7405-eng-26 with the Union Carbide Corporation.

temperature by measuring the energy distribution of escaping H^0 atoms and relating this distribution to plasma ion temperatures through a knowledge of atomic collision cross sections or rates. Active diagnostics involve the probing of a plasma with a neutral or charged atomic beam and observing the results of the beam particle interaction with the plasma particles. Ion temperatures can be determined by using the same detection and analysis technique as with the passive mode but with the added advantage that local plasma properties can be obtained. By sweeping the line of sight of the detector assembly along the probing beam path, spatial profiles of the ion temperature are obtainable.

Methods using atomic interactions or collisions within a high temperature plasma have been applied to quantitatively determine a wide variety of plasma parameters such as plasma ion density and temperatures along with their spatial profiles, space potentials, plasma fluctuations, and the ohmic heated current flux in tokamak plasmas. A few of these plasma parameters (e.g., space potential) have been accessible only to beam probing techniques.

Many advantages are inherent in using particle diagnostics to obtain properties of high temperature plasmas. A listing of some of the more obvious ones follows.

   (i) Spatial and temporal profiles of plasma parameters are obtainable. Using beam probing methods, local plasma properties with 1-cm spatial resolution have been measured. Interfacing with computers permits time resolutions from microseconds to milliseconds on a continuous basis throughout the plasma discharge.

   (ii) Particle velocity or energy distribution functions for plasma ions have been determined.

   (iii) Neither passive nor active techniques perturb the gross plasma properties.

   (iv) The flux of escaping particles or electromagnetic radiation arising from the interaction is sufficient to produce a large signal-to-noise ratio at the detector.

   (v) For plasmas heated by neutral beams, particle diagnostics permit the study of energy transfer from fast ions to plasma ions and electrons.

   (vi) Since most of the relevant atomic cross sections are known or calculable, analysis and unfolding of the raw data are unambiguous.

Although the advantages of particle diagnostics far outweigh the disadvantages, difficulties are encountered in the use of these techniques. Foremost are the particle attenuation of the neutrals escaping the plasma when both passive and active methods are used and the added attenuation of the probing beam when active methods are used. To accurately assess the effects of beam attenuation one must know electron, proton, H-atom, and impurity density and temperature profiles. In tokamak plasmas, toroidal asymmetries are present in the H^0 density distribution due to the trapping by resonant charge exchange of the neutral heating beam, desorption of H_2 from the limiter, and gas puffing in various toroidal locations to sustain the discharge. Care must be taken to ensure that ion temperatures measured during neutral-beam heating are not influenced by distortions produced in the high-energy tail of the ion distribution function. The neutral-particle spectrum may be strongly affected by ions trapped in the small ripple field produced by the toroidal field coils or in trapped "banana" orbits that occur when the ratio of the particle toroidal velocity to the total velocity is less than some critical value. Thus, the particle orbits may produce effects that result in measurements not representative of the average plasma particle. Finally, added to the many difficulties in interpreting the measurements is the problem of costs in constructing ion sources and accelerators capable of projecting a beam of atoms or ions across a plasma without excessive attenuation. The plasma dimensions, temperature, and density will increase in the next generation of plasma experiments. Required will be intense probing beams up to mega-electron-volt energies which will demand development of new techniques for production of compact sources of ion and atom beams.

In the second section of this brief review the atomic collisional cross section or reaction rate data relevant to the measurements of particles escaping the plasma boundary and the probing of a plasma with a neutral beam will be discussed. This will be followed by an apparatus description and a discussion of the interpretation and implications of data obtained in measuring ion temperatures and H^0 atom density in tokamak- and mirror-confined plasmas. Subsequent sections will treat the use of neutral- and charged-particle beams to determine plasma proton and impurity densities, space potential, plasma fluctuations, and the ohmic current flux distribution in tokamak plasmas. Emphasis will be placed on the atomic physics relating to the measurements and interpretation. The isotopic species of hydrogen will be used interchangeably throughout the discussion.

II. Particle Diagnostic Atomic Physics

In utilizing the passive technique to determine the plasma ion temperature or ambient neutral-particle density, an understanding is required of the origin of the neutral atoms in the plasma interior and the collisional processes that determine the transport of neutral H atoms into the plasma interior where they act as charge centers to form H atoms representative of the plasma. Collisional processes determine the attenuation of these atoms as they escape the plasma and are subsequently detected. Hydrogen molecules incident on the plasma edge originate from two sources: (1) the desorption by particles and photons at the plasma wall and, for tokamaks, from the limiter; and (2) the injection of H_2 to sustain the plasma density. In addition to these sources energetic H atoms enter the plasma from the reflection of H^+ or H^0 at surfaces. This H component is a small fraction of the H_2 molecular component that feeds the plasma. As the H_2 molecule enters the plasma edge, dissociative collisions result in H^0 fragments with energies of a few electron volts. Table I summarizes the principal reactions or processes leading to dissociation and tabulates the reaction thresholds and energies of the H^0 fragments. The values listed are derived on the assumption that the H_2 molecule is in the ground vibrational state; the collision can be described as a Franck–Condon transition. Electron impact dissociation is the dominant collisional process leading to the formation of H^0. For vibrationally/rotationally excited H_2, the dissociation cross section increases as the quantum number increases. For example, the cross section for dissociative attachment increases by approximately four orders of magnitude as the vibrational level of H_2 increases from 0 to 4 (Wadehra and Bardsley, 1978).

A fraction of the 2–4 eV H^0 fragments is directed inward toward the plasma center. Through repetitive resonant charge exchange collisions, the energetic hydrogen atoms penetrate to the plasma center with an energy distribution representative of the plasma proton or deuteron distribution. One usually assumes that the bulk plasma is in thermal equilibrium such that the distribution can be described as Maxwellian. Thus, this H^0 distribution acts as target atoms for production by resonant electron capture collisions of the outwardly directed H^0 atoms. A typical H^0 density profile calculated for

TABLE I

Summary of the Principal Reactions or Processes Leading to Dissociation

Reaction	Process	Energy threshold from ground state (eV)	H^0 energy (eV)
$e + H_2 \rightarrow H + H + e$	Molecular dissociation	8.6	2.1–4.2
$e + H_2 \rightarrow H^* + H + e$	Dissociative excitation	18.2	
$e + H_2 \rightarrow H^+ + H + 2e$ $\rightarrow 2H^+ + 3e$	Dissociative ionization	>18.2	6.2–9.5
$e + H_2 \rightarrow H_2^- \rightarrow H + H^-$	Dissociative attachment	<1.0	2.1–4.2
$e + H_2 \rightarrow H_2^+ + 2e$	Molecular ionization	15.4	
$e + H_2^+ \rightarrow H + H^+ + e$	Molecular dissociation	24.7	4.7–7.3
$H_2^+ + H_2 \rightarrow H_2^0 \rightarrow H + H$	Resonant charge exchange	0	2.1–4.2

5D. Particle Plasma Diagnostics

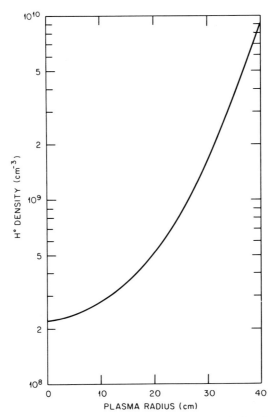

Fig. 1. H^0 neutral density profile generated by computer code for the Princeton Large Torus (PLT) tokamak; $n_e = 3 \times 10^{13}$ cm^{-3}, $T_i(0) = 3.7$ keV, $T_e = 2.7$ keV. [From Medley and Davis (1979).]

a tokamak plasma is shown in Fig. 1 (Medley and Davis, 1979). The neutral density decreases approximately two orders of magnitude across the 40-cm-radius plasma. The profile is dependent on the electron density and temperature.

Observations on the high temperature, high-density Alcator tokamak revealed that as the average electron density $\bar{n}_e$ was increased from 1×10^{14} to 4×10^{14} cm^{-3}, the flux of H^0 escaping the plasma periphery decreased to a level below the detector sensitivity. By modifying the detecting apparatus, energetic neutral atoms escaping the plasma were again measurable. The escaping flux was indicative of a central H^0 density of 10^8–10^9 cm^{-3}, whereas calculations using known cross sections predicted the central density to be 10^2–10^3 cm^{-3} (Gaudreau et al., 1978). Further analysis showed that by including electron–ion recombination collisions in the computations, the H^0 density was increased four to five orders of magnitude. Hydrogen recombination cross sections have been formulated by adding to the exact quantum calculations for radiative recombination into the 1s level of H an additional empirical term to account for capture into all nl levels (Gordeev *et al.*, 1977). The results take the form

$$\sigma_{rr} = \sigma_{rr}(1s)(1.2 + 0.28\alpha), \tag{1}$$

where $\alpha = Ze^2/hv$, with v the relative velocity between the ion and electron, Z the atomic number, and h Planck's constant. The validity of this formula is restricted to those plasmas with electron temperature $T_e > 1.5$ eV. The dependence of the e–H$^+$ recombination rate averaged over a Maxwellian distribution is shown in Fig. 2. As the plasma temperature increases from 1 to 10 keV, the rate decreases by four and a half orders of magnitude. Application of these rates to the high-density Alcator plasma is shown in Fig. 3, where the H^0 radial profile calculated with radiative recombination included is compared to the calculated profile, assuming no recombination collisions. Inclusion of the radiative recombination process in the calculations results in good agreement with the experimentally determined profile and the magnitude of escaping H^0 flux.

An additional source of neutral particles is present when intense neutral beams are injected to heat the confined plasma. The predominant collisional process that traps or ionizes the incoming energetic H^0 or D^0 is again resonant charge exchange. Reaction products include thermal neutrals with the same thermal velocity or energy distribution as the plasma protons or deuterons. Since the injected neutral beam penetrates the plasma, a radial distribution of thermal H^0 is produced by charge exchange collisions and

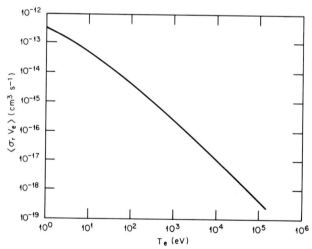

Fig. 2. Dependence of H$^+$ recombination rate on plasma electron temperature. [From Gordeev *et al.* (1977).]

5D. Particle Plasma Diagnostics

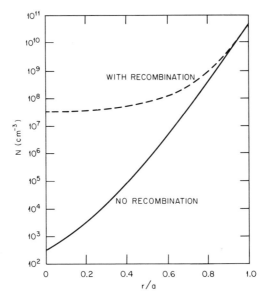

Fig. 3. Computed H^0 neutral density radial profile over a plasma column with and without radiative recombination for an average plasma density of 5×10^{14} cm^{-3}. [From Dnestrovskij et al. (1979).]

beams injected at different spatial positions in the plasma, producing large toroidal asymmetries in tokamak plasmas. For example, the thermal H^0 density is increased by a factor of 100 along the injected beam path.

In active-beam diagnostics of a plasma, the collisional processes that determine the penetration of the energetic probing beam and the escape of the thermal H^0 from the plasma center are identical to those found when the passive technique is used. The principal collisional process that attenuates a neutral particle beam traversing a plasma is the electron capture collisions with plasma ions and ionization by electrons and positive ions. Plasma ions include both the principal components H^+ and impurity ions of C, O, Fe, Cr, etc. Pertinent cross sections leading to H^0 attenuation by protons and electrons are shown in Fig. 4 as a function of the H^0, H^+, or electron energy. To obtain an "effective" electron ionization cross section for attenuation of an H atom in a plasma, it is necessary to average the electron ionization cross section over the Maxwellian electron velocity distribution at plasma temperature T_e. This effective cross section $\langle \sigma_{ei} v_e \rangle / v_{H^0}$ is plotted in Fig. 4 for two plasma temperatures at 1 and 5 keV. The "effective" cross section must be used when $v_e > v_{H^0}$. These cross sections have been approximated by Riviere (1971) with the following analytic expressions:

Electron capture:

$$\sigma_{ex} = 0.6937 \times 10^{-14}(1 - 0.155 \log_{10} E)^2/(1 + 0.112 \times 10^{-14} E^{3.3}). \quad (2)$$

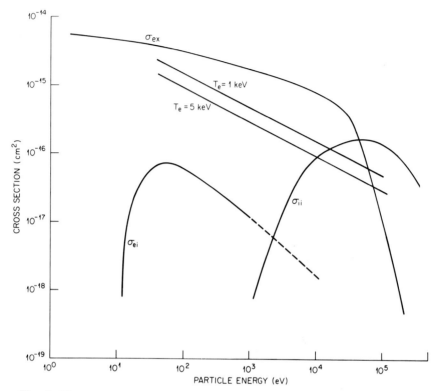

Fig. 4. Electron and hydrogen ion cross sections relevant to H^0 attenuation in a plasma; $\sigma_{ex} - H^+ + H \to H^0 + H^+$, $\sigma_{ii} - H^+ + H \to 2H^+ + e$, $\sigma_{ei} - e + H \to H^+ + 2e$. For electron ionization when $v_e > v_{H^0}$, the quantity $\langle \sigma_{ei} v_e \rangle / v_0$ is plotted, where σ_{ei} has been averaged over a Maxwellian electron velocity distribution at plasma temperatures of 1 and 5 keV. [From Barnett et al. (1977).]

Ionization by positive ions:

$$\log_{10}(\sigma_{ii}/Z^2) = 0.8712(\log_{10} E)^2 + 8.156 \log_{10} E - 34.833, \quad (3)$$

when $E < 150$ keV.

$$\sigma_{ii} = Z^2/E(3.6 \times 10^{-12}) \log_{10}(0.1666E),$$

when $E > 150$ keV.

Electron ionization:

$$\sigma_{ei} = 6.153 \times 10^{-14} g(x)/E^2,$$

$$g(x) = \frac{1}{x}\left(\frac{x-1}{x+1}\right)\left[1 + \frac{2}{3}\left(1 - \frac{1}{28}\right) \ln(2.7 + \sqrt{x-1})\right],$$

$$x = E/13.605, \quad (4)$$

5D. Particle Plasma Diagnostics

when E is the relative energy of H^0 in electron volts and cross sections are in square centimeters.

For H^0 energies less than 30–40 keV, the attenuation is dominated by charge exchange; above 40 keV, proton ionization dominates. For electron temperatures of 100 eV found in the plasma outer region, the electron ionization cross section is approximately equal to the cross section of a 300-keV H^0 that is ionized by a plasma proton. The reader is referred to papers by Riviere (1971) and Galbraith and Kammash (1979) for a discussion of the numerical integration of the rate coefficients $\langle \sigma v \rangle$ for a monoenergetic beam interacting with a plasma with a Maxwellian distribution.

Interactions of hydrogen neutrals with multicharged impurity ions should also be considered in computing the transport of an H^0 beam in a plasma. The total cross section for electron loss of an energetic H^0 beam passing through a plasma is the sum of the electron capture cross section and the ionization or stripping cross section. For relative particle velocities less than 4–5×10^8 cm/s corresponding to energies of 60–80 keV/amu, impurity electron capture cross sections dominate the total loss cross section. Thus, in considering the attenuation of thermal H^0 by impurities, one needs only to include the electron capture cross sections. Typical results for high-energy electron capture and ionization cross sections for $H^0 + C^{q+}$ are shown in Fig. 5 (Phaneuf et al., 1978; Olson and Salop, 1977). For hydrogen energies greater than 50 keV, the total electron loss cross section can be scaled with energy and is given by the empirical formula (Olson et al., 1978)

$$\sigma_T = 4.6q(32q/E)[1 - \exp(-E/32q)] \times 10^{-16} \text{ cm}^2, \quad (5)$$

where E is the H^0 energy in kilo-electron-volts and q is the particle charge. The electron capture cross section scales as $q^{1.46}$. At lower energies where collisions of impurity ions with H^0 are important, scaling does not exist (Fig. 6). In Fig. 5 calculations were made using Monte Carlo classical methods, while calculation of the cross sections in Fig. 6 was performed using full quantum close-coupling methods. In Fig. 6 results of the calculations of electron capture cross sections are shown for fully stripped C, O, and Fe, H-like C^{5+}, and He-like O^{6+} (Shipsey et al., 1981; Salop and Olson, 1977, 1979a,b). Experimental data have been obtained which substantiate the close-coupling theoretical calculations for charge exchange cross sections of O and C ions in atomic hydrogen (Phaneuf et al., 1982).

III. Neutral-Particle Spectrometers Used in Determining Ion Temperatures

Excellent reviews of the design and use of neutral-particle analyzers have been published by Afrosimov and Gladkovskii (1967), Petrov (1976), and Kislyakov and Krupnik (1981). A schematic diagram illustrating the experi-

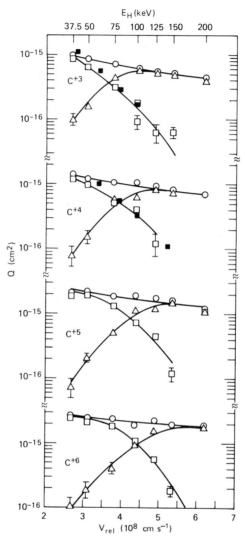

Fig. 5. Cross sections for $C^{q+}(q = 3 - 6)$ + H collisions. The open circles denote the total electron loss cross section (i.e., ionization plus charge exchange). The theoretical cross sections for charge exchange are shown as open squares and should be compared to experimental measurements, denoted by the solid squares. Theoretical cross sections for impact ionization are shown as open triangles. [From Olson and Salop (1977).]

mental arrangement commonly used to determine ion temperatures and other plasma parameters is shown in Fig. 7. In active particle diagnostics an ion source and neutralizer produces a beam of energetic neutral particles which is projected across the plasma to be studied. Detector 1, usually a thermal- or secondary-electron emission detector, measures the attenuation

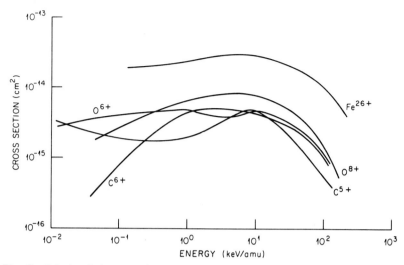

Fig. 6. Calculated charge exchange cross sections for plasma impurity ions with atomic hydrogen. For energies less than 25–50 keV/amu, close-coupling methods were used to compute the cross section. At energies greater than 50 keV/amu, classical methods were used. C^{5+}, O^{6+} + H, Shipsey et al. (1981); O^{8+} + H, Salop and Olson (1979a); C^{6+} + H, Salop and Olson (1977); Fe^{26+} + H, Salop and Olson (1979b). [Adapted by permission of *J. Phys. B,* © 1981, The Institute of Physics.]

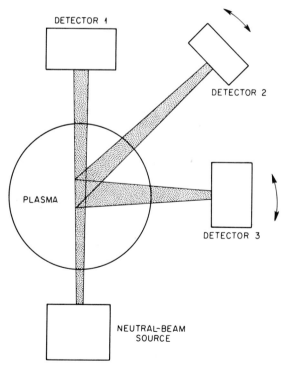

Fig. 7. Schematic diagram illustrating the geometrical configuration in the active-particle diagnostic technique.

of the beam, which provides information as to the plasma density and impurity content. Located at angle θ is a second detector that can be scanned across the neutral-beam trajectory. This detector can be a simple secondary-electron-emission-type detector to detect the H^0 scattered by the plasma ions or it can be a more sophisticated detector. Located usually at an angle of 90° is detector 3, which consists of a gas cell used to ionize or strip the H^0 atoms leaving the plasma, an electric and magnetic field analyzer to determine the energy and momentum distribution of the stripped H^0, and a particle detector. Using a neutral probing beam, detector 3 measures only those events occurring in the volume bounded by the intersection of the beam and the solid angle subtended by the detector. Without the probing beam, detector 3 operates in the passive mode and detects all those events in the detector line of sight. Ideally, detector 3 is arranged to rotate in both the poloidal and toroidal directions. Replacement of the particle analyzer detector system with an optical spectrometer permits the measurement of impurity ion density through optical radiation that arises from electron capture into excited states of the impurity ion. The use of detectors 2 and 3 to determine other plasma parameters will be discussed in the following sections.

In the diagnostic passive mode the measured spectrum of neutral particles is a superposition of electron capture events along the radius r. Use of a probing beam provides local plasma parameters. However, in the design of the source four requirements must be satisfied. First, the beam intensity must be sufficient to provide an H^0 density for charge exchange with plasma ions that is greater than the ambient plasma H^0 density. This condition can be expressed as follows:

$$J_B/J_0 \cong n_B l/n_0 L, \tag{6}$$

where J is the neutral flux escaping the plasma, n is the neutral density, l is the thickness or diameter of the injected neutral beam, L is the length along the direction of sight of detector 3, and subscripts B and 0 refer to beam and residual particles, respectively. In designing an experiment, attempts are made to obtain a flux ratio of 10 or better, which results in an excellent detector signal-to-noise ratio. To increase the detector signal-to-noise ratio, the beam can be modulated with phase-sensitive detection. The technology of modern ion sources and neutralizers is capable of producing H^0 beams with densities of 10^8-10^9 cm^{-3}, which may be compared to the 10^7-10^8 cm^{-3} plasma neutral density found on axis in tokamak plasmas. Second, the neutral beam energy must be such that it will readily penetrate the plasma. Third, the beam intensity must not perturb the plasma. Care must be taken to provide differential pumping between source and plasma to prevent the neutralizer cell gas from entering the plasma region. Finally, the collimation of the neutral beam must provide a well-defined interaction volume to achieve spatial resolution of 2–3 cm. To obtain high current densities in the neutral probing beams, mass analysis has not been used; the result is that H^0

beam particles have not only the initial accelerating energies E_0, but $\frac{1}{2}E_0$ and $\frac{1}{3}E_0$ formed from H_2^+ and H_3^+ dissociative collisions in the gas-neutralizing cell. Also, an H_2^0 component may be present. These added mass and energy components complicate data analysis.

The complexity of a neutral-particle spectrometer design and its use depend on the plasma parameters. For measurements of a one-component, dc or long-pulse plasma, the ion temperature can be determined to the required accuracy with a single-channel analyzer. If the plasma is composed of two or more species and if ion temperatures as a function of time and position are required, it is necessary to use sophisticated multichannel analyzers.

Schematic diagrams of various neutral-particle analyzers are shown in Figs. 8–10. The simplest is the single-channel analyzer shown in Fig. 8 (Barnett and Ray, 1972). H^0 particles escaping the plasma are stripped in the gas cell; the protons formed enter a parabolic electrostatic analyzer at 45° and are detected by a channel electron multiplier. The geometry is compact to minimize the adverse effects of scattering from the stripping collisions. The energy resolution ΔE is given by

$$\Delta E/E = (\Delta x_1 + \Delta x_2)/x, \qquad (7)$$

where Δx_1 and Δx_2 indicate the analyzer entrance and exit slit width separated by distance x. Thus, the entrance and exit aperture can be chosen for

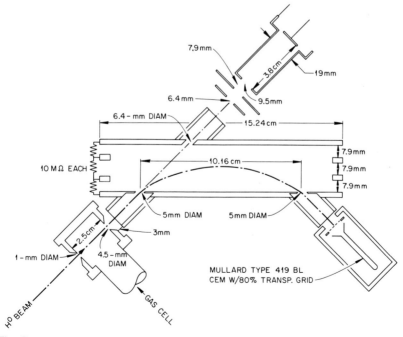

Fig. 8. Line drawing of a single-channel neutral-particle analyzer. [From Barnett and Ray (1972).]

optimum energy resolution and at the same time provide a sufficient counting rate at the detector. The positive potential on the back electrode can be ramped to sweep across the particle energy range during a few milliseconds. This geometry has been extended to multichannels (Berry *et al.*, 1974; Cordey *et al.*, 1974; Becker, 1970; Summers *et al.*, 1978; Takeuchi *et al.*, 1977). The principal disadvantage of the multichannel energy analyzers is that the minimum to maximum energy measurable for one voltage setting is one-half.

A design that is more flexible is shown in Fig. 9 (Afrosimov *et al.*, 1975). Neutrals are stripped in a gas cell, momentum analyzed by a magnet, energy analyzed by a five-channel cylindrical analyzer, and detected by open-dynode electron multipliers. An additional advantage of the magnetic-energy tandem analyzer is that mass discrimination permits the measurement of a H^+ thermal plasma heated by D^0 injection. A similar analyzer has been used to measure the energy distribution of ions in the 2XIIB mirror-contained plasma at Lawrence Livermore Laboratory (Nexsen *et al.*, 1979). The analysis and detection system consists of 15 channels that cover the 0.5–50 keV energy range for D^+.

To minimize scattering problems encountered by the analyzer shown in Fig. 10, the TFR group (Equipe TFR, 1978) has used a compact, single-channel analyzer which has been expanded to seven channels on the ISX plasma (Neilson, 1980). A schematic diagram of the analyzer is shown in Fig. 10. The short path length between gas cell and channeltron multiplier detectors minimizes the loss of ions due to inelastic scattering in the gas cell.

From the simple analyzers in use in the 1960s, developments have evolved to the highly sophisticated and complex neutral-particle diagnostic system to be installed on the TFTR plasma facility (Kaita and Medley, 1979; Medley *et al.*, 1980). Planned are 12 analyzers for perpendicular ion temperature measurements and 8 for tangential measurements of the slowing-down

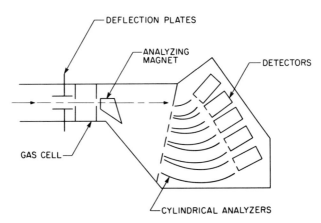

Fig. 9. Five-channel momentum-energy neutral-particle analyzer. [From Afrosimov *et al.* (1975).]

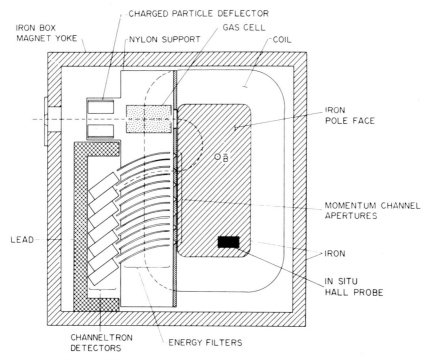

Fig. 10. Diagram of the 180° momentum analyzer and the seven-channel cylindrical analyzer used on the impurity study experiment (ISX). [From Neilson (1980).]

spectra of beam-injected ions. The analysis section consists of an electric field parallel to the magnetic field direction. Particles entering the combined E and H fields will be deflected in the direction of E with a displacement proportional to m/e and independent of particle velocity, where m is the ion mass and e the charge. The magnetic field deflects the entering ions through 180°. Thus, at the collector position the H^+, D^+, and T^+ will be separated into three columns of ions with each column being energy dispersed. The collector will be three 60-cm semicontinuous channel multiplier arrays. Only through interfaces with high-speed computers is it possible to record and analyze the massive quantities of data from systems such as this.

In reducing the raw data to obtain the energy or velocity distribution function, corrections must be applied for the energy resolution of the analyzer. For analysis of energy only, the analyzer dispersion is expressed as

$$\frac{dx}{dE} = \text{const}, \qquad (8)$$

where x is the particle displacement along the direction of the beam deflection. If dx is fixed or all channels have identical aperture widths, the dispersion can be written as

$$\frac{\Delta E}{E} = \frac{\Delta x}{x} \propto \frac{1}{E}. \tag{9}$$

For a given energy distribution the collector intensity of each channel increases with energy. For momentum energy analyzers the dispersion is

$$\frac{dx}{d(mv)} = \text{const} \tag{10}$$

and if Δx is fixed,

$$\frac{\Delta E}{E} \propto \frac{1}{E^{1/2}}. \tag{11}$$

The conversion efficiency of an analyzer depends on the gas-cell stripping coefficient, on scattering that results in loss of particles between gas cell and collector, and on the detector efficiency. To obtain an absolute calibration including these three factors, the analyzer is calibrated using monoenergetic beams. Care must be taken to ensure that the detector sensitivity remains constant over a long period of time. The major problem encountered in the calibration procedure is the absolute detection of neutral hydrogen particles. In previous work, the common practice has been to detect the electrons ejected when energetic atoms impinge on a metallic target and to assume that the secondary emission coefficient was the same for H^+ and H^0. By determining the secondary emission coefficient for H^+, the flux of H^0 could be determined. According to the theory of Hagstrum (1954, 1956, 1961) secondary-electron emission is dominated at high impact energies by the kinetic emission process. For H^+ or H^0 energies of 100–200 eV, potential emission dominates kinetic emission with the secondary emission coefficient independent of energy below 100–200 eV. Thus, for low-energy impact the H^0 secondary emission may be vastly different from that for H^+. Stier et al. (1954) found that the ratio H^0/H^+ of the secondary emission coefficient γ_0^-/γ_+^- for a gas-covered surface increased from 1.1 at 20 keV to 1.3 at 200 keV. To within an uncertainty of 25%, a value of 1.18 for λ_0^-/γ_+^- was found by Kislyakov et al. (1975) for H^0 and H^+ impact on Cu, CuBe, Mo, and Al in the energy range 0.2–8 keV. Ray et al. (1979) produced a known H^0 flux by photodetaching H^- within the cavity of a YAG laser ($\lambda = 1064$ nm). The ratio λ^0/λ^+ is plotted as a function of the impact energy in Fig. 11. The ratio was 1.15 ± 0.08 over the particle energy range 30–1600 eV. For a gas-covered surface, the secondary yield was independent of the metal substrate. Comparison of the yield as a function of energy with the cross section for ionization of O_2 by H indicates the same energy dependence for the negative yields and cross sections. Indications are that negative ions and electrons emitted from a gas-covered surface are a consequence of binary

5D. Particle Plasma Diagnostics

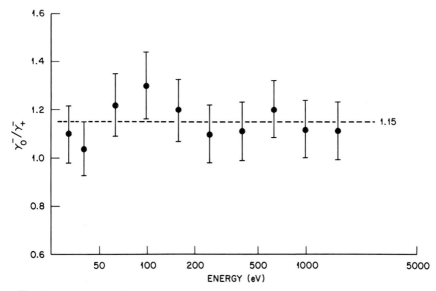

Fig. 11. Ratio of the H^0 secondary-electron emission to that of H^+ in the energy range 30–1600 eV. The surface was gas-covered Cu. [From Ray *et al.* (1979).]

collisions between the incoming low-energy ion or atom and the molecules absorbed on the surface.

The design of the analyzer and choice of a stripping gas for the conversion cell depend critically on a gas or vapor with a large stripping cross section and small elastic and inelastic scattering cross section. With a knowledge of the total scattering the conversion cell and analyzer geometry can be optimized for maximum detector signal. Inelastic scattering cross sections have been measured for 1–10 keV stripping of H^0 in several gases. The results of this investigation (Fleischman *et al.*, 1974) can be summarized as (1) the scattering cross section at a given energy increased as the target particle mass increased, (2) the scattering cross sections for H and D stripping were identical at the same center-of-mass energy, and (3) the scattering resulting from electron capture collisions is small compared to scattering from stripping collisions.

Using the neutral-particle analyzer as shown in Fig. 8, the conversion efficiencies for seven gases were determined (Barnett and Ray, 1972). Results obtained are shown in Fig. 12, with N_2 having the greatest conversion efficiency for gas-cell areal density less than 1.5×10^{15} cm^{-2}. With N_2 admitted to the conversion cell, the conversion efficiencies were obtained in the energy range 0.1–10 keV for cell pressures 1–10 mtorr as shown in Fig. 13. At 10 keV the conversion efficiency was 0.01–0.05, decreasing to 10^{-4}–10^{-3} at 0.1 keV. These efficiencies are typical of other analyzer designs that have been used. As the particle energy was increased above 10 keV, the

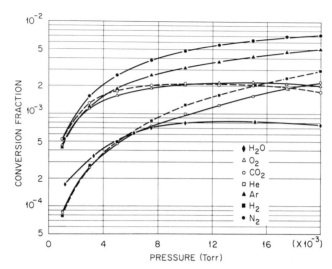

Fig. 12. Neutral-particle analyzer conversion efficiency as a function of gas-cell pressure for seven target gases; the analyzer is shown in Fig. 8. [From Barnett and Ray (1972).]

maximum conversion efficiency occurred at approximately 50 keV. In Fig. 13 the D^0 results are plotted at one-half the incident energy or at equivelocity H^0. At energies >400 eV the conversion efficiencies for H and D are identical, while at lower energies the H^0 and D^0 efficiencies diverge. This is a consequence of the stripping cross section for H^0 and D^0 being identical at equivelocity and the scattering cross section being energy dependent. The energy at which the divergence occurs is determined by the geometry of the conversion cell and analyzer. To avoid contamination of the plasma by the stripping gas, H_2 gas is often used. The conversion efficiency was a factor 8–10 less for H_2 than that found for N_2 gas.

To increase the conversion efficiency, a pulsed gas cell has been used to increase the target density to an equilibrium charge state density ($nl \sim 10^{16}$ cm^2) (Bezlyudnyi *et al.*, 1975). The gas cell consisted of a cylinder 9 cm long in which internal baffles with 2-cm apertures were placed to impede the gas flow and act as a gas delay line. A fast-acting mechanical valve admitted gas into the cell to an equilibrium pressure in 15 ms. The gas pressure remained at its equilibrium value for 15 ms and thereafter decayed to its $1/e$ point at the end of 60 ms. Using helium as the stripping gas the conversion efficiency was increased by a factor of 20 over that obtained with a static pressure cell. For 2–7 keV H^0, the conversion efficiency to H^+ was nearly constant between 10 and 15%. For H^0 energies less than 2 keV, where inelastic scattering is important, the efficiency decreased to 0.08% at 500 eV. The pulsed gas cell has been used to increase the sensitivity of the analyzer on the Alcator high-density tokamak plasma facility (Dnestrovskij *et al.*, 1979). The increased

5D. Particle Plasma Diagnostics

conversion efficiency has permitted the measurement of ion temperatures from plasmas with average densities as great as $1\text{–}5 \times 10^{14}$ cm^{-3}. In using the analyzer in the pulsed mode for plasma of 1 s or longer duration, the gas cell must be repeatedly pulsed to obtain temporal resolution of the ion temperature.

In the outer regions of tokamak plasmas the ion temperatures are less than 100 eV. Since the conversion efficiency for a gas stripping cell for H^0 energies less than 100 eV is of the order of 10^{-4}, other techniques must be employed to provide a detector signal above the background noise. Also, most of the particles escaping the plasma are in the low-energy region. An absolute measurement of the low-energy particles being ejected at the plasma surface will provide important data in assessing wall erosion by

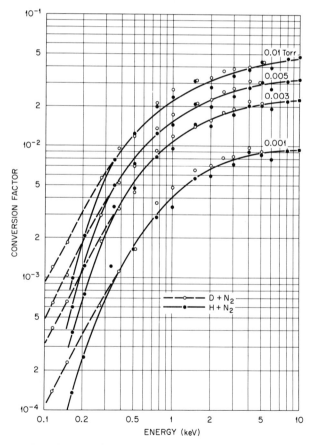

Fig. 13. Neutral-particle analyzer conversion efficiency for H and D incident on a N$_2$ gas cell. Results for D atoms are plotted at one-half the incident energy. [From Barnett and Ray (1972).]

particle sputtering. Two apparatuses have been used to measure low temperature plasmas and the flux of escaping H^0: (1) a neutral-particle analyzer that uses a cesium vapor cell to convert H^0 to H^- (Brisson *et al.*, 1980) and (2) a time-of-flight spectrometer (Voss and Cohen, 1980, 1982).

The overall efficiency of a neutral-particle analyzer for measuring the energy distribution of low-energy H^0 has been increased by replacing the gas cell with a cesium vapor cell in which electron capture collisions produce H^-, which is subsequently energy analyzed. Not only is the electron capture cross section to form H^- much larger in this energy region than the H^0 stripping cross section but, the scattering is markedly less. A schematic diagram of the apparatus is shown in Fig. 14. A heat pipe (Bacal and Reichelt, 1974; Bacal *et al.*, 1974) was used to contain the Cs vapor and prevent contamination of the plasma. At a temperature of 140°C the Cs vapor pressure was 1×10^{-3} Torr, which produced the optimum Cs density for maximum conversion of a 100-eV H^0 beam. The conversion efficiency of the analyzer is shown in Fig. 15 by the dashed curve for 0.1–6 keV H^0. For a 100-eV H^0 beam the efficiency was two and a half orders of magnitude greater than that for the N_2 cell as shown by the solid curve. The analyzer has been used to determine the ion temperature on the ELMO Bumpy Torus (EBT) plasma whose ion temperatures are less than 100 eV. More recently a four-channel analyzer has been calibrated and used successfully on a tokamak plasma (Thomas and Neilson, 1981).

A time-of-flight spectrometer has been used to determine the energy spectrum of neutral particles escaping a tokamak plasma in the energy range 20–1000 eV (Voss and Cohen, 1980, 1982). Neutral particles ejected from the plasma are gated on a 1-μs time scale by a slotted, rotating, chopper disk. Particle detection was by means of secondary emission from a BeCu surface. Particle emission spectra of D^0 from the PLT discharge during the initial phase of the discharge is shown in Fig. 16. Two spectra are shown—one at 30 ms into the discharge and the other at 50 ms, at which time the plasma central ion temperature is near its steady-state value. The plasma edge tem-

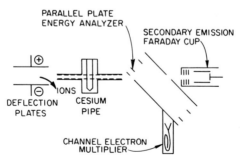

Fig. 14. Schematic diagram of the Cs heat-pipe conversion cell and single-channel neutral-particle spectrometer. [From Brisson *et al.* (1980).]

5D. Particle Plasma Diagnostics

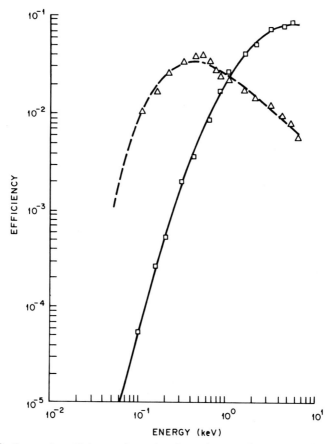

Fig. 15. Conversion efficiency of a cesium heat pipe as a function of H⁰ energy. Open triangles are the experimental data, the dashed curve is a fourth-order polynomial least-squares fit to the data, and the solid curve and open squares are data for an equivalent N_2 conversion cell. [From Brisson et al. (1980).]

perature has been approximated to be 18 eV at the early time rising to 52 eV at 50 ms.

In mirror-confined plasmas a gridded, retarding potential analyzer has proved useful in providing the energy distribution of ions escaping out the mirror ends. A schematic diagram of the analyzer is shown in Fig. 17. The analyzer is designed to meet the following criteria: (i) the entrance grid must have low transmission such that the Debye length $(kT_e/4\pi n_e e^2)^{1/2}$ inside the analyzer is larger than the grid spacing; (ii) the acceptance angle must be greater than $\pm 25°$ to accept ions lost by scattering from the mirror loss cone; and (iii) the radius of the ion and electron repeller grid must be greater than the radius of the entrance aperture by two ion gyroradii so that all ions at the entrance grid have the same probability of reaching the grids of the collector.

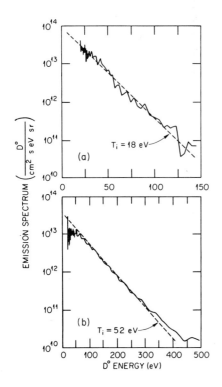

Fig. 16. D^0 emission spectra from the PLT discharge during the initial phase of the discharge. [From Voss and Cohen (1980, 1982).]

Gridded detectors designed according to these criteria have been used successfully to measure ion energies escaping along the mirror magnetic field lines in the range of 50–1000 eV.

In analyzing the measured ion spectrum obtained with a stripping cell and electrostatic analyzer, not only must corrections be made for the energy

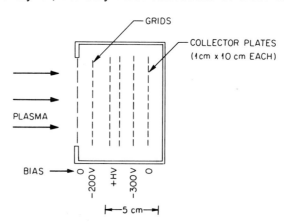

Fig. 17. Schematic diagram of a gridded energy analyzer. The potential arrangement has a double analyzer grid to assure a uniform retarding potential. [From Osher (1979).]

5D. Particle Plasma Diagnostics

dependence of the analyzer conversion efficiency and resolution but also corrections must be applied to the particle attenuation as the particles traverse the plasma from center to edge. Eubank (1979) has written a general expression for the attenuation of the neutral particle flux diffusing through a plasma:

$$\text{Attenuation} = \exp\left(-\sum_{j,k} \int_r^a \sigma_{kj}(E_j) f(E_j, r) \, dE_j \, dr'\right), \tag{12}$$

where j denotes the plasma particle species—electron, proton, impurity ion, etc.; k is the sum over all collisional interactions; a is the plasma radius; E_j is the energy of particle j; and r is the plasma radius at which the neutral particle was formed. This formula can be expressed in explicit form as

$$\text{Attenuation (from plasma center)} = \exp[-an_i(\sigma_{ex} + \sigma_{ii} + \sigma_s) + an_e\langle\sigma_{ei}v_e\rangle/v_b] \tag{13}$$

where n_i and n_e are average values of plasma ion and electron density, σ_{ex} is the cross section for charge exchange with plasma ions, σ_{ii} is the ionization cross section by protons and impurity ions, σ_s is the scattering cross section, σ_{ei} is the electron ionization cross section, v_e and v_b are the electron and H^0 particle velocities, respectively, $\langle\sigma_{ei}v_e\rangle$ is the electron ionization rate coefficient averaged over the velocity distribution, and a is the plasma radius.

It is seen in Fig. 4 that the dominant collisional processes leading to particle or beam attenuation are resonance charge exchange and electron ionization. Thus, a convenient expression is introduced to signify the plasma capture cross section,

$$\sigma_{pc} = \sigma_{ex} + \langle\sigma_{ei}v_e\rangle/v_b, \tag{14}$$

which for typical tokamak plasmas is $1-2 \times 10^{-15}$ cm^2.

In addition to providing data needed to obtain plasma ion temperatures, knowing the cross sections involved in the beam attenuation calculations permits an estimate of particle lifetime in a plasma and the equilibrium time. Particle lifetimes are given by

$$\tau_{ex} = 1/n_a\langle\sigma_{ex}v_r\rangle, \tag{15}$$

where n_a is the H^0 density and v_r is the relative velocity between ions and atoms. Typical lifetimes in tokamak plasmas are 0.01–0.1 s. In similar manner the equilibrium time can be estimated by assuming that the hydrogen flux leaving the plasma is equal to the incoming flux. With this assumption the equilibrium time is

$$\tau_{eq} \simeq (n_i\sigma_{ex}v_r)^{-1}, \tag{16}$$

which for tokamak plasmas is a few microseconds.

With the above formulas approximate corrections can be made to the neutral-particle analyzer data. For more precise corrections consideration must be given to the density and temperature profile, which for tokamak plasmas is adequately described by a profile of the form

$$T(r) = T(0)[1 - (r/a)^u]^m. \quad (17)$$

Elaborate computer codes have been developed to determine the ion temperature from the particle energy distribution as measured by the analyzer (Goldston, 1978).

The spectrum of charge exchange neutrals obtained from the PLT high temperature tokamak plasma is shown in Fig. 18 (Eubank *et al.*, 1979; Medley and Davis, 1979). Plotted is the natural log of the charge exchange flux as a function of the particle energy. The charge exchange flux was corrected in Fig. 18 for the analyzer conversion efficiency and energy resolution. If a Maxwellian energy distribution, which for one dimension is given by

$$f_n(E) = K[E^{1/2}/(kT)^{3/2}] \exp(-E/kT\ dE), \quad (18)$$

is assumed, the plasma ion temperature is given by the slope of the plot represented by the straight line. For the data of Fig. 18, 1.6 MW of H^0 was injected into a D^+ plasma, heating the plasma to 4.5 keV. The particle analyzer detected those thermal D^0 which escaped the plasma. At energies less than 3–4 keV the distribution departs from a Maxwellian, and the parti-

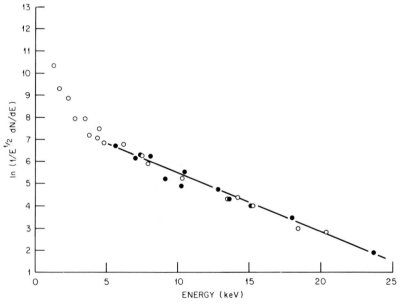

Fig. 18. PLT charge exchange neutral energy distribution. The solid line is a least-squares fitting to the data points between 1.5 and 6.0 kT_i resulting in an ion temperature of 4.5 keV. The symbols represent data taken during four discharges. [From Medley and Davis (1979).]

5D. Particle Plasma Diagnostics

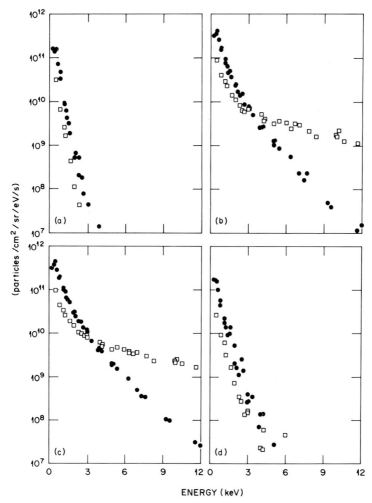

Fig. 19. Neutral-flux spectra at different times during a neutral-beam-heated discharge in the ISX tokamak. The low-energy thermal and fast-neutral components were measured with a seven-channel momentum energy analyzer. Closed circles are the D^0 spectrum; open squares are the fast H^0 spectrum. (a) $t = 60$ ms; (b) $t = 100$ ms; (c) $t = 150$ ms; (d) $t = 186$ ms. [From Neilson (1980).]

cles in this energy region are representative of the particles in the plasma outer region.

To illustrate the necessity of using a combined momentum–energy analyzer for a plasma heated by neutral-beam injection, the neutral spectra of particles from a D^+ plasma heated by an H^0 beam are shown in Fig. 19 (Neilson, 1980). The data were taken with the mass analyzer shown in Fig. 10 for a tokamak plasma with 1 MW of injected H^0 beam. The four energy spectra of Fig. 19 portrays the time history of the plasma heating. For particle energies greater than 7 keV the H^0 fast flux is one to two orders of magnitude greater than D^0 from the thermal plasma. Without the capability of mass discrimination it would be impossible to obtain accurate ion temperatures with neutral-beam heating.

Ion temperature measurements in plasmas confined by magnetic mirror configurations are basically the same as the measurements in toroidal containment devices in that the charge exchange flux is representative of the plasma ion energy distribution. However, owing to particles being lost from the containment volume by scattering into the loss cone, a thermal distribution is not obtainable. Instead an average energy is used to characterize the energy distribution. This quantity is defined in the usual manner as

$$\overline{E} = \int Ef(E) \, dE \bigg/ \int f(E) \, dE, \tag{19}$$

where the integrals can be represented as sums over all the analyzer channels. The energy spectra from the 2XIIB mirror plasma is shown in Fig. 20 at four times during the discharge (Nexsen et al., 1979). For these data a 430-A neutral D^0 beam was injected into the confining field. The mean energy $\overline{E}$ was 10–11 keV until the injected beam was interrupted at 6.6 ms. After turn-off the mean energy decayed with an e-folding time of 360 μs.

Many questions have been raised as to the validity of assuming that the escaping neutral-particle flux energy distribution is representative of the true plasma ion energy distribution. Foremost is the question of the distortion of the high-energy tail of the distribution under neutral-beam heating conditions when the beam power is equal to or exceeds the ohmic heating power of a tokamak plasma. In the process of unfolding the neutral-particle spectrum, the procedure has been to consider that section of the particle energy curve in which $E \geq 2kT_i$. Two separate studies have been made to investigate any distortion produced by the heating beam as it thermalizes through Coulomb collisions with the confined ions and electrons. In one of the studies the Fokker–Planck equation for the equilibrium thermal ion distribution was solved for both low and high particle velocities (Bittoni et al., 1980). The study disclosed that the high-energy tail is distorted and non-Maxwellian for neutral-beam heating when particle loss is dominated by thermal conduction. When particle loss occurs principally by charge exchange or diffusion, the high-energy tail should be Maxwellian. The distortion increases with beam power. As a guide to experiment the theory predicts that reliable ion temperatures should be obtained for $v_{\text{plasma neutrals}} \ll v_{\text{injected neutrals}}$.

The second investigation involved the solution of the nonlinear Fokker–Planck equation (Killeen and Marx, 1970). In this study it was found that the particles in the high-energy tail thermalize among themselves. Thus, the high-energy tail is representative of the actual ion temperature.

In laboratory high temperature plasmas, the departure of the velocity or energy distribution from a Maxwellian is not known. In tokamak plasmas the

5D. Particle Plasma Diagnostics

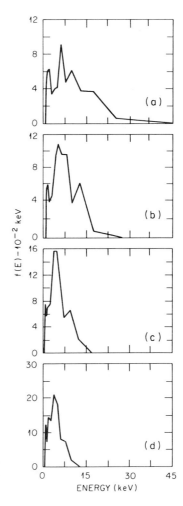

Fig. 20. Energy distribution of neutral particles escaping the 2XIIB mirror-contained plasma. [From Nexsen et al. (1979).]

current flow, potential gradients, and other perturbations result in nonequilibrium conditions. To obtain a degree of confidence in ion temperature measurements several experimental methods have been compared. Results from three measurements are shown in Fig. 21, where the ion temperature is plotted as a function of time during PLT discharges (Eubank et al., 1979). A 1.6-MW H^0 beam was injected into an ohmically heated D^+ discharge 450 ns after the discharge was initiated. An ion temperature of 4 keV was measured by the passive charge exchange technique as shown in Fig. 21a. Figure 21b shows the ion temperature deduced from the D–D neutron yield. Large errors are introduced by this method since the absolute on-axis D^+ density must be known to calculate the reaction rate at temperature T_i. The presence of multicharged impurity ions makes it difficult to obtain the D^+ den-

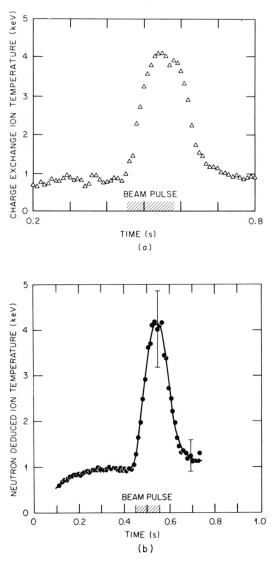

Fig. 21. Variation of ion temperature during a PLT tokamak discharge. (a) Ion temperature determined by charge exchange measurements. Peak temperature was 4 keV obtained with injection of 1.6 MW of H^0 into a D^+ plasma. (b) Ion temperature as measured by total neutron emission. Ion temperature was 4 keV with 1.4 MW H^0 into a D^+ plasma. (c) Ion temperature of 4 keV obtained from Doppler broadening of the Fe(XX) 2665-Å line with 1.6 MW H^0 into a D^+ plasma. [From Eubank et al. (1979).] (*Continued on next page.*)

5D. Particle Plasma Diagnostics

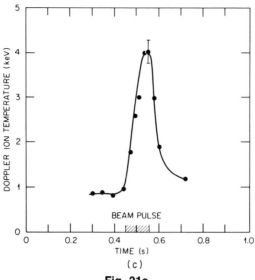

Fig. 21c.

sity. The results of the third method are shown in Fig. 21c, where the Doppler broadening of the impurity ion Fe(XX) was used to obtain the ion temperature. A quadrupole "forbidden" transition of Fe(XX) at 2665 Å was detected with a grazing-incidence spectrometer. Observing "forbidden" transitions permits the use of conventional spectrometers, which are easier to calibrate in the UV spectral region. Increasing the beam power from 1.6 to 2 MW increases the ion temperature from 4 to 6 keV where the K_α line of Fe(XXV) was used to measure the 2p-1s transition at 1.85 Å using a crystal spectrometer. For the higher temperatures where the ion charge increases, additional "forbidden" transitions need to be identified. In the Doppler broadening measurements made from tokamak plasmas the line profiles can be described by a Gaussian.

Both the charge exchange and neutron yield techniques are subject to high-energy tail distortion. The Doppler broadening technique is not influenced by this uncertainty but has an uncertainty introduced in that the measured impurity ion temperature is not the same as the D^+ temperature for beam-heated plasmas. Energy lost to the impurity ions by the injected particle as the ions thermalize in the plasma is proportional to $n_i Z_i^2 / m_i$, where Z_i is the nuclear charge of particle of mass m_i. Thus, the energy transfer to impurity ions increases faster with the ion nuclear charge than with the mass, resulting in the preferential heating of impurity ions. Corrections taking into account this classical energy transfer must be applied to the Doppler broadening measurements. Comparing the data of Figs. 21a–21c indicates that ion temperatures measured by the three techniques give satisfactory agreement and are within the experimental errors.

IV. Plasma Ion Density and Effective Charge by Neutral-Beam Attenuation

Plasma electron densities are accurately determined either by laser Thomson scattering or interferometry using microwaves or submillimeter laser beams. If impurity ions are not present in the plasma, $n_e = n_{H^+}$. However, with impurity ions constituting a part of the positive charge fraction, the H^+ ion density is no longer equal to the electron density but to $n_e - n_i Z_i$, where n_i is the impurity ion density and Z_i is the electronic charge.

One of the methods that have been utilized to determine the proton density in a plasma consists of projecting an H^0 beam across several chords of a plasma and measuring the resulting attenuation. By performing an Abel inversion, a radial profile is obtained if the plasma geometric dimensions are known. This technique has proven to be successful on several high temperature plasmas during the past two decades. Eubank et al. (1965) projected a 2.5–5.0 keV H^0 across both a plasma blob and a stellarator plasma and determined the H^+ density in the range 10^{13}–10^{15} cm^{-3}. In the Soviet Union the method has been used extensively to determine the ion density in a toroidal plasma compression facility (Berezovskii et al., 1974), in a tokamak (Kislyakov and Petrov, 1970), and in the Alpha device (Afrosimov et al., 1966). In these measurements the attenuation due to the presence of impurity ions was assumed to be negligible. More recently the TFR group (Equipe TFR, 1978, 1979) have made beam attenuation measurements on the TFR tokamak plasma taking into account Mo^{q+} and O^{q+} impurities in the plasma. Mirror-machine experiments have relied almost exclusively on H^0 beam attenuation to determine ion densities.

The diagnostic apparatus installed on the TFR facility consisted of an ion source, an accelerator to accelerate the ions to energies of 2–40 keV, and a gas neutralizing cell to convert H^+ to H^0. On the opposite side of the plasma, a 2-cm aperture defined the H^0 beam after traversing the plasma. Passing the beam through a 100-Å C foil converted the H^0 to H^+. The H^+ beam was passed through a magnetic field to select only those protons with full accelerating energy. After momentum analysis the ions were detected with a continuous-strip electron multiplier. The source-detector was scanned across the plasma on a shot-to-shot basis.

As discussed in Section II the collisional processes causing beam attenuation are charge transfer with protons; ionization by plasma protons and electrons; elastic scattering by plasma electrons and protons; charge transfer, ionization, and elastic scattering by impurity ions; and elastic scattering and stripping by neutral gas surrounding the plasma.

To calculate the beam attenuation across a single plasma chord, Eq. (13) can be rewritten as

5D. Particle Plasma Diagnostics

$$\frac{I}{I_0} = \exp\left[-\left(\sigma_c \int_{l_p} n_p \, dl + \frac{\langle \sigma_{ei} v_e \rangle}{v_b} \int_{l_p} n_e \, dl \right.\right.$$
$$\left.\left. + \sigma_s \int_{l_0} n_0 \, dl + \sum_{i,q} \sigma_{iq} \int n_{iq} \, dl\right)\right]. \quad (20)$$

In this expression the first term is the attenuation due to charge exchange, ionization, and elastic scattering collisions with plasma protons integrated over the length of the plasma chord l_p. The second term is the attenuation by plasma electrons with $\langle \sigma_{ie} v_e \rangle$ being the reaction rate coefficient averaged over the electron thermal distribution. The third term arises from the stripping and scattering of the probing beam with the neutral gas particles outside the plasma. Finally, the last term is the collisional interaction with plasma impurity ions, and the cross section σ_{iq} is the sum of the charge exchange, ionization, and scattering cross sections of impurity ions with D^0. Attenuation of the beam due to gas scattering and charged-particle scattering depends on the diameter of the probing beam in relation to the acceptance angle of the detector. If the beam and detector designs are optimized, such that the detector accepts all those particles scattered through $\pm 2°$ (on the TFR experiment), the elastic scattering cross section is much less than the other cross sections leading to attenuation and can be neglected. In practice this condition is difficult to achieve, and an estimate of the elastic scattering is obtained by filling the plasma container with H_2 at a pressure of approximately 1×10^{-4} Torr. Measurement of the beam attenuation by H_2 not only gives the attenuation by elastic scattering but also evaluates the third term of Eq. (20). In tokamak plasmas with densities $\sim 3 \times 10^{13}$ cm^{-3}, the attenuation due to scattering and stripping has been found to be small and can be neglected.

In comparing the cross section of Fig. 4, it is seen that for an H^0 probing beam with energies less than 20 keV (40 keV for D^0), the resonant charge exchange cross section dominates the H^+ ionization cross section. Thus, for H^0 beam energies less than 20 keV one needs only to consider charge exchange with protons, electron ionization, and collisions with impurity ions in analyzing attenuation measurements. From Fig. 5 the same behavior is expected for impurity-ion collisions such as C^{q+}. The impurity ionization cross section is equal to the electron capture cross section in the energy region 60–100 keV. Generally, the energy at which the $\sigma_{iq} \cong \sigma_{ex}$ increases with increasing charge state.

The attenuation of a 17-keV H^0 beam transmitted through a H^+ TFR tokamak plasma as a function of electron line density is shown in Fig. 22. On the semilog plot, the experimental data lie on a straight line represented by

$$I/I_0 = \exp(-\sigma_c [n_e L]) \quad (21)$$

or

$$\sigma_c = (\ln I/I_0)/[n_e L]. \quad (22)$$

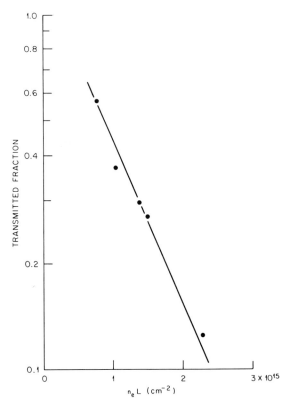

Fig. 22. Fraction of a 17-keV H⁰ beam transmitted through a H⁺ TFR tokamak plasma as a function of the electron line density obtained using an HCN interferometer. [From Equipe TFR (1979).]

The electron line density $n_e L$ was obtained using HCN laser interferometry. Since σ_c was dominated by the proton charge exchange collisions, n_e is directly proportional to n_{H^+}. For the TFR plasma the measured σ_c was $8.9 \pm 1.3 \times 10^{-16}$ cm² at 17 keV D⁰. The experimental value of σ_c can be compared to a calculated $\sigma_c = 9.9 \times 10^{-16}$ cm² assuming a pure-H plasma. To determine line densities and the effective Z, it was necessary to determine σ_c as a function of the probing D⁰ beam energy. Results of the attenuation as a function of the D⁰ energy are shown in Fig. 23. The full circles are experimental data for a 300-kA ohmic-heated discharge. Open squares are the data obtained when the plasma current was decreased to 200 kA. The solid line is a theoretical fit to the full circles and provides a line-integrated deuterium density given by $[n_D L]/[n_e L] = 0.80 \pm 0.09$. Thus, impurity ions make up 20% of the positive charge density in the plasma. The dashed curve is the calculated fit obtained when the plasma impurity concentration is assumed to be zero. The open circles are for the case when the ratio of deuteron to electron line density is 0.8, but the impurity cross sections are neglected in

5D. Particle Plasma Diagnostics

the calculation. Modeling of the impurity concentration and radial distribution indicated a plasma capture or attenuation cross section of 9.2×10^{-16} cm² when the impurity concentration was 3% C and 0.1% Mo. Effective Z values defined as

$$Z_{\text{eff}} = \sum n_i Z_i^2 \Big/ \sum n_i Z_i \tag{23}$$

ranged from 2.4 to 3.0, which was consistent with Z_{eff} obtained from plasma conductivity measurements. The computations have shown that the plasma capture cross section was insensitive to the uncertainty in the assumed impurity concentration. However, the plasma capture cross section is sensitive to the impurity density through the relation $[n_D L]/[n_e L]$. The TFR experiment indicates the importance of cross section data in determining the ion density that is not amenable to other diagnostic techniques.

Beam attenuation methods have been used extensively to determine ion density profiles in the tandem mirror experiment TMX (Cohen, 1980). Fifty-

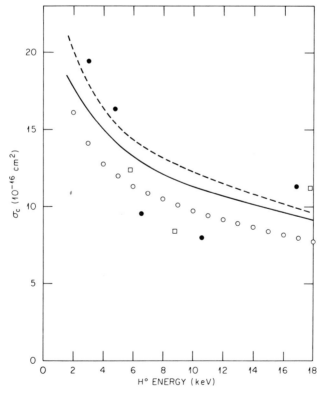

Fig. 23. Plasma capture cross section as a function of injection H⁰ probing beam energy: (●) 300-kA ohmically heated D⁺ plasma, (□) 200-kA ohmically heated D⁺ plasma, (—) theoretical curve for best fit to solid circles, (- - -) theoretical curve for $Z_{\text{eff}} = 1$, (○) calculated curve assuming $D = 0.8 n_e$. [From Equipe TFR (1979).]

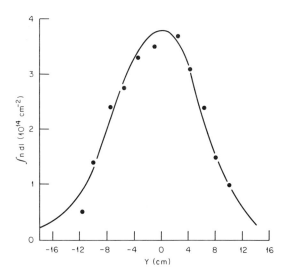

Fig. 24. Spatial profile of the tandem mirror experimental (TMX) plasma. Plotted is the H^+ line density as a function of the distance below and above the plasma center line. [From Cohen (1980).]

four tightly collimated secondary-emission-type detectors have been utilized to measure the attenuation of a neutral heating beam in the central cell and end plug regions of the mirror-confined plasma. The effective plasma capture cross section was calculated in a manner similar to that described for the TFR tokamak plasma. This cross section was 1.54×10^{-15} cm², which is approximately 70% greater than that found for the tokamak plasma. The line density distribution across the plasma is shown in Fig. 24. Peak line density observed was 3.9×10^{14} cm^{-2}. These measurements introduce large uncertainties in the measurement of absolute densities. The probing beam has three energy components resulting from the acceleration and neutralization of H^+, H_2^+, and H_3^+. Without beam analysis prior to the detector, there is no way to discriminate and correct for the attenuation of the three components of energy. In the experiment the detector "looks" directly at the plasma, which results in a detector noise arising from the electromagnetic radiation. Finally, the acceptance angle of the detector is $\pm 0.2°$, which implies that elastic scattering must be included in the data analysis.

Beam attenuation methods have limited usefulness. With the increased size and higher density plasmas anticipated in the future, the beam energies must be increased to hundreds of kilovolts in order for the beam to penetrate the plasma.

V. Beam Scattering Diagnostics

In 1960 Russek (1960) proposed that if a beam of particles were scattered by plasma particles, the differential scattering cross section $\sigma(T_i, \theta)$ would

5D. Particle Plasma Diagnostics

be a function of the thermal motion of the plasma. For both hard-sphere and Coulomb scattering, the number of particles scattered per unit solid angle was computed as a function of the scattering angle, plasma temperature, and density. It was shown that by measuring the particles scattered at two angles, the temperature and density could be determined for those collision partners where the masses were equal or when the plasma target mass was greater than the incident particle mass.

Later Abramov *et al.* (1972) considered injecting a beam into a plasma and measuring the energy distribution of the beam scattered at a fixed angle. The calculations indicated that the energy or velocity distribution of the scattered particles was determined by the plasma particle velocity distribution and, in fact, that the width of the measured distribution was directly proportional to the plasma ion temperature for a Maxwellian distribution.

It is well known that the differential elastic scattering cross section for particles with several kilo-electron-volts energy decreases rapidly with increasing scattering angle and increasing particle energy. Thus, in designing an experiment, the scattering angle at which the observations are to be made must be small ($\sim 5°$–$10°$) and the incident beam energy as low as possible to permit plasma penetration. These two requirements limit the application of this technique to plasmas with T_i in the range 100–500 eV (found in the outer layers of tokamak plasmas), theta pinches, field-reversed pinches, and other low temperature plasmas.

The energy distribution of particles of mass m_1 elastically scattered by a target particle of mass m_2 at zero target particle energy is given by the classical binary collision formula

$$E = E_0 \frac{m_1 \cos\theta + m_2[1 - (m_1/m_2)^2 \sin^2\theta]^{1/2}}{m_1 + m_2}, \quad (24)$$

where E_0 is the incident energy and θ is the scattering angle. This expression is valid only for small scattering angles. If $m_1 = m_2$, the distribution can be represented as

$$E = E_0 \cos\theta. \quad (25)$$

For small scattering angles and impurity ions in a plasma, one would expect that each impurity ion mass would be displayed as a peak in the scattered energy distribution. However, as the impurity ion mass becomes large (e.g., Fe^{q+}), the energy distribution would be centered near the incident energy, thereby requiring an energy analyzer with excellent resolution.

The expected energy spectrum of an elastically scattered beam by a plasma with ion temperature T_i has been given by Abramov *et al.* (1972) as

$$\Delta E \simeq 4\theta[E_0 k T_i (m_i/m_2) \ln 2]^{1/2}, \quad (26)$$

where ΔE is defined as the full width at half-height. This equation is valid only at small scattering angles and for $E_0 \geq 5$–$10 T_i$. Using this expression the

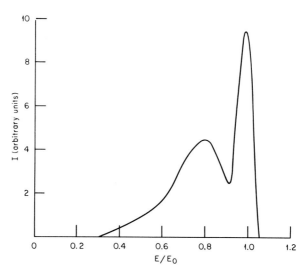

Fig. 25. Computed spectrum of He atoms scattered from a 200-eV plasma. The 10-keV He atoms were scattered through an angle of 13.5° by a plasma which was 98% H^+ and 2% O^{8+}. [From Afrosimov et al. (1976).]

expected spectrum has been computed (Afrosimov et al., 1976) for a 10-keV He beam elastically scattered from a plasma whose constituents were 90% H^+ and 2% O^{8+}. The results are shown in Fig. 25 for a plasma whose ion temperature was 200 eV. The oxygen peak is located at $0.99E_0$. For higher masses difficulties would be encountered in resolving the distribution.

The only instance of applying the beam scattering technique to measure ion temperature and density in high temperature plasma has been a tokamak (Aleksandrov et al., 1979; Berezovokii et al., 1980). To minimize the background signal, an 8-keV He neutral beam was injected into an H^+ plasma. The detector was at an angle of 9°. With the detector viewing the 1-cm² cross-sectional beam area, the longitudinal resolution was 9–10 cm. (A great disadvantage of viewing the plasma scattering of a particle beam or photon beam at small angles is the inherent bad spatial resolution along the beam.) The result of the measurements is shown in Fig. 26, where dN/dE is plotted as a function of the scattered particle energy. Curve 1 is the measured instrumental broadening as determined from the monoenergetic He beam. Curve 2 is the distribution obtained from gas scattering when $T_i = 0$. The open circles are the experimental points obtained for plasma scattering, which are in excellent agreement with the solid curve computed for a plasma with a 117-eV ion temperature.

An additional method employing elastic scattering of particles involves selecting the probing particle mass m_1 to be greater than the target mass m_2

(Afrosimov and Kislyakov, 1971). If $m_1 > m_2$, a sharp limiting scattering exists such that

$$\sin \theta_{\lim} = m_2/m_1. \quad (27)$$

For a gas or plasma with $T_i = 0$ the particles cannot be elastically scattered at angles given by $\theta_{\lim}$. The differential scattering cross section for He^{2+} ions elastically scattered by plasma protons has been calculated for several plasma temperatures. The resulting scattering curves are shown in Fig. 27 for various values of the ratio kT_i/E_0. The values of kT_i/E_0 range from 2.5×10^{-6} to 2.5×10^{-1}. As the plasma temperature increases, scattering occurs at angles greater than $\theta_{\lim}$. Curve 1 represents the scattering for $kT_i = 0$. The limiting scattering angle for He^{2+} on H^+ is 14°. Note that $d\sigma/d\omega$ decreases four orders in magnitude as the scattering angle is increased.

Below 14°, the differential scattering cross section is essentially independent of kT_i/E. To obtain a measurement of T_i, it would only be necessary to determine the ratio of the scattered particle intensity at a scattering angle greater than $\theta_{\lim}$ to that of the scattered intensity less than $\theta_{\lim}$. The same

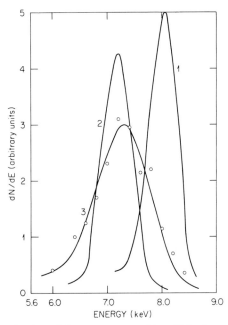

Fig. 26. Energy distribution of He atoms scattered by a T-4 tokamak H^+ discharge. Curve 1 is the energy resolution of the He beam without a plasma, curve 2 is the energy distribution of the He beam from gas scattering, curve 3 is the calculated energy distribution scattered from the plasma assuming an ion temperature of 117 eV; open circles are the experimental values obtained at a scattering angle of 9°. The energy of the incident He^0 beam was 8.05 keV. [From Berezovskii et al. (1980).]

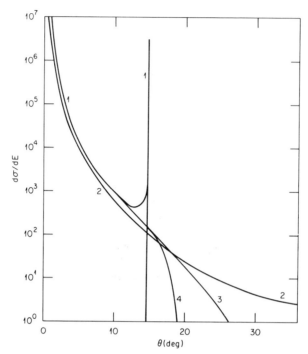

Fig. 27. Calculated differential scattering cross section for scattering of He^{2+} ions by H^+ for various ratios of the H^+ temperature kT and the incident He^+ energy E: (1) $kT/E = 0$, (2) 2.5×10^{-1}, (3) 2.5×10^{-2}, (4) 2.5×10^{-3}. [From Afrosimov and Kislyakov (1971).]

analysis can readily be extended from He^{2+} to He atoms since the elastic differential scattering cross section will be altered slightly by the electron screening of the He atom. The feasibility of this technique must be confirmed in future experiments.

VI. Impurity Ion Density

Usually, impurity ion concentrations in plasmas are determined by measuring the spectral line intensities of transitions from excited states. For high temperature plasmas, light ions such as C and O are completely stripped of electrons in the first few centimeters upon entering the plasma. As the ions diffuse into the plasma the only radiation emitted is the small amount resulting from radiative recombination collisions or charge exchange into excited states by collisions with H atoms. To overcome the difficulty in determining the impurity ion density, a technique has been developed to inject into the plasma an energetic beam of H^0 atoms and observe the radiation from capture into the ion excited state (Afrosimov *et al.*, 1977).

5D. Particle Plasma Diagnostics

The intensity of the radiation from state $n_2 l_2$ into state $n_1 l_1$ is given by

$$I(n_2 l_2 - n_1 l_1) = n_e n_i (A^{q+}) \langle \sigma_{rr} v_e \rangle + n(H^0) n_i (A^{q+}) \langle \sigma_{ex} v_i \rangle \alpha \beta$$
$$+ n_e n [A^{(q-1)+}] \sigma_{exc} v_e \rangle, \quad (28)$$

where n_e is the electron density, n_i is the impurity ion density, $n(H^0)$ is the hydrogen ion density at the point of observation, σ_{rr} is the radiative recombination cross section to level $n_2 l_2$, σ_{exc} is the electron excitation cross section, σ_{ex} is the electron capture cross section into state $n_2 l_2$, α is the fraction cascading into level $n_2 l_2$, and β is the branching ratio to go to state $n_1 l_1$.

The first term of Eq. (28) is the contribution to the radiation intensity from radiative recombination collisions, the second term arises from charge exchange collisions, and the third term results from electron excitation. For high temperature plasmas the radiative recombination is small compared to electron capture and excitation and may be neglected. Thus, if the charge exchange cross sections into a particular excited state and the branching ratios are known, the spectral line intensity can be determined by subtraction of the electron excitation contribution when the probing beam is turned off.

The electron capture cross section for fully stripped ions into principal quantum states by collisions with H^0 have been calculated by Salop and Olson (1977). Later these calculations were extended by Olson (1981) to include capture into the angular momentum quantum states l. In these calculations the Monte Carlo classical trajectory method was used. These computations have been confirmed by experiment for total electron capture cross sections for interaction energies greater than 30 keV/amu. The theoretical predicted cross sections for capture into various n levels for H^0 colliding with C^{6+} are shown in Fig. 28 for energies 25–100 keV/amu. Shown in parentheses beside each energy is the total electron capture cross section in units of 10^{-16} cm^2. This reaction can be characterized as follows: (1) the total cross section decreases rapidly with energy, (2) the cross sections are maximum for capture into the $n = 4$ level, and (3) the n level distribution broadens with increasing impact energy. Extension of the calculations to higher-Z fully stripped ions suggests that the most probable n level increases with Z being $n = 12$ for Ne^{20+}. The results of the calculations for capture into nl levels for He^{2+} to C^{6+} are shown in Fig. 29 for a 50-keV/amu collision. The n levels are shown by the numbers on the graph, with the abscissa denoting the orbital angular momentum quantum number l. For C^{6+} the capture probability is most probable into the $n = 4$, $l = 3$ level.

Capture into fully stripped impurity ions of O^{8+} (Isler *et al.*, 1981; Zinov'ev *et al.*, 1980) and C^{6+} (Afrosimov *et al.*, 1978, 1979) have been observed in tokamak plasmas by injecting a beam of H^0 atoms. The intensity of the observed spectral line can be expressed as

$$\Delta I = n_i J(r) \sigma_{ex} V \beta \alpha Y, \quad (29)$$

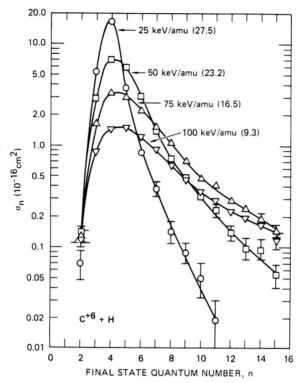

Fig. 28. Electron capture cross sections into level n for the C^{6+} + H reaction. The values in parentheses are the total cross sections (in cm^2) for the given energy. [From Olson (1981).]

where $J(r)$ is the hydrogen flux density of H^0 at the point of observation, V is the volume of the probing beam subtended by the spectrometer solid angle, Y is the spectrometer efficiency, and other terms are as defined in Eq. (28).

If the Lyman-α transition (hydrogenlike 2p–1s) is observed, the $n_2 l_2$ branching ratio becomes 1. The H^0 flux density at the volume under observation is determined by calculating the attenuation of the incident beam as a function of the electron temperature and density and the proton density. Under this procedure, calculations must be made to ensure that the impurities after capturing an electron are not reexcited by electron impact. Otherwise, the observed signal will be too large. A 12-keV, 3.5-mA/cm^2 H^0 beam has been used to determine the O^{8+} radial profile in a tokamak plasma (Zinov'ev et al., 1980). Radiation from the 2p–1s Lyman-α transition (19 Å) was observed with a photoelectron detector. A cascade factor of 0.55 was assumed for the 2p level. The results of the measurements are shown in Fig. 30, where the O^{8+} density is plotted as a function of the normalized radius. The O^{8+} density was constant out to $0.6a$, which is 0.25% of the electron

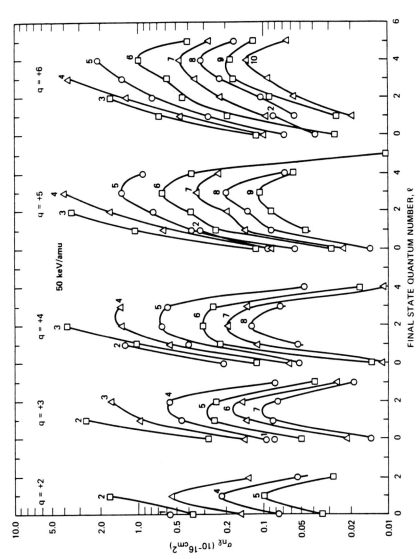

Fig. 29. Theoretical electron capture cross sections into nl levels for 50 keV/amu of fully stripped ions in charge states 2–6. The numbers identifying each curve represent the n level, while the ordinate denotes the l level captured into. [From Olson (1981).]

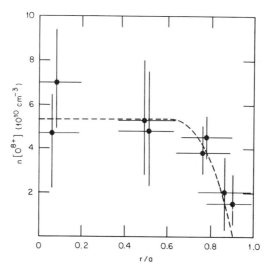

Fig. 30. Calculated radial profile of O^{8+} ions in the T-10 tokamak plasma at 300–400 ms into the discharge. The dashed curve is the computed steady-state, coronal approximation of the O^{8+} ions under the assumption that the oxygen was uniformly distributed in the plasma. The curve is normalized to the experimental data at small r/a. [From Zinov'ev et al. (1980).]

density. The large error bars are the result of large plasma oscillations that induced radiation from the plasma, making it difficult to distinguish the background from the beam-produced signal.

VII. Magnetic Field Measurements

In ohmically heated tokamak plasmas the current flow around the torus creates a poloidal magnetic field that provides plasma stability and confinement. Stability is achieved when the safety factor q, defined as $q = (r/R)(B_t/B_p)$ is greater than 1, where r is the plasma radius, R the torus major radius, B_t the toroidal magnetic field, and B_p the poloidal field. In steady state ohmically heated plasmas, the current density j is inversely proportional to the plasma resistivity, such that $j \propto T_e^{3/2}/Z_{\text{eff}}$. With auxiliary heating, high beta, fast transients in the plasma current, density or temperature can result in distortions in the current distribution (i.e., skin currents), leading to an instability or disruption. In such situations a diagnostic technique is required to sense the current distribution or the internal magnetic field.

Internal magnetic fields of high temperature plasmas may be inferred indirectly from external fields measured by magnetic pickup loops placed external to the plasma boundary, or directly by making use of Faraday rotation, or the Zeeman effect. Inserting probes or loops inside the plasma

5D. Particle Plasma Diagnostics

distorts the plasma parameters; also, the probes have limited lifetime due to the erosion of the protective sheath surrounding the loop. By placing magnetic loops with different orientations around the plasma, techniques have been developed in which moments of the plasma current distribution are determined from the loop signals and combined with other measurements providing electron and ion temperature and density. Arrays of soft x-ray detectors determine the magnetic surface where $q = 1$. Incorporating these data into the magnetohydrodynamic (MHD) equilibrium theory predicts the internal magnetic field. These measurements are straightforward but are limited in that profile information is available for only one or two time points.

The other technique used has been to observe the optical level splitting from the Zeeman effect. In one of the earliest measurements the Zeeman effect was observed of the C and O ion transitions in a theta-pinch plasma (Jahoda et al., 1963). More recently techniques have been developed to inject a neutral or charged beam of particles into a plasma and observe the shifted wavelengths in the presence of a magnetic field.

When an atom or ion is subjected to an external field, the energy levels are split into $2J + 1$ sublevels, where J is the total angular momentum quantum number. This splitting is the result of the interaction of the magnetic field with one of the orbital electrons. The resulting shift of the energy levels is the normal Zeeman effect. The energy shift of the level from the original energy is stated

$$\Delta E = M_j g \mu_B B \tag{30}$$

where M_j is the magnetic quantum number taking on values $-J, \ldots, +J$, μ_B is the Bohr magneton, g is the Lande g factor, and B is the magnetic field. The displacement or shift of a spectral line in a transition from state M_2 to M_1 is

$$\Delta E_2 - \Delta E_1 = (M_2 g_2 - M_1 g_1) \mu_B B, \tag{31}$$

or, expressed in terms of wavelength,

$$\beta \lambda = 4.668 \times 10^{-10} \lambda^2 B (M_2 g_2 - M_1 g_1). \tag{32}$$

The selection rule for M, which is valid for LS coupling in weak fields, is $\Delta M_j = \pm 1$ for σ transitions and $\Delta M_j = 0$ for π transitions, where σ transitions are those in which the electric field vector is parallel to B and in π transitions, the vector is perpendicular to B. The Zeeman pattern for Ba^+ ion is shown in Fig. 31. Both the excited state $^2P_{1/2}$ and ground state $^2S_{1/2}$ are split into two sublevels when the Ba^+ is placed in a magnetic field.

The analysis for the level splitting is only valid for weak magnetic fields where LS coupling holds. As the magnetic field increases, the ratio of the magnetic energy to the spin–orbit energy approaches 1 and LS coupling breaks down. This breakdown is known as the Paschen–Back effect with the critical magnetic field at which breakdown occurs being dependent on the

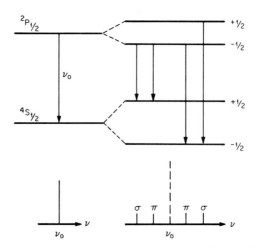

Fig. 31. Energy-level diagram of the Zeeman splitting of the $^2P_{1/2}$ and $^2S_{1/2}$ levels of the Ba$^+$ ion.

level structure of the atom. Weak-field splitting and LS coupling are valid when the total spread or shift of the spectral lines is small relative to the level spacing. Li^0 atoms commonly used as probing beams have magnetic field limits of 3–4 T, whereas Ba$^+$ ions have a much greater critical field.

Basically, there are two ways in which the magnetic field can be determined by the Zeeman effect: (1) measure the direction of the polarized components which are parallel and perpendicular to the total magnetic field, (2) determine the splitting or shift of the $\Delta M = \pm 1$ or 0 transitions. The plasma current giving rise to the poloidal magnetic field B_p causes the total magnetic field to be at an angle $\theta = B_p/B_t$ with respect to the toroidal field B_t. By measuring the direction of polarization parallel to the field and knowing B_t, B_p is readily calculable.

McCormick was the first to demonstrate the feasibility of injecting a beam of atoms and measuring the internal field by determining the direction of the total field (McCormick *et al.*, 1977; McCormick and Olivain, 1978; Fujita and McCormick, 1973). The experimental apparatus is shown in Fig. 32. Ions were produced by a surface ionization ion source, accelerated to 5–10 keV, and passed through a gas neutralizer and into the plasma. The source was mounted on a carriage such that the Li^0 beam could be scanned across the plasma. Li atoms were collisionally excited by electrons to the $^2P_{1/2}$ state, giving rise to the resonance line at $\lambda = 6708$ Å. The optical system viewed the resulting radiation in a direction perpendicular to the toroidal field, which results in the σ components being polarized perpendicular to the local total magnetic field and the π component parallel to B_{total}. After passing through a lens system to increase the light-gathering power, the π component is separated from the σ component by a Fabry–Perot interferometer.

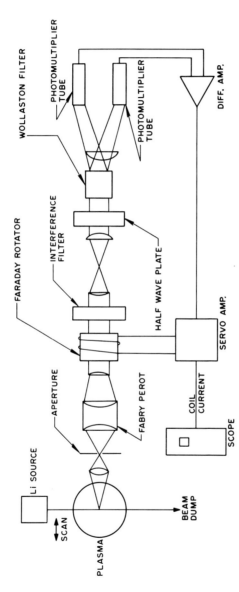

Fig. 32. Schematic diagram of apparatus used to measure the magnetic field in a plasma utilizing the Zeeman effect. [From McCormick *et al.* (1977).]

The orientation of the π component is measured by a polarimeter consisting of a Faraday rotator and a Wollstrom prism. The prism separates the beam into two parts which are detected by two photomultiplier tubes. The output signals from the photomultipliers are fed into a differential amplifier whose function is to feed back a signal to the Faraday rotator. The polarization vector is rotated until a null is produced at the amplifier output. The rotation angle is directly proportional to the feedback current, which can be calibrated. Radial profiles were obtained by scanning the beam on a shot-to-shot basis.

Magnetic fields have been measured on a low-intensity tokamak discharge (Pulsator) whose density was 3×10^{13} cm^{-3} in a toroidal magnetic field of 2.7 T. The experimental results are shown in Fig. 33, where the ratio of B_P/B_T is plotted as a function of the plasma radius. Data points are shown as solid circles, with the error bars representing the greatest and smallest values of all data taken at the given radius. Also plotted are three curves computed for Z_{eff} profiles, as shown in the inset. All three curves were derived assuming the electron temperature profile to be given by $T_e = 800$ eV $[1 - (r/11.5)^2]^2$. Although the ratio B_P/B_T is very sensitive to the assumed electron temperature profile, the method provides a B_P/B_T profile, in reasonable agreement to that computed. Further effort is being directed toward increasing the Li beam energy to decrease the beam attenuation in the plasma.

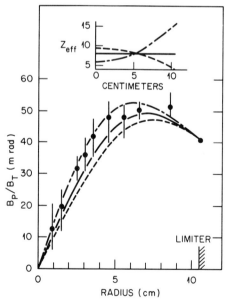

Fig. 33. Radial profile of the Pulsator tokamak poloidal field. [From McCormick *et al.* (1977).]

5D. Particle Plasma Diagnostics

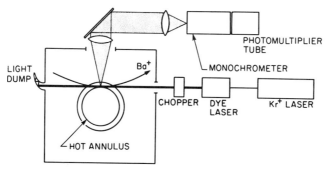

Fig. 34. Schematic diagram of the laser-induced resonance fluorescence of a Ba^+ beam experiment to measure both the internal magnetic fields and the local electric potential. [From Wickham and Lazar (1982).]

Cobble and Glowienka (1979) are developing a system in which a Ba^+-ion beam is injected into an EBT plasma with the spacing of the σ components (4934 Å) determined directly by a filter and photomultiplier.

The method of measuring the beam shift or splitting has been unsuccessful in the past owing to line broadening, which masks out the separation of the shifted components. Line broadening arises from the natural linewidth, instrumental broadening, Stark effects, Doppler broadening due to the energy dispersion of the probing beam, hyperfine structure, and isotope effects. Beam energy dispersion results from the finite ion-source temperature, ripple on the ion-source accelerating power supply, and beam divergence. These effects have been evaluated for the Ba^+ beam in EBT and found to increase the natural linewidth by 60%. With this small increase in linedwidth, the splitting of 0.18 Å should be observable.

One of the constraints in the experiments discussed above is the low signal level with the available beam intensities. The data shown in Fig. 33 are from a series of 36 discharges. Developments are in progress to enhance the light intensity by optical pumping of the Ba^+ and Li^0 beams by a tunable dye laser (Wickham *et al.*, 1982; Rynn and Fornaca, 1980; Weber and Erickson, 1981; Baur *et al.*, 1980). In addition to enhancing the signal level, the Doppler broadening can be minimized by taking advantage of the effective cooling that occurs in the beam direction. The cooling reduces the Doppler broadening from beam energy dispersion by a factor v/c, where v is the beam velocity. A Ba^+-ion beam system is being developed to determine the magnetic field produced by the hot annulus in a plasma similar to the EBT plasma (Wickham *et al.*, 1982). A schematic diagram of the apparatus is shown in Fig. 34. A Kr^+ laser pumps a tunable dye laser whose output is chopped at 2.5 MHz. The laser beam intersects the Ba^+ beam at the innermost point of the Ba^+ orbit. Fluorescence arising from resonance absorption of the laser light is viewed by a monochrometer and a photomultiplier in a direction perpendicular to both the laser and ion beam. Precise alignment is required to minimize the Doppler shift by preventing the Ba$^+$ beam and laser beam from intersecting at an angle.

In addition to determining the magnetic field, the technique will also be also to determine the local electrostatic potential. The energy of the incident Ba$^+$ ion is $E_0 = \frac{1}{2}mv_0^2$. As the ion enters the region of electrostatic potential, the energy will change to $E = E_0 - e\phi$, where ϕ is the plasma potential. If $e\phi \ll E_0$ the velocity will change by

$$\Delta v = v_0 e\phi/2E_0, \tag{33}$$

or the shift in wavelength of an emitted photon is

$$\frac{\Delta\lambda}{\lambda} = \frac{1}{2}v_0 e\phi/cE_o. \tag{34}$$

Each Zeeman component will shift by the same amount. Thus, by measuring the shift from λ_0, the space potential is determined.

A similar method is being developed for application to a tokamak plasma (TEXT) (Baur *et al.*, 1980). In this development a tunable dye laser is injected colinearly with a 70–100 keV Li^0 beam into a plasma. The polarization vector of the laser is rotated at 50 kHz, and phase-sensitive detection is used to measure the fluorescence induced by the laser. The phase angle between the fluorescence signal and a reference signal containing the laser polarization direction gives the direction of the total magnetic field. Nine phototubes view nine channels across the plasma along the path the Li^0 beam traverses.

As the size, density, and temperature of fusion plasmas increase in the future, probing beams must have energies of hundreds of kilo-electron-volts to permit beam penetration into the plasma. The high cost of producing beams at the high energies may restrict future use.

VIII. Heavy-Ion Beam Probe

During the past decade a heavy-ion beam probe diagnostic has been developed which has proven to be invaluable in measuring many properties of magnetically confined plasmas. These parameters, which include space potential, electron density, electron temperature, electrostatic fluctuations, and magnetic fields, can be determined by injecting an energetic, singly ionized ion beam into a plasma and observing the double-charged ions resulting from electron ionization collisions. The first measurements in the development of the technique consisted of injecting 1.8-meV H$_2^+$ into a small-diameter, He-arc plasma and detecting H$_2^+$ ions produced by collisional dissociation (Hickok, 1967; Jobes and Hickok, 1967; Jobes *et al.*, 1969). Since the Larmor radius of both the incident H$_2^+$ ion and the H$^+$ ion (formed in the dissociating collision) is small in the magnetic fields found both in tokamak

5D. Particle Plasma Diagnostics

and mirror plasmas, it was found to be necessary to increase the H_2^+ energy to several tens of mega-electron-volts in order for the H^+ ion to escape the plasma and be detected. To overcome the difficulty encountered with molecular ions, the H_2^+ was replaced with a high-mass, single-charged ion beam such as Cs^+ or Tl^+ (Jobes and Hickok, 1970).

A schematic diagram of the beam probe system used to study the space potential, electron density, electron temperature, and potential fluctuations in the ST tokamak is shown in Fig. 35 (Hosea et al., 1973). Tl^+ ions were produced in a thermionic emitter source and accelerated to 200 keV. The Tl^+ ions were passed through a set of sweep plates whose function was to sweep the ion beam across the plasma diameter. Focusing was achieved by a strong-focusing, linear-field, electrostatic lens. Ions passed through the plasma interior where electron collisions produced Tl^{2+} ions. Since the radius of curvature of the Tl^{2+} ions is one-half that of the primary beam, the Tl^{2+} beam was separated from the Tl^+ trajectory. An electrostatic energy analyzer determined the Tl^{2+} energy. Ions were detected with a split-plate detector.

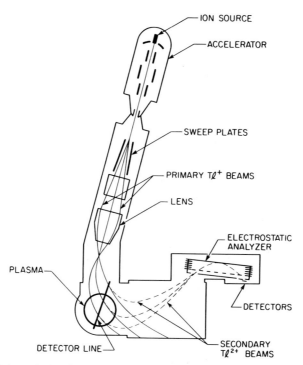

Fig. 35. Schematic drawing of the heavy-ion beam probe apparatus on the ST tokamak. Three primary and secondary beam paths are shown. These paths correspond to maximum positive, zero, and maximum negative voltages applied to the beam sweep plates. [From Hickok and Jobes (1972).]

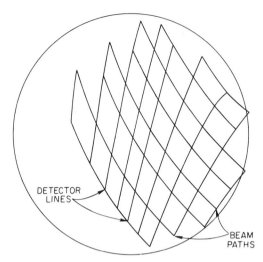

Fig. 36. Grid of detector lines and primary beam paths for the heavy-ion beam probe. [From Hickok and Jobes (1972).]

A detector fixed at one location will only intercept ions formed at some point p along the Tl$^+$ path in the plasma. If the beam is moved, the detector sees another point p_2, and if the beam is swept across the plasma, the detectors will intercept the Tl^{2+} ions formed in a line across the plasma. This line is termed a detector line. By choosing the beam geometry carefully, the detector line can be made to lie nearly in the horizontal plane. To move the detector line vertically, the beam energy must be increased (decreased) or the detector position can be moved. For convenience the beam energy is usually changed. Sweeping the beam and varying the energy allows one to map out the plasma and make measurements over the entire cross-sectional area. A typical grid of detector lines and beam paths is shown in Fig. 36. Optimum spatial resolution is obtained when the detector lines are at right angles to the beam path.

After leaving the plasma boundary, ions were energy analyzed by a plane parallel-plate electrostatic energy analyzer and detected by a split detector plate (Fig. 35). The ion signals from the two plates were amplified by a differential amplifier, and a feedback circuit controlled the voltage to the negative deflection plate of the analyzer. Without plasma, the deflection plate voltage was adjusted so that equal quantities of current were detected by the two plates and produced a null at the amplifier output. The energy of the Tl$^+$ ions at point p is $e(V - \phi)$, where ϕ is the plasma potential at p. The Tl^{2+} ions on leaving the point p have an additional energy of $2e\phi$, resulting in a net gain in energy of $e\phi$. Thus, the particle energy is slightly greater and will unbalance the differential amplifier and produce a corrective voltage to the analyzer plate. This corrective voltage is just the space potential at point p. The ingenious use of split detector plates makes it possible to measure space potentials as low as 10–20 V with temporal resolution ~1 μs.

The ohmic heating current in a tokamak plasma produces a poloidal field which deflects the beam in the Z direction (direction of applied toroidal field). This beam momentum produced in the Z direction is proportional to the magnetic vector potential A_z. By measuring the Z deflection or change in Z momentum, the vector potential can be mapped, with the derivative of A_z giving the magnetic field strength. To perform this measurement it was necessary to divide the split detector into four quadrants and move the beam in the Z direction with and without the plasmas. The Z sweep-plate voltage was again directly proportional to the beam deflection.

Electron densities can be determined from the total current to the plates if the electron temperature is known. The current of doubly charged ions was proportional to $n_e f(T_e)$, where $f(T_e)$ is the functional dependence of the effective ionization rate for Tl^+ on the plasma electron temperature. In principal $n_e f(T_e)$ can be determined separately by determining the secondary ion current using two different reactions, which could be Cs^+ and Tl^+ or one primary beam Cs^+ and secondary beams of Cs^{2+} and Cs^{3+}.

Results obtained for several plasma configurations using the heavy-ion beam probe in apparatus similar to that described are shown in Figs. 37–41. Shown in Fig. 37 is the space potential measured as a Cs^+ beam was scanned across the horizontal plane of the EBT plasma (Bieniosek, 1981; Bieniosek

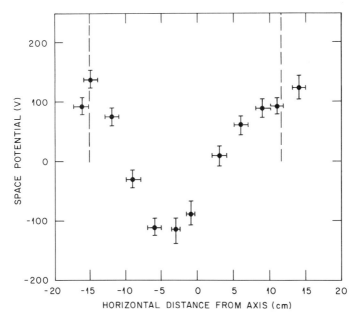

Fig. 37. Space potential in the EBT plasma as measured by a Cs^+ ion beam. [From Bieniosek (1981).]

et al., 1980; Colestock et al., 1978). The vertical dashed lines correspond to the position of the hot-electron annulus of the plasma. The depth of the potential well was comparable to that of the electron temperature. By increasing the gas pressure by a factor of 2, the operating mode of the plasma was changed, resulting in the space distribution as shown in Fig. 38. This stark comparison of the space potential of the same plasma configuration operating in a different mode indicates the usefulness of the technique and data in interpreting the plasma behavior. Space potential measurements have also been made on a magnetically confined arc (Jobes and Hickok, 1970).

A heavy-ion Tl^+ beam probe has been used to probe the ST tokamak plasma. Low-frequency oscillations which may lead to MHD instabilities were observed as shown in Fig. 39. The top trace is the signal obtained from a magnetic pickup loop located outside the plasma. Density or potential oscillations of approximately $\pm 9\%$ in the Tl^{2+} signal correspond in both phase and frequency to the magnetic field oscillations. The fall-off of the Tl^{2+} signal at $y = -5$ cm was caused by the vacuum wall intercepting the beam. Other measurements of potential fluctuations have been made on a hollow-cathode discharge (Glowienka et al., 1976) and the EBT plasma (Colestock et al., 1978).

The results of the measurements separating the electron density and temperature from the $n_e f(T_e)$ value are shown in Figs. 40 and 41 for Na^+ and K^+ beams probing a He discharge (Reinovsky et al., 1973, 1974). The secondary ion current measured by the detector is given by

$$I_s = \gamma I_p n_e f(T_e), \tag{35}$$

where I_p is the primary ion current, γ is a geometric factor, and $f(T_e)$ is the effective cross section. If two beams of different masses are used to probe the plasma, the ratio of the secondary currents to the primary currents is

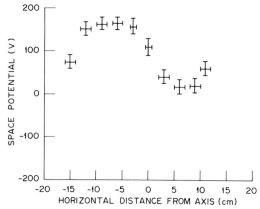

Fig. 38. Space potential in the EBT plasma. Operating conditions as for data in Fig. 37 except for increase in gas pressure by a factor of 2. [From Bieniosek et al. (1980).]

5D. Particle Plasma Diagnostics

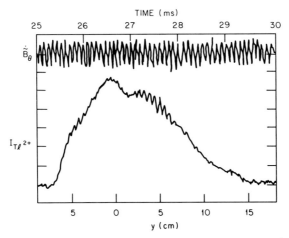

Fig. 39. Density profile measurements on the ST tokamak plasma using Tl^+ ion beam probe. The top trace is from a magnetic pickup probe located external to the plasma. The bottom trace is the Tl^{2+} current showing oscillations in density corresponding in phase and frequency to the oscillations in the magnetic probe signal. [From Hickok (1980).]

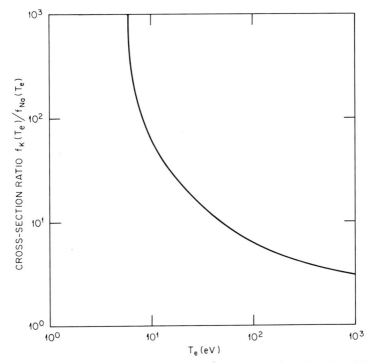

Fig. 40. Ratio of the effective electron ionization cross sections for K^+ and Na^+ as a function of a plasma electron temperature. [From R. E. Reinovsky, J. C. Glowienka, A. E. Seaver, W. C. Jennings, and R. L. Hickok, Ion Beam Probe Measurements of Electron Temperature, *IEEE Transactions on Plasma Science*, © 1974 IEEE.]

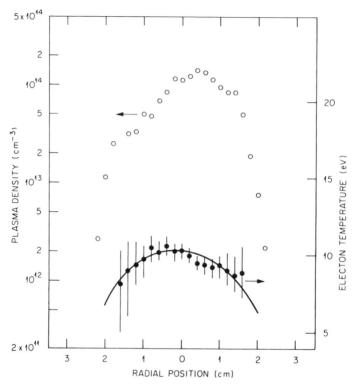

Fig. 41. Radial profile of the electron temperature and density in a hollow-cathode discharge using a K$^+$ and Na$^+$ ion beam. [From R. E. Reinovsky, J. C. Glowienka, A. E. Seaver, W. C. Jennings, and R. L. Hickok, Ion Beam Probe Measurements of Electron Temperature, *IEEE Transactions on Plasma Science*, © 1974 IEEE.]

$$\frac{(I_s/I_p)_1}{(I_s/I_p)_2} = \frac{\gamma n_e f_1(T_e)}{\gamma n_e f_2(T_e)} = \frac{f_1(T_e)}{f_2(T_e)}. \tag{36}$$

If the ratios of the effective cross sections are known, T_e may be found. The ratio of the effective ionization cross section for K$^+$ to Na$^+$ is shown in Fig. 40 for plasma electron temperatures up to 1 keV. Above 1 keV the ratio is relatively insensitive to the electron temperature. With these cross-section data the electron density and temperature were determined for a hollow-cathode helium discharge. The results are shown in Fig. 41. In the well-diagnosed plasma facilities, the electron temperature and density are more easily and precisely determined as measured by laser Thomson scattering.

In tokamak plasma the displacement of the beam in the Z direction has been measured. In taking derivatives of the experimental data, large errors are involved. It does not seem that the beam probe technique is a viable means of measuring the poloidal fields arising from the ohmic heating current. However, the beam probe technique does provide accurate and reliable measurements of local space potential and fluctuations.

References

Abramov, N. G., Afrosimov, V. V., Gladkovskii, I. P., Kislyakov, A. I., and Perel, V. I. (1972). *Zh. Tekh. Fiz.* **41**, 1924–1932; *Sov. Phys.—Tech. Phys. (Engl. Transl.)* **16**, 1520–1525.
Afrosimov, V. V., and Glakkovskii, I. P. (1967). *Zh. Tekh. Fiz.* **37**, 1557–1597; *Sov. Phys.—Tech. Phys. (Engl. Transl.)* **12**, 1135–1168.
Afrosimov, V. V., and Kislyakov, A. I. (1971). *Zh. Tekh. Fiz.* **41**, 1933–1935; *Sov. Phys.—Tech. Phys. (Engl. Transl.)* **16**, 1526–1528 (1972).
Afrosimov, V. V., Gladkovskii, I. P., Gordeev, Y. S., Kalinkevich, I. F., and Fedorenko, N. V. (1960). *Zh. Tekh. Fiz.* **30**, 1456–1468; *Sov. Phys.—Tech. Phys. (Engl. Transl.)* **5**, 1378–1388.
Afrosimov, V. V., Ivanov, B. A., Kislyakov, A. I., and Petrov, M. P. (1966). *Zh. Tekh. Fiz.* **36**, 89–101; *Sov. Phys.—Tech. Phys. (Engl. Transl.)* **11**, 63–71.
Afrosimov, V. V., Berezovskii, E. L., Gladkovskii, I. P., Kislyakov, A. I., Petrov, M. P., and Sadovnikov, V. A. (1975). *Zh. Tekh. Fiz.* **45**, 56–63; *Sov. Phys.—Tech. Phys. (Engl. Transl.)* **20**, 33–37.
Afrosimov, V. V., Gladkovskii, I. P., and Kislyankov, A. I. (1976). *Pis'ma Zh. Tekh. Fiz.* **3**, 10–13; *Sov. Phys.—Tech. Phys. (Engl. Transl.)* **3**, 3–5 (1977).
Afrosimov, V. V., Gordeev, Y. S., and Zinov'ev, A. N. (1977). *Pis'ma Zh. Tekh. Fiz.* **3**, 97–101; *Sov. Phys.—Tech. Phys. (Engl. Transl.)* **3**, 39–40.
Afrosimov, V. V., Gordeev, Y. S., Zinov'ev, A. N., and Korotkov, A. A. (1978). *Pis'ma Zh. Eksp. Teor. Fiz.* **28**, 540–543; *JETP Lett. (Engl. Transl.)* **28**, 500–502.
Afrosimov, V. V., Gordeev, Y. S., Zinov'ev, A. N., and Korotkov, A. A. (1979). *Fiz. Plazmy* **5**, 987–995; *Sov. J. Plasma Phys. (Engl. Transl.)* **5**, 551–556.
Aleksandrov, E. V., Afrosimov, V. V., Berezovskii, E. L., Izvozchikov, A. B., Marasev, V. I., Kislyakov, A. I., Mikhailov, E. A., Petrov, M. P., and Rosylakov, G. V. (1979). *Pis'ma Zh. Eksp. Teor. Fiz.* **29**, 3–7; *JETP Lett. (Engl. Transl.)* **29**, 1–4.
Bacal, M., and Reichelt, W. (1974). *Rev. Sci. Instrum.* **45**, 769–772.
Bacal, M., True, A., Doucet, H. J., Lamain, H., and Chretien, M. (1974). *Nucl. Instrum. Methods* **114**, 407–409.
Barnett, C. F., and Ray, J. A. (1972). *Nucl. Fusion* **12**, 65–72.
Barnett, C. F., Dunlap, J. L., Edwards, R. S., Haste, G. H., Ray, J. A., Reinhardt, R. G., Schill, W. J., Warner, R. M., and Wells, E. R. (1961). *Nucl. Fusion* **2**, 264–272.
Barnett, C. F., Ray, J. A., Ricci, E., Wilker, M. I., McDaniel, E. W., Thomas, E. W., and Gilbody, H. B. (1977). *Oak Ridge Natl. Lab. [Rep.] ORNL (U.S.)* **ORNL-5207 and ORNL-5206**.
Baur, J. F., West, W. P., and Ensberg, E. S. (1980). *Bull. Am. Phys. Soc.* **25**, 684.
Becker, G. (1970). *Z. Phys.* **234**, 6–16.
Berezovskii, E. L., Kislyakov, A. I., and Mikhailov, E. A. (1974). *Pis'ma Zh. Eksp. Teor. Fiz.* **19**, 283–287; *JETP Lett. (Engl. Transl.)* **19**, 166–168.
Berezovskii, E. L., Kislyakov, A. I., Petrov, S. Y., and Roslyakov, G. V. (1980). *Fiz. Plazmy* **6**, 1385–1395; *Sov. J. Plasma Phys. (Engl. Transl.)* **6**, 760–766.
Berry, L. A., Clarke, J. F., and Hogan, J. T. (1974). *Phys. Rev. Lett.* **32**, 362–365.
Bezlyudnyi, S. V., Berezovskii, E. L., Kislyakov, A. I., and Khudoleav, A. V. (1975). *Fiz. Plazmy* **1**, 749–756; *Sov. J. Plasma Phys. (Engl. Transl.)* **1**, 410–412.
Bieniosek, F. M. (1981). *Oak Ridge Natl. Lab. [Rep.] ORNL-TM (U.S.)* **ORNL-TM-7487**.
Bieniosek, F. M., Colestock, P. L., Conner, K. A., Hickok, R. L., Kuo, S. P., and Dandl, R. A. (1980). *Rev. Sci. Instrum.* **51**, 206–212.
Bittoni, E., Cordey, J. G., and Cox, M. (1980). *Nucl. Fusion* **20**, 931–938.
Brisson, D., Baity, F. W., Quon, B. H., Ray, J. A., and Barnett, C. F. (1980). *Rev. Sci. Instrum.* **51**, 511–515.
Cobble, J. A., and Glowienka, J. C. (1979). *IEEE Trans. Plasma Sci.* **PS-7**, 147–151.
Cohen, B. I., ed. (1980). "Status of Mirror Fusion Research 1980." *Lawrence Livermore Lab. Rep.* [UCAR Rep.] **10049-80- Rev-1**, B-1–B-36.

Colestock, P. L., Conner, K. L., Hickok, R. L., and Dandl, R. A. (1978). *Phys. Rev. Lett.* **40,** 1717–1720.
Cordey, J. G., Hugill, J., Paul, J. W. P., Sheffield, J., Speth, E., Stott, P. E., and Tereshin, V. I. (1974). *Nucl. Fusion* **14,** 441–444.
Dnestrovskij, Y. N., Lysenko, S. E., and Kislyakov, A. I. (1979). *Nucl. Fusion* **19,** 293–299.
Equipe TFR (1978). *Nucl. Fusion* **18,** 647–731.
Equipe TFR (1979). *Nucl. Fusion* **19,** 1261–1267.
Eubank, H. P. (1979). *In* "Diagnostics for Fusion Experiments" (E. Sindoni and C. Wharton, eds.), pp. 7–15. Pergamon, Oxford.
Eubank, H. P., Noll, P., and Tappert, F. (1965). *Nucl. Fusion* **5,** 68–72.
Eubank, H. P., *et al.* (1979). *Plasma Phys. Controlled Nucl. Fusion Res., Proc. Int. Conf., 7th, Innsbruck, 1978* 167–198.
Fleischman, H. H., Barnett, C. F., and Ray, J. A. (1974). *Phys. Rev. A* **10,** 569–583.
Fujita, J., and McCormick, K. (1973). *Eur. Conf. Controlled Fusion Plasma Phys., Proc., 6th, Moscow* **1,** 191.
Galbraith, D. L., and Kammash, T. (1979). *Nucl. Fusion* **19,** 1047–1060.
Gaudreau, M. P. J., Kislyakov, A. I., and Sokolov, Y. A. (1978). *Nucl. Fusion* **18,** 1725–1729.
Glowienka, J. C., Jennings, W. C., and Hickok, R. L. (1976). *Appl. Phys. Lett.* **28,** 485–487.
Goldston, R. J. (1978). *Princeton Plasma Phys. Lab., Rep.* **PPPL-1443.**
Gordeev, Y. S., Zinov'ev, A. N., and Petrov, M. P. (1977). *Pis'ma Zh. Eksp. Teor. Fiz.* **25,** 223–227; *JETP Lett. (Engl. Transl.)* **25,** 204–207.
Hagstrum, H. D. (1954). *Phys. Rev.* **96,** 325–335.
Hagstrum, H. D. (1956). *Phys. Rev.* **104,** 317–320.
Hagstrum, H. D. (1961). *Phys. Rev.* **122,** 83–113.
Hickok, R. L. (1967). *Rev. Sci. Instrum.* **38,** 142–143.
Hickok, R. L. (1980). "Heavy Ion Beam Probing," Rep. RPDL-80-14. Rensselaer Plasma Dyn. Lab., Rensselaer Polytech. Inst., Troy, New York.
Hickok, R. L., and Jobes, F. C. (1972). *Air Force Off. Sci. Res. Rep.* **AFOSRTR-72-0018.**
Hosea, J. C., Jobes, F. C., Hickok, R. L., and Dellis, A. N. (1973). *Phys. Rev. Lett.* **30,** 839–842.
Isler, R. C., Murray, L. E., Kasai, S., Dunlap, J. L., Bates, S. C., Edmonds, P. H., Lazarus, E. A., Ma, C. H., and Murakami, M. (1981). *Phys. Rev. A* **24,** 2701–2712.
Jahoda, F. C., Ribe, F. L., and Sawyer, G. A. (1963). *Phys. Rev.* **134,** 24–29.
Jobes, F. C., and Hickok, R. L. (1967). *Rev. Sci. Instrum.* **38,** 928–931.
Jobes, F. C., and Hickok, R. L. (1970). *Nucl. Fusion* **10,** 195–197.
Jobes, F. C., Marshall, J. F., and Hickok, R. L. (1969). *Phys. Rev. Lett.* **22,** 1042–1045.
Kaita, R., and Medley, S. S. (1979). *Princeton Plasma Phys. Lab., Rep.* **PPPL-1582.**
Killeen, J., and Marx, K. D. (1970). *In* "Methods of Computational Physics" (B. Alder, ed.), Vol. 9, pp. 421–489. Academic Press, New York.
Kislyakov, A. I., and Krupnik, L. I. (1981). *Fiz. Plazmy* **7,** 866–906; *Sov. J. Plasma Phys. (Engl. Transl.)* **7,** 478–498.
Kislyakov, A. I., and Petrov, M. P. (1970). *Zh. Tekh. Fiz.* **40,** 1609–1614; *Sov. Phys.—Tech. Phys. (Engl. Transl.)* **15,** 1252–1256 (1971).
Kislyakov, I. A., Stockel, J., and Jukaubka, K. (1975). *Zh. Tekh. Fiz.* **45,** 1545–1547; *Sov. Phys.—Tech Phys. (Engl. Transl.)* **20,** 986–987.
McCormick, K., and Olivain, J. (1978). *Rev. Phys. Appl.* **13,** 85–92.
McCormick, K., Kick, M., and Olivain, J. (1977). *Eur. Conf. Controlled Fusion Plasma Phys., Conf. Proc., 8th, Prague* **1,** 140.
Medley, S. S., and Davis, S. L. (1979). *Princeton Plasma Phys. Lab., Rep.* **PPPL-1507.**
Medley, S. S., Goldston, R. J., and Towner, H. H. (1980). *Princeton Plasma Phys. Lab., Rep.* **PPPL-1673.**
Neilson, G. H., Jr. (1980). *Oak Ridge Natl. Lab.* [*Rep.*] **ORNL-TM** (*U.S.*) **ORNL-TM-7333,** p. 33.

5D. Particle Plasma Diagnostics

Nexsen, W. E., Jr., Turner, W. C., and Cummins, W. F. (1979). *Rev. Sci. Instrum.* **50,** 1227–1235.
Olson, R. E. (1981). *Phys. Rev.* **24,** 1726–1733.
Olson, R. E., and Salop, A. (1977). *Phys. Rev. A* **16,** 531–541.
Olson, R. E., Berkner, K. H., Graham, W. G., Pyle, R. V., Sachlacter, A. S., and Stearns, J. W. (1978). *Phys. Rev. Lett.* **41,** 163–166.
Osher, J. E. (1979). *In* "Diagnostics for Fusion Experiments" (E. Sindoni and C. Wharton, eds.), pp. 47–76. Pergamon, Oxford.
Petrov, M. P. (1976). *Fiz. Plazmy* **2,** 371–389; *Sov. J. Plasma Phys. (Engl. Trans.)* **2,** 210–211.
Phaneuf, R. A., Meyer, F. W., and McKnight, R. H. (1978). *Phys. Rev. A* **17,** 534–545.
Phaneuf, R. A., Alvarez, I., Meyer, F. W., and Crandall, D. H. (1982). *Phys. Rev. A* **26,** 1892–1906.
Ray, J. A., Barnett, C. F., and Van Zyl, B. (1979). *J. Appl. Phys.* **50,** 6516–6519.
Reinovsky, R. E., Jennings, W. C., and Hickok, R. L. (1973). *Phys. Fluids* **16,** 1772–1773.
Reinovsky, R. E., Glowienka, J. C., Seaver, A. E., Jennings, W. C., and Hickok, R. L. (1974). *IEEE Trans. Plasma Sci.* **PS-2,** 250–256.
Riviere, A. C. (1971). *Nucl. Fusion* **11,** 363–369.
Russek, A. (1960). *Phys. Rev.* **120,** 1536–1542.
Rynn, N., and Fornaca, S. (1980). *Bull. Am. Phys. Soc.* **25,** 897.
Salop, A., and Olson, R. E. (1977). *Phys. Rev. A* **16,** 1811–1816.
Salop, A., and Olson, R. E. (1979a). *Phys. Rev. A* **19,** 1921–1929.
Salop, A., and Olson, R. E. (1979b). *Phys. Rev. A* **71A,** 407–410.
Shipsey, E. J., Browne, J. C., and Olson, R. E. (1981). *J. Phys. B* **14,** 869–880.
Stier, P. M., Barnett, C. F., and Evans, G. E. (1954). *Phys Rev.* **96,** 973–982.
Summers, D. D. R., Gill, R. D., and Stott, P. E. (1978). *J. Phys. E* **11,** 1183–1190.
Takeuchi, H., Funahashi, A., Takahashi, K., Shirakata, H., and Yano, S. (1977). *Jpn. J. Appl. Phys.* **16,** 139–147.
Thomas, D. M., and Neilson, G. H., Jr. (1981). *Bull. Am. Phys. Soc.* **26,** 879.
Voss, D. E., and Cohen, S. A. (1980). *Princeton Plasma Phys. Lab., Rep.* **PPPL-1884.**
Voss, D. E., and Cohen, S. A. (1982). *Rev. Sci. Instrum.* **53,** 1227–1235.
Wadehra, J. M., and Bardsley, J. N. (1978). *Phys. Rev. Lett.* **41,** 1795–1798.
Weber, P. G., and Erickson, R. N. (1981). *Bull. Am. Phys. Soc.* **26,** 966.
Wickman, M. G., Lazar, N. H., and Rynn, N. (1982). *Bull. Am. Phys. Soc.* **27,** 1053.
Zinov'ev, A. N., Korotko, A. A., Krzhizhanovskii, E. R., Afrosimov, V. V., and Gordeev, Y. S. (1980). *Pis'ma Zh. Eksp. Teor. Fiz.* **32,** 557–560; *JETP Lett. (Engl. Transl.)* **32,** 539–542.

5E
The Electron Bremsstrahlung Spectrum from Neutral Atoms and Ions

R. H. Pratt and I. J. Feng[*]

Department of Physics and Astronomy
University of Pittsburgh
Pittsburgh, Pennsylvania

I. Introduction.	307
II. Bremsstrahlung in a Plasma: Observables and Assumptions.	308
III. Coulomb Spectrum.	311
IV. Atomic Electron Screening Effects for an Isolated Atom or Ion.	313
V. End Points of the Spectrum: Elastic Scattering and Direct Radiative Recombination.	315
VI. Angular Distributions and Polarization Correlations.	317
VII. Bremsstrahlung Emission in Hot Dense Plasmas.	318
References.	319

I. Introduction

Electron bremsstrahlung, the radiation of a photon as an electron is scattered by an atom or ion, becomes an increasingly important energy loss mechanism as energies increase (see, e.g., Merts *et al.*, 1976). The spectrum of bremsstrahlung radiation also provides a diagnostic for plasma temperature (Tucker, 1975). These facts illustrate the practical interest in the study of bremsstrahlung. Yet the study of bremsstrahlung is also a fundamental problem of theoretical physics, illustrating the connections between classical mechanics, quantum mechanics, and quantum electrodynamics. The theory must unify the radiation from an acclerated charge, as predicted by Maxwell (1881), with the finite end point of the bremsstrahlung spectrum (tip) due to the photon as particle and conservation of energy in the free–free transition process (Kramers, 1923), with structures in the spectrum associated with resonances in electron–atom scattering (Lee and Pratt, 1975a; Olsen and

---

[*] Present address: AT&T Bell Laboratories, Murray Hill, New Jersey 07974.

Maximon, 1978; Dyachkov, 1981). Closely related, and also an important radiative process in hot plasmas, is direct radiative recombination (DRR), the radiative capture of a fast continuum electron into an empty bound state of an ion.

Our purpose here is to give a brief report on our current understanding of the electron bremsstrahlung process, focusing on aspects of concern for the hot plasma environment. We have in mind particularly the range of incident electron energies from 100 eV to 1 MeV; both low-Z and high-Z elements are of interest. The viewpoint is largely theoretical, mainly focusing on the radiation resulting when an electron is incident on an isolated atom or ion and primarily on the energy spectrum of that radiation. For other aspects one may consult the still useful review of Koch and Motz (1959), the discussion by Nakel (1980) of the triply differential cross section, texts on radiation processes in astrophysics (Tucker, 1975; Sobelman, 1979), and the various articles [for discussions of the spectrum, see Lee *et al.* (1976) and Pratt and Tseng (1975), tip region; Pratt and Lee (1977), soft photon region; Tseng and Pratt (1979), high energies, above 2 MeV; Lee *et al.* (1977), ions; for discussions of the angular distribution, see Tseng and Pratt (1971) and Tseng *et al.* (1979)], tabulations (Pratt *et al.*, 1977, note the 1-keV values of the original tables are incorrect; Kissel *et al.*, 1981a), and previous accounts (Pratt, 1980, 1981, 1982) on which the present subchapter is based.

In Section II we summarize the observables of the process and the main physical and model assumptions, and discuss the characteristic distances associated with the process and the insight this provides for the formulation of appropriate descriptions. In Section III we discuss the Coulomb spectrum, identifying classical features, modifications which result from quantum mechanics, further modifications of relativistic origin, and discussing when these modifications become significant. In Section IV we examine modifications which result from screening, comparing the Coulomb, neutral, and ionic (intermediate) cases in classical, quantum, and relativistic circumstances. In Section V we discuss the soft-photon end point of the spectrum, related through the low-energy theorem to elastic electron scattering, and we devote greater attention to the hard-photon endpoint of the spectrum, related to the DRR process, which is also of interest in hot plasmas. In Section VI we briefly note some aspects of angular distributions and polarization correlations which may be of some interest for plasma applications. Finally, in Section VII we discuss the special situation of bremsstrahlung in a hot but also dense plasma, for which many of the assumptions of the preceding discussion no longer apply.

II. Bremsstrahlung in a Plasma: Observables and Assumptions

We assume that our basic process of concern, and capable of observation, is the radiation of a single photon in the scattering of an electron by an

5E. Electron Bremsstrahlung Spectrum from Neutral Atoms and Ions

isolated atom or ion at rest which, within the precision of the measurement of the electron and photon, remains in the same state after the scattering. The process is then characterized by the incident electron momentum and spin ($\mathbf{p}_i$, $\mathbf{s}_i$), the final electron momentum and spin ($\mathbf{p}_f$, $\mathbf{s}_f$), and the radiated photon momentum and polarization ($\mathbf{k}$, $\boldsymbol{\epsilon}$). The momentum transfer $\mathbf{q}$ to the atom is then given through momentum conservation as

$$\mathbf{q} = \mathbf{p}_i - \mathbf{p}_f - \mathbf{k}.$$

Energy conservation imposes the constraint

$$(1 + \mathbf{p}_i^2)^{1/2} = k + (1 + \mathbf{p}_f^2)^{1/2}$$

(with units $\hbar = m_e = c = 1$) in the approximation that the mass m of the atom is large enough so that the energy transfer to the atom ($q^2/2m$ when small enough that the final atom has nonrelativistic velocity) is negligible in comparison to electron and photon energies. It is often convenient to specify the fraction k/T_i of the incident electron kinetic energy $T_i \equiv (1 + \mathbf{p}_i^2)^{1/2} - 1 \sim \frac{1}{2}\mathbf{p}_i^2$ (at nonrelativistic velocities) which is radiated; for bremsstrahlung this fraction can vary from zero (soft-photon end point) to one (hard-photon end point), while for direct radiative recombination the fraction is greater than one. In most of our subsequent discussion we shall assume that none of the polarization variables ($\mathbf{s}_i$, $\mathbf{s}_f$, $\boldsymbol{\epsilon}$) are observed.

Implicit in this description are crucial physical assumptions regarding both the process, as it is to be observed, and the environment in which it takes place. More complex processes (including multiphoton emission, excitation or ionization of the atom in the course of the bremsstrahlung event) are to be distinguished in observation, neglected as providing a small contribution to the observation, or summed over together with the basic process in interpreting the observation. Further, any process takes place in an environment; the description of a process as isolated is always an idealization. For such a description of bremsstrahlung to be useful we must be able to neglect effects of the environment, whether from other charged particles in a plasma, the presence of an intense background radiation field, the channeling and coherence effects of a lattice, etc.

The bremsstrahlung process as we have defined it depends on the kinetic energy T_i of the incident electron and the target atom (often characterized by its nuclear charge Z, its ionic charge Z_i, and assumed to be in its ground state). The energies and momenta of the final electron and photon may be observed, subject to the constraint of energy conservation. Consequently the most detailed possible independent observation is of the triply differential cross section $d^3\sigma \equiv d^3\sigma/d\Omega_{pf}\, d\Omega\, dk$ corresponding to coincidence measurement of directions of photon emission and recoil electron and to the sharing of incident electron kinetic energy between them. If only the radiated photon is observed this is described by the angular distribution (doubly differential cross section) $d^2\sigma \equiv d^2\sigma/d\Omega_k\, dk = \int d^3\sigma\, d\Omega_{pf}$, or the shape function $s = d^2\sigma/d\sigma$, with $d\sigma$ the energy spectrum as defined below. If only the sharing of energy is considered this will be described by $d\sigma \equiv d\sigma/dk \equiv$

$\int d^2\sigma \, d\Omega_k$. The total cross section $\int (d\sigma/dk) \, dk$ diverges, since, owing to the zero mass of the photon, $\sigma \propto 1/k$ for small k. However, $\int k(d\sigma/dk) \, dk$ exists and, since $d\sigma/dk$ gives the probability unit time to radiate a photon of energy k per unit flux of beam particle incident on the atom, the integral gives the average energy-radiated per unit time per unit flux of beam particles, the so-called total energy loss.

In a plasma we may expect to have a distribution of incident electron kinetic energies T_i, (and perhaps also a distribution of target atoms and ions). If the electron distribution is Maxwellian $M(v) \, dv$ (as in LTE or coronal equilibrium) the observables of the bremsstrahlung process can be characterized by temperature T rather than kinetic energy T_i, corresponding to weighted Maxwellian averages over the T_i bremsstrahlung observables. Of particular interest is the emissivity (energy of given frequency emitted per unit volume per second), corresponding to a weighting of $d\sigma/dk$ with $vM(v) \, dv$ (the extra v entering to determine the rate at which collisions are occurring in the plasma) and the total bremsstrahlung emission (radiated power loss), obtained by integrating the emissivity over all frequencies. In this subchapter we shall confine our attention to the basic bremsstrahlung process, considered as a function of incident electron kinetic energy T_i.

Even the best current calculations of bremsstrahlung involve a large number of model assumptions. The process is separated into radiation in inelastic scattering from bound electrons (Haug, 1975a,b; Nakel and Pankau, 1972, 1973, 1975) (often neglected, or calculated neglecting binding) and radiation in scattering from the nuclear charge as screened by an effective central potential due to the bound electrons. In this separation other many-electron effects, including excitation and ionization and exchange, are being omitted. Radiative corrections from quantum electrodynamics (McEnnan and Gavrila, 1977; Mork and Olsen, 1965, 1968), as well as finite nuclear mass and size (Berg and Linder, 1958) and nuclear magnetic moments (Ginsberg and Pratt, 1964, 1965; Goldemberg and Pratt, 1966), have been considered in gamma radiation, but do not appear to be of concern in current plasma environments. Under these assumptions bremsstrahlung is described as a free–free transition of a single continuum electron in a screened self-consistent relativistic central potential. Cross sections are obtained by evaluating the matrix element M proportional to

$$\int \psi_f^* \boldsymbol{\alpha} \cdot \boldsymbol{\epsilon} \exp(-i\mathbf{k} \cdot \mathbf{r}) \, \psi_i \, d^3r,$$

where the electron wave functions ψ are solutions of the Dirac equation in the potential v. Both the ψ and the integral for M can be obtained numerically (Tseng, 1970) through expansions in partial waves and in photon multipoles. Simpler approaches are available in various cases, as we shall discuss, and we may use our best available results to confirm the circumstances of their validity. It is more problematic to argue for the circumstances in which the "best available" results are not good enough.

5E. Electron Bremsstrahlung Spectrum from Neutral Atoms and Ions

One situation which clearly requires special attention is bremsstrahlung in a hot dense plasma, for which it is clear that the process cannot be entirely isolated from its environment. In addition, the description of the projectile and the target become more problematic. We will explore some aspects of these matters in Section VII.

The distance probed in a particular bremsstrahlung event is generally measured by the inverse q^{-1} of the momentum transfer to the atom, which provides an oscillatory cutoff on the matrix element M at larger distances. For a given incident energy values of q range from a small q_{min} obtained in a colinear configuration with soft-photon emission to a q_{max} corresponding to a head-on elastic scattering. Bremsstrahlung from low-energy electrons is sensitive to the large distance region; bremsstrahlung from high-energy electrons can sample *either* small or large distance regions, depending on the choice of kinetics in a coincidence measurement. The differential cross section is largest in the case of small q, where a larger volume (larger distances) contributes to the matrix element. Thus in integrated cross sections the regions probed by the smallest momentum transfers included in the integration dominate. For the spectrum $q_{min} \sim p_i - p_f$ (nonrel), so that large distances are important in the soft-photon region of the spectrum, whereas small distances are important in the hard-photon region of the spectrum from relativistic electrons.

III. Coulomb Spectrum

When an electron is scattered in a point-Coulomb potential the main classical, quantum mechanical, and relativistic features of the bremsstrahlung process can be understood by discussing the predictions for the spectrum obtained with three types of theories: (a) the classical Landau–Lifshitz formula (Landau and Lifshitz, 1971), (b) the Sommerfeld formula (nonrelativistic dipole approximation) (Sommerfeld, 1931, 1939), and (c) the relativistic Born approximation calculation of Bethe and Heitler (1934; see also Sauter, 1934; Racah, 1934). The full relativistic results of a numerical partial wave calculation in the point-Coulomb potential can be reproduced to a fairly good accuracy with these approximations, since the validity regions of the three theories overlap (Feng and Pratt, 1981).

(a) Landau and Lifshitz describe the calculation of the classical spectrum resulting from the acceleration of a charge e in a trajectory without energy loss. The result, expressed in Hankel functions of imaginary arguments and indices, is a function only of the one variable $\mu = \gamma k/2T_i$, and is fairly well represented by its large-μ and small-μ limits. The simple Kramers formula $\sigma = 16\pi\alpha^3/3\sqrt{3} \simeq 5.61$ mb (the large-μ limit) away from the soft-photon end of the spectrum, together with the small-μ limit $\sigma = \frac{16}{3}\alpha^3 \ln(2/\gamma\mu)$ ($\gamma = e^c$; c is Euler's constant) at the soft-photon end, may be combined to give a semiquantitative representation of the full classical result. The soft-photon end-point behavior is the same in all theories of radiation from a charge, but at high energies the differences among theories are large even quite near that end point. In the bulk of the spectrum at high energies the Kramers formula remains qualitatively more useful than the full classical result, corresponding to the observation that Gaunt factors (ratios of full calculations to Kramers) remain of order unity. Further terms in the soft-photon (small-μ) and the hard-photon (large-μ) expansions have been obtained by Florescu and Costescu (1978). The expression for small μ is

$$\sigma(k) = \frac{16\alpha^3}{3} (1 + \pi\mu) \ln \frac{2}{\gamma\mu} + O(\mu^2). \qquad (1)$$

The expression for large μ is

$$\sigma(k) = (16\pi\alpha^3/3\sqrt{3})[1 + d_1\mu^{-2/3} + d_2\mu^{-4/3} + d_1d_2\mu^{-2} + O(\mu^{-8/3})], \qquad (2)$$

where $d_1 = 0.217747$, $d_2 = -0.0131214$, and $d_1d_2 = -1/350$. Combining these two expressions, the exact classical result can be reproduced with extremely good accuracy (Feng and Pratt, 1981; Florescu and Costescu, 1978).

(b) Quantum mechanical features are illustrated in the nonrelativistic dipole result of Sommerfeld, expressed in hypergeometric functions, which depends on the two variables ν_i and ν_j. For low energies (large ν_i and ν_j) this reduces to the classical result, while with increasing energy the validity of the classical form is increasingly restricted to the soft-photon region. Quantum mechanical effects associated with the quantization of angular momentum in electron motions increasingly depress the bulk of the spectrum below the predictions of the Kramers formula. For high energies (where ν_i and ν_j are both small) the Sommerfeld formula reduces to the nonrelativistic Born approximation result

$$\sigma(k) = \tfrac{16}{3} \alpha^3 \ln[(p_i + p_f)/(p_i - p_f)]$$

away from the tip, while at the tip (ν_i small, ν_j large) it reduces to

$$\sigma = (16\pi/3)\alpha^3\nu_i.$$

The value at the tip of the spectrum is zero in Born approximation and within nonrelativistic dipole approximation it does indeed go to zero with increasing energy. The Elwert factor (Elwert, 1939)

$$\frac{\nu_f}{\nu_i} \left[\frac{1 - \exp(-2\pi\nu_i)}{1 - \exp(-2\pi\nu_f)} \right]$$

($\rightarrow 1$ at the soft-photon end point of the spectrum and in Born approximation) may be used to modify Born approximation results. This factor was derived from the Sommerfeld formula as the ratio of Sommerfeld to Born under conditions that $2\pi(\nu_f - \nu_i) \ll 1$, but in fact it has been found to have greater validity, extending to the tip region where ν_f becomes large. The Elwert factor in these cases is correcting the Born approximation normalization of the out-going slow electron and converts the vanishing Born approximation prediction at the tip to a finite prediction. At low energies, there is substantial cancellation among relativistic, retardation, and higher multipole effects, with the consequence that even in high-Z elements the Sommerfeld formula remains fairly good to 50 keV.

(c) The relativistic Born approximation of Bethe and Heitler, exhibiting separate dependence on p_i and p_f, still agrees with classical predictions for soft photons and still vanishes at the tip; multiplying by the Elwert factor again gives improved results, quantitative for low Z and qualitative for high Z. The finite value at the tip does not continue to decrease as T_i grows, since now v_i remains finite in the high-energy limit; this agrees with the behavior (but not the value) to be expected for an exact calculation in the Coulomb potential. In the mega-electron-volt range the tip value is increased for high-Z elements by contribution from final p states, while in the high-energy limit both the final s- and p-state contributions are significantly suppressed. In the ultrarelativistic regime (above energies of concern here) σ begins to grow, rising even above the Kramers formula value. Z-dependent corrections in this regime obtained by Bethe and Maximon (1954) do not modify these qualitative features.

Feng and Pratt (1981) have examined the regions of validity of these theories and have given a prescription for the characterization of the Coulomb spectrum in terms of simple expressions. The full relativistic result for all Z's, for energies 1–200 keV, can be reproduced within 15% general accuracy with four simple formulas. When $v_i(1 - k/T_i) > 0.7$ classical results may be used, namely the large-μ expansion for $\mu \geq 0.3$ and the small-μ expansion for $\mu \leq 0.3$. (The full Landau–Lifshitz result is not needed.) When $v_i(1 - k/T_i) < 0.7$ quantum effects contribute and can be included through use of the Elwert factor and Born approximation. (The full Sommerfeld formula is not needed.) Owing to the cancellation of relativistic and retardation effects when $T_i \geq (2a)^4 mc^2$ the nonrelativistic Born approximation should be used, and in this case the exponential factors may be omitted from the Elwert factor, leading once again to a one-variable expression. For $T_i \geq (2a)^4 mc^2$ the full Elwert factor should be used together with the Bethe–Heitler formula, and in high-Z elements the accuracy in the tip region will deteriorate as mega-electron-volt energies are reached.

IV. Atomic Electron Screening Effects for an Isolated Atom or Ion

Screening of the nucleus by atomic electrons substantially reduces the bremsstrahlung cross sections of a neutral atom from the Coulomb case, particularly at low energies (see Fig. 1). In the nonrelativistic quantum re-

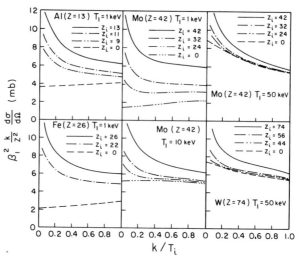

Fig. 1. Bremsstrahlung energy spectrum from Al and Fe at $T_i = 1$ keV; from Mo at $T_i = 1$, 10, and 50 keV; and from W at $T_i = 50$ keV.

gime an accurate form-factor approximation can give a fairly good estimate of the reduction due to screening. In this approximation $d^3\sigma$ is multiplied by $|F(q)|^2$, where

$$F(q) = 1 - \frac{1}{Ze} \int \rho(\mathbf{r}) \exp(i\mathbf{q} \cdot \mathbf{r}) \, d^3r, \quad \int \rho(\mathbf{r}) \, d^3r = Ze,$$

with $\rho(\mathbf{r})$ the electron charge density of the atomic electrons. We may understand $|F(q)|^2$ as reducing $d^3\sigma$ corresponding to a reduced effective charge seen by the incident electron penetrating the atom. At large q (small r) $F(q) = 1$ and the electron has penetrated the atomic electron charge and is seeing the nuclear charge; for small q (large r) $F(q) = 0$ and the electron is outside the neutral atom and sees no charge to scatter from. Form-factor approximation removes the logarithmic divergence of the soft-photon end point of the spectrum: $k \, d\sigma/dk$ is now finite as $k \to 0$ (but the Born approximation is no longer exact at the end point). Generally in the tip region of the spectrum (particularly at high energies) screening has little effect, since small distances dominate. However, within tens of electron volts of the tip a low-energy screened-continuum normalization differs substantially from Coulomb, so that the spectrum vanishes at the end point as in the Born approximation. For lower incident energies Green (1979), Collins and Merts (1981), Scofield (1981), and Geltman (1973) have calculated the free–free Gaunt factor within the nonrelativistic dipole approximation in model potentials. Full relativistic calculations have been reported by Lee and Pratt (1975a) as low as 500 eV. Born approximation is poor (too big) at these energies, shape resonance features are observed, the exchange effect be-

5E. Electron Bremsstrahlung Spectrum from Neutral Atoms and Ions

comes important, and it is believed that below about 10 eV elaborate multi-state close-coupling calculations become necessary. As the incident energy increases into the mega-electron-volt range, more of the spectrum is determined at smaller distances where screening is unimportant. At very high energy use of form factors together with Bethe–Maximon Coulomb corrections can be justified (Bethe and Maximon, 1954).

Lee et al. (1977) have calculated the bremsstrahlung spectrum from isolated atomic ions. Figure 1 shows the bremsstrahlung energy spectrum from Al, Fe, Mo, and W ions of various degree of ionization, at different incident energies. The ionic cross sections are intermediate between the point-Coulomb and the screened neutral-atom results. Note that the larger portion of an ionic spectrum remains closer to the screened (not the Coulomb) spectrum until a high degree of ionization is reached. By defining the "ionization factor"

$$I(k/T_i, T_i, Z, Z_i) = \frac{\sigma(k/T_i, T_i, Z, Z_i) - \sigma(k/T_i, T_i, Z, 0)}{\sigma(k/T_i, T_i, Z, Z_i) + \sigma(k/T_i, T_i, Z, 0)}$$

and using the fact that I is a smoothly varying function of Z_i/Z, k/T_i, and T_i (it is observed that the ionization factor is simply a function of Z_i/Z and not of the two parameters separately), one can reproduce the ionic spectrum with fairly good accuracy by a simple interpolation between the Coulomb and neutral-atom results (Feng and Pratt, 1981).

V. End Points of the Spectrum: Elastic Scattering and Direct Radiative Recombination

The soft-photon and hard-photon end-point regions of the bremsstrahlung spectrum are related to other atomic processes. The soft-photon end point is related to the cross section $(d\sigma/d\Omega_f)_{el}$ for elastic electron scattering through the low-energy theorem (Jauch and Rohrlich, 1954; Rohrlich, 1955; Burnet and Kroll, 1968)

$$\lim_{k \to 0} d^3k = \frac{\alpha}{4\pi^2} \left(\frac{\boldsymbol{\epsilon} \cdot \mathbf{p}_i}{\mathbf{k} \cdot \mathbf{p}_i} - \frac{\boldsymbol{\epsilon} \cdot \mathbf{p}_f}{\mathbf{k} \cdot \mathbf{p}_f} \right)^2 \left(\frac{d\sigma}{d\Omega_f} \right)_{el},$$

where the dot products are of 4-vectors; the next term in the expansion for small k may also be related to the elastic amplitudes. This relationship has been of practical use in obtaining d^3k in the soft-photon region (Pratt, 1979) where partial wave expansions converge very slowly. In the Coulomb case the forward angle divergence of elastic scattering corresponds to the $\ln k$ divergence of $k\, d\sigma/dk$ which we have seen; in the screened case both are finite.

The hard-photon end point (tip) of the bremsstrahlung spectrum is related to direct radiative capture of a continuum electron into an unoccupied bound state, also an important atomic process in plasmas. The reduced matrix element (wave function normalizations removed) can be continued smoothly across the tip limit into the capture region. For low incident energies the Kramers formula for bremsstrahlung becomes the Kramers formula for radiative capture into outer levels. For much higher incident energies, even the K shell is near the tip on a k/T_i scale, and so one gets the relationship between tip bremsstrahlung and atomic photoeffect (or its inverse) first noticed by Fano (1959; see also McVoy and Fano, 1959; Fano et al., 1958; Jabbur and Pratt, 1963). For large T_i (quantum region of the spectrum) the final low-energy electron s and p states dominate, while in the classical Kramers region all final angular momentum states contribute. This is why shape resonance effects in higher-*l final* states become visible only for low-energy *initial* states (Lee et al., 1976, 1977; Pratt and Tseng, 1975; Pratt and Lee, 1977; Tseng and Pratt, 1971, 1979; Tseng et al., 1979).

Earlier work on direct radiative recombination (DRR) cross sections and rate coefficients, culminating in the papers of Seaton (1959), concentrated on obtaining corrections to the Kramers formula within the hydrogenic approximation via expansion of the Gaunt factor as a function of energy and principle quantum number *n*. This Coulomb free–bound Gaunt factor ordinarily does not lead to more than 20% corrections, and is often set to unity. Subsequently, and particularly with concern for impurity elements in fusion plasmas, there has been increasing interest in screening effects in DRR when some inner shells are filled (see, e.g., Merts et al., 1976). Merts estimated that in a thin iron-seeded plasma, for temperatures above 1.5 keV DRR would dominate over dielectronic recombination. Kim (1981; see also Kim and Pratt, 1983) has utilized the connection between the bremsstrahlung tip and DRR, following methods outlined by Lee and Pratt (1975b, 1976), to obtain DRR cross sections (by subshell and summed) for several choices of Z, Z_i, T_i. (This approach exploits the continuity of the reduced matrix element as a function of final state energy, so that it is not necessary to perform separate calculations for each state into which capture occurs.) Kim demonstrated that the dominant cross sections are well characterized by a modified Kramers formula with an effective charge $Z_{eff} = \frac{1}{2}(Z + Z_i)$, so that alternative uses of the Kramers formula will tend to fail except in the case of totally ionized atoms. This choice of Z_{eff} indicates that for x-ray energies the dominant captures (into unoccupied states just at the edge of the ion) are characterized not by distances exterior to the ion (Z_i, as they would be at low energies) or far inside the ion (Z, as they would be at very high energies), but rather by intermediate distances—the Z_{eff} appropriate for the capture is not the Z_{eff} appropriate for the energy level of the state into which capture occurs. Kim (1981) has also shown that this dependence leads to simple scaling features of total DRR cross sections and also to simple analytic expressions for the rate coefficients obtained as weighted averages over

5E. Electron Bremsstrahlung Spectrum from Neutral Atoms and Ions

Maxwellian distributions of incident electron kinetic energies. For the total cross section of DRR Kim and Pratt (1983) obtain the expression

$$\sigma_{tot} = \frac{\gamma \rho \alpha}{3\sqrt{3}} \frac{a^2}{T_i} \ln\left(1 + s\frac{a^2}{T_i}\right),$$

where $a = Z_{eff}\alpha$ and $s = (2n_{eff}^2)^{-1}$, with $n_{eff} = n_0 - 0.3 + (1 - \omega_0)$, n_0 the principal quantum number of the valence shell of the ion, and ω_0 the fraction of unoccupied states in the valence shell. For the three basic DRR plasma rate coefficients, weighted Maxwellian averages, they obtain for the recombination rate

$$\alpha = 2a^2 A' kT \left[\exp\left(\frac{sa^2}{kT}\right) E_1\left(\frac{sa^2}{kT}\right) + c + \ln\left(\frac{sa^2}{kT}\right)\right],$$

for the rate of electron kinetic energy loss

$$\beta = kT\left[\alpha - 2a^4 A's \exp\left(\frac{sa^2}{kT}\right) E_1\left(\frac{sa^2}{kT}\right)\right],$$

and for the radiated power loss

$$\gamma = 2A'a^4 skT,$$

where $A' = \sqrt{\frac{2}{3\pi}} \frac{8}{3}\alpha (kT)^{-3/2}$, $c = 0.577$ is Euler's constant, and $E_1(Z) = \int \exp(-T) dt/t$ is the first exponential integral. These expressions were in good agreement with more detailed calculations of Barfield (1980) and Merts et al. (1976).

VI. Angular Distributions and Polarization Correlations

As the description of a hot plasma becomes more realistic, the averaged information regarding atomic processes represented by integrated cross sections, such as the spectrum σ or the total energy loss, becomes less adequate. Transport codes go beyond one dimension. The polarization of emitted radiation can be used as a diagnostic for the velocity distribution of electrons in the plasma (Sauthoff, 1980; Milchberg and Weisheit, 1983). We therefore briefly discuss here what is known about the photon angular distribution (Tseng et al., 1979; Kissel, 1982) $d^2\sigma$ (which is of course far from isotropic, and becomes increasingly forward peaked at high energies) and about the polarization correlation C_{03}, describing the polarization properties of the radiation emitted in scattering from an unpolarized electron beam incident on a collection of target atoms not characterized by any net polarization (Tseng and Pratt, 1973). The main message is that, unlike the spectrum σ, the nonrelativistic dipole approximation is *not* adequate for the angular distribution $d^2\sigma$ and the polarization correlations, either for bremsstrahlung or in DRR. Higher multipole effects are important throughout the x-ray regime, so that the simple nonrelativistic dipole forms which prove so useful for the spectrum do not characterize the distribution.

A tabulation of the bremsstrahlung angular distribution shape $d^2\sigma/d\sigma$ in terms of a small set of coefficients B_n is now available (Kissel et al., 1983); the shape can be parametrized as $(1 - \beta_i \cos \theta)^{-4} \Sigma_n B_n P_n (\cos \theta)$, where the B_n depend on Z_i, T_i, k/T_i, and the incident velocity β_i enters a factor which displays the high-energy forward peaking of the distribution (exact in Born approximation). This may be contrasted with the nonrelativistic dipole shape proportional to $1 + \frac{1}{2}a_2 P_2 (\cos \theta)$, characterized by one shape parameter a_2 (Thaler et al., 1956), which (as for photoeffect or DRR) is already inadequate by the kilo-electron-volt range. The Bethe–Heitler shape (Bethe and Heitler, 1934), which includes forward peaking, is complicated; it agrees moderately well with numerical calculations (within 15% for high-Z elements, except at forward and backward angles), even when the magnitudes are poorly predicted. With increasing energy only forward angles remain sensitive to the large distances where screening matters. The connection between bremsstrahlung and DRR implies that the tip bremsstrahlung shape $d^2\sigma$ can be identified as a weighted sum of DRR shapes over capture into different angular momentum substates for fixed (high) n (Kissel et al., 1981b). For high incident electron energy T_i the sum is dominated by s states (with an important p-state contribution in high-Z elements), while for low T_i all angular momenta contribute.

Limited information is thus far available (Motz and Placious, 1960; Sheer et al., 1968; Kuckuck, 1972, 1972; Kuckuck and Ebert, 1973) on the polarization correlation C_{03}, which reflects the fact that beginning with unpolarized electrons one will in general produce photons which are linearly polarized parallel or perpendicular to the emission plane. In dipole approximation C_{03} is characterized by the shape parameter a_2. However, deviations from a dipole distribution are already noticeable in the x-ray regime. For the bremsstrahlung tip C_{03} can be identified with a weighted sum of corresponding DRR correlations (Feng et al., 1983).

VII. Bremsstrahlung Emission in Hot Dense Plasmas

Bremsstrahlung emission by superthermal electrons from laser-irradiated targets is a potentially useful diagnostic for the measure of hot electrons produced by laser–plasma interaction. The connection of the bremsstrahlung intensity and spectrum with the number and energy of superthermal electrons was analyzed by Brueckner and Lee (1979). The calculations of free–free Gaunt factors within the Born approximation for the Debye–Hückel ion–sphere, Thomas–Fermi, and average atom potentials have been summarized by Weisheit (Chapter 8, this volume). Green (1979) has reported numerical results for a gold target in the nonrelativistic dipole approximation for the Thomas–Fermi (TF) potential. A full relativistic calculation within the TF potential for Cs for incident e⁻ energies $T_i \geq 1$ keV was reported by Lamoureux et al. (1982) and Feng et al. (1983b).

Their major conclusions are (1) the Born approximation is inadequate for most cases (especially at low incident electron energies). (2) The finite-temperature TF potential is quite different from an isolated neutral-atom potential (HFS); hence the energy spectra calculated within these two potentials are significantly different. (3) In some cases the spectrum in a hot dense plasma is similar to that from an isolated ion of charge corresponding to the average degree of ionization appropriate to an atom at that temperature and density. However, this is a bad approximation at extreme density, where the spectrum is more screened (suppressed) than for a neutral atom even if there is substantial ionization.

References

Barfield, W. D. (1980). *J. Phys. B* **13**, 931.
Berg, R. A., and Linder, C. N. (1958). *Phys. Rev.* **112**, 2072.
Bethe, H. A., and Heitler, W. (1934). *Proc. R. Soc. London, Ser. A* **A146**, 83.
Bethe, H. A., and Maximon, L. C. (1954). *Phys. Rev.* **93**, 768.
Brueckner, K. A., and Lee, Y. T. (1979). *Nucl. Fusion* **19**, 1431.
Burnet, T. H., and Kroll, N. M. (1968). *Phys. Rev. Lett.* **20**, 86.
Collins, L. A., and Merts, A. L. (1981). *Quant. Spectrosc. Radiat. Transfer* **26**, 443.
Dyachkov, L. G. (1981). *J. Phys. B* **14**, L695.
Elwert, G. (1939). *Ann. Phys. (Leipzig)* **34**, 178.
Fano, U. (1959). *Phys. Rev.* **116**, 1156.
Fano, U., Koch, H. W., and Motz, J. W. (1958). *Phys. Rev.* **112**, 1679.
Feng, I. J., and Pratt, R. H. (1981). *Univ. of Pittsburgh Rep.* **266**.
Feng, I. J., Goldberg, I., Kim, Y. S., and Pratt, R. H. (1983a). *Phys. Rev. A* **28**, 609.
Feng, I. J., Lamoureux, M., Pratt, R. H., and Tseng, H. K. (1983b). *Phys. Rev. A* **27**, 3209.
Florescu, V., and Costescu, A. (1978). *Rev. Roum. Phys.* **23**, 131.
Geltman, S. (1973). *J. Quant. Spectrosc. Radiat. Transfer* **13**, 601.
Ginsberg, E. S., and Pratt, R. H. (1964). *Phys. Rev.* **134**, B773.
Ginsberg, E. S., and Pratt, R. H. (1965). *Phys. Rev.* **137**, B1500.
Goldemberg, J., and Pratt, R. H. (1966). *Rev. Mod. Phys.* **38**, 311.
Green, J. M. (1979). Rep. RDA-TR-108600-003. R & D Associates, Santa Monica, California.
Haug, E. (1975). *Z. Naturforsch., A* **30A**, 1099.
Haug, E. (1975b). *Phys. Lett. A.* **54A**, 339.
Jabbur, R. J., and Pratt, R. H. (1963). *Phys. Rev.* **129**, 184.
Jauch, J. M., and Rohrlich, F. (1954). *Helv. Phys. Acta* **27**, 613.
Kim, Y. S. (1981). Ph.D. Thesis, Univ. of Pittsburgh.
Kim, Y. S., and Pratt, R. H. (1983). *Phys. Rev. A* **27**, 2913.
Kissel, L. (1982). *Sandia Lab.* [*Tech. Rep.*] *SAND* **SAND81-2154**.
Kissel, L., MacCallum, C., and Pratt, R. H. (1981a). *Sandia Lab.* [*Tech. Rep.*] *SAND* **SAND81-1337**.

Kissel, L., Kim, Y. S., and Pratt, R. H. (1981b). Abstracts of contributed papers, Gatlinburg. Tennessee (Sheldon Datz, ed.). *Abstr., ICPEAC, 12th.*
Kissel, L., Quarles, C. A., and Pratt, R. H. (1983). *At. Data Nucl. Data Tables* **28**, 381.
Koch, H. W., and Motz, J. W. (1959). *Rev. Mod. Phys.* **31**, 920.
Kramers, H. A. (1923). *Philos. Mag.* **46**, 836.
Kuckuck, R. W. (1972). Ph.D. Thesis, Univ. of California.
Kuckuck, R. W. (1972). *Lawrence Livermore Lab.* [*Rep.*] *UCRL* **UCRL-51188**. (Unpubl.)
Kuckuck, R. W., and Ebert, P. J. (1973). *Phys. Rev. A* **7**, 456.
Lamoureux, M., Feng, I. J., Pratt, R. H., and Tsang, H. K. (1982). *J. Quant. Spectrosc. Radiat. Transfer* **27**, 227.
Landau, L. D., and Lifshitz, L. M. (1971). "The Classical Theory of Fields," 3rd ed. Pergamon, New York.
Lee, C. M., and Pratt, R. H. (1975a). *Phys. Rev. A* **12**, 707.
Lee, C. M., and Pratt, R. H. (1975b). *Phys. Rev. A* **13**, 1325.
Lee, C. M., and Pratt, R. H. (1976). *Phys. Rev. A* **14**, 990.
Lee, C. M., Kissel, L., Pratt, R. H., and Tsang, H. K. (1976). *Phys. Rev. A* **13**, 1714.
Lee, C. M., Pratt, R. H., and Tseng, H. K. (1977). *Phys. Rev. A* **16**, 2169.
McEnnan, J., and Gavrila, M. (1977). *Phys. Rev. A* **15**, 1557.
McVoy, K. W., and Fano, U. (1959). *Phys. Rev.* **116**, 1168.
Maxwell, J. C. (1881). "Treatise on Electricity and Magnetism," 3rd ed., Vols. 1 and 2. Dover, New York. (Orig. publ., 1959.)
Merts, A. L., Cowan, R. D., and Magee, N. H., Jr. (1976). *Los Alamos Sci. Lab.* [*Rep.*] *LA* **LA-6220MS**.
Milchberg, H. M., and Weisheit, J. C. (1982). *Phys. Rev. A* **26**, 1023.
Mork, K., and Olsen, H. (1965). *Phys. Rev.* **140**, B1661.
Mork, K., and Olsen, H. (1968). *Phys. Rev.* **166**, 1862.
Motz, W., and Placious, R. C. (1960). *Nuovo Cimento* **15**, 571.
Nakel, W. (1980). *In* "Coherence and Correlation in Atomic Collisions" (H. Kleinpoppen and J. F. Williams, eds.), pp. 187–203. Plenum, New York.
Nakel, W., and Pankau, E. (1972). *Phys. Lett. A* **38A**, 307.
Nakel, W., and Pankau, E. (1973). *Phys. Lett. A* **44A**, 65.
Nakel, W., and Pankau, E. (1975). *Z. Phys. A* **274**, 319.
Olsen, H. A., and Maximon, L. C. (1978). *Phys. Rev. A* **18**, 2517.
Pratt, R. H. (1981). *Proc. Int. Conf. X-Ray Processes Inner Shell Ioniz., August 25–29, 1980, Stirling, Scotland* (D. J. Fabian, H. Kleinpoppen, and L. Watson, eds.), p. 367. Plenum, New York.
Pratt, R. H. (1981). *Comments At. Mol. Phys.* **10**, 121.
Pratt, R. H. (1982). *In* "Advances in X-Ray Spectroscopy" (C. Bonnelle and C. Mande, eds.), p. 411. Pergamon, Oxford.
Pratt, R. H., and Lee, C. M. (1977). *Phys. Rev. A* **16**, 1733.
Pratt, R. H., and Tseng, H. K. (1975). *Phys. Rev. A* **11**, 1797.
Pratt, R. H., Tseng, H. K., Lee, C. M., Kissel, L., MacCallum, C., and Riley, M. (1977). *At. Data Nucl. Data Tables* **20**, 175; erratum published **26**, 477 (1981).
Racah, G. (1934). *Nuovo Cimento* **11**, 461, 467.
Rohrlich, F. (1955). *Phys. Rev.* **98**, 181.
Sauter, F. (1934). *Ann. Phys. (Leipzig)* **20**, 404.
Sauthoff, N. R. (1980). Personal communication.
Scofied, J. (1981). Personal communication.
Seaton, M. J. (1959). *Mon. Not. R. Astron. Soc.* **119**, 82.
Sheer, M., Trott, E., and Zahs, G. (1968). *Z. Phys.* **209**, 68.
Sobelman, I. I. (1979). "Atomic Spectra and Radiative Transitions." Springer-Verlag, Berlin and New York.
Sommerfeld, A. (1931). *Ann. Phys. (Leipzig)* **11**, 257.

Sommerfeld, A. (1939). "Atombau und Spektrallinien, 2." Au fl., Bd. 2. Braunschweig: Vieweg & Sohn.
Thaler, R. M., Goldstein, M., McHale, J. L., and Biedenharn, L. C. (1956). *Phys. Rev.* **102**, 1567.
Tseng, H. K. (1970). Ph.D. Thesis, Univ. of Pittsburgh.
Tseng, H. K., and Pratt, R. H. (1971). *Phys. Rev. A* **3**, 100.
Tseng, H. K., and Pratt, R. H. (1973). *Phys. Rev. A* **7**, 1502.
Tseng, H. K., and Pratt, R. H. (1979). *Phys. Rev. A* **19**, 1525.
Tseng, H. K., Pratt, R. H., and Lee, C. M. (1979). *Phys. Rev. A* **19**, 187.
Tucker, W. H. (1975). "Radiation Processes in Astrophysics." Chap. 5. MIT Press, Cambridge, Massachusetts.

6
Heating of Plasma by Energetic Particles

6A
Introduction

M. F. A. Harrison

Culham Laboratory
Abingdon, Oxfordshire
England

The temperature of a magnetically confined toroidal plasma can be raised by ohmic heating (i.e., by inducing a plasma current to flow in the direction of the confining field). Nevertheless, most forms of toroidal device will require auxiliary heating (see Harrison, Chapter 2, Sections IV and V) in order to achieve ignition. Auxiliary heating by neutral-beam injection has been successfully demonstrated in a number of laboratory scale experiments and the technique has potential for reactor applications. Its principal aspects are (a) formation of intense beams of energetic neutral atoms (e.g., currently envisaged beams consist of ~50 A of D^+ at ~150 keV), (b) transport of these beams through the confining field and into the plasma, (c) ionization of the energetic atoms and the subsequent trapping of these ions within the magnetic field, and (d) transfer of energy from the trapped ions to the thermalized electrons and ions of the confined plasma by means of collisions.

The simplest concept of an ignition scenario in a toroidal device is based upon ohmic heating to a modest temperature (~10^2 eV), followed by neutral injection until ignition is attained; the neutral beams are then switched off and the plasma temperature is self-sustained by α-particle heating. In reality the scenario is more complex but the underlying physical principles can be indicated by considering the ignition curve (i.e., $n\tau_E$ plotted against T) in Fig. 8 of Chapter 2. In the case of a pure hydrogenous plasma, the gradient $d(n\tau_E)/dT = 0$ when $T \approx 20$ keV; at lower temperature the gradient is negative but at a higher temperature it is positive. If the value of $n\tau_E$ is low, then the plasma must be heated to a temperature in excess of 20 keV in order to attain ignition; the plasma is then thermally unstable and its temperature will continue to rise when the beam is switched off. Conversely, if $n\tau_E$ is large, then ignition is attained in a thermally stable regime of temperature and there is the possibility of controlling temperature by feedback to the neutral injector system. Thus controlled heating is best performed at moderately high plasma density, but it is also important that the neutral beam can penetrate deeply into the plasma and so a balance in n and T must be sought throughout the heating phase. A recent discussion of the procedure for heating to ignition can be found in the INTOR—Phase One report (1982).

Heating by confined α-particles is a central issue in the concept of a D–T fusion reactor and has many similarities with neutral-beam heating. However, atom beams have directed momentum that can be controlled, whereas the initial motion of the α-particles depends upon the properties of the fusion collisions. The momentum transferred from the atom beams to the bulk plasma has significant consequences (see Cordey, Subchapter 6B, Section V.B), one such being the ability to drive a plasma current around the torus. It should be noted that, in the tokamak, the plasma current is induced by transformer action so that the device must be operated in a pulsed mode but the application of beam-driven current offers the prospect of a DC tokamak. This would substantially reduce the engineering problems associated with pulsed operation.

In this chapter the theoretical analyses of heating that are presented in Cordey (Subchapter 6B) and Post (Subchapter 6D) place emphasis upon toroidal devices and hence the motion of energetic ions is considered in relation to the particular properties of the toroidal magnetic field. However, from the technological standpoint, it is worth remembering that neutral-beam injection evolved initially from the requirements of mirror machines and that its present extensive application to toroidal devices is of relatively recent origin. The essential components in an injector are discussed in detail in Green (Subchapter 6C); basically they consist of (a) the source of energetic ions and (b) the neutralizer in which these ions are converted into energetic atoms. The highest stage of development has been attained using ion beams of H^+ and D^+ but conversion into H^0 or D^0 (which relies predominantly upon charge capture) becomes increasingly inefficient as the ion energy is increased. This has stimulated research into sources of H^- or D^- ions where more efficient neutralization can be achieved at high energies by charge stripping; the remarkable progress in this direction is considered in Subchapter 6C, Section V.

To date most fusion studies are performed in plasmas of H or at best D, so that the diagnostics of energetic α-particles have received relatively little attention; this aspect is covered in Subchapter 6D, Section III. A further facet of α-particle heating is that, in steady state, it is necessary to exhaust helium in order to maintain a low concentration of unreactive ^{4}He within the plasma. This topic is grouped together with the α-particle heating in Subchapter 6D but it should be noted that exhaust of helium as neutral gas relates strongly to conditions in the plasma boundary and these are discussed in detail in Harrison (Chapter 7 of this volume).

Reference

International Tokamak Reactor—Phase One (1982). Report of the International Tokamak Reactor Workshop, 1980 and 1981, IAEA, Vienna, p. 130.

6B
Trapping and Thermalization of Fast Ions

J. G. Cordey

Joint European Tokamak
Culham Laboratory
Abingdon, Oxfordshire
England

I. Introduction	327
II. Fast-Ion Deposition	328
III. The Slowing Down of the Fast Ions	332
IV. Energy and Momentum Transfer Rates	334
V. Effect on Plasma Temperature, Current, and Rotation	335
A. Plasma Temperature	335
B. Beam-Driven Currents	336
C. Plasma Rotation	337
References	338

I. Introduction

The basic theory of neutral-injection heating is now well established and has been checked on several experiments (Bol *et al.*, 1974; Cordey *et al.*, 1974; Berry *et al.*, 1975; Equipe TFR, 1978). The injected fast neutrals are trapped in the plasma by the combined effects of charge exchange, proton ionization, and electron ionization. The fast ions lose their energy through Coulomb collisions with the background plasma electrons and ions. As well as heating the plasma, the fast ions can also generate currents and increase the plasma rotation. In this section the theory and its experimental confirmation, where possible, will be reviewed. For convenience the subject will be split into four sections: Section II on the ionization of the fast neutrals and the deposition profile; Section III on the slowing down of the fast ions through Coulomb collisions with the background plasma; Section IV on the energy and momentum transfer rates; and Section V on the effect on the plasma temperature, current, and rotation.

II. Fast-Ion Deposition

The objective here, of course, is to deposit the fast ions as close to the center of the plasma as possible so as to maximize the efficiency of the heating. The deposition profile is determined by ionization of the fast neutrals that are trapped by three separate atomic processes: charge exchange on the hydrogenic species and the impurities, impact ionization by the hydrogenic species, and impurities and impact ionization by electrons. The cross sections for the ionization by the hydrogenic species and the electrons have been reviewed by Riviere (1971) and are reproduced in Fig. 1. For injection energies below 40 keV/amu charge exchange is the dominant process, whereas above 40 keV/amu ion impact ionization is dominant. Electron impact ionization is small for the range of energies of interest. A reasonable fit to the total cross section for ionization by the hydrogenic species is given by the function

$$\sigma_H = 2 \times 10^{-14}[1 - \exp(-\mathscr{E}/10)]/\mathscr{E} \quad (cm^2), \tag{1}$$

where $\mathscr{E}$ is the energy of the fast neutrals (keV/amu).

The question of ionization of the fast neutrals by the impurities has received considerable discussion recently. At first it was thought that impact

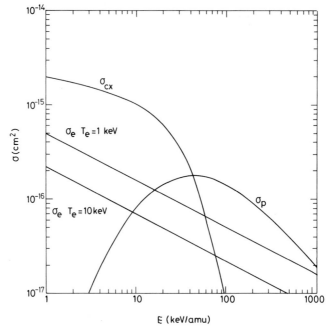

Fig. 1. Cross sections for the ionization of the fast neutral by hydrogenic ions and electrons as a function of the fast neutral energy per atomic mass $\mathscr{E}$ (keV/amu). σ_{cx} is charge exchange, and σ_e and σ_p are impact ionization by electrons and ions.

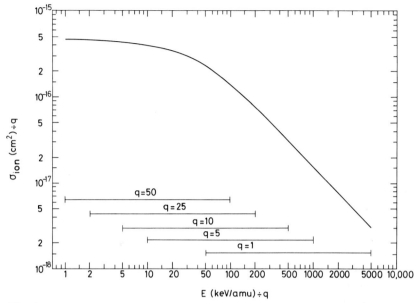

Fig. 2. Cross section for the ionization of fast neutrals of energy $\mathscr{E}$ (keV/amu) by an ion of charge state q. [From Olsen *et al.* (1978).]

ionization would be the dominant process and that the cross section would scale as q^2 (q is the charge of the impurity). However, Olsen *et al.* (1978) have shown that impact ionization is only dominant at very high energies [$>32q$ (keV)] and for lower energies charge exchange is the dominant ionization process. In Fig. 2 the total cross section for ionization by an impurity of charge state q given by Olsen *et al.* (1978) is reproduced. The analytic fit to the cross section is

$$\sigma_{\text{imp}} = 4.6q \times 10^{-16} \{32q/\mathscr{E}[1 - \exp(-\mathscr{E}/32q)]\}. \tag{2}$$

A mean free path λ for the fast neutral can be defined as

$$\lambda = 1/(n_H \sigma_H + n_{\text{imp}} \sigma_{\text{imp}} + n_e \sigma_e), \tag{3}$$

where n_H and n_{imp} are, respectively, the densities of the hydrogenics and impurity components of the plasma, n_e is the electron density, and the cross sections are given in Figs. 1 and 2. The number of ions S deposited on a magnetic surface per second can then be calculated as a function of the mean free path λ. For a pencil beam of current I_{inj} at an angle θ to the magnetic axis and passing through the center of the machine the source rate S at radius r from the plasma axis is

$$S = \frac{I_{\text{inj}}}{4\pi^2 erR\lambda \sin \theta} \left[\exp\left(-\frac{1}{\sin \theta} \int_r^a \frac{dr}{\lambda}\right) + \exp\left(-\frac{1}{\sin \theta} \int_{-r}^a \frac{dr}{\lambda}\right) \right], \tag{4}$$

where R and a are the major and minor radii of the torus. Now, of course, the actual source geometry cannot usually be accurately represented by a pencil beam since the radius of the source is often a large fraction of the minor radius of the torus. In this case it is more accurate to use a set of pencil beams to represent the beam width and then sum them. An alternative method is to use a Monte Carlo technique (Lister et al., 1976) to select the fast ions from the source; using this technique allows one to more easily take into account beam focusing.

Examples of deposition profiles in the DITE tokamak are given in Fig. 3 where the normalized source rate is shown, $H(r) = (2\pi^2 a^2 Re/I_{inj})S$, for three different ratios of the minor radius to the mean free path (λ_0 at the central density) and two different injection geometries "co" (whose beam is parallel to the direction of flow of toroidal plasma current) and "counter" (where the beam is antiparallel to the current). From the figure it can be seen that a/λ_0 should be of order unity in order to achieve reasonable deposition profile peaked in the center with minimum losses. For a/λ_0 large the fast ions are all trapped on the outside of the torus, as in the case $a/\lambda_0 = 6$ in the figure, and when a/λ_0 is small the fast ions pass right through the torus and only a few are trapped, as in the case $a/\lambda_0 = \frac{1}{10}$.

From this we see that it is fairly important to match the penetration length λ to the radius of the device. Since λ is approximately proportional to the energy and inversely proportional to the density, this means that for the larger, higher density machines of the future high injection energies of the order of 500 keV may be required. At this energy one would have to go to negative ion technology, which is not as well developed as positive ion

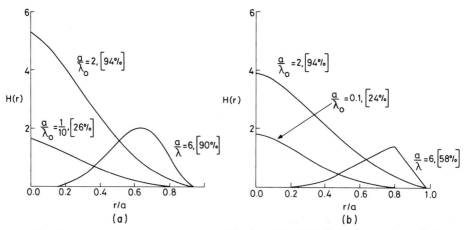

Fig. 3. Spatial shape factor $H(r) [= (2\pi^2 a^2 Re/I_{inj})S]$ in DITE for various values of a/λ_0, where λ_0 is the mean free path of a neutral for the central density. The percentage of beam atoms that are trapped in the plasma as ions is also labeled on each curve. (a) coinjection, (b) counterinjection.

6B. Trapping and Thermalization of Fast Ions

technology. Thus the key question here is whether positive ion technology with an injection energy less than, say, 160 keV will be sufficient to achieve ignition in the larger devices. This depends critically on the radial dependence of the thermal conductivity χ, as can be seen from the solution of the following thermal transport equation.

The one-dimensional thermal transport equation for the plasma temperature T may be written in the form

$$\frac{1}{r}\frac{\partial}{\partial r}\left(r\chi\frac{\partial T}{\partial r}\right) + \frac{P}{4\pi^2 eR\lambda \sin\theta}\left[\exp\left(-\frac{1}{\sin\theta}\int_r^a \frac{dr}{\lambda}\right)\right.$$

$$\left. + \exp\left(-\frac{1}{\sin\theta}\int_{-r}^a \frac{dr}{\lambda}\right)\right] = 0, \tag{5}$$

where the second term is the heat deposition, i.e., Eq. (4), and P is the injected power. Equation (5) may be trivially integrated twice to give

$$T(r) = \frac{P}{4\pi^2 eR}\int_r^a \frac{dr'}{r'K}\left[\exp\left(-\frac{1}{\sin\theta}\int_{r'}^a \frac{dr''}{\lambda}\right)\right.$$

$$\left. - \exp\left(-\frac{1}{\sin\theta}\int_{-r'}^a \frac{dr''}{\lambda}\right)\right]. \tag{6}$$

To evaluate the above a suitable expression is required for the thermal conductivity χ and the form

$$\chi = \alpha/(1 - r^2/a^2)^\rho \tag{7}$$

will be used so that the strength of the radial dependence can be changed by varying ρ. The case $\rho = 0$, $\alpha = 5 \times 10^{19}$ m^{-1} s^{-1} is the INTOR or ALCATOR scaling assumption used in many reactor studies. The integral in Eq. (6) can be evaluated analytically by use of the method of steepest descent for large values of $\int_0^a dr/\lambda$ (poor penetration) and numerically for other values. The analytic expression for the central temperature $T(0)$ in the limit of poor penetration is

$$T(0) = \frac{P}{4\pi^2 eR\alpha}\Gamma(\rho + 1)\left(\frac{2\lambda \sin\theta}{a}\right)^{\rho+1} \tag{8}$$

for a uniform plasma density, i.e., a flat density profile (here Γ is the gamma function). Thus for the INTOR scaling where $\rho = 0$ the central temperature is proportional to the product of the power and the mean free path λ.

The numerical solution of Eq. (6) for a plasma of uniform density $\bar{n}$ is given in Fig. 4 where the ratio of the central temperature to the injected power P is shown as a function of both a/λ and $a\bar{n}/\mathscr{E}$ (i.e., the line density normalized to beam energy). Thus we see at large values of ρ that the heating is only efficient when the mean free path for ionization is equal to the plasma

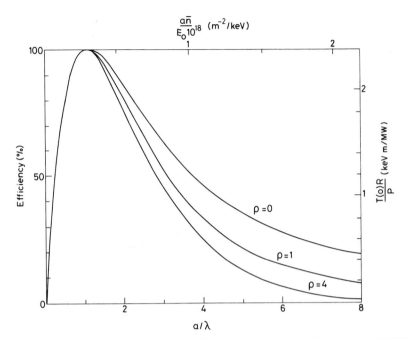

Fig. 4. Heating efficiency and the ratio of central temperature to input power $[T(0)R/P]$ plotted versus the ratio of the plasma radius to the mean free path and the ratio of plasma line density to beam energy $[a\bar{n}/\mathcal{E}\,10^{18}]$.

radius (i.e., $a/\lambda = 1$). However, when the thermal conductivity is independent of radius ($\rho = 0$), 50% heating efficiency can be obtained with injection energies a factor of 4 less than optimum ($a/\lambda = 4$). If the density profile is taken to be parabolic rather than flat, even lower injection energies can give rise to reasonably efficient heating.

The theoretical ion deposition profiles of Fig. 3 have not been checked experimentally in any great detail as yet. There is some evidence from PLT (Eubank *et al.*, 1979) that there is definitely an optimum density at which heating is most efficient and curves similar to those of Fig. 4 have been produced.

III. The Slowing Down of the Fast Ions

The fast ions lose their energy and are scattered by Coulomb collisions with the thermal ions and electrons. This process is best described by a Fokker–Planck equation for the fast ions. Analytical and numerical (Cordey *et al.*, 1974; Callen *et al.*, 1975) as well as Monte Carlo (Lister *et al.*, 1976) methods have all been used to obtain a solution of this equation. In this Subchapter I will concentrate on the analytic solutions since I think the

6B. Trapping and Thermalization of Fast Ions

physics is more transparent here. Our starting point is the Fokker–Planck equation (Cordey et al., 1974). This equation for the fast ion velocity distribution function f may be written in terms of the velocity variables v and $\zeta = v_\parallel/v$ ($v_\parallel$ is the velocity parallel to the magnetic field) in the form

$$\tau_s \frac{\delta f}{\delta t} = v^{-2} \frac{\delta}{\delta v}\underbrace{[(v_c^3 + v^3)f]}_{\substack{\text{ion} \\ \text{drag}}} + \underbrace{\frac{\beta v_c^3}{v^3} \frac{\delta}{\delta \zeta}\left[(1-\zeta^2)\frac{\delta f}{\delta \zeta}\right]}_{\substack{\text{electron} \quad \text{scattering} \\ \text{drag}}}$$

$$- \underbrace{\tau_s n_0 \sigma_{cx} vf}_{\substack{\text{charge} \\ \text{exchange}}} + \underbrace{\frac{eZ_f}{m_f} \mathbf{E} \cdot \frac{\partial f}{\partial \mathbf{v}} + \frac{T_e}{m_f}}_{\text{electric field}}$$

$$- v^{-2} \frac{\partial}{\partial v}\underbrace{\left[(v^2 + \gamma v_i^2 v_e/v)\frac{\partial f}{\partial v}\right]}_{\text{energy diffusion}} + \underbrace{SK(\zeta)\tau_s \,\delta(v - v_0)}_{\text{source}}. \quad (9)$$

Here $\mathbf{E}$ is the electric field within the plasma, m and Z are the particle mass and charge state (fast ion denoted by f and plasma ions by i), $K(\zeta)$ is the angular distribution function of the source of strength S, and δ is the delta function. The characteristic time is

$$\tau_s = 3m_e v_e^3 m_f/16\pi^{1/2} e^4 Z_f^2 \ln \Lambda n_e \sim 10^{12} A_f T_e^{3/2}(\text{keV})/nZ_f^2 \quad (10)$$

and the critical energy

$$\mathscr{E}_c = \tfrac{1}{2} m_f v_c^2 = 14.8 T_e A_f/A_i^{2/3}.$$

The remainder of the parameters are defined as follows: $\gamma = 0.75\pi^2$, $\beta = Z_{\text{eff}} A_i/2A_f$, Z_{eff} (see Chapter 4, Section II) is the effective charge of the plasma, A is the atomic weight, and $\ln \Lambda$ is the Coulomb logarithm.

In the analysis that follows we neglect the electric field, energy diffusion, and charge exchange terms since these are usually small and have been considered elsewhere (Cordey et al., 1974). The rate of loss energy can be obtained from the velocity characteristic of Eq. (9), which can be written in the form

$$\frac{dv}{dt} = (v_c^3 + v^3)/v^2 \tau_s. \quad (11)$$

Integrating this equation and writing the solution in terms of energy rather than velocity, the time to slow down to energy $\mathscr{E}$ from the injection energy $\mathscr{E}_0$ is given by

$$t = \frac{\tau_s}{3} \log\left[\frac{\mathscr{E}_0^{3/2} + \mathscr{E}_c^{3/2}}{\mathscr{E}^{3/2} + \mathscr{E}_c^{3/2}}\right], \quad (12)$$

and thus the time to slow down to thermal energies is approximately $t = \tau_s/3 \log[(\mathscr{E}_0/\mathscr{E}_c)^{3/2} + 1]$. For large injection energies $\mathscr{E}_0 \gg \mathscr{E}_c$ the energy loss of the fast ions is mainly through collisions with electrons and the slowing down

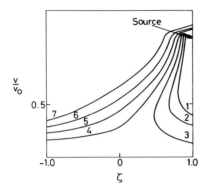

Fig. 5. Contours of the fast ion distribution in v, $\zeta (= v_\parallel/v)$ space, the contours decrease by 14% per contour, 1 being the highest.

time is essentially τ_s, whereas for low injection energies $\mathscr{E}_0 \ll \mathscr{E}_c$ the slowing down is due to collisions with plasma ions and the time to thermalize is shorter: $\tau_s \mathscr{E}_0^{3/2}/3\mathscr{E}_c^{3/2}$.

Returning now to the complete solution of Eq. (9) since the equation is separable in v and ζ, the solution may be expressed as a series of Legendre polynomials $P_j(\zeta)$ in the form

$$f = \sum_{j=0}^{\infty} a_j(v) P_j(\zeta). \tag{13}$$

The $a_j(v)$ are obtained by integrating the subsidiary equation in v:

$$a_j(v) = S\tau_s K_j St(1 - v/v_0) \frac{(v_0^3 + v_c^3)^{\beta \Lambda_j/3}}{(v^3 + v_c^3)^{1+\beta \Lambda_j/3}} \left(\frac{v}{v_0}\right)^{\beta \Lambda_j}, \tag{14}$$

where $\Lambda_j = j(j+1)$.

A typical contour plot of the distribution function f in v, ζ space is shown in Fig. 5. These distributions have been compared with the measured energy spectra on several experiments (Bol et al., 1974; Cordey et al., 1975; Berry et al., 1974; Equipe TFR, 1978), and the close agreement suggests that the fast ions do indeed slow down classically.

IV. Energy and Momentum Transfer Rates

The power transfer to the plasma can be obtained by multiplying Eq. (9) by $\frac{1}{2}mv^2$ and integrating over velocity space. The resulting transfer rates are

$$Q_i = m_f v_c^3 \int v a_0 \, dv \qquad Q_e = m_f \int v^4 a_0 \, dv \tag{15}$$

to the ions and electrons, respectively. Substituting the expression for a_0 from Eq. (5), we can then calculate the fraction of energy F going to the electrons and to the ions as a function of the injection energy and this is given in Fig. 6.

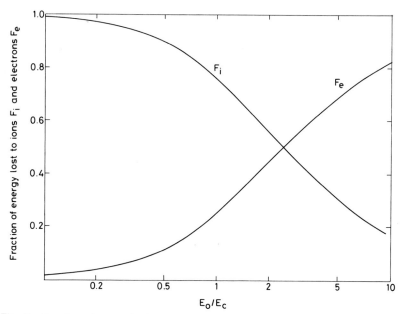

Fig. 6. Fraction of energy by loss to the electrons F_e and the ions F_i versus $\mathscr{E}_0/\mathscr{E}_c$. $\mathscr{E}_0$ is the injection energy, $\mathscr{E}_c = 14.8 A_f/A_i^{2/3}$, A_f is the fast ion atomic weight, and A_i is the thermal ion atomic weight.

The toroidal momentum balance is more complicated. After multiplying the guiding centre Fokker–Planck equation by $mv\zeta$ and integrating over velocity space and poloidal angle θ, the resulting momentum balance equation may be expressed in the form

$$\frac{1}{\tau_s}\int dv \int \zeta \, d\zeta \left[v \frac{\partial}{\partial v}[v_c^3 + v^3)f] - \frac{(1-\zeta^2)}{\zeta} v_c^3 \beta \frac{\partial f}{\partial \zeta} + v^3 S \tau_s \right] = 0. \quad (16)$$

The first term of Eq. (16) is the loss of fast-ion momentum through friction with thermal ions and electrons, the second term is the loss of momentum by angular scattering with the thermal ions, and the final term is the fast-ion source of toroidal momentum.

We now move on and see what effect the heating and momentum transfer has on the plasma.

V. Effect on Plasma Temperature, Current, and Rotation

A. *Plasma Temperature*

In all the experiments the heating gives a substantial rise in the ion temperature, which increases linearly with power and has not been observed to saturate. The highest ion temperatures ~7 keV has been obtained in PLT where 3 MW of 40-keV D^0 neutrals were injected into a deuterium plasma (Eubank et al., 1979). The increase in electron temperature was quite modest, going from 1.8 to around 3 keV. This is actually a common feature of most of the experiments because at low densities ($\bar{n} = 10^{19}$ m^{-3} in PLT) the electrons and ions become decoupled and ion temperature substantially exceeds the electron temperature. The basic reason for the decoupling is that there is a greater loss of energy from the electrons than the ions, although roughly equal amounts of power go into each species, hence the ions achieve a higher temperature. At high densities the ion–electron thermal equilibration time becomes much shorter than the electron loss time and the ion temperature then approaches the electron temperature. High density discharges with intense neutral injection heating have been obtained on ISXB. Although the ion temperatures are lower than PLT, the energy density and in particular the plasma β is very high ($\beta^* \sim 3\%$), which has enabled detailed studies of MHD stability limits to be made (Murakami et al., 1981).

B. Beam-Driven Currents

It was first pointed out by Ohkawa (1970) that the injected fast ions would induce a current in the plasma. The physical explanation for the current is as follows. In the time it takes the fast ions to thermalize they circumnavigate the torus many times and thereby form a set of stacked current loops. The transfer of momentum from the fast ions to the electrons also generates an electron current that tends to cancel out the fast ion current. The calculation of the fast ion contribution to the current is quite straightforward, using the distribution given by Eq. (4) and, taking the velocity moment, one readily arrives at an expression for the fast ion current. The back electron current is normally obtained by solving the electron Fokker–Planck equation using the technique described by Connor and Cordey (1974). For the limit $v_e > v_0$, which is usually appropriate for most tokamak injection schemes, the expression for the beam-driven current density can be written

$$j_{bd} = \bar{n}_f v_{\parallel} e Z_f \left[1 - \underbrace{\frac{Z_f}{Z_{eff}}}_{\text{fast ion current}} + \underbrace{1.46 \frac{\varepsilon^{1/2} Z_f}{Z_{eff}} \mathcal{A}(Z_{eff})}_{\text{electron current}} \right], \qquad (17)$$

where $\mathcal{A}$ is of order unity and given by (Connor and Cordey, 1974). The first term in Eq. (17) is the fast-ion current, and the second and third terms are the electron current driven by the fast ions. The third term that reduces the back electron current is due to the trapping of the electrons in the toroidal field gradient. In tokamaks the trapped electrons play an important role by (a) reducing the number of current-carrying electrons and (b) by increasing the frictional drag on the passing electrons.

Moving on now to the effect of this current on a tokamak, the Ohm's law with a beam-driven current is

$$\sigma E = j - j_{bd}, \qquad (18)$$

where j is the toroidal current density and σ the Spitzer electrical conductivity. In tokamaks in which the total current is kept constant by the iron core, the toroidal electric field E, and hence the loop volts, change to accommodate the beam-induced current. Integrating the Ohm's law over the minor radius yields the following expression for the change in loop volts $\delta\phi = \phi_F - \phi_I$:

$$\frac{\delta\phi}{\phi_I} = \frac{\int \sigma_I r \, dr}{\int \sigma_F r \, dr} - 1 - \frac{I_{bd}}{I_p} \frac{\int \sigma_I r \, dr}{\int \sigma_F r \, dr}, \qquad (19)$$

where I_{bd} and I_p are the total beam-driven and plasma currents and the subscripts I and F stand for initial and final values. This new value of the potential is set up in the magnetic field diffusion time.

The existence of the beam-driven current was first demonstrated experimentally on the Culham Superconducting Levitron (Start et al., 1978), and more recently its effect on the loop voltage has been identified in the DITE tokamak (Clarke et al., 1980).

C. Plasma Rotation

The fast ions lose momentum to the background plasma, which will then rotate. The toroidal rotation velocity u is given by the following momentum balance equation:

$$m_i n_i u / \tau_m = \Gamma_f, \qquad (20)$$

where Γ_f is the momentum input from the fast ions, which are the first two terms of Eq. (17), and τ_m is the momentum loss of the background plasma. A more sophisticated form of Eq. (22) that includes the effect of trapped particles may be found in the paper by Stacey and Sigmar (1979). There are several classical momentum loss mechanisms that contribute to the τ_m. These are perpendicular viscosity (Callen and Clarke, 1971), convection (Kovrizhnykh, 1971), and charge exchange and ripple viscosity (Conner and Cordey, 1974). Perpendicular viscosity is very small indeed and hence can be neglected. The last three are more significant; in particular, loss by charge exchange is usually the largest in present experiments.

In experiments on PLT (Suckewer et al., 1979) an increase in the toroidal velocity of the background plasma during injection was observed. The rotational velocity was smaller than would be expected if the only loss mechanisms were classical, and so the authors suggest that the toroidal momentum loss may be nonclassical.

References

Berry, L. A., Bush, C. E., Dunlap, J. L., Edmonds, P. H., Jernigan, T. C., Lyon, J. F., Murakami, M., and Wing, W. R. (1975). *Plasma Phys. Controlled Nucl. Fusion Res., Proc. Int. Conf., Tokyo, 1974* **1,** 113.

Bol, K., Cecchi, J. L., Daughney, C. C., Ellis, R. A., Eubank, H. P., Furth, H. P., Goldston, R. J., Hsuan, H., Jacabsen, R. A., Mazzucato, E., Smith, R. R., and Stix, T. H. (1974). *Phys. Rev. Lett.* **32,** 661.

Callen, J. D., and Clarke, J. F. (1971). *Bull. Am. Phys. Soc.* **16.**

Callen, J. D., Colchin, R. J., Fowler, R. H., McAlees, D. G., and Rome, J. A. (1975). *Plasma Phys. Controlled Nucl. Fusion Res., Proc. Int. Conf., 5th, Tokyo, 1974* **1,** 645.

Clarke, W. H. M., Cordey, J. G., Cox, M., Gill, R. D., Hugill, J., Paul, J. W. M., and Start, D. F. H. (1980). *Phys. Rev. Lett.* **45** (13), 1101.

Connor, J. W., and Cordey, J. G. (1974). *Nucl. Fusion* **14,** 185.

Cordey, J. G., and Core, W. G. F. (1974). *Phys. Fluids* **17,** 1626.

Cordey, J. G., Hugill, J., Paul, J. W. M., Sheffield, J., Speth, E., Stott, P. E., and Tereshin, V. I. (1974). *Nucl. Fusion* **15,** 441.

Equipe TFR (1978). *Nucl. Fusion* **18,** 1271.

Eubank, H., Goldston, R., Arunasalam, V., Bitter, M., Bol, K., and the PLT team (1979). *Plasma Phys. Controlled Nucl. Fusion Res., Proc. Int. Conf. 7th, Innsbruck, Austria, 1978.*

Kovrizhnykh, L. M. (1971). *Plasma Physics Controlled Nucl. Fusion Res., Proc. Int. Conf., 4th, Madison, Wisc.* p. 399.

Lister, G. G., Post, D. E., and Goldston, R. (1976). *Proc. Symp. Plasma Heat. Toroidal Devices, 3rd, Varenna, Italy, 1976* p. 303.

Murakami, M., Swain, D. W., Bates, S. C., Bush, C. E., Charlton, L. A., and the ISX-B team (1981). *Plasma Phys. Controlled Nucl. Fusion Res., Proc. Int. Conf. 8th, Brussels, Belgium.*

Ohkawa, T. (1970). *Nucl. Fusion* **10,** 185.

Olsen, R. E., Berkner, K. H., Graham, W. G., Pyle, R. V., Schlachter, A. S., and Stearns, J. W. (1978). *Phys. Rev. Lett.* **41,** 163.

Riviere, A. C. (1971). *Nucl. Fusion* **11,** 363.

Stacey, W. M., Jr., and Sigmar, D. J. (1979). *Phys. Fluids* **22,** 2000.

Start, D. F. H., Collins, P. R., Jones, E. M., Riviere, A. C., and Sweetman, D. R. (1978). *Phys. Rev. Lett.* **40,** 1497.

Suckewer, S., Eubank, H. P., Goldston, R. J., Hinnov, E., and Sauthoff, N. R. (1979). *Phys. Rev. Lett.* **43,** 207.

6C

Neutral-Beam Formation and Transport

T. S. Green

Euratom/UKAEA Fusion Association
Culham Laboratory
Abingdon, Oxfordshire
England

I. Introduction	339
II. Ion Beam Extraction and Acceleration	341
A. Analytical Treatment of Single-Gap Extraction	342
B. Multigap Systems	344
C. Beam Steering	346
D. Computational Studies of Beam Extraction	346
E. Electrode Power Loading	348
III. Plasma Sources for Positive Ions	352
A. Ionization Efficiency in Arc Discharge Sources	353
B. Ion Species	357
C. Plasma Uniformity	358
D. Ion Source Types	360
IV. Beam Transport in a Gas Neutralizer	363
A. Neutralizer Efficiency	363
B. Slow Hydrogen Ion Production in Neutralizers	367
C. Beam Particle Trajectories	367
D. Space-Charge Compensation	368
V. Negative-Ion Beams	372
A. Double-Charge Capture	372
B. Direct Extraction	374
C. Volume Production of Negative Ions	376
References	378

I. Introduction

The advantages of neutral-beam injection as a method for heating plasmas arise from its relatively simple physical basis. Ions are created in a plasma source and are electrically accelerated through an electrode structure to the required energy; some are then converted (by charge exchange collisions) to atoms that can traverse the magnetic field surrounding the target plasma

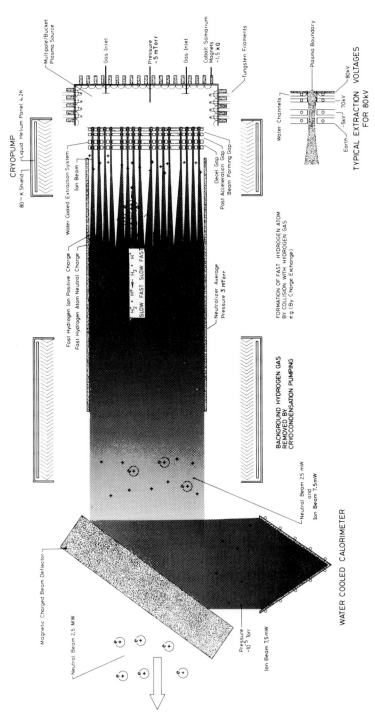

Fig. 1. Schematic diagram of neutral atom injector for multimegawatt beams based on positive-ion source with multiaperture extraction system.

(Fig. 1). In the target plasma the atoms are ionized and slowly transfer energy to the plasma by collisions.

The application of neutral-beam injection as a heating method at the megawatt power level has depended on the extension of beam technology from the subampere level used in nuclear physics to the multiampere level. This development has rested on two main factors:

(a) the recognition of the need to subdivide the ion beam into small beamlets in the acceleration stage by the use of multiaperture extraction electrodes, as illustrated in Fig. 1 (Kaufmann, 1961; Hamilton *et al.*, 1968).

(b) the use of a close coupled neutralizer cell so that the ions are converted into atoms in the shortest possible distance.

The use of multiaperture electrodes has led in turn to the need to develop plasma sources of large area (≥ 1000 cm^2) with a high level of plasma uniformity over the area of the extraction electrode.

In this subchapter we discuss the development of different components of the outline neutral injection system: (a) plasma source, (b) accelerating structure, (c) neutralizer cell, and (d) beam transport. The emphasis is on positive-ion systems, as outlined in Fig. 1. The developing area of negative-ion sources is included in Section V because of the importance of these sources for the high injection energies discussed in Cordey (Subchapter 6B, Section II).

The development of the technology of the beam line system, which has been essential for the achievement of high power injection, is not reviewed in this discussion, being extensively covered in the literature (see, e.g., Cooper *et al.*, 1981).

II. Ion Beam Extraction and Acceleration

Ions are extracted from a plasma source by the application of an electric potential between electrodes and the source plasma. A single-gap accelerator with three electrodes is shown in Fig. 2. The first electrode, at positive potential V^+, defines the equipotential plane close to the source plasma; the third is at earth potential. An intermediate electrode held $\simeq -0.1 V^+$ prevents electrons formed in the neutralizer cell from returning to the source; it is called a suppressor electrode. Its influence on the beam extraction is small but it does play a significant role in the motion of secondary ions and electrons that may produce power loading in the electrode system.

Limitation in performance of three-electrode (triode) systems due to electrical breakdowns have been discussed extensively in the literature (Green, 1976; Thompson, 1974; Cooper *et al.*, 1974). These discussions indicated that high voltage, high current sources could best be operated with four-electrode (tetrode) designs. More recently, extended operational experience

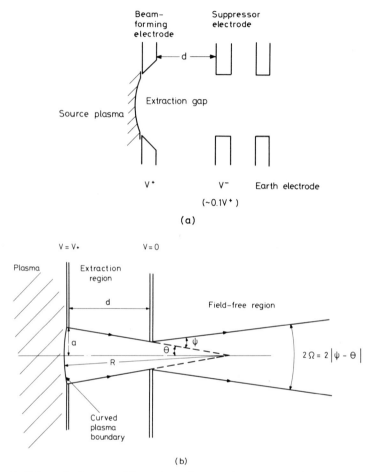

Fig. 2. Schematic diagram of three-electrode extraction system (triode): (a) geometry, (b) beam trajectories.

at high voltage has indicated that the situation is less clearcut than originally proposed. In this discussion the design of both three- and four-electrode structures will be considered without reference to their limitations.

A. Analytical Treatment of Single-Gap Extraction

In the present discussion we are concerned with the problem of the acceleration of ions extracted from a plasma surface, which is analogous to that of electron acceleration in a space-charge limited diode.

The theoretical evaluations (i.e., analytical and computational) of this problem differ in detail in their treatment of the plasma surface. In the analytical approach it is normally assumed that the plasma boundary is

6C. Neutral-Beam Formation and Transport

curved (either as a section of a cylinder or a section of a sphere) with constant current density over its surface. The subsequent analysis follows closely that presented by Brewer (1967; see also Pierce, 1949) for an electron gun with curved emitter surfaces. It is based on the treatment by Langmuir and Blodgett (1924), of a spherically converging, space-charge limited flow. Figure 2b shows schematically the trajectories of the ions.

The current of ions I from an aperture, radius a, accelerated across a gap d is given approximately by

$$\frac{I}{V^{3/2}} = 1.72 \times 10^{-7} \frac{a^2}{d^2} \left(1 - \frac{1.6d}{R}\right) \quad (\text{A V}^{-3/2}), \tag{1}$$

where V is the acceleration voltage and R the radius of curvature of the plasma surface.

The angle of convergence of the beam from the plasma surface is θ and equals a/R. The ions pass through the aperture in the negative suppressor electrode, which behaves like an electrostatic lens. The focal length is calculated from the Davisson–Calbrick formula allowing for the influence of space charge on the electric field (Davisson and Calbrick, 1931; Coupland et al., 1972), i.e.,

$$f = 3d. \tag{2}$$

The equation for the trajectory of a beam particle can be written using the matrix transfer formulation:

$$\begin{vmatrix} r' \\ r \end{vmatrix} = \begin{vmatrix} 1 & -1/f \\ 0 & 1 \end{vmatrix} \begin{vmatrix} 1 & 0 \\ d & 1 \end{vmatrix} \begin{vmatrix} r'(0) \\ r(0) \end{vmatrix}, \tag{3}$$

where r is the radius of the particle trajectory and r' is the angle made by the particle trajectory to the beam axis, i.e., $r' = dr/dz$.

The equation for the beam envelope is obtained by equating $r'(0)$ to θ and $r(0)$ to a. The condition that the emergent beam be parallel is R equals $4d$, to a first approximation, but $R = 3d$ is more accurate.

Inserting this relation into Eq. (1) for the perveance leads to the result that for a parallel or zero divergence beam

$$P_0 = \frac{I}{V^{3/2}} \simeq 0.76 \times 10^{-7} \frac{a^2}{d^2} \quad (\text{A V}^{-3/2})$$

$$= 0.44 P_c, \tag{4}$$

where P_c is the perveance for a plane diode. The exact value of the constant depends on the approximation introduced and can vary from 0.44 to 0.60. This analysis also leads to an expression for the variation of the angular divergence Ω with perveance

$$\Omega = 0.5 \, a/d [(P/P_0) - 1]. \tag{5}$$

Corresponding relations can be derived for slit apertures.

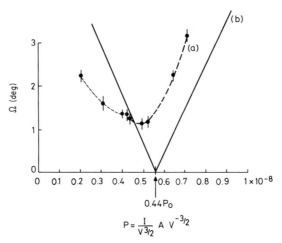

Fig. 3. Variation of beamlet divergence Ω with perveance P in a three-electrode system: curve (a) experimental data, curve (b) theoretical curve [Eq. (5)].

These predictions agree reasonably well with the experimental observations (Fig. 3), except that they indicate zero divergence at optimum perveance rather than the nonzero values observed experimentally, due to neglect of the finite transverse energies of the ions.

B. Multigap Systems

In principle the addition of more electrodes can be analyzed straightforwardly by an extension of the transfer matrix formulation:

$$\begin{vmatrix} r' \\ r \end{vmatrix} = M_{11} \cdot M_{12} \cdot M_{22} \cdot M_{23} \cdot \begin{vmatrix} r'(0) \\ r(0) \end{vmatrix}, \qquad (6)$$

where M_{11}, M_{22}, etc., represent the transfer matrices for the drift spaces between electrodes and M_{12}, M_{23} the transfer matrices of the lenses in the electrodes.

As a specific example we can consider a two-gap (or tetrode) system (Fig. 4). M_{11} is calculated assuming straight line trajectories (as above) and equals $\begin{vmatrix} 1 & 0 \\ d_1 & 1 \end{vmatrix}$. M_{22} is calculated assuming that space-charge effects are small and ions move in the constant electric field in this gap (Pierce, 1949; Holmes and Thompson, 1981; Okumura et al., 1980; Gardner et al., 1978):

$$M_{22} = \begin{vmatrix} A^{-1} & 0 \\ \dfrac{2(A-1)V_1}{E_2} & 1 \end{vmatrix}, \qquad (7)$$

6C. Neutral-Beam Formation and Transport

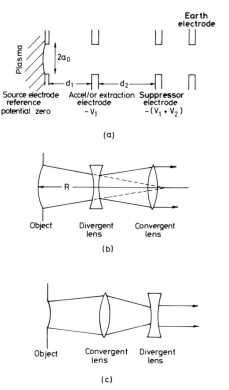

Fig. 4. Schematic diagram of four-electrode extraction system (tetrode): (a) geometry of electrodes, (b) equivalent optical system ($V_2 \leq V_1$), (c) equivalent optical system ($E_2 \geq E_1$).

where $A = [1 + (V_2/V_1)]^{1/2}$, V_1 is the voltage across the first gap, and V_2 is that across the second. E_2 is the electric field in the second gap V_2/d_2. The lens transfer matrices are

$$\begin{vmatrix} 1 & -1/f_1 \\ 0 & 1 \end{vmatrix} \quad \text{and} \quad \begin{vmatrix} 1 & -1/f_2 \\ 0 & 1 \end{vmatrix},$$

where the focal lengths are calculated using the Davisson–Calbrick formula. Various degrees of sophistication may be used in calculating the electric fields used in these formulas to allow for space charge and geometry. The final trajectory equation may be written as

$$\begin{vmatrix} r' \\ r \end{vmatrix} = \begin{vmatrix} 1 & -1/f_2 \\ 0 & 1 \end{vmatrix} \begin{vmatrix} A^{-1} & 0 \\ \dfrac{2(A-1)V_1}{E_2} & 1 \end{vmatrix}$$

$$\times \begin{vmatrix} 1 & -1/f_1 \\ 0 & 1 \end{vmatrix} \begin{vmatrix} 1 & 0 \\ d_1 & 1 \end{vmatrix} \begin{vmatrix} r'(0) \\ r(0) \end{vmatrix}. \tag{8}$$

The emergent beam has zero divergence ($r' = 0$) when

$$\frac{r(0) + d_1 r'(0)}{d_1 r'(0)} \left(\frac{2V_1}{E_2} \frac{A-1}{f_1 f_2} - \frac{1}{f_2} - \frac{1}{f_1 A} \right) = \frac{2V_1}{E_2} \frac{A-1}{d_1 f_2} - \frac{1}{A d_1}. \quad (9)$$

This rather complicated expression relates the possible values of the voltage ratio (V_2/V_1 designated as Γ) and the ratio of the gap distances (d_2/d_1 designated by γ) to allowable values of the radius of curvature of the plasma boundary ($r(0)/r'(0)$). Particular solutions for a flat boundary, giving the highest perveance, have been presented by Holmes and Thompson (1981). These correspond to high values of Γ/γ, i.e., of E_2/E_1.

C. Beam Steering

It follows from the treatment presented above that the beams extracted from a source may be deflected by offset of one aperture relative to the others because of the electrostatic lenses formed at the apertures. The geometry for beam deflection (or steering) in a single-gap system is shown in Fig. 5a together with experimental data on the angle of deflection produced by offset in Fig. 5b.

For both a single- and a two-gap system the deflection can be calculated using the matrix formulation. The calculated values are in good agreement with experiment, as can be seen from data for a single-gap system shown in Fig. 5b. Similar experiments have been performed with two-gap systems and the data compared with theory (Holmes and Thompson, 1981; Okumura *et al.*, 1980; Gardner *et al.*, 1978).

D. Computational Studies of Beam Extraction

Analytical treatments suffer from a number of defects: idealization of plasma profile, neglect of electrostatic field perturbations, and neglect of finite temperature of the ions. Numerical simulations of extraction systems have been developed since the early work of Bates (1966) by a number of authors (a detailed bibliography is given in Whealton and Whitson, 1980).

The computer programs have varied in sophistication and the degree to which they can take account of the effects omitted from the analytical treatments. All programs allow for the effect of geometry on the electrostatic fields, and some for the finite transverse ion energy. Treatment of the plasma density at the aperture in the beam-forming electrode is the most difficult task since it requires that the computation be carried out over a volume within the plasma in which a uniform flux of ions towards the aperture can be defined. Typical ion trajectories calculated by Whealton and Whitson (1980) using such a program are shown in Fig. 6. To minimize the computing required to obtain this result, Raimbault (1979) has used an analytical model

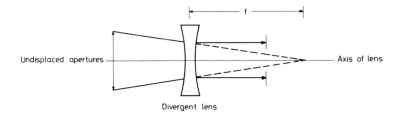

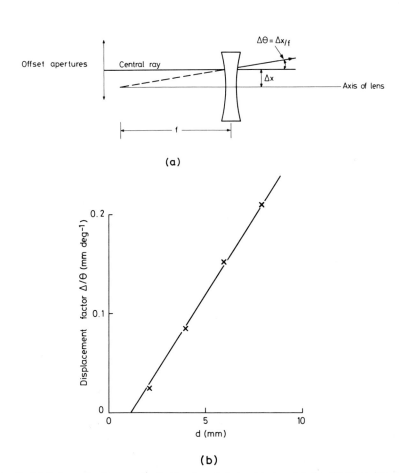

Fig. 5. (a) Schematic diagram illustrating beamlet steering in a triode. (b) Experimental data for beamlet steering in slit aperture extraction system compared with theory. Displacement factor Δ/θ versus gap separation d (θ being angle steering due to an aperture displacement of Δ). Crosses represent experimental points: the fitted line has a slope of 0.031 deg^{-1} compared with theoretical value of 0.026 deg^{-1}.

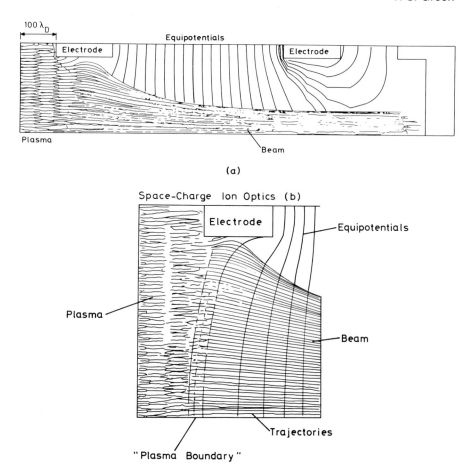

Fig. 6. Computed ion trajectories in an extraction system: (a) trajectories between electrode, (b) trajectories near plasma surface. [After Whealton and Whitson (1980).]

to calculate flow towards the aperture matching the output to the computation of ion trajectories from the plasma boundary.

One of the most important applications of the computer programs has been to the calculation of the motion of secondary particles created in the gap, either by ion–molecule collisions or by collisions of particles with the electrodes. These determine the power loading as discussed below.

E. *Electrode Power Loading*

The extraction of ion beams with high power densities (≥ 20 kW/cm^2) for long pulses (≥ 1 s) requires integral cooling of the electrode structures, the design of which is based on known or estimated power deposition rates. This

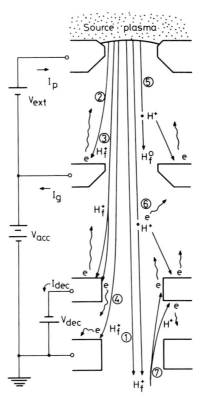

Fig. 7. Schematic diagram indicating channels for power loading in extraction systems. The following aspects are indicated: (1) transmitted beam of H_f^+ ions; (2)–(4) fast ions (H_f^+) with aberrated ion trajectories that intercept electrodes and give rise to secondary electrons; (5) charge exchange and also slow ions (H^+) hitting electrode, giving rise to secondary electrons; (6) ionization of background gas giving slow (H^+) ions and backstreaming electrons; (7) ions returning from neutralizer.

had led to an intensive theoretical and experimental investigation of power deposition, an important result of which has been the recognition of the significance of particles created in the extraction gap: electrons that stream back into the source, and ions and neutrals with degraded energies and different trajectories.

1. Analysis of Power Loading

The analysis of the factors that contribute to power loading naturally breaks into four components (Fig. 7):

i. *Volume ionization and charge exchange.* Gas is introduced into the plasma source and possibly into the neutralizer cell. As a result there is a finite density of molecular hydrogen in the extraction gap (n_0). The rate of formation of ions in this gap is given by $(J_B/e)n_0(\sigma_i + \sigma_{cx})$. The cross sections σ_i (ionization cross section) and σ_{cx} (charge exchange cross section) are energy dependent, and hence position dependent (the relation between energy and position is derived from the computer simulation). The neutral gas density profile is generally calculated from the conductance of each electrode for room temperature gas. [However, Holmes and Green (1980) propose instead that the gas streams through the structure at high temperature, giving a different density profile.] Figure 8 shows a typical set of calculations for the fractional currents and powers carried by the electrons, slow ions, and neutrals deriving from collisions due to H^+ ions with H_2 molecules in the extraction gap.

ii. *Trajectories of volume produced particles.* The trajectories of these particles can be derived using the computational techniques described above, which, following the original studies of Cooper *et al.* (1974), are now universally applied. Typical trajectory tracing is shown in Fig. 9. The results indicate that volume electrons mainly return through the aperture in the beam-forming electrode to the source, and volume ions and neutrals dominantly pass into the neutralizer but a few may impact on the negative suppressor electrode.

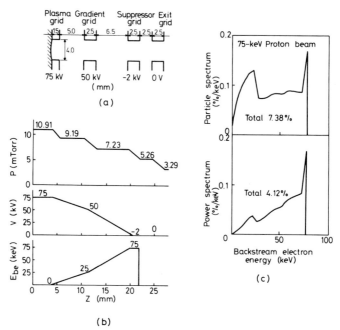

Fig. 8. Calculated values of electron currents and powers due to collisional processes in extraction system: (a) extraction geometry; (b) assumed axial variation of pressure P, potential V, and calculated energy of electrons created by collisions E_b; (c) particle and power spectra of backstreaming electrons that are created by collisions. [After Ohara *et al.* (1980).]

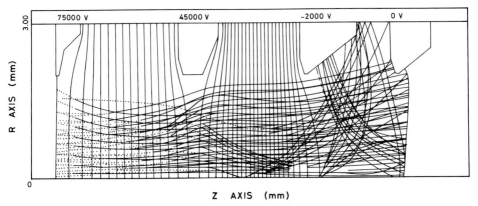

Fig. 9. Trajectories of secondary particles produced by ionization processes in an extraction system. (The solid lines denote H_2^+ ions, the dotted lines electrons. The vertical scale is three times the size of the horizontal scale for clarity.) [After Ohara *et al.* (1980).]

iii. *Trajectories of slow ions from neutralizer.* A number of authors compute the trajectories of slow ions from the neutralizer, assuming a plasma boundary exists in the aperture in the earth electrode (Fig. 9) whose position depends on the potentials in the beam. However, the Debye length in the beam plasma is approximately equal to the beamlet radius, so that the definition of the plasma boundary is unclear and the calculation of current density uncertain.

iv. *Trajectories of secondary electrons.* The major problem in design of extraction systems is to reduce the fraction of secondary electrons originating on the suppressor electrode that may return to the source or beam-forming electrode. Computations of the trajectories show it is important to shape the suppressor electrode, as shown in Fig. 10 due to Ohara *et al.* (1980). These are typical of other published results.

2. Experimental Data

There is a wide range of experimental data corresponding to different optimization of electrode design in both circular and slit geometries. It is difficult at this stage to present a unified discussion of these data.

As an example of the discussion in the literature we consider the contribution to backstreaming electrons due to secondary electrons from the suppressor electrode. This has been examined semiempirically by Ohara *et al.* (1980). The power in these electrons expressed as a percentage of the beam power F is given by

$$F = (I_{\text{sup}}/I_+)\gamma/(\gamma + 1)\beta, \qquad (10)$$

where I_{sup} is the current from the negative suppressor electrode, I_+ the beam current, γ the average secondary emission coefficient, and β the fraction of secondary electrons that returns to the source.

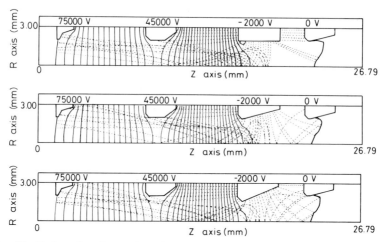

Fig. 10. Trajectories of secondary electrons emitted from the negative suppressor grid for three different grid shapes. Solid lines are equipotentials and dashed lines trajectories. The calculations are for circularly symmetrical ion beams (R is the radial, and Z the axial, displacement). [After Ohara *et al.* (1980).]

Experimentally I_{sup}/I_+ is observed to be ~5% per mTorr of source pressure in the experiment; γ is taken to be 0.5 for low energy ion incidence, and β is estimated to be 0.3–0.5 from the computed trajectories. Thus, F is estimated to be ~5–8 × 10^{-3} per mTorr (Ohara *et al.*, 1980).

Holmes and Green (1980) have shown that it is possible to derive a more detailed insight into the factors determining F by measurement of the variation of power loading on electrode and the neutralizer with the voltage applied to the suppressor electrode.

III. Plasma Sources for Positive Ions

The plasma sources now used in neutral-beam injection systems have in general evolved from those originally used in nuclear physics, principally the rf sources, the duoplasmatron, and reflex sources. The status of such sources, mainly dominated by arc discharge sources, will be discussed below, but first it is relevant to consider the underlying physical principles. All sources require that energy is coupled into a plasma to maintain a steady rate of ionization, so the first consideration is that of ionization efficiency. Uniformity of the plasma density at the extraction plane is required, as discussed above, and the conditions for achieving this will be considered. Finally the need for high fractions of H^+ and D^+ in the extracted ion beams has led to studies of the collisional processes in sources that determine these fractions and this will be discussed in Section III.B.

6C. Neutral-Beam Formation and Transport

A. *Ionization Efficiency in Arc Discharge Sources*

In arc discharge sources electrons are injected from a cathode into the plasma volume and ionize the gas introduced into the source. The current flow to the anode is comprized of some of these injected electrons and also thermal electrons. The cathode may be a directly heated filament or may be a virtual cathode formed at a magnetic restriction (as in a duoplasmatron and its variants).

The electrons are accelerated across the cathode sheath.[†] In the following it will be assumed that the sheath potential V_c equals the arc voltage, i.e., the anode to plasma potential is small (Goede and Green, 1982; Hershkowitz *et al.*, 1979). The treatments of the subsequent development of the energy spectra of the electrons differ considerably; unfortunately not all the treatments have been discussed in the open literature, some being reported only in internal laboratory discussion notes.

Some discussions, deriving from the early work of Wiesemann (1972), treat the problem on the basis of single-particle collisions in which the electrons are degraded in energy by inelastic collisions. This type of calculation for a hydrogen (or deuterium) plasma is extremely complex, requiring complete categorizing of the inelastic processes for neutral atoms and molecules and molecular ions; it can only be treated by computer modeling.

A number of analytical models have been developed that introduce approximations to facilitate the treatment. Three of these are described below; they represent different types of approximation.

(a) *Ionization by thermal electrons.* This model, as presented here, was first derived by Von Goeler (1970) for ionization in an rf source. Several authors (Cooper, 1975; Kulygin *et al.*, 1976) propose that even in an arc discharge it is the thermal electrons that are responsible for ionization. The electrons from the filaments, accelerated across the cathode sheath, provide the input energy to maintain thermal equilibrium.

The conservation equation for the ion density is

$$n_+/\tau_+ = n_e n_0 \langle \sigma v \rangle_\text{ion}, \qquad (11)$$

where n_+ is the ion density, n_e the electron density, n_0 the neutral gas density, τ_+ the ion containment time, and $\langle \sigma v \rangle_\text{ion}$ the reaction rate coefficient for ionization of the neutral gas by thermal electrons; it is a function of electron temperature.

Charge neutrality requires that

$$n_+ = n_e. \qquad (12)$$

[†] The plasma sheath is discussed by Harrison (Chapter 7, Section III, this volume).

Hence

$$\langle\sigma v\rangle_{\text{ion}} = 1/n_0\tau_+ . \tag{13}$$

Since $\langle\sigma v\rangle_{\text{ion}}$ is a function of temperature, it follows that the electron temperature is determined by the product $n_0\tau_+$. Table I gives data for a source operating with low discharge voltage and without magnetic field (Kulygin et al., 1976) that show that the product of pressure in the source (proportional to n_0) and $\langle\sigma v\rangle_{\text{ion}}$ (calculated from the measured electron temperature) is constant over a range of operating pressures.

(b) *Ionization by primary electrons allowing for stepwise degradation of energy.* This model, proposed by Friedman et al. (1978) to explain ionization rates in sources operating in helium, is based on the assumption that the electrons from the cathode are slowed down in steps by inelastic collisions.

One may then write

$$I_+ = e \int_0^{\tau_p} n_e n_0 \langle\sigma v\rangle_{\text{ion}} \, dt \times \text{volume},$$

where I_+ is the total ion current produced and τ_p the lifetime of primary electrons.

Two cases are investigated. One is a reflex discharge in which the electrons are efficiently contained by the magnetic field and the lifetime is the slowing-down time. The ion production equation then becomes

$$I_+ \simeq I_e \int_{\phi_i}^{V_c} \frac{\langle\sigma v\rangle_{\text{ion}}}{\Delta_{\text{ion}}\langle\sigma v\rangle_{\text{ion}} + \Delta_{\text{in}}\langle\sigma v\rangle_{\text{in}}} \, d\varepsilon, \tag{14}$$

where I_e is the injected current of primary electrons, $\langle\sigma v\rangle_{\text{in}}$ the reaction rate coefficient for inelastic scattering, Δ_{ion} the average energy lost in ionization collisions, Δ_{in} the average energy lost in inelastic collisions, ε the electron energy, and ϕ_i the ionization potential χ_i divided by the electron charge. The

TABLE I

Ionization Rates in Field-Free Source[a,b]

P_0 (mTorr)	T_e (eV)	$\langle\sigma v(T)\rangle_{\text{ion}}$ (cm^3 s^{-1})	$P_0\langle\sigma v(T)\rangle$
35	4.4 → 5.5	3.5 × 10^{-10} → 1.4 × 15^{-9}	1.2 → 4.9
15	5.5 → 6.5	1.4 × 10^{-9} → 2.6 × 10^{-9}	2.1 → 3.9
5	9.5 → 10.5	7 × 10^{-9} → 7.4 × 10^{-9}	3.5 → 3.7

[a] From Kulygin et al. (1976).
[b] Theory predicts $P_0\langle\sigma v(T)\rangle = $ const.

6C. Neutral-Beam Formation and Transport

values of these parameters in helium given by the authors are

$$\Delta_{ion} = 40 \text{ eV}, \quad \Delta_{in} = 20 \text{ eV},$$

$$\frac{\langle \sigma v \rangle_{ion}}{\langle \sigma v \rangle_{in}} = 0.4.$$

Hence

$$I_+ \sim I_e(V_c - \phi_i)/50.$$

Note that I_+/I_e is predicted to be independent of gas density. This relation is in good agreement with the experimental data at low arc voltages, as shown in Fig. 11. The departure from this linear relation is ascribed to changes in electron containment that depend on arc voltage, source pressure, and magnetic field.

The second case considered by Friedman *et al.* (1978) is for a source with cusp magnetic fields. Friedman *et al.* assume that the electron confinement time is not significantly greater than the slowing-down time and consequently allow for both electron loss and slowing down. The problem is treated using Monte Carlo computational methods.

(c) *Ionization by primary electrons assuming an energy distribution that is independent of gas density.* This model proposed by Green *et al.* (1975) is based on the assumption that the primary electrons are degraded by beam plasma instabilities into a broad energy distribution whose form is independent of the gas density.

The ion production rate is calculated from the relation

$$I_+ = n_p n_0 \langle \sigma v \rangle_{ion} e v,$$

where n_p is the density of primary electrons v is the plasma volume.

Now $\langle \sigma v \rangle_{ion}$ is the reaction rate coefficient for ionization averaged over the electron energy distribution function.

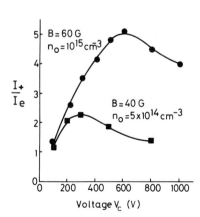

Fig. 11. Variation of ionization efficiency I_+/I_e with arc voltage in reflex ion source. [After Friedman *et al.* (1978).]

n_p is given by

$$I_e = en_p[(1/\tau_p) + n_0\langle\sigma v\rangle_{in}]v.$$

$\langle\sigma v\rangle_{in}$ is now defined as the reaction rate coefficient for the primary electrons to be inelastically scattered to energies below the ionization threshold.

Hence

$$\frac{I_e}{I_+} = \frac{\langle\sigma v\rangle_{in}}{\langle\sigma v\rangle_{ion}} + \frac{1}{n_0\langle\sigma v\rangle_{ion}\tau_p} \quad (15)$$

This relation between I_e/I_+ and n_0 has been confirmed experimentally for a number of sources (magnetic field free; reflex and magnetic multipole; Green et al., 1975; Leung et al., 1978; Goede et al., 1978). Typical data are shown in Fig. 12. Values of $\langle\sigma v\rangle_{ion}/\langle\sigma v\rangle_{in}$ obtained at different are voltages agree well with earlier data by Langmuir and Jones (1928).

(d) *Boundary conditions.* To these analyses, it is necessary to add consideration of the boundary conditions. The cathode condition, first discussed by Tonks and Langmuir (1929), requires that the fraction of ions returning to the cathode αI_+ is related to the emission current I_e by the relation (see also Gabovich, 1972)

$$\alpha I_+/I_e = |m_e/m_i|^{1/2}, \quad (16)$$

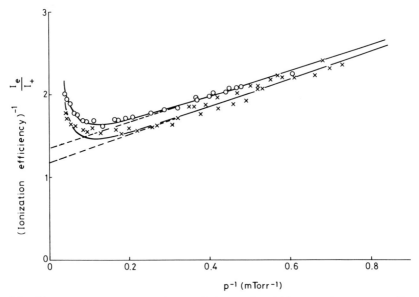

Fig. 12. Variation of inverse ionization efficiency I_e/I_+ with inverse of gas pressure. Experimental data are compared with straight line predicted by Eq. (15). O, $V_c = 110$ V; X, $V_c = 130$ V. [After Green et al. (1975).]

6C. Neutral-Beam Formation and Transport

m_e and m_i being the electron and ion masses, respectively. A simple application of this relation has been discussed by Lejeune (1971) for a duoplasmatron, in which case I_+/I_e is given by $n_0 \sigma_{ion} l$, where l is the discharge length. Hence

$$n_0 \sigma_{ion} = (1/\alpha l)|m_e/m_i|^{1/2}. \qquad (17)$$

Now σ_{ion} is a function of the electron energy, i.e., of the arc voltage. Thus the voltage at which the source can operate is a function of n_0. (Note the similarity to the relation for thermal ionization.) Lejeune also shows that the decrease of n_0 due to gas burnout leads to an increase in arc voltage. Generalization of this treatment has been discussed by Green (1974).

The boundary condition at the anode derives from the requirement that the electron flux density be sufficiently high as to carry the anode current. Different treatments of this problem are still being formulated in the literature and vary with the details of source operation.

B. Ion Species

The need to produce ion beams with high ratios of H^+ or D^+ ions for neutral injectors has led to experimental and theoretical studies of the factors influencing the ion species. Several authors indicate that ~80% of the extracted current can be in the H^+ component rather than H_2^+ and H_3^+ (Stirling et al., 1979; Goede and Green, 1979). There is, however, still considerable uncertainty in predicting how these ratios may be improved.

The range of reactions occurring in a hydrogen plasma is very wide and the relative importance of different processes varies according to the electron energy spectrum. Further complications arises from uncertainties in the recombination of the H_0 (or D_0) atoms at the wall and in the levels of vibration states excited in the molecules that influence the possible reactions (Bacal et al., 1980a).

Several studies (not published in the open literature) have proceeded by computing all reaction rates for certain boundary conditions with a minimum of fitting parameters (e.g., Cooper et al. at Berkeley use the recombination coefficient as one parameter and calculate total energy balance leading to prediction of electron energy spectrum, current density, and all reaction rates). It is difficult, however, to derive physical insight from such computational modeling.

Martin and Green (1976) in an alternative approach have shown that the ratio of currents of the different species have reasonably straightforward functional dependence on the source pressure and electron density, and attempt to use these dependencies to evaluate the dominating processes in the sources.

As a simple example, H_3^+ ions are formed by collision of H_2^+ ions with H_2

molecules. In the limit of low current density at which H_3^+ ions are not destroyed by collisions one finds

$$\frac{N(H_3^+)}{\tau_3} = N(H_2^+)N(H_2)\langle\sigma v\rangle_3, \tag{18}$$

where the values of N are the relevant particle densities, τ_3 is the containment time of the H_3^+ ion, and $\langle\sigma v\rangle_3$ is the reaction rate coefficient for production of H_3^+ in collisions between H_2^+ ions and H_2 molecules. The experimental data show an approximately linear variation of the ratio of H_3^+ and H_2^+ currents with gas pressure as predicted by this equation (the pressure dependence of the ion containment time due to the change in ion mobility causes a departure from strictly linear dependence).

Such a treatment shows that the fraction of H^+ ions increases with the product of the electron density and the ion containment time (Fig. 13a). This is in general agreement with the observations that the H^+ fraction increases as the current density in the source increases, shown by typical data plotted in Fig. 13b.

C. Plasma Uniformity

The physical phenomena that control the distribution of plasma density in an ion source have been discussed in general by Kohlberg and Nablo (1966). They are similar to those discussed for a positive plasma column (see, e.g., Franklin, 1976). Figure 14 shows results of the calculation of the variation of plasma density in a plane, one-dimensional source as presented by Franklin. In the calculation the basic equations solved are (a) particle balance, (b) ion momentum, and (c) electron momentum. The ions are treated as a fluid and it is assumed that there is charge neutrality.

The resulting equations can be solved subject to the boundary conditions: that the ion velocity and the potential are zero at the source center, and that the ion velocity increases monotonically until $du/dx \to \infty$ at the sheath (u is the velocity and x the distance from the center). Also shown in Fig. 14 is the distribution of density for the case in which the ions do not behave as a fluid but move freely in the potential gradient (Harrison and Thompson, 1959; Tonks and Langmuir, 1929).

The various results indicate that in the regimes of quasicollisionless plasmas the density profile is insensitive to the details of the calculation, essentially because the potential developed has to accelerate the ions to the ion acoustic speed by the time they enter the sheath (Bohm, 1949), and because the potential relates to the density via the electron equation, which to a good approximation is $\eta = -\log N$ (η is the normalized potential and N the normalized density). There is still a need for analyses in two or three dimensions, including some magnetic field effects, to allow direct comparison with experiment.

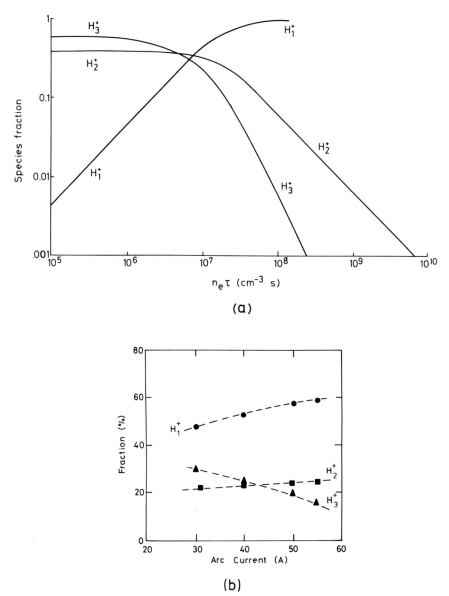

Fig. 13. (a) Theoretical prediction of $H^+ : H_2^+ : H_3^+$ fractions in an ion source as a function of the product $n_e\tau$ (n_e is the electron density and τ the containment time of H^+ ions). (b) Experimental data for the variation of H^+ fraction with arc current in hydrogen ion source.

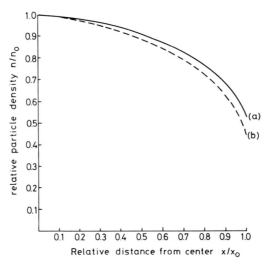

Fig. 14. Calculated density profile in plasma discharge. One-dimensional solution: curve (a) fluid motion of ions, curve (b) free fall motion of ions. [After Franklin (1976).]

D. Ion Source Types

1. *The Duopigatron Source*

The duopigatron ion source initially developed at Oak Ridge established itself as a "workhorse" source in the fusion community and was built in various forms for use on many injection experiments on tokamaks. Most of these sources were capable of yielding ~10 A (Davis et al., 1972). This source, developed from the duoplasmatron, has two plasma regions separated by an intermediate electrode. The first plasma forms a plasma cathode emitting electrons into the second reflex discharge (or PIG) region (Fig. 15).

At the Oak Ridge National Laboratory the source has been developed to larger diameters with modifications to the intermediate electrode design to optimize plasma uniformity, and by use of a multipole magnetic field configuration in the PIG region. In other laboratories similar developments are being or have been undertaken to extend the application to different geometries (Becherer, 1975; Ohara et al., 1979). Beam currents of several tens of amperes have been produced in the latest forms of these sources.

2. *The Periplasmatron*

In a duopigatron the localized electron emission leads to plasma nonuniformities. Fummelli and Valckx (1976) proposed to overcome this by using a peripheral primary plasma volume (Fig. 16). This gives the added advantage of a large magnetic field free volume in the center of the source with local-

6C. Neutral-Beam Formation and Transport

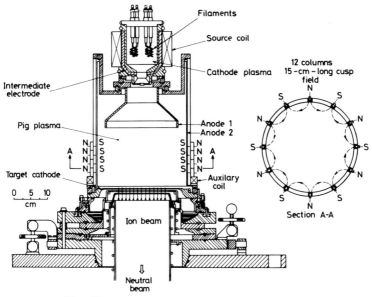

Fig. 15. Schematic diagram of duopigatron source.

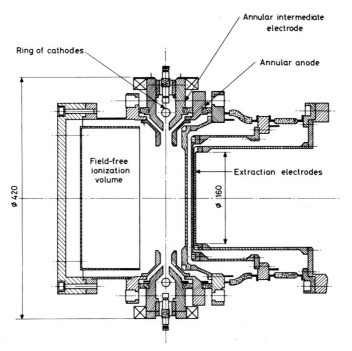

Fig. 16. Schematic diagram of periplasmatron ion source. (Note: source is symmetric about horizontal axis.)

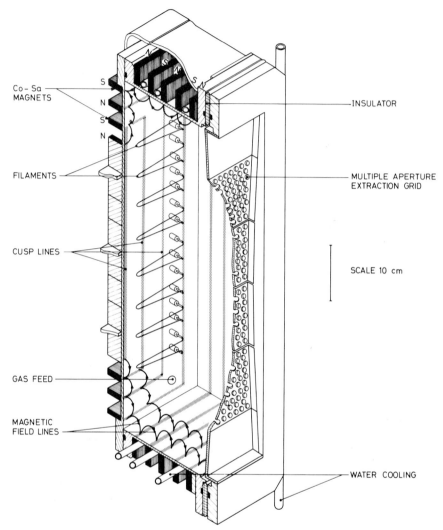

Fig. 17. Schematic diagram of magnetic multipole (bucket) ion source.

ized magnetic shielding of the anode. The circular and rectangular forms of this source have been operated at currents of 40 A (Bariaud *et al.*, 1979).

3. *Magnetic Field Free Sources*

As an alternative approach a source with zero magnetic field, but a peripheral anode of limited area was developed at the Berkeley Laboratory (Baker *et al.*, 1971). This source gives reasonable uniformity. However, it is not very efficient electrically due to the small value of the containment time

6C. Neutral-Beam Formation and Transport

of the ionizing electrons τ_p and as a consequence of the use of low arc voltages. Modified versions of this source have been designed for long pulse operation, particular emphasis being placed on anode and cathode design.

4. *Magnetic Multipole Sources*

The latest forms of high current sources are based on the use of a multipole magnetic field to shield the anode, thus providing electron and ion containment; they are similar to devices introduced by Limpaecher and MacKenzie (1973) as generators of plasmas for investigation of the properties of plasmas. Figure 17 shows one type of this source built at Culham Laboratory (Hemsworth *et al.*, 1978). The multipoles create a localized magnetic field around a large volume with low magnetic field, within which the plasma is uniform.

A magnetic multipole source was first tested at Culham on the DITE injector test line and yielded 30–40 A extracted current (Hemsworth *et al.*, 1978). Larger versions are now being tested to yield 60–80 A. Similar sources are being tested in many laboratories.

IV. Beam Transport in a Gas Neutralizer

The two basic aspects of beam transport in a gas neutralizer (i.e., a charge exchange) cell are that of the continuous change from a pure ion beam to a mixed ion–neutral beam, and that of the trajectories of the beam particles that determine the efficiency of transport of the beam through a beam line. There is a very strong coupling between these two aspects because the trajectories can be strongly influenced by space charge in the ion beam, which in its turn depends on the production rate of slow ions and electrons by collision of beam particles with the gas in the neutralizer.[†]

A. *Neutralizer Efficiency*

1. *Positive Hydrogen Ions*

The spatial evolution of hydrogen ions and fast neutrals along a gas neutralizer has been calculated by several authors (e.g., Berkner *et al.*, 1975). For a single beam of H^+ ions passing through hydrogen gas (assumed to be molecular) the steps are

$$H^+ + H_2^0 \rightarrow H^0 + \text{(slow atom or ion products)} - \text{(cross section } \sigma_{10})$$

$$H^0 + H_2^0 \rightarrow H^+ + \text{(slow atom or ion products)} - \text{(cross section } \sigma_{01}).$$

[†] We will use the term neutralization to describe charge exchange, and space-charge compensation to describe equalization of space charge in the beam.

Signifying I^+ as the current of fast ions, I^0 as the equivalent current of fast neutrals, and n_0 as the density of molecular hydrogen, one has

$$\frac{dI^+}{dz} = -\frac{dI^0}{dz} = -n_0\sigma_{10}I^+ + n_0\sigma_{01}I^0, \quad (19)$$

the solution of which is

$$I^+ = I^+(0)\left\{\frac{\sigma_{01}}{\sigma_{10} + \sigma_{01}} + \frac{\sigma_{10}}{\sigma_{10} + \sigma_{01}}\exp[-(\sigma_{10} + \sigma_{01})\Pi]\right\}, \quad (20)$$

where

$$\Pi = \int_0^L n_0\, dz$$

and

$$I^0 = I^+(0)\left\{\frac{\sigma_{10}}{\sigma_{01} + \sigma_{10}} - \frac{\sigma_{10}}{\sigma_{01} + \sigma_{10}}\exp[-(\sigma_{01} + \sigma_{10})\Pi]\right\}, \quad (21)$$

L being the length of the neutralizer. The limiting value of $I^0/I^+(0)$ as Π tends to infinity is the thick target yield F_0^∞ and equals $\sigma_{10}/(\sigma_{10} + \sigma_{01})$. It is the rapid decrease in F_0^∞ at high energies (Fig. 18a) due to the decrease of σ_{10} relative to σ_{01}, which leads to low efficiency for the production of neutral beams from H^+ (or D^+) ion beams. A further implication for the design of a neutralizer is the rapid increase with energy in the value of Π required to achieve 80% of the value of F_0^∞, as shown in Fig. 18b.

The equation for the evaluation of atomic and ionic species from the breakup of H_2^+ and H_3^+ are more complex and are not included here. They have been treated in the papers referred to above.

2. *Negative Hydrogen Ions*

When negative ions are used to provide a beam of neutral atoms the situation is simpler in that only one energy component is involved. The resulting set of equations can be solved. Typical plots of $I^0/I^-(0)$ are shown in Fig. 19. The optimum value is ~0.6–0.7. This is much less sensitive to energy due to the constant and high value of the cross section for electron detachment in collision of a negative ion with a gas molecule.

3. *Positive Impurity Ions*

The observations of impurity ions in neutral beams raises the important question of their neutralization. For simple impurities such as the metallic ions the situation is relatively clear, though the experimental data are limited. However, for more complex ions derived from water vapor and hydrocarbons the possible breakup channels need to be identified.

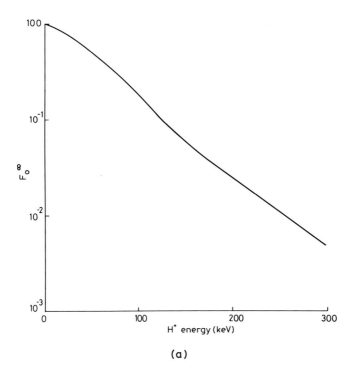

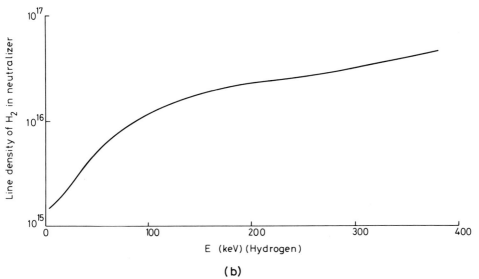

Fig. 18. (a) Variation with energy of thick target yield (F_0^∞) of atoms due to H$^+$ ions passing through molecular hydrogen gas. (b) Variation with energy of target thickness (cm^{-2}) required to achieve $0.8 F_0^\infty$ in a neutralizer.

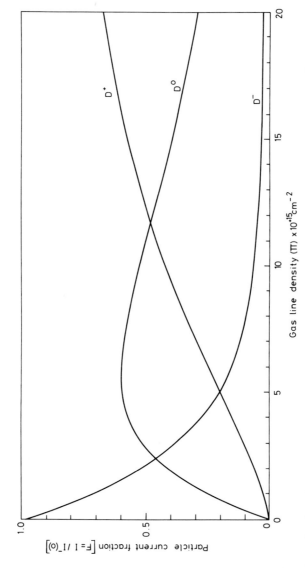

Fig. 19. Neutralization of negative-ion beam in deuterium gas. Calculated values of particle current fractions $F = I^0/I^-(0)$ versus line density f in neutralizer. Data for 300-keV D$^-$ beam.

B. Slow Hydrogen Ion Production in Neutralizers

Slow ions are produced in neutralizers, both by charge exchange collisions (cross section σ_{10}) and by ionization of the background gas by fast ions and fast neutrals (cross sections σ_{ei} and σ_{eo}, respectively). Both these processes produce predominantly H_2^+ ions though a significant yield of H^+ ions and H^0 neutrals is also obtained. The relevant reactions may be coupled into those describing the evolution of the ion–neutral beam to give a complete calculation of the spatial variation of slow ion production (Martin, private communication), which can in principle be compared either with the measurement of the local current of slow ions to the wall of the neutralizer or with the intensity of radiation from the dissociated H^0. These studies have been initiated but are, to the author's knowledge, relatively undeveloped. They may be important in understanding the behavior of gas in a neutralizer because of the high probability that the gas molecules are ionized during their transit through the neutralizer cell.

C. Beam Particle Trajectories

1. Single Beamlets

The multiampere beams utilized in neutral injection are formed from many smaller beamlets and the trajectories of the beam particles derive, to a first order, from the properties of these single beamlets. The transport of a single beamlet, either ions or mixed ions/neutrals, may be usefully described by the Kapchinsky–Vladimirsky equation (Lapostelle, 1972; Lawson, 1974). Though this equation is only strictly valid for a particular physical situation (that in which the density of the particles in phase space is uniform and bounded by an ellipse), it gives reasonable guidance in evaluating the influence of different terms on the beam transport. The equation may be written in the form

$$\frac{d^2a}{dz^2} - \frac{K_{sp}}{a} - \frac{K_M}{a} - \frac{\varepsilon^2}{a^3} = 0 \tag{22}$$

for a beam of radius a moving along the z axis. The "force" factors are as follows.

Emittance factor: The factor ε^2/a^3 represents the effect of the transverse energy of the beam ions, i.e., of the beam emittance ε (ε is the area in phase space divided by π).

Space-charge factor:

$$K_{sp} = 6.4 \times 10^5 (I/E^{3/2})(1 - f) \quad \text{(cm)}, \tag{23}$$

where f is the fraction of space-charge compensation, I the current in amps, and E the beam energy in electron volts.

Magnetic constriction factor:

$$K_M = 6.4 \times 10^5 \, (I/E^{3/2})\beta^2 \quad \text{(cm)}, \tag{24}$$

where β equals v/c.

Normally for nonrelativistic beams the magnetic constriction term is relatively small. However, for beams with a high degree of space-charge compensation this may not be the case.

It is normal in discussing the transport of ion beams to introduce a parameter representing the ratio of the space charge factor to the emittance factor (Evans and Warner, 1971). We may derive the result

$$\delta_1 \equiv \frac{\text{space-charge factor}}{\text{emittance factor}} = \frac{K_{sp}}{\varepsilon^2} a^2 = 6.4 \times 10^5 \frac{I}{E^{1/2}} \frac{(1-f)}{T_i}, \tag{25}$$

where T_i is the energy of the ion beam transverse to the direction of propagation expressed as a temperature [T_i is in electron volts in Eq. (25)]. Similarly we may define a second parameter (Cottrell 1981):

$$\delta_2 \equiv \frac{\text{magnetic constriction factor}}{\text{emittance factor}} = \frac{K_M a^2}{\varepsilon^2} \simeq 0.7 \times 10^{-3} \frac{IE^{1/2}}{T_i}. \tag{26}$$

Figure 20 shows the variation of $I(1-f)$ and I with beam energy E for δ_1 and δ_2 equal to unity. Of these factors which may affect beam transport, only space-charge effects have been discussed significantly and are discussed in Section IV.D.

There are relatively few experimental studies of emittance of high current ion beams. Some measurements of emittance of low current beams have been made. These do not yet give an adequate description of the factors that determine the emittance of the H^+, H_2^+, and H_3^+ ions and the neutrals into which they are converted.

D. Space-Charge Compensation

The space-charge compensation of high current ion beams has been reviewed in articles by Osher (1977) and Green (1976), being extensions of the earlier work on low current beams by Gabovich *et al.* (1975). The calculation of the potential in an ion beam proceeds by solution of coupled equations:

(i) Poisson's equation relating the potential to differences in charge density,
(ii) momentum and continuity equation for slow ions produced by beam–gas collisions,
(iii) continuity equation for electrons produced by beam–gas collisions,

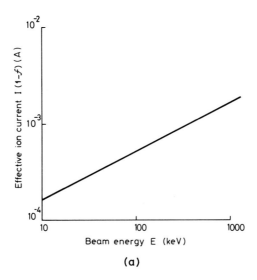

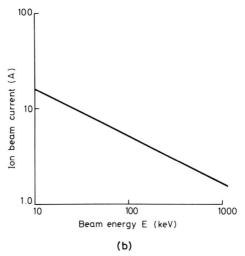

Fig. 20. Beam transport parameters as a function of beam energy. (a) $I(1 - \Pi)$ versus beam energy E for δ_1 [Eq. (25)] = 1 and $T_i = 1$ eV. (b) I versus beam energy E for δ_2 [Eq. (26)] = 1 and $T_i = 1$ eV.

(iv) equation relating electron density to potential involving calculation of the electron energy spectrum.

It is remarkable that there is not one of these equations for which there is a unique treatment in the literature. Below we consider one particular treatment and indicate where others differ from it.

1. *Plasma Limit*

This treatment, developed by Dunn and Self (1964) for the case of a plasma produced by an electron beam, may be applied to negative-ion beams or modified to the case of positive-ion beams (Green, 1977). It is assumed that charge neutrality exists throughout the beam, except in a thin sheath at the wall. For a positive-ion beam

$$n_b + n_i = n_e, \tag{27}$$

where n_b is the beam ion density, n_i the slow ion density, and n_e the electron density.

The electrons are assumed to be in thermal equilibrium such that

$$n_e = n_e(0)\exp(-e\phi/kT_e), \tag{28}$$

$n_e(0)$ being the electron density on axis where the potential ϕ is zero. T_e is the electron temperature that is not calculated in this treatment.

The slow ions are assumed to move without collisions, being accelerated by the potential. The density at x is calculated as the integral of the contribution from all points x' between zero and x:

$$n_i(x) = \int_0^x \frac{n_b(x')n_0\sigma_{bi}v_b\,dx'}{(2e/m_i)^{1/2}[\phi(x) - \phi(x')]^{1/2}}, \tag{29}$$

where n_0 is the neutral gas density, σ_{bi} the cross section for slow ion production, v_b the beam ion velocity, and m_i the slow ion mass.

The neutrality condition (plasma equation) becomes

$$n_b(x) = \int_0^x \frac{n_b(x')n_0\sigma_{bi}v_b\,dx'}{(2e/m_i)^{1/2}[\phi(x) - \phi(x')]^{1/2}} = n_e(0)\exp\left[-\frac{e\phi(x)}{kT_e}\right]. \tag{30}$$

This equation can be solved in certain cases using the method developed by Harrison and Thompson (1959), subject to the boundary condition that $d\phi/dx \to \infty$ at the wall where the plasma blends into the sheath. An approximate value for the ratio $n_e(0)/n_b(0)$ is given by the equation

$$n_b(0) + n_b(0)\frac{n_0\sigma_{bi}v_b a}{v_i} = n_e(0), \tag{31}$$

where v_i equals $(2kT_e/m_i)^{1/2}$ and a is the half-width of the beam. This equations shows the importance of the beam parameter $(n_0\sigma_{bi}v_b a/v_i)$. When it is small the slow ions do not contribute significantly and the electron density balances the beam ion density. The potential is then given by

$$\frac{e\phi(x)}{kTe} = \log\left[\frac{n_b(0)}{n_b(x)}\right]. \tag{32}$$

If the beam parameter is large, the slow ions dominate and the potential is that calculated for a plasma by Harrison and Thompson (1959). For a gas pressure of 1 mTorr in a 10-cm-wide beam line, a beam energy of 80 keV (H^+) and a temperature of 10 eV, the beam parameter would be ~10, indicating that the slow ions dominate.

Under these conditions the beam density for 200 mA/cm² of 100-keV ions is $\sim 2 \times 10^9$ cm^{-3} and the electron density 2×10^{10} cm^{-3}. The corresponding Debye length is 1.0×10^{-2} cm, which is much less than the beam width, so the plasma approximation is generally valid.

The model is one dimensional, and is thus limited in its application because an important factor in neutralizer physics is the motion of slow ions in the potentials that develop near the earth electrode through which the beam ions enter the neutralizer. Here it is important that two dimensions (or three) should be considered. Anderson and Hooper (1977) have developed a two-dimensional model for negative ions using a fluid approximation for the ions, though this approximation is hardly valid at the densities involved.

The model also excludes discussion of energy balance which determines the electron temperature. Energy balance has been included by Hamilton (1971) and Holmes (1979). However, both these authors use a quasi-neutrality model, assuming that $n_b + n_i = n_e$ only on the axis (see also Gabovich *et al.*, 1975).

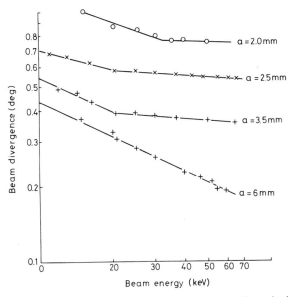

Fig. 21. Variation of beamlet divergence with beam energy (data obtained in He). Decrease of divergence with increasing beamlet radius indicates decreasing space-charge effect. [After Holmes (1979).]

The lack of neutrality in the beam edges is equivalent to saying that the sheath extends through the beam. It then becomes important to define the electron loss rate, and all these authors use energy distributions that have cutoff at the energy corresponding to the potential well depth. Electrons are assumed to be lost by gaining energy to take them beyond the cutoff.

An important implication of these studies is that the space-charge forces in a beam decrease as the beam diameter increases; as a consequence the divergence of a beam due to space-charge decreases with beam diameter (Fig. 21; Holmes, 1979).

The literature also includes theoretical and experimental studies of the influence of beam fluctuations on space-charge compensation (see Osher, 1977). However, there is still a need for a more thorough unification of the steady state models in the parameter range relevant to high current beams in gas neutralizers before these stability problems can be tackled.

V. Negative-Ion Beams

The need for high efficiency neutral beam systems at high energies (≥ 100 keV per nucleon) has led to an investigation of the possibility of building systems based on negative ions. Two approaches are evolving that will be considered below: double-charge capture by low energy positive ions in alkali vapor cells; and direct extraction of ions from a source in which the negative ions are produced either by surface interaction or by volume collisional processes.

A. Double-Charge Capture

This proposal is based on the known cross-section data for the process of the capture of electrons by positive ions to form neutral atoms and subsequently negative ions (Fig. 22). Since positive-ion beams also consist of molecular ions, the production of negative ions via the breakup of the molecules is also important.

The beam line required for such a system is shown schematically in Fig. 23.

The major components are (a) positive-ion source, (b) drift space, (c) charge exchange cell, (d) drift space, and (e) accelerator.

The main purpose of the two drift spaces is to provide pumping and to prevent material from the charge exchange cell reaching high voltage insulators.

The current density at the accelerator stage (J_-^∞) has been calculated by Hooper and Poulson (1980) to be

$$J_-^\infty = F_-^\infty \frac{J_s A_s}{\pi} \left(\frac{180}{\pi L_f}\right)^2 \frac{1}{\theta^2}, \tag{33}$$

6C. Neutral-Beam Formation and Transport

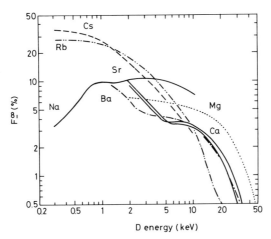

Fig. 22. Energy dependence of yield of D^- (F_-^∞) from double-charge capture (D^+ ions in alkali vapors). [After Schlacter (1980).]

where F_-^∞ is the thick target yield of the negative ion; J_s the current density of positive ions at the source of area A_s; L_f the focal length of the extraction array (equal to the distance between source and accelerator, i.e., ~1–2 m); and θ the beam divergence in degrees (for slit geometry extraction θ^2 is replaced by $\theta_\parallel \times \theta_\perp$, i.e., the product of the divergences in the planes perpendicular and parallel to the slit axis).

Strictly speaking this expression should be summed over the ionic and molecular species taking account of scattering in the molecular ion breakup. Hooper and Poulsen introduce a normalized negative-ion current defined by the equation

$$\frac{J_-^\infty \cdot L_f^2}{J_s A_s} = \frac{F_-^\infty}{\pi}\left(\frac{180}{\pi\theta}\right)^2. \tag{34}$$

The beam divergence is energy dependent. Hooper and Poulsen quote that $\theta_\perp$ is equal to $7.0 E^{-1/2}$ deg and $\theta_\parallel$ is equal to $2.1 E^{-1/2}$ deg for the beam energy E in keV. As a consequence of this energy dependence the maximum normalized current density is obtained at higher energies. It is advantageous

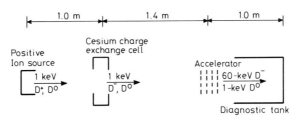

Fig. 23. Schematic diagram of beam line for negative ion beam based on double-charge capture by D^+ ions in cesium. [After Hooper et al. (1977).]

to use the charge capture process in Na, which peaks as 10 keV, rather than the process in Cs, which peaks at 1.2 keV, even though the latter has a higher cross section. Figure 24 shows the calculated values of $J^\infty L_f^2/J_s A_s$ versus D^+ energy for different alkali vapors and also the experimental values obtained by Hooper and Poulsen (1980).

The energy dependence of the divergence arises from space-charge forces in the low energy beams. Geller et al. (1980) (see also Delauney et al., 1980) have proposed a method to reduce this space-charge divergence by the use of an axial magnetic field that links the ion source to the accelerator region. (This field limits radial loss of ions and forces the electric potential to be axial rather than radial.) Consequently these authors are able to use low energy ions and Cs vapor.

Charge exchange cells (Bacal et al., 1980b; Semashko et al., 1979) using both Na and Cs have been developed based on supersonic nozzles. The major requirements are to produce adequate axial line density over a defined area (~5 cm × 20 cm) without axial diffusion of the alkali onto the accelerators. Problems of plasma production in the cell and its implication for design have been discussed in the literature.

The problems of beam transport for the negative ions are also being analyzed (Green, 1977; Anderson and Hooper, 1977) to define parameter regions in which space-charge neutralization is adequate, without overfocusing of the negative ions in the positive potential well and without a neutral particle density so high as to lead to electron stripping.

B. Direct Extraction

Early experiments with Calutron and PIG sources had indicated the possibility of extracting negative ions from a plasma source, but at limited currents of a few milliamps and with a relatively high gas throughput. [Bacal (1982) reviews these early experiments and the recent work discussed below.]

Dimov and colleagues later developed a 1-A H^- source using direct extraction for nuclear accelerators. The geometry shown in Fig. 25 relies on the conversion of positive ions to negative ions by collisions on a cesiated cathode. This interpretation of the data was supported by experimental measurement of the energy spectra of the negative ions reported by Bel'chenko et al. [The work at the Novosibirsk Laboratory is reviewed by Dudnikov (1980). Similar work at the Brookhaven Laboratory is also reviewed by Prelec (1980).]

More detailed studies of the ion spectra in direct extraction sources of a different geometry have been reported by Ehlers and Leung (1980). The spectra show features corresponding to (a) desorption and (b) backscattering of ions.

6C. Neutral-Beam Formation and Transport

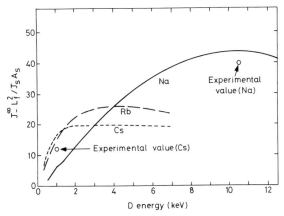

Fig. 24. Energy dependence of normalized current density of D$^-$ ions produced by double-charge capture in Cs, Rb, and Na. [The normalized current density $J_-^\infty L_f^2/J_s A_s$ is defined in Eq. (34).] [After Hooper and Poulson (1980).]

In desorption, positive ions incident on the surface transfer the energy they have gained in acceleration across the plasma sheath in collisions. The average energy of a desorbed H$^-$ ion is usually small as it leaves the surface. It is then accelerated away from the converter surface through the potential difference V_s across the sheath. Such ions show up as a peak in the energy spectrum at an energy close to eV_s.

The backscattered ions are formed by the process of reflection of the positive ion during which the positive ion is first neutralized to an atom and then captures an electron to become an H$^-$ ion. For incident H$^+$ ions the energy of the emerging H$^-$ ion is $2eV_s$. Experimentally it is observed that H$_2^+$

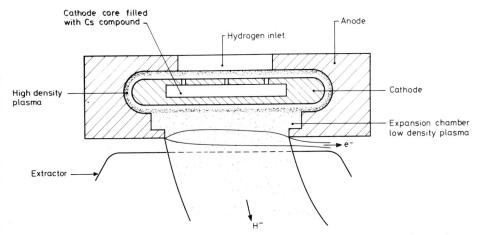

Fig. 25. Schematic diagram of magnetron ion source for negative ions. [After Prelec (1977).]

and H_3^+ ions play an important role: these break up into fractional energy H_0 atoms in the surface and give rise to H^- ions of $\frac{2}{3}eV_s$ and $\frac{1}{3}eV_s$, respectively.

Limitations to the output of surface interaction sources have been discussed by Green (1975) and Dudnikov (1980). The current of negative ions emitted from a surface is equal to α times the incident positive-ion current density (J_+). However, the negative-ion current is attenuated between the conversion surface and the extraction region due to recombination collisions $(H^- + H^+$ or $D^- + D^+)$. The attenuation rate depends on the positive-ion density and can hence be written as $\exp(-\beta J_+)$. The extracted current of negative ions J_- is therefore given by

$$J_- = \text{const} \times \alpha J_+ \exp(-\beta J_+). \tag{35}$$

Present studies are concerned with optimization of α and β.

C. Volume Production of Negative Ions

Bacal et al. (1980a) have reported a detailed study of the processes leading to negative-ion production in plasma sources by collisional processes in the source volume. Early experiments with probes indicated densities greater than those predicted on the basis of usually considered reaction processes

$$e + H_2 \rightarrow H^- + H.$$

Subsequent measurements using a laser beam for photodissociation of the H^- substantiated the high densities ($n_-/n_e \sim 0.35$ at $n_e \sim 10^{10}$ cm^{-3}) and have led to further discussion of the production mechanisms for the H^- ions (references for the theoretical papers on mechanisms are given by Bacal et al., 1980a, and Bacal, 1982). As a consequence, it is assumed that the dominant processes involve hydrogen molecules in an excited vibrational state

$$e + H_2(v^*) \sim H^- + H.$$

The population of $H_2(v^*)$ molecules is determined either by the reactions

$$H_2^+(v^*) + H_2(v = 0) \sim H_2(v^*) + H_2(v^*)$$

and

$$H_2^+(v^*) + e(\text{wall}) \sim H_2(v^*),$$

or the electron excitation of H_2

$$e + H_2 \rightarrow h\nu + H_2[X^1 \Sigma_g^+ (v'')] + e,$$

or by processes involving H_3^+.

If the process of electron excitation of the molecule dominates (Hiskes et al., 1982; estimate $\sigma \sim 3 \times 10^{-17}$ cm^2), then the density of negative ions is related to the discharge parameters by the expression

6C. Neutral-Beam Formation and Transport

$$n_- = \frac{n_e n_+ \sigma v(\text{DA})}{n_+ \sigma v(\text{ii}) + 1/\tau_d} b \frac{v_+}{v^*} \frac{\sigma(\text{EV})}{\sigma_i}, \qquad (36)$$

where n_e is the density of thermal electrons, n_+ the ion density, $\sigma v(\text{DA})$ the reaction rate coefficient for dissociative attachment to the excited molecules, $\sigma v(\text{ii})$ the reaction rate coefficient for loss of H^- by ion–ion recombination ($= 2 \times 10^{-7}$ cm^3 s^{-1}), τ_d the diffusion time of H^- to the walls, v_+ the velocity of the positive ions, v^* the velocity of the vibrationally excited molecules, $\sigma(\text{EV})$ the cross section for vibrational excitation, σ_i the cross section for ionization of H_2, and b the number of collisions made by a molecule with the wall before it is deexcited.

Experimental data for the product $n_-/n_e n_+ \times [n_+ \sigma v(\text{ii}) + (1/\tau_d)]$ are shown in Fig. 26 as a function of electron temperature. Also shown is the theoretical estimate of $\sigma v(\text{DA})$ for $v = 8$, from Wadehra (1979). Comparison of the absolute values indicates that $b \sim 10$, but this estimate depends on the relative contribution of H_3^+ to the formation of the vibrationally excited molecules.

The investigation of these processes still continues because this may lead to a simple source of H^- or D^- ions for future injectors.

Acknowledgment

All the figures in this chapter have been reproduced courtesy of Culham Laboratory.

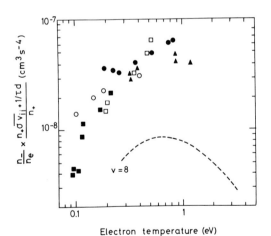

Fig. 26. Experimental data for a plasma source with H^- ions. The parameter

$$\frac{n_-}{n_e} \times \frac{n_+ \langle \sigma v \rangle_{ii} + 1/\tau_D}{n_+}$$

is plotted as against the measured electron temperature. The symbols indicate data taken at different gas pressures in the source. This parameter should be proportional to the reaction rate coefficient for dissociation attachment of thermal electrons to vibrationally excited molecules. Shown for comparison is the coefficient for this process calculated by Wadhera (1979) for the case that the molecule is in the state $v = 8$. [After Bacal et al. (1980a).]

References

Anderson, A. O., and Hooper, E. B. (1977). *Proc. Symp. Prod. Neutral. Negat. Hydrogen Ions Beams, Brookhaven Natl. Lab. Rep.* **BNL-50727**, p. 205.
Bacal, M. (1982). *Phys. Scripta* **T2/2**, 467.
Bacal, M., Bruneteau, A. M., Doucet, H. J., Graham, W. G., and Hamilton, G. W. (1980a). *Proc. Int. Symp. Prod. Neutral. Negat. Hydrogen Ions Beams, 2nd, Brookhaven Natl. Lab. Rep.* **BNL-51304**, p. 90.
Bacal, M., Buzzi, J. M., Doucet, H. J., Labaune, G., Lamain, H., Stephan, J. P., Delauney, M., Jacquot, C., Ludwig, P., and Verney, S. (1980b). *Proc. Int. Symp. Prod. Neutral. Negat. Hydrogen Ions Beams, Brookhaven Natl. Lab. Rep.* **BNL-51304**, p. 270.
Baker, W. R., Ehlers, K. W., Kunkel, W. B., and Lietzke, A. F. (1971). *Proc. Symp. Ion Sources Form. Ion Beams, Brookhaven Natl. Lab. Rep.* **BNL-50310**.
Bates, D. G. (1966). *Culham. Lab. Rep.* **CLM-R53**.
Bariaud, A., Becherer, R., Desmons, M., Fumelli, M., Raimbault, P., and Valckx, F. P. G. (1979). *Proc. Symp. Eng. Prob. Fusion Res., 8th, San Francisco, Calif., 1979* **2**, p. 685.
Becherer, R. (1975). *Fontenay-aux-Roses Int. Rep.* **EUR-CEA-FC-788**.
Berkner, K. H., Pyle, R. V., and Stearns, J. W. (1975). *Nucl. Fusion* **15**, 249.
Bohm, D. (1949). *In* "The Characteristics of Electrical Discharges in Magnetic Fields" (A. Guthrie and R. K. Wakerling, eds.), p. 77. McGraw-Hill, New York.
Brewer, G. R. (1967). *In* "Focussing of Charged Particles" (A. Septier, ed.), Vol. 2, p. 23. Academic Press, New York.
Cooper, W. S., Halbach, K., and Magyary, S. B. (1974). *Proc. Symp. Ion Sources Form. Ion Beams, 2nd., Berkeley, Calif., 1973* Paper II-1.
Cooper, W. S. (1975). *Bull. Am. Phys. Soc.* **19**, 857.
Cooper, W. S., Elischer, V. P., Goldbey, D. A., Hopkins, D. B., Jacobson, V. L., Lores, K. H., and Tanabe, J. T. (1981). *Lawrence Berkeley Lab. Rep. LBL* **LBL-12383**.
Cottrell, G. A. (1981). *Rev. Sci. Instrum.* **52**, 55.
Coupland, J. R., Green, T. S., Hammond, D. P., and Riviere, A. C. (1972). *Culham Lab. Rep.* **CLM-P312**.
Davis, R. C., Morgan, O. B., Stewart, L. D., and Stirling, W. L. (1972). *Rev. Sci. Instrum.* **43**, 278.
Davisson, C. J., and Calbrick, C. J. (1931). *Phys. Rev.* **28**, 585.
Delauney, M., Fourcher, J. L., Geller, R., Jacquot, C., Ludwig, P., Mazhari, F., Ricard, E., Rocco, J. C., Sermet, P., and Zadworney, F. (1980). *Proc. Int. Symp. Prod. Neutral. Negat. Hydrogen Ions Beams, 2nd, Brookhaven Natl. Lab. Rep.* **BNL-51304**, p. 255.
Dudnikov, V. G. (1980). *Proc. Int. Symp. Prod. Neutral. Negat. Hydrogen Ions Beams, 2nd, Brookhaven Natl. Lab. Rep.* **BNL-51304**, p. 137.
Dunn, D. A., and Self, S. A. (1964). *J. Appl. Phys.* **35**, 113.
Ehlers, K. W., and Leung, K. N. (1980). *Proc. Int. Symp. Prod. Neutral. Negat. Hydrogen Ions Beams, 2nd, Brookhaven Natl. Lab. Rep.* **BNL-51304**, p. 198.
Evans, L. R., and Warner, D. J. (1971). *CERN* **CERN/MPS/LIN**, 71-3.
Franklin, R. N. (1976). "Plasma Phenomena in Gas Discharges." Oxford University Press, London and New York.
Friedman, S., Jerde, L., Carr, W., and Seidl, M. (1978). *J. Appl. Phys.* **49**, 3209.
Fumelli, M., and Valckx, F. P. G. (1976). *Nucl. Instrum. Methods* **135**, 203.
Gabovich, M. D. (1972). "Physics and Technology of Plasma Ion Sources," Transl. FTDH.T 23-1690, U.S. Air Force.
Gabovich, M. D., Katsubo, L. P., and Solochenko, I. A. (1975). *Sov. J. Plasma Phys. (Engl. Transl.)* **1**, 162.
Gardner, W. L., Kim, J., Menon, M. M., and Whealton, J. H. (1978). *Rev. Sci. Instrum.* **49**, 1214.
Geller, R., Jacquot, C., Ludwig, P., Sermet, P., Gustavson, H. G., Pauli, R., and Rocco, J. C.

6C. Neutral-Beam Formation and Transport

(1980). *In* "Low Energy Ion Beams" (K. G. Stephens and I. H. Wilson, eds.), Conference Series, No. 54, p. 143. Inst. Physics, London.
Goede, A. P. H., Green, T. S., Martin, A. R., Mosson, A. G., and Singh, B. (1978). *Proc. Int. Symp. Heat. Toroidal Plasmas, Grenoble, Fr.* p. 77.
Goede, A. P. H., and Green, T. S. (1979). *Proc. Symp. Eng. Probl. Fusion Res., 8th, San Francisco, Calif., 1979* **2**, 680.
Goede, A. P. H., and Green, T. S. (1982). *Phys. Fluids* **25** (10), 1797.
Green, T. S. (1974). *Proc. Symp. Ion Sources Form. Ion Beams, Berkeley, Calif., 1973* Paper II-80.
Green, T. S. (1975). *Nucl. Instrum. Methods* **125**, 345.
Green, T. S., Goble, C., Inman, M., and Martin, A. R. (1975). *Proc. Eur. Conf. Controlled Fusion Plasma Phys., 7th, Lausanne, Switz.* p. 93.
Green, T. S. (1976). *IEEE Trans. Nucl. Sci.* **NS-23** (2), 918.
Green, T. S. (1977). *Proc. Symp. Prod. Neutral. Negat. Hydrogen Ions Beams, Brookhaven Natl. Lab. Rep.* **BNL-50727**, 228.
Hamilton, G. W., Hilton, J. L., and Luce, J. S. (1968). *Plasma Phys.* **10**, 687.
Hamilton, G. W. (1971). *Proc. Int. Symp. Ion Sources Form. Ion Beams, Brookhaven Natl. Lab. Rep.* **BNL-50310**, p. 171.
Harrison, E. R., and Thompson, W. B. (1959). *Proc. Phys. Soc., London* **72**, 214.
Hemsworth, R. S., Aldcroft, D. A., Allen, T. K., Bayes, D. V., Burcham, J. N., Cole, H. C., Cowlin, M. C., Coultas, J. C., Hay, J. H., and McKay, W. J. (1978). *Proc. Int. Symp. Heat. Toroidal Plasmas, Grenoble, Fr.* p. 83.
Herschkowitz, N., Smith, J. R., and Kozima, H. (1979). *Phys. Fluids* **22**, 122.
Hiskes, J. R., Karo, A. M., Bacal, M., Bruneteau, A. H., and Graham, W. G. (1982). *J. Appl. Phys.* **53** (5), 3469.
Holmes, A. J. T. (1979). *Phys. Rev. A* **19**, 389.
Holmes, A. J. T., and Green, T. S. (1980). *In* "Low Energy Ion Beams" (K. G. Stephens and I. H. Wilson, eds.), Conference Series No. 54, pp. 163–170. Inst. Phys., London.
Holmes, A. J. T., and Thompson, E. (1981). *Rev. Sci. Instrum.* **52**, 172.
Hooper, E. B., Jr., Anderson, O. A., Orzechowski, T. J., and Poulson, P. (1977). *Proc. Int. Symp. Prod. Neutral. Negat. Hydrogen Ions Beams, Brookhaven Natl. Lab. Rep.* **BNL-50727**, p. 163.
Hooper, E. B., Jr., and Poulson, P. (1980). *Proc. Int. Symp. Prod. Neutral. Negat. Hydrogen Ions Beams, Brookhaven Natl. Lab. Rep.* **BNL-51304**, p. 247.
Kaufmann, H. R. (1961). *NASA Tech. Note* **NASA TN D-585**.
Kohlberg, I., and Nablo, S. (1966). *In* "Physics and Technology of Ion Motors" (F. E. Marble and J. Surugue, eds.), pp. 155–206. Gordon & Breach, New York.
Kulygin, V. M., Panacenkov, A. A., and Semashko, N. N. (1976). *All Union Conf. Plasma Accel., 3rd, Minsk, 1976* p. 2.
Langmuir, I., and Blodgett, K. R. (1924). *Phys. Rev.* **24**, 49.
Langmuir, I., and Jones, H. A. (1928). *Phys. Rev.* **31**, 357.
Lapostelle, P. (1972). *Proc. Int. Conf. Ion Sources, 2nd, Vienna, 1972* p. 133.
Lawson, J. D. (1974). *Plasma Phys.* **17**, 567.
Lawson, J. D. (1977). "The Physics of Charges Particle Beams." Oxford Univ. Press (Clarendon), London and New York.
Lejeune, C. (1971). *Proc. Symp. Ion Sources Form. Ion Beams, Brookhaven Natl. Lab. Rep.* **BNL-50310**, p. 21.
Leung, K. N., Kribel, R. E., Goede, A. P. H., and Green, T. S. (1978). *Phys. Lett. A* **66A**, 112.
Limpaecher, R., and Mackenzie, K. R. (1973). *Rev. Sci. Instrum.* **44**, 726.
Martin, A. R., and Green, T. S. (1976). *Culham Lab. Rep.* **CLM-R159**.
Ohara, Y., Akiba, M., Arakawa, Y., Horiike, H., Kawai, M., Matsuda, S., Mizutani, Y., Ohga, T., Okumara, Y., Sakuraba, J., Shibata, T., Shirakata, H., and Tanaka, S. (1979). *Proc. Symp. Eng. Probl. Fusion Res., San Francisco, Calif., 1979* **2**, 198.

Ohara, Y., Akiba, M., Arakawa, Y., Okumura, Y., and Sakuraba, J. (1980). *J. Appl. Phys.* **51** (7), 3614.

Okumura, Y., Mizutani, Y., and Ohara, Y. (1980). *Rev. Sci. Instrum.* **51**, 471.

Osher, J. E. (1977). *In* "Low Energy Ion Beams" (K. G. Stephens, I. H. Wilson, and J. L. Moruzzi, eds.), p. 201. Instit. Phys., London.

Pierce, J. R. (1949). "Theory and Design of Electron Beams." Van Nostrand, New York.

Prelec, K. (1980). *Proc. Int. Symp. Prod. Neutral. Negat. Hydrogen Ions & Beams, 2nd, Brookhaven Natl. Lab. Rep.* **BNL-51304**, p. 145.

Raimbault, P. R. (1979). Private communication.

Schlacter, A. S. (1980). *Proc. Int. Symp. Prod. Neutral. Negative Ions Beams, 2nd, Brookhaven Natl. Lab. Rep.* **BNL-51304**, p. 42.

Semashko, N. N., Kuznetsov, V. V., and Krylov, A. I. (1979). *Proc. Symp. Eng. Probl. Fusion Res., San Francisco, Calif.* p. 853.

Stirling, W. L., Tsai, C. C., Haselton, H. H., Schechter, D. E., Whealton, J. H., Dagenhart, W. K., Davis, R. C., Gardner, W. L., Kim, J., Menon, M. M., and Ryan, P. M. (1979). *Rev. Sci. Instrum.* **50**, 523.

Thompson, E. (1974). *Proc. Symp. Ion Sources Form. Ion Beams, 2nd, Berkeley, Calif., 1973* II-3.

Tonks, L., and Langmuir, I. (1929). *Phys. Rev.* **34**, 87.

Von Goeler, S. (1970). *Proc. Int. Conf. Isot. Separators, Marburg, Fed. Rep. Ger.* **BMBW-FB70-28**, p. 399.

Wadehra, J. M. (1979). *Appl. Phys. Lett.* **35**, 917.

Whealton, J. H., and Whitson, J. C. (1980). *Part. Accel.* **10**, 235.

Wiesemann, K. (1972). *Proc. Int. Conf. Ion Sources, 2nd, Vienna, 1971* p. 325.

6D

Alpha-Particle Heating

D. E. Post

Plasma Physics Laboratory
Princeton University
Princeton, New Jersey

    I. Alpha-Particle Production and Heating 381
   II. Fast-Alpha-Particle Diagnostics 387
  III. Alpha-Particle Ash 390
        References 393

I. Alpha-Particle Production and Heating

Alpha particles are produced primarily from the reaction D + T → α (3.5401 MeV) + n(14.048 MeV) in most commonly discussed fusion schemes. The reaction rate for D–T fusion is illustrated in Fig. 1 of Chapter 2. Convenient formulas for these rates have been computed (Hively, 1977). The 14-MeV neutrons produced by D–T fusion leave the plasma and are used either to breed tritium and heat a blanket or else to breed Pu^{239} from U^{238}. The alpha particles are confined by the magnetic field, and heat the plasma ions and electrons through Coulomb collisions. If the confinement of plasma heat is sufficiently "good," the alpha heating may equal the plasma energy loss rate, the plasma will be "ignited," and the fusion "burn" can be self-sustaining. Defining a heat loss rate as the plasma thermal energy divided by a confinement time, we can balance the losses and the heating and, assuming $T_e = T_i$, obtain

$$3n_e T/\tau_E = n_D n_T E_\alpha \langle \sigma v \rangle_{\text{D-T}} \tag{1}$$

or with $n_D = n_T = \frac{1}{2} n_e$, $n_e \tau_E = 12 T/E_\alpha \langle \sigma v \rangle_{\text{D-T}}$ as a condition for ignition, where n_e, n_D, and n_T are the electron, deuterium, and tritium densities; T is the temperature; $\langle \sigma v \rangle_{\text{D-T}}$ is the fusion reactivity; and E_α is the alpha energy, 3.5 MeV.

The heating rate of the plasma by alpha particles due to Coulomb collisions of the fast alphas and plasma ions and electrons can be calculated from the Fokker–Planck equation for fast alphas (Cordey *et al.*, 1981). Neglecting pitch-angle scattering (Mikkelsen and Post, 1979), and defining $f(\mathbf{v})$ as the alpha-particle density in phase space, we have

$$\frac{\partial f}{\partial t} = \frac{1}{v^2 \tau_s} \frac{\partial}{\partial v} (v^3 + v_c^3) f + S \frac{\delta(\mathbf{v} - \mathbf{v}_0)}{4\pi v_0^2}, \quad (2)$$

where S is the total alpha-particle source rate (in cm^3 s^{-1}), v_0 is the velocity of the alpha particle at birth, and

$$\tau_s = \frac{3 m_e m_\alpha v_e^3}{16 \sqrt{\pi} Z_\alpha^2 e^4 n_e \ln \Lambda_e} \approx \frac{1.17 \times 10^{12}}{n_e} [T_e \text{ (keV)}]^{3/2}, \quad (3)$$

$$v_c = \left(\frac{3\sqrt{\pi} \, m_e}{4 m_p \ln \Lambda_e} \sum \frac{Z_j^2 \ln \Lambda_j n_j}{A_j n_e} \right)^{1/3} v_e, \quad (4)$$

$$\ln \Lambda_e = \ln \left[\frac{kT_e}{\hbar e} \left(\frac{m_e}{2\pi n_e} \right)^{1/2} \right] - \frac{1}{2}$$

$$= 23.9 + \ln \left(\frac{T_e \text{ (eV)}}{n_e^{1/2}} \right), \quad (5)$$

$$\ln \Lambda_i = \ln \left[\left(\frac{kT_e}{\pi n_e} \right)^{1/2} \frac{m_p v_\alpha A_\alpha A_j}{\hbar e (A_\alpha + A_j)} \right] - \frac{1}{2}$$

$$= 14.2 + \ln \left[\left(\frac{T_e \text{ (eV)}}{n_e} \right)^{1/2} \frac{A_\alpha A_j v_\alpha}{A_\alpha + A_j} \right], \quad (6)$$

$$\tfrac{1}{2} m_\alpha v_\alpha^2 = 3.5 \text{ MeV}, \quad \tfrac{1}{2} m_e v_e^2 = kT_e.$$

In the above equations, Z and A are respectively the particle charge state and atomic number, m_p and m_e are the proton and electron masses, and the subscript j indicates the plasma ions. The steady state solution for constant T_e, n_e is

$$f(\mathbf{v}) = S\tau_s / 4\pi (v^3 + v_c^3). \quad (7)$$

The alphas split their energy between the electrons and ions. In (2), the slowing down is due to collisions between the alpha particles and electrons, and alpha particles and ions. The first part, $(1/v^2\tau_s)(\partial/\partial v)(v^3 f)$ is due to electron collisions and the second part $(1/v^2\tau_s)(\partial/\partial v)(v_c^3 f)$, to ions; we can see that the heating rate of the electrons and ions is equal when v equals the critical velocity v_c. We can transform (1) to energy coordinates and calculate the fraction of the alpha energy that goes to the ions as the alpha slows down. This yields

$$f_{\text{ion}} = \frac{2}{\varepsilon^2} \left[\frac{1}{\sqrt{3}} \tan^{-1} \left(\frac{2\varepsilon - 1}{\sqrt{3}} \right) - \frac{1}{6} \ln \left(\frac{(1+\varepsilon)^2}{1 - \varepsilon + \varepsilon^2} \right) + \frac{1}{\sqrt{3}} \frac{\pi}{6} \right] \quad (8)$$

for the ion heating fraction, where $\varepsilon \equiv v_\alpha / v_c$ ($v_\alpha = 13 \times 10^8$ cm s^{-1} is the initial velocity of the alpha particle). For usual plasma conditions for fusion

6D. Alpha-Particle Heating

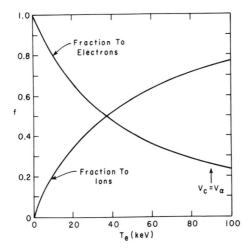

Fig. 1. Fraction of energy deposited in the ions and electrons by Coulomb collisions as a 3.5-MeV alpha particle slows down.

$T_e \sim 10$ keV, $n_e \sim 10^{14}$ cm^{-3}, so that $v_\alpha \gg v_c$, and the alphas are slowed down primarily by the electrons (Fig. 1) and thus most of their heat is transferred to the electrons.

Alpha heating requires that the alpha particles be confined as they slow down (Kolesnichenko, 1980; McAlees, 1974; Mikkelsen and Post, 1979; Hively and Miley, 1980). In axisymmetric toroidal systems such as tokamaks, the alpha particles drift because of curvature and gradients in the magnetic field. These drifts must be averaged out by rapid motion along the field lines.

We can derive a set of equations which describe the orbit of an alpha particle in an axisymmetric field. With no collisions, the energy $E = \tfrac{1}{2}mv^2$ and magnetic moment $\mu = mv_\perp^2/2B$ are constants of motion. In addition, the condition of axisymmetry implies that the toroidal angular momentum of an alpha particle going around the torus is constant; i.e.,

$$p_\phi \approx m_\alpha R v_\parallel + Z_\alpha(e/c) R_0 A_\phi(r) = \text{const}, \qquad (9)$$

where we have written p_ϕ as the guiding center momentum, R as the major radius of the particle trajectory, R_0 as the major radius of the magnetic axis, and A_ϕ as the toroidal vector potential. We can apply these constants to an axisymmetric tokamak with concentric circular toroidal magnetic surface, using $B_\phi = B_0 R_0/R$, and $B_\theta = B_\theta(r) R_0/R$ for the toroidal and poloidal magnetic fields, with r being the distance from the magnetic axis. The vector potential A_ϕ can be computed from the current distribution.

From the three constants of the motion, we can derive an algebraic equation for the spatial trajectory:

$$R(r) = \tfrac{1}{2}[R^* + (R^{*2} + 4\{[p_\phi - Z_\alpha(e/c)R_0 A_\phi(r)]/m_\alpha v\}^2)^{1/2}],$$
$$Z(r) = \{r^2 - [R(r) - R_0]^2\}^{1/2},$$
(10)

where $R^* = \mu R_0 B_0/E$ is a constant of the motion and the range in r is limited to those values which produce real values of $Z(r)$, the height above the toroidal plane. Using Eq. (10), we can draw a typical set of orbits for alpha particles with different pitch angles $v_\parallel/v$ (Fig. 2) for the standard TFTR-sized tokamak described in Table I. The orbital excursions are large. Some of the orbits are unconfined, and those alphas will strike the wall or limiter and give rise to prompt losses from the plasma. The models described in Eqs. (2)–(6), (9), and (10) have been used to perform detailed Monte Carlo calculations to determine the conditions necessary for good confinement of alpha particles in tokamaks (Mikkelsen and Post, 1979; McAlees, 1974). These calculations indicate that the figure of merit for alpha confinement is $I_p \mathcal{A}$, the product of the plasma current and the aspect ratio R_0/a. The effects of increasing plasma current for the TFTR tokamak in Table I are illustrated in Fig. 3, which shows the fraction of the total alpha content which suffers prompt loss as a function of the plasma minor radius and of the plasma current I_p. Figure 4 shows the effects of increasing the aspect ratio for improving the alpha confinement. Since the confinement of alpha particles is a function of $I_p \mathcal{A}$, a good rule of thumb is that $I_p \mathcal{A} \gtrsim 7$ MA confines more than 90% of the alphas (Mikkelsen and Post, 1979; McAlees, 1974).

The slowing down of the particles changes their orbits. Combining μ and E as $R^* = \mu R_0 B_0/E$, we obtain $R^* = (R_0 B_0/B)[1 - (v_\parallel/v)^2]$ as a constant of

TABLE I

Standard-Case TFTR Tokamak Configuration

$E_\alpha = 3.54$ MeV
$I_p = 2.5$ MA
$B_0 = 50$ kG
$R_0 = 255$ cm
$a = 85$ cm
$n(r) = n(0)[1 - (r/a)^2]$
$T_e(r) = T_e(0)[1 - (r/a)^2]$
$J(r) = J(0)[1 - (r/a)^2]^{1.5}$
$n_B(r) = [n^*(r)]^2 \exp\{-20/[T^*(r)]^{1/3}/T^*(r)^{2/3}\}$
  where the auxiliary functions n^* and T^* are
  $n^*(r) = n^*(0)[1 - (r/a)^2]$
  $T^*(r) = T^*(0)[1 - (r/a)^3]$
  $T^*(0) = 20$ keV

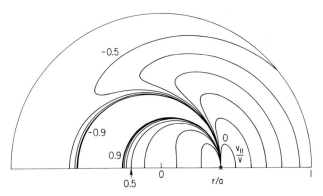

Fig. 2. Typical orbits for a 3.5-MeV alpha particle for the standard TFTR plasma conditions of Table I. All orbits start at the point X and the labels are the cosine of the pitch angle at the birth point. [After Mikkelsen and Post (1979).]

motion. The particles are reflected by magnetic mirrors when $v_\parallel = 0$, and $R^*B(R^*) = R_0 B_0$. R^* is the "mirror" radius, the major radius at the turning point. R^* is not a function of v, only of $v_\parallel/v$, the cosine of the pitch angle. Since alpha particles slow down primarily by collisions with the electrons, there is very little pitch-angle scattering, and the mirror radius is very nearly a conserved quantity for alpha particles as they thermalize. This is illustrated in Fig. 5, which shows ten successive orbits of an alpha particle as it slows down from 3.5 MeV to 350 keV. The net result is that the orbit shrinks

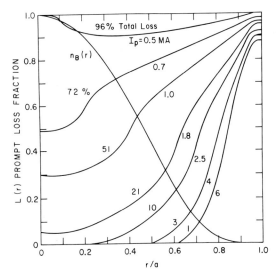

Fig. 3. Prompt loss fraction $L(r)$ and percent of total alpha particles lost during the first orbit as a function of minor radius for a range of plasma currents. Plasma conditions relate to Table I ($\mathcal{A} = 3$). $n_B(r)$ is the alpha birth distribution. [After Mikkelsen and Post (1979).]

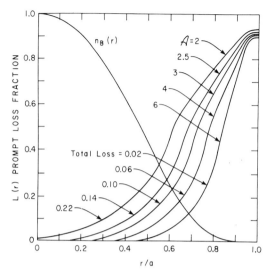

Fig. 4. Effect of aspect ratio $\mathcal{A}$ on the prompt losses of 3.5-MeV alpha particles for the plasma in Table I. The plasma current was 2.5 MA and the major radius was varied from 170 to 510 cm. $n_B(r)$ is the alpha birth distribution. [After Mikkelsen and Post (1979).]

in width as it slows down. The large orbital excursions of the alphas mean that the energy they deposit will be spread around the plasma rather than localized in the flux surface on which they were born. This is illustrated in Fig. 6, which compares $n_B(r)$, the initial birth density of the alphas; $n_s(r)$, the density after the alphas on unconfined orbits have been subtracted out; and $n_D(r)$, the energy deposition profile for the confined alphas including the spread due to orbits. The energy deposition profile is considerably broadened by the orbital spreading. For larger plasma currents, the orbital spread-

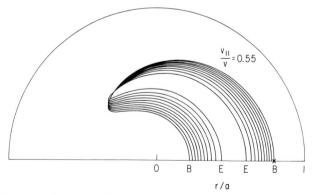

Fig. 5. Evolution of a typical alpha particle orbit as it slows down from 3.5 MeV to 350 keV. The particle starts at X. The beginning and ending bounces are labeled B and E, respectively. [After Mikkelsen and Post (1979).]

6D. Alpha-Particle Heating

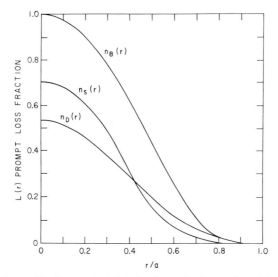

Fig. 6. Density profiles for a typical alpha birth profile, the density of alphas after first orbit losses, and the energy deposition profile including orbits and slowing down for the 1-MA plasma described in Table I. Loss = 51%.

ing of the energy deposition and orbital losses will be reduced, and in very large-current devices with $I_p \sim 6$–10 MA, deposition of the alpha energy locally at the birth radius is an adequate approximation.

Confinement of alpha particles in other magnetic geometries has not received extensive study. Almost all other schemes are not axisymmetric, which means that p_ϕ is not a good constant of the motion. This violation of axisymmetry occurs to a small extent in tokamaks owing to the discrete nature of the toroidal field coils, which introduces a "ripple" in the toroidal field. The basic effect of this nonaxisymmetry is to destroy the ability of particles on a flux surface to average out the grad-B and curvature drifts. This occurs because the lack of axisymmetry can introduce local wells in the magnetic field in which particles can be trapped and then drift out, or the particles effectively undergo small net drifts over each periodic motion which add instead of canceling.

The confinement of alpha particles in mirror systems has been studied by Driemeyer *et al.* (1979) and Miley *et al.* (1979) for reversed-field mirrors, and by Devoto *et al.* (1980) for tandem mirrors.

II. Fast-Alpha-Particle Diagnostics

The present estimates of alpha-particle heating assume that the alpha particles drift as single particles undergoing collisions with the background plasma. Such behavior will have to be verified in large fusion experiments. There exist a number of theoretically predicted instabilities (Kolesnichenko,

1980) that could lead either to anomalous diffusion and loss of fast alphas, or to nonclassical coupling to the plasma. In near-term fusion experiments such as JET and TFTR, the alpha heating will be about one-fifth of the total heating, and knowledge of the alpha behavior will be difficult to extract from the behavior of the gross plasma parameters such as the electron temperature, etc. Thus, for near-term as well as long-term (ignition) fusion experiments, there will be intense interest in the fast-alpha heating and the fast-alpha distribution function, which requires a method for measuring the fast-alpha distribution as it heats the plasma.

Techniques have been proposed for measuring the fast alphas which have escaped (Hendel and Seiler, 1978). Unconfined 14-MeV protons from the $D + He^3$ reaction in PLT have also been measured (Strachan, 1980). More preferable is an elucidation of the complete distribution function similar to the kind of data obtained for neutral hydrogen beam injection and heating, where the energy distribution of the escaping high-energy neutrals formed by charge exchange between the fast beam ions and background neutral hydrogen atoms (Goldston, 1975) is measured.

A technique has been proposed (Post *et al.*, 1981) in which some He^{2+} present in the plasma is neutralized by charge exchange with an injected beam of Li^0 and the energy and angular distribution of the escaping helium neutrals is measured. The scheme is sketched in Fig. 7. An energetic neutral lithium beam is passed through the plasma; the alpha particles are neutralized, and analyzed as they escape from the plasma. The double-charge-exchange cross section for alphas and neutral helium is given in Fig. 8. The cross section is $\sim 10^{-16}$ cm^2 for low relative velocities, and drops off rapidly for large relative velocities (Okuno *et al.*, 1978). Thus to have a significant

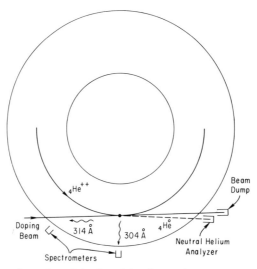

Fig. 7. Schematic outline of the fast alpha diagnostic. [From Post *et al.* (1981).]

6D. Alpha-Particle Heating

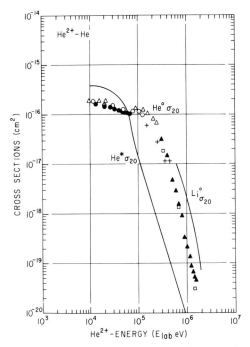

Fig. 8. Cross sections for double charge exchange of He^{2+} and He^0 and Li^0. He^0 is ground state neutral helium, He^* labels an estimate of the metastable $He^0 + He^{2+}$ cross section, and Li^0 is a calculation of $Li^0 + He^{2+} \rightarrow He^0 + \cdots$. (+) Allison (1958), (△) Berkner (1968), (○), Bayfield (1975), (□) Nikolaev (1962), (▲), Pivovar (1962), (●) Shah (1974). [Adapted from Okuno (1978). He^0 data from Okuno (1978); He^* data and Li^0 estimate from Olson (1981, private communication).]

reaction rate, a technique equivalent to that of "merged beams" is necessary, where one "beam" is in effect the fast alphas and the other is the injected lithium beam. Since a 3.5-MeV alpha particle has a velocity of $\sim 1.3 \times 10^9$ cm s^{-1}, the "doping" neutral Li beam must have a similar velocity (~ 6 MeV). Practical considerations of count rates, etc., require that the beam current be in the 10-mA range. Since the detection of fast He^0 atoms is relatively straightforward, the key problem is the production of a 6-MeV, 10-mA beam of Li^0. An efficient way to produce such a beam might be to form negative ions, accelerate them with an RF accelerator, and neutralize them in a thin gas cell or by laser photodetachment. H^- is ruled out since H^0 has only one electron and is not a good candidate for double charge exchange. Although it is possible to produce sufficiently large He^- currents (Hooper *et al.*, 1980) the neutralization of He^- at 3 MeV produces a large fraction of metastable helium, which would have a low $\langle \sigma v \rangle$ for double charge exchange and would be ionized before penetrating to the center of the plasma. The neutralization of Li^- is expected to produce primarily ground state lithium neutrals. The double-charge-exchange cross section for Li^0 and

He^{2+} has not been measured, but some preliminary calculations (Olsen, 1980) indicate that it is as large as the He^0 + He^{2+} → He^{2+} + He^0 cross section (Fig. 8). At low relative velocities one would expect some capture of both a 1s and a 2s electron from lithium by the alpha particle. As the relative velocity increases above the orbital speed of the 2s electron on lithium, capture of the two 1s electrons should be the dominant channel. However, measurements of the cross section still need to be done.

In principle, the 6-MeV neutral lithium beam offers a practical method for measuring fast-alpha distribution in a reacting plasma. A similar technique has been proposed (Burrell, 1979) to measure the low-energy alpha particle (ash) density at the plasma edge using a low-energy neutral beam to populate the $n = 3$ level of He$^+$ by the reaction Li + He^{2+} → (He$^+$)* + Li$^+$. The 3 → 2 transition would have a wavelength of 1640 Å and could be detected by optical techniques. It is relatively easy to make a 100-keV neutral lithium beam. The usefulness of the technique is limited to the plasma edge since a 100-keV Li^0 beam would be ionized in the first 10–20 cm of a large plasma.

III. Alpha-Particle Ash

Alpha-particle production is inherent in alpha heating. The thermalized alpha particle "ash" must be removed in a long-pulse fusion experiment. Considering a large tokamak such as INTOR with 100 MW of alpha heating ($R \sim 530$ cm, $a \sim 120$ cm, $\bar{n} \sim 10^{14}$ cm^{-3}), the alpha production rate is $\sim 2 \times 10^{20}$/s, or $0.01 n_e$/s, so that in 20 or 30 s the discharge would be largely alpha particles if the D–T continued burning. If the helium to be pumped is allowed to accumulate to $\sim 3\%$, and the neutral density at the pumping aperture is $\sim 10^{13}$ cm^{-3} (10^{16}/liter), then the required pumping speed is $2 \times 10^{20}/(0.03 \times 10^{16}) \sim 7 \times 10^5$ liter/s. This is a high pumping speed, especially for helium. There is therefore great interest in schemes for increasing helium pumping efficiencies.

Several schemes have been proposed (Shimomura et al., 1979) which postulate that helium ions could be concentrated within a divertor plasma so that helium pumping problems could be reduced. This concept was based on the assumption that hydrogen and helium atoms formed on the neutralizer plate of the divertor might transport differently in a plasma since the hydrogen could undergo charge exchange. This was investigated quantitatively by two Monte Carlo calculations using somewhat different assumptions (Takizuka et al., 1980; Callen et al., 1980; Seki et al., 1981). The physics of the calculation is illustrated in Fig. 9. Plasma ions (D$^+$, T$^+$, and He^{2+}) flow along the field lines into the divertor (described in Section 7.II.A). They are accelerated across the electrostatic sheath (described in Section 7.III) at the neutralizer plate and some are reflected as fast neutrals, while the remainder are desorbed as slow neutrals (described in Section 7.IV.A). The hope was

6D. Alpha-Particle Heating

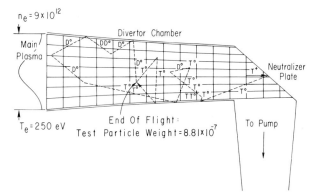

Fig. 9. Schematic diagram of a divertor chamber with a D, T, and He plasma showing a sample path for a test particle. Plasma = 5% He^{2+}, 47.5% D^+, 47.5% T^+.

that many of the slow and fast helium atoms would be ionized before they could return to the main plasma, whereas the hydrogen (D, T) could undergo charge exchange collisions with the divertor plasma ions, and any slow D, T atoms would thus be raised to the local plasma temperature. There would thus be a greater probability that the D, T atoms would return to the main plasma in the toroidal tokamak chamber. The calculation reported by Seki *et al.* (1981) and Takizuka *et al.* (1980) assumed that the low-energy D–T was molecular but that it dissociated immediately into two hydrogen atoms (with energy 3 eV) as it left the wall. It also had a prescription for following the paths of the subsequently ionized helium and hydrogen atoms. Their results are given in terms of a backflow fraction R (fraction of particles that return to the main plasma) in Fig. 10 (Takizuka *et al.*, 1980). At very low plasma

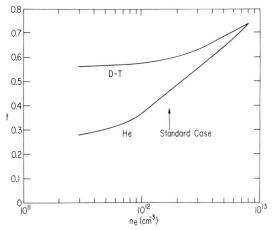

Fig. 10. Backflow fraction R for D–T and He as a function of divertor plasma density [Data from Takizuka *et al.* (1980) and Seki *et al.* (1981).]

densities in the divertor the fraction of hydrogen flowing back to plasma is greater than the fraction of helium flowing back. The "helium enrichment" of a divertor R_{D-T}/R_{He}, varies from 2.0 at $n_e = 3 \times 10^{11}$ cm^{-3} to 1.0 at $n_e \sim 8 \times 10^{12}$ cm^{-3}, and is 1.3 for the "standard case" (i.e., $T_e = 250$ eV, $n_e = 1.7 \times 10^{12}$ cm^{-3}) (Takizuka et al., Seki et al., 1981).

The large enhancement expected from simple arguments for hydrogen backflow was not found in the detailed calculations primarily because the charge exchange of hydrogen is both an advantage and a disadvantage. It is an advantage in that it can increase the average energy of neutrals flowing back to the plasma, thereby increasing the ionization mean free path of neutrals that have lost energy by colliding with the wall. However, charge exchange decreases the total collision mean free path by about a factor of 3. This means that neutral hydrogen atoms heading toward the plasma have their direction randomized, thereby increasing the average distance a neutral atom travels in the plasma, and thus increasing the probability that the atom will be ionized. The calculations of Seki et al. (1981) and Takizuka et al. (1980) indicate that the charge exchange replacement of neutral energy lost to the wall is a slightly stronger effect but not by much.

The calculation of Callen et al. (1980) included a comprehensive assessment of hydrogen molecules in that details of their ionization and dissociation by electrons (see Section 7.V.B) of the divertor plasma were taken into account. It was assumed that the slow neutrals were desorbed as molecules at ~300 K and as can be seen from Fig. 11 of Chapter 7 the dominant electron collision process at $T_e \gtrsim 50$ eV is ionization of H_2 to H_2^+ followed by dissociation into a proton and a neutral with a dissociation energy of ~4 eV. Assuming that $T_e \sim 100$–500 eV for typical expected cases, and that the sheath potential is ~$3T_e$, hydrogen ions striking the neutralizer plate will have energies of ~300–1500 eV or more, and from the reflection data in Fig. 6 of Chapter 7 it is clear that 60% or more of the neutralized ions leave the surface as molecules, not fast atoms. Thus 50% of the hydrogen atoms in molecules become protons (deuterons or tritons) in their first collision. The effectiveness of charge exchange in increasing the slow-particle velocity is reduced by up to 50% compared to assuming that all the hydrogen leaves the surface as H^0 (or D^0, T^0). Inclusion of these molecular effects in the model of Callen et al. turned helium enrichment into deenrichment ($R \sim 0.72$). The validity of both calculations is uncertain since the reflection data is largely extrapolated from high-energy experiments (>1 keV) or based on theoretical models for the interaction of the neutral and wall material. Another serious failing is that the plasma in both calculations is assumed fixed. In a self-consistent calculation, the plasma would probably be denser and colder than assumed since the ionized neutrals will raise the density.

Thus it does not appear that divertors can significantly (factors of 5–10) enrich the helium in the pump and help solve the exhaust problem. However, work described by Callen et al. points out that helium may be "enriched" in the plasma edge. The penetration of helium is less than that of hydrogen because of the absence of charge exchange for helium and the unimportance of molecules since the neutrals in the main plasma do not strike the walls as often as they do in the divertor.

Interpreted correctly, the results of Callen *et al.* indicate that helium pumping will not be a real problem. The presence of the plasma significantly retards the backflow of both H and He by as much as a factor of 10–20. This is because a large fraction of the backflowing neutrals are ionized and thus returned as ions to the neutralizer plate, and the neutrals going into the pump begin at high energy (~10–50 eV) and thus have higher flow rates down the pump than room-temperature neutrals [by $(E_0/300 \text{ K})^{1/2}$]. So even with no enrichment, this model predicts that the required pumping speed is ~10^4–5×10^4 L/s to provide for He exhaust.

Acknowledgments

The author is grateful to D. Mikkelsen, C. Singer, D. Heifetz, and J. Weisheit for discussions and contributions. This work was supported by U.S. Department of Energy Contract No. DE-AC02-76-CHO-3073.

References

Burrell, K. (1979). Private communication.
Callen, J. D. *et al.* (1980). *Plasma Phys. Controlled Nucl. Fusion Res., Proc. Int. Conf., 8th, Brussels, 1980* **1**, 775.
Cordey, J. G., Goldston, R. J., and Mikkelsen, D. R. (1981). *Nucl. Fusion* **21**, 581.
Devoto, R. A., Ohnishi, M., Kerns, J., and Woo, J. T. (1980). *Alpha Particle Confinement in Tandem Mirrors*, ANS Top. Meet., 4th Techn. Controlled Nucl. Fusion, King of Prussia **A.12**, 36.
Driemeyer, D., Miley, G. H., and Condit, W. C. (1979). *A Monte Carlo Method for Calculating Fusion Product Behavior in Field-Reversed Mirrors*, ANS Top. Meet. Comput. Methods Nucl. Engin., Williamsburg **2**, 7–37.
Goldston, R. J. (1975). *Nucl. Fusion* **15**, 651.
Hendel, H., and Seiler, S. (1978). Measurement of Charged Fusion Reaction Products, *TFTR Diagnostics Rev.* PPL, November 16–17, 1978, p. 118.
Hively, L. (1977). *Nucl. Fusion* **17**, 873.
Hively, L., and Miley, G. H. (1980). *Nucl. Fusion* **20**, No. (8), 969–983.
Hooper, E. B. *et al.* (1980). *Rev. Sci. Instrum.* **15**, 1066.
Jones, E. (1978). *Culham Lab. Rep.* **CLM-R-175**.
Kolesnichenko, Ya. I. (1980). *Nucl. Fusion* **20**, 727.
Maeda, H., and Okabayashi, M. (1979). *Bull. Am. Phys. Soc.* **24**, 993.
McAlees, D. G. (1974). *Oak Ridge Nat. Lab. [Rep.] ORNL-TM (U.S.)* **ORNL-TM-4661**.
Mikkelsen, D. and Post, D. (1979). *Proc. Workshop Phys. Plasmas Close Thermonucl. Condition, Varenna.* **DOE CONF-790866**, 41.
Miley, G. H., Gilligan, J. G., Driemeyer, D. E., Morse, E. C., and Condit, W. C. (1979). *Studies of Finite Gyro-Radius Effects in a Field Reversed Mirror Configuration*, Eur. Conf. Controlled Fusion Plasma Phys., Oxford, 1979 **AP34**, 41.
Okuno, K. (1978). *Res. Rep.—Nagoya Univ., Inst. Plasma Phys., 1978* **IPPJ-AM-9**.

Olson, R. E. (1980). Private communication.
Post, D. E. *et al.* (1981). *J. Fusion Energy* **1,** 129; also Grisham, L. *et al.* (1981). *Princeton Univ. Plasma Phys. Lab., Rep.* **PPPL-1661.**
Seki, Y. *et al.* (1981). *Nucl. Fusion* **20,** 1213.
Shimomura, Y. *et al.* (1979). Research Rep. JAERJ M-8294, Tokai, Japan.
Strachan, J. (1980). *Plasma Phys. Controlled Nucl. Fusion Res., Proc. Int. Conf., 8th, Brussels, 1980* **2,** 95.
Takizuka, T. *et al.* (1980). *Plasma Phys. Controlled Nucl. Fusion Res., Proc. Int. Conf., 8th, Brussels, 1980* **1,** 679.

7
Boundary Plasma

M. F. A. Harrison

Culham Laboratory
Abingdon, Oxfordshire
England

I. Description of the Boundary Region	. . . . . .	395
II. The Boundary of a Toroidal Device	. . . . . .	399
A. Divertors	. . . .	402
B. Plasma Transport	. . . .	406
C. Impurity Control	. . . .	411
III. The Sheath and Long-Range Electric Field Regions		413
IV. Particle–Surface Interactions	. . . .	416
A. Backscattering due to Ion and Atom Impact	. .	417
B. Sputtering	. . . .	420
V. Atomic Processes in the Boundary Plasma	. . .	423
A. Energy Losses due to Collisions of Electrons with "Hydrogen" Atoms	. . . .	424
B. Effects of Molecular Hydrogen	. . . .	427
C. Recycling and Trapping of "Hydrogen" Atoms	.	430
D. Collisions of Electrons with Impurity Ions	. .	432
VI. Significance of the Boundary Plasma	. . . . .	436
References	. . . .	437

I. Description of the Boundary Region

The term "boundary" or "edge" is used in fusion research to describe a rather indeterminate region that links a magnetically confined plasma to the walls of its containment vessel. The vessel is exposed to fluxes of both particles and energy which escape from the hot plasma, and the manner in which its walls respond to this irradiation is influenced by local conditions within the boundary of the plasma and by the detailed nature of plasma–surface interactions that take place at the wall. Charged particles can most readily leave the confined plasma by drifting parallel to the magnetic field so that surface interactions tend to be concentrated in regions where magnetic flux tubes intersect the wall, i.e., in regions of "open" magnetic field. A linear containment device must inevitably be prone to some loss of particles along open flux tubes but, in the nominally closed field of a toroidal device such as a tokamak, the equivalent losses must be initiated by the outward diffusion of plasma across the inner region of closed magnetic field. This flow of charged particles and energy feeds the boundary plasma through which it is transported predominately in the direction parallel to the magnetic field until it reaches the wall. The spatial distributions of the particle fluxes and power loading at the wall are thus determined by the topology of the field and the walls and also by the transport properties of the plasma along and across the magnetic field.

The boundary is defined in this chapter as that region where transport is dominated by plasma flow parallel to the magnetic field. It will be seen later that the containment time $\tau_\parallel$ for plasma in such a boundary is generally much shorter than the containment time $\tau_\perp$, that characterizes outward transport from the bulk of the plasma. The present treatment also emphasizes conditions where atoms released from the wall due to plasma bombardment return to the wall without traversing (to any significant extent) regions of closed magnetic field. This condition can be defined as $\Delta_\perp \sim \Delta_0$ where $\Delta_\perp$ is the scale length that characterizes the outward flow of plasma during its transit through the boundary and Δ_0 characterizes the penetration range of atomic species released from the wall. Clearly these are limiting conditions and in practice they must be coupled to the mechanisms that control the outward transport of plasma particles and energy. However, it is argued here that the simplification is of value insofar as it facilitates a coherent discussion of the manifold processes that take place within the boundary.

Plasma interactions with the walls are sensitive to a wide range of parameters such as the energy and atomic species of incident particles as well as the bulk and surface properties of wall material. The subject is briefly discussed in Section IV and for the present it is adequate to accept that plasma–surface interactions cause (i) backscattering or "reflection" of atoms and ions and that most of the returning particles are neutral and some retain an appreciable fraction of their incident energy, (ii) desorption of "hydrogen"[†] (and of impurity elements such as oxygen) that are trapped on the surface, and (iii) sputtering of atoms of the wall material which may subsequently contaminate the plasma with radiating ions of metallic elements. The relative contributions from these various processes are sensitive to the parameters of the plasma adjacent to the surface; for example, the local temperature of electrons and ions influences the energy with which plasma particles strike the surface and the localized fluxes of energy and particles transported by the plasma affect the power loading and fluence of particles at the surface.

[†] Most present-day fusion experiments are conducted using a hydrogen or deuterium plasma although the longer-term objective is to develop a reactor based upon the nuclear fusion of D and T. Parentheses are used when it is unnecessary to distinguish between the properties of the hydrogen isotopes.

7. Boundary Plasma

The electron temperature within a fusion plasma is generally far too high for electron–"proton" recombination to occur at a significant rate within the plasma volume, and the principal sources of neutral particles are backscattering and desorption at the vessel wall. The density of neutral particles is thus peaked in the boundary region. Motion of these atoms and molecules is not directly affected by the magnetic field and they can therefore travel inward from the wall to hotter regions of the plasma until such time as they become ionized by the plasma electrons. The resulting ion pairs form a component of the boundary plasma, but they eventually diffuse back to the vessel wall where they are neutralized again. This recirculation process is termed "recycling" and it affects the gradients in plasma temperature and density and thereby (as discussed in Section II) influences the diffusive transport processes of plasma near the wall. When the surface of the wall is heavily loaded with trapped gas, each incident ion pair may release more than one atom and the recycling process will be correspondingly enhanced until the surface concentration of trapped gas is reduced to a steady-state value such that the release rate of neutral particles equals the incident rate of ions. Ion impact also causes desorption of trapped impurity species (e.g., O and CO) that are often observed during the initial phases of a pulsed discharge; this has prompted extensive research into the preparation of clean walls.

An important atomic mechanism associated with recycling arises from charge exchange collisions between the released "hydrogen" atoms and the plasma "protons." The rate coefficients for charge exchange and electron impact ionization are such that, in a homogeneous plasma, the frequency of charge exchange collisions is greater than that of ionizing collisions (see Section V). During each charge exchange collision a plasma proton captures an electron and the resultant atom moves freely across the magnetic field in a direction which is almost randomly distributed around the collision site. In effect, the collision scatters the plasma "proton" so that it moves (as a "daughter" atom with energy corresponding to the ion temperature) either outward toward the wall or inward toward the region of hotter plasma. A sequence of such charge exchange collisions may take place before the transported atom is ionized and at each collision the daughter atom has the local temperature of the plasma "protons." The mean free path for ionization is longer for the faster atoms produced by charge exchange in the hotter region of the plasma and these atoms can either penetrate rather deeply, or else escape rather easily. Molecules of "hydrogen" do not play a direct role in this cycle because their cross section for charge exchange is relatively small; nevertheless, they are readily dissociated in collisions with electrons and their product atoms subsequently undergo charge exchange. These atomic processes govern the distribution of energy among those "hydrogen" atoms that return to the wall.

An essential property of any plasma is that it must on average be electrically neutral and, to maintain this neutrality, it is necessary for the boundary plasma to adjust its transport properties so that the average rate at which charge is lost to the wall by ions equals the average rate due to electrons. If it is accepted that plasma recycling predominates in the region of parallel flow, then the requirement for ambipolar transport forces the local potential of the plasma to be positive with respect to the surface. Mobile electrons that leave the plasma must then traverse a region of electron-repelling electric field which impedes their flow to such a degree that it becomes equal to that of the less mobile ions. The potential U is approximately equal to $3kT_e/e$ for a hydrogen plasma and is spatially distributed in a region adjacent to the surface and which is characterized by the Debye shielding length λ_D. This restricted region is called a "plasma sheath" and ions that enter it are accelerated to the surface. In general λ_D is much smaller than the mean free paths for atomic processes so that the sheath is collisionless. A more detailed description is given in Section III and it is sufficient to note here that both the sheath potential and hence the energy of ions incident upon the surface are sensitive to the temperature T_e of plasma electrons at the sheath edge, and thus to any process which cools these electrons. Cooling due to free–bound atomic collisions is likely to occur in the vicinity of the wall because most ions incident upon the surface return as atoms which are

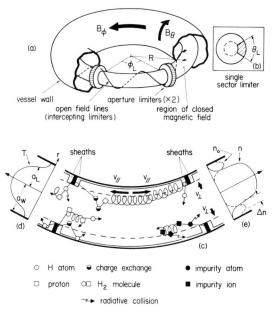

Fig. 1. Boundary regime in a toroidal containment vessel. The configuration of the magnetic fields together with the vessel wall and aperture limiters is shown in (a) and a single-sector limiter is illustrated in the inset (b). Various mechanisms by which atomic reactions interact with the plasma are illustrated schematically in (c) and the radial profile of plasma temperature is sketched in (d); the corresponding density profiles of atoms and of plasma are shown in (e).

subsequently excited or ionized by collisions with plasma electrons. In the case of a "hydrogen" plasma, comparable radiative processes may also arise due to the release of molecules from the surface. Collisions with atoms and ions which have several bound electrons can dissipate appreciable amounts of energy so that the presence of relatively small concentrations of impurity elements can cool the boundary plasma electrons.

Atomic processes are discussed in more detail in Section V and in Fig. 1. Their relevance to the boundary region is illustrated schematically for the specific case of a toroidal confinement device such as a tokamak.

II. The Boundary of a Toroidal Device

Consider a confined plasma which in bulk must drift with a velocity that is predominantly parallel to the direction of the confining magnetic field. Nevertheless, a variety of scattering mechanisms (such as those described in Harrison, Chapter 2 and Hogan, Chapter 4, this volume) cause ions and electrons to move across the field so that they eventually reach the wall; here electron capture takes place and most of the ions return to the plasma as neutral particles. The wall can thus be regarded as a sink for charged particles. It is also a sink for energy which must be lost from the hot plasma by convective and conductive mechanisms whenever the plasma can come into contact with the wall. Radiation emitted due to free–free and free–bound atomic collision processes within the plasma provides further channels for energy loss which, in an optically thin plasma, are not directly affected by the confining magnetic field. The maximum values of the density n and temperature T of the plasma therefore tend to occur along the geometric axis of the magnetic field and losses of charge and energy to the walls of the vessel set up gradients in both density and temperature which drive plasma outward from the magnetic axis. The outward flux $\Gamma_\perp(r)$ at an intermediate distance r between the axis and the wall can be expressed in terms of a "radial" diffusion coefficient $D_\perp$ for motion perpendicular to the field; it is given by

$$\Gamma_\perp(r) = - D_\perp(r) \left(\frac{dn}{dr}\right)_r, \qquad (1)$$

and the corresponding flow velocity is

$$v_\perp(r) = \Gamma_\perp(r)/n(r). \qquad (2)$$

The magnitude of $D_\perp$ is uncertain because the detailed nature of these "radial" flow mechanisms has not yet been clearly established. They are also likely to be dependent upon the particular containment device but it is often accepted for the boundary plasma of a tokamak that D_B, the semiempirical

Bohm diffusion coefficient,

$$D_B = \tfrac{1}{16} kT_e/eB = 6.25 \times 10^6 T_e/B \quad (\text{cm}^2\ \text{s}^{-1}),^\dagger \tag{3}$$

is indicative of radial transport in a magnetic field of intensity B near the wall. This contention is in general agreement with experimental evidence (see, e.g., Proudfoot and Harbour, 1980; Staib, 1982) and is taken as a guideline throughout the subsequent discussion. Insertion of typical values for B ($\approx 3 \times 10^4$ G) and T_e (≈ 30 eV) into Eq. (3) yields $D_\perp = D_B \approx 6 \times 10^3$ cm^2 s^{-1} and $\Gamma_\perp \approx 1.2 \times 10^{16}$ ion pairs cm^{-2} s^{-1} when the density is about 10^{13} cm^{-3} and the scale length of the density gradient is taken to be 5 cm. The corresponding flow velocity given by Eq. (2) is $v_\perp \approx 1.2 \times 10^3$ cm s^{-1}, which is much smaller than the thermal velocity of the particles involved.

The spatial distribution of open magnetic flux tubes at the wall of the confinement vessel governs the distribution of energy and charged particles incident upon the wall. This in turn influences the construction of the vessel, which must be compatible with the thermal loading produced by the plasma and the surface erosion concomitant upon processes such as sputtering. It is customary to shield the torus wall of a tokamak by means of a "limiter" which is a plate or comparable structure that projects from the wall and into the plasma. Interactions between the plasma and its boundary surfaces are thereby concentrated in the vicinity of the limiter. A highly simplified form of this configuration is adopted in Fig. 1 to provide a framework for illustrating the description of the boundary plasma that has so far been evolved. The major radius of the torus is R and the minor radius a_w lies in the direction r. The confining magnetic field consists of two components, B_θ in the poloidal direction and B_ϕ in the toroidal direction. The magnetic surfaces are nominally coaxial with the wall of the vessel but the flux tubes are twisted in a helical manner around the magnetic axis, their pitch being dependent upon criteria for confinement. This twisting of the field lines in the poloidal (θ) direction generates a rotational transform ι, and a "safety factor"

$$q(r) \equiv \frac{2\pi}{\iota} = \frac{r}{R}\left(\frac{B_\phi}{B_\theta}\right)_r \tag{4}$$

can be defined such that each magnetic field line makes $q(r)$ transits of the torus before being twisted back to its original poloidal angle. The effective length of the line is thus $2\pi R q(r)$.

The configuration shown in Fig. 1 has two limiters in the form of poloidally symmetric aperture plates which are so spaced that they subtend an angle ϕ_L in the toroidal plane. It is assumed that all magnetic field lines are

---

† Throughout Chapter 7 the symbol T is used whenever kT is expressed in electron volts and in general Gaussian units are used for other parameters. The symbols e, k, etc., have their conventional meaning. The electron and ion masses are m_e and m_i and their temperatures are T_e and T_i. For a D–T mixture m_i is taken as $2.5m_i$ (proton). The ion charge state is denoted by Z.

closed inboard of the limiter apertures (i.e., where $r < a_L$), whereas for $r > a_L$ all field lines intersect the limiters. The length L_B of each of the intercepting flux tubes in the cut-away sector illustrated between the limiters can thus be expressed as

$$L_{B1} \approx \phi_L R q(a_L),$$

and that in the other sector of the torus is

$$L_{B2} \approx (2\pi - \phi_L) R q(a_L).$$

Radial diffusion outward at velocity $v_\perp$ across the magnetic field carries plasma into these flux tubes. The plasma particles also move readily in the direction parallel to the magnetic field and the bulk flow of plasma drifts along the flux tube with a velocity $v_\parallel$. Plasma flow to the limiter surface must on average be ambipolar in order that overall charge neutrality can be maintained, and so it is reasonable to express the flow velocity by

$$v_\parallel = \mathcal{M} C_s = \mathcal{M}[(ZkT_e + kT_i)/m_i]^{1/2} \qquad (5)$$

where C_s is the ion sound speed and $\mathcal{M}$ the Mach number of the flow. In general the flow in the flux tube is subsonic, i.e., $\mathcal{M} < 1$, except at the sheath edge adjacent to the limiter, where $\mathcal{M} \approx 1$. If the average Mach number is taken as 0.3,[†] then the average value of $v_\parallel$ for a 30-eV hydrogen plasma is about 2.3×10^6 cm s^{-1}, which is about 10^3 times the comparable value of $v_\perp$. The drift time of the plasma in this region of flow parallel to the open magnetic field is

$$\tau_\parallel = L_\parallel / v_\parallel, \qquad (6)$$

where $L_\parallel$ is the effective length of the flow. In the absence of a net flow of current it is assumed that the plasma drifts with equal probability in opposite directions around the torus so that on average

$$L_\parallel \approx \tfrac{1}{2} L_B. \qquad (7)$$

The simple geometry of the two aperture limiters shown in Fig. 1 is not likely to be encountered in practice and a more general appreciation of the boundary can be gained from the concept of a single limiter plate that is not fully symmetric but which, as shown inset in Fig. 1, subtends an angle θ_L in the poloidal plane. The flow length for this geometry is given by

$$L_\parallel(\theta_L) \approx [(2\pi - \theta_L)/\theta_L] \pi R q(a_L). \qquad (8)$$

Typical values in present-day experiments are $R = 150$ cm and $q(a_L) \approx 4$, so, if θ_L is taken to be 60°, then $L_\parallel^{60} \approx 9.4 \times 10^3$ cm and for the preceding plasma conditions $\tau_\parallel^{60} \approx 4 \times 10^{-3}$ s.

[†] The Mach number of the flow is discussed in Section II.B.

An indication of the scale length Δ_n for the radial gradient of the plasma density in the boundary region[†] can be obtained from

$$\Delta_n \approx (D_\perp \tau_\parallel)^{1/2}. \tag{9}$$

The value $\Delta_n^{60} \approx 5$ cm for the preceding conditions is applicable only in those cases where recycling is insignificant (see Section II.B), but it does provide a useful indication of a lower limit to the radial extent of the boundary region for this particular configuration.

The plasma flux tube carries plasma particles to the limiter; this can be envisaged in the simplified form shown in Fig. 2, namely as a ribbon of approximately rectangular cross section that is wrapped over the toroidal boundary surface of the closed field region. The ribbon is closely wound so that it completely covers the surface (the pitch of this winding is much exaggerated in Fig. 2), and its linear direction is denoted by z; its length for any configuration equals $2L_\parallel$. The thickness of the ribbon that corresponds to the "channel" for particle transport is determined by Δ_n and width l_B (normal to z) is governed by the geometry of the limiter. The cross-sectional area $A_\parallel$ of the flow channel normal to z will be uniform throughout the channel length (provided that magnetic flux is conserved) and can be expressed as

$$A_\parallel \approx 2\pi^2 R a_L \Delta_n / L_\parallel, \tag{10}$$

which for the preceding case of a single sector limiter becomes

$$A_\parallel(\theta_L) = \frac{2\pi a_L \Delta_n}{q(a_L)} \frac{\phi_L}{2\pi - \phi_L}. \tag{11}$$

Taking $a_L \approx 50$ cm yields $A_\parallel^{60} \approx 80$ cm², which is typical of present experiments.

There is a comparable channel for transport of plasma energy in the boundary but its thickness Δ_Q may be somewhat smaller than Δ_n (see Section II.B).

A. Divertors

The wall of the toroidal vessel of minor radius a_w should be effectively shielded from direct impact of plasma whenever the limiter geometry is such that $a_w - a_L \gtrsim \Delta_n$, but the flow of boundary plasma is then concentrated upon the limiter surface, which is liable to release impurity ions into the adjacent region of confined plasma. This potential source of impurities within the torus can, in some degree, be avoided by the use of a "divertor" which is a device that deflects or "diverts" the plasma flow channel into a

[†] In effect Δ_n is equivalent to $\Delta_\perp$ (discussed in Section I) but Δ_n applies to a specific plasma configuration.

7. Boundary Plasma

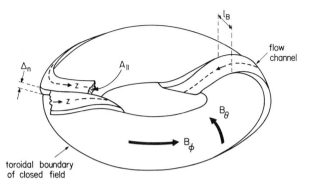

Fig. 2. Topology of a flow channel for plasma transported parallel to the magnetic field lines in a toroidal device. The channel can be compared to a ribbon which completely covers the toroidal surface; its thickness Δ_n, width l_B, and cross-sectional area are discussed in the text. Direction z lies parallel to the magnetic field and the linear extent in direction z is given by $L_B \approx 2L_\|$. The winding pitch of the ribbon is much exaggerated to illustrate the effect of rotational transform.

separate chamber appended to the containment vessel. The viability of the concept was first demonstrated for the stellarator (see, e.g., Burnett *et al.*, 1958) but is now more generally applied to tokamaks; a bibliography of recent literature dealing with divertors and related plasma physics can be found in Harbour (1981), and an appreciation of the current status of this subject can be gained from the IAEA Technical Committee Meeting on Divertors and Impurity Control (Keilhacker and Daybelge, 1981). The present discussion is limited to an outline of the properties of two systems, namely the "poloidal" and the "bundle" divertor.

The poloidal field in a tokamak is produced by a current that is induced to flow within the plasma; this field can be opposed, at least over a small range of poloidal angle θ, when a complementary current is passed in the same direction through a nearby external conductor. A poloidal divertor relies upon such local annulment of the poloidal field; its principles are illustrated in Fig. 3 where a single divertor winding is shown lying above but parallel to the magnetic axis of the plasma. Location of zero B_θ, or "null point" in the poloidal plane, is dependent upon the spacing between the winding and the plasma and also upon the relative magnitude of the currents. The walls of the vessel are necked in around the single null point so that communication between the top divertor chamber and the lower containment vessel is restricted; the system is symmetrical in the toroidal plane and additional divertor windings can be provided to increase the number of null points. The surface that bounds the region of closed field is called the "separatrix" and the null point lies on this surface. Magnetic flux tubes outboard of the separatrix are open and the rotational transform forces the flux tubes to twist (in the manner shown in Fig. 2) so that plasma at position A is rotated to B

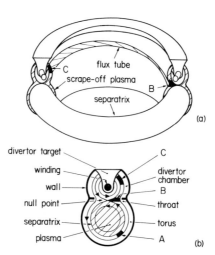

Fig. 3. Schematic illustration of a single–null poloidal divertor. The perspective view (a) illustrates the toroidal symmetry and the effects of rotational transform. The view in the poloidal plane (b) shows the configuration of poloidal field, the boundary surfaces of the containment vessel, and the combination of divertor winding and plasma current appropriate to a tokamak. The divertor chamber communicates with the torus vessel only via the restricted throat and recycling plasma, and impurity ions released by plasma-target interactions are thereby constrained to backflow along the magnetic field before entering the torus vessel.

and then to C as it moves in the toroidal direction around the torus. In effect, plasma that diffuses outward across the separatrix into region A is scraped out of the torus and deposited upon the divertor target; indeed, the boundary is also termed the "scrape-off" region and the diverted field configuration serves the role of a "magnetic limiter." The effective length of flow in the scrape-off plasma is $\pi R q(a_s)$ and the flow area $A_{\|,\,\text{pol}}$, is given by

$$A_{\|,\,\text{pol}} \approx 2\pi a_s \Delta_n / q(a_s), \tag{12}$$

where a_s is the radius of the separatrix (typically ≈ 50 cm in present-day experiments) so that $A_{\|,\,\text{pol}} \approx 400$ cm^2. It should be noted that $A_{\|,\,\text{pol}}$ is appreciably smaller than the geometric area of the throat of the divertor chamber and so the ability of the divertor to retain "hydrogen" gas as well as impurity atoms released at the divertor target is strongly dependent upon ionization of these neutral species before they can flow back into the torus. Another aspect of the poloidal divertor is that its flux tubes graze the divertor target; plasma is thus spread over a large area and power loading is minimized. The local magnetic field is strong and, because of its orientation to the target, it can affect plasma surface interactions; for example, it is likely to suppress secondary electrons that are emitted from the target (see Section III).

The configuration of a bundle or "local" divertor is quite different; the type evolved by Stott *et al.* (1977, 1978) is illustrated in Fig. 4. It consists of two external coils which are located almost tangential to the torus wall and

7. Boundary Plasma

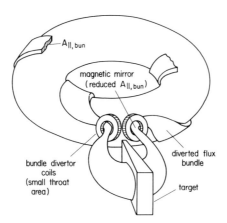

Fig. 4. Illustration of the principles of a bundle divertor. The divertor coils and target are sketched as well as the flow channel, which is reduced in area as it traverses the magnetic mirror region in the throat of each divertor coil. The downstream channels expand prior to intersecting the divertor target.

whose windings are energized so that their combined magnetic field acts locally in opposition to the toroidal field. A bundle of the exterior magnetic flux is thereby deflected outwards through the narrow throat of one coil and into the external divertor chamber from whence it returns to the torus in a comparable manner through the other throat. The flux lines are concentrated upon a relatively small area of divertor target but atomic species released at the target can communicate with the torus only via the restricted throat regions. The divertor coils generate a local magnetic field that is stronger than the average field in the scrape-off region so that plasma particles are reflected by the magnetic mirrors which exist close to each throat. The flow of plasma then relies strongly on ion–ion scattering to fill the mirror loss cone and the Mach number of the flow in the scrape-off-region tends to be lower than that for a poloidal divertor. The area of the flow channel is given by

$$A_{\parallel,\text{bun}} \approx 2\pi a_s \Delta_n / q_{\text{bun}} \tag{13}$$

where q_{bun} (the "diversion frequency") accounts for the rotational transform of the diverted flux bundle. Typically $q_{\text{bun}} \approx 20$ so that $A_{\parallel,\text{bun}}$ in present experiments is about 100 cm². Within the scrape-off region, $A_{\parallel,\text{bun}}$ is always less than $A_{\parallel,\text{pol}}$ but this difference is greatest in the throats of the bundle coils.

Experiments show that divertors reduce the concentration of impurity ions in tokamaks. This is particularly evident for large and powerfully heated ($\approx$2 MW of neutral-beam injection) devices (see, for example, Wesley et al., 1982 and Keilhacker et al., 1982). This fact is attributable firstly to the localized cooling of the boundary plasma in the vicinity of the divertor target, which thereby reduces the yield of impurity atoms from the target and secondly to the ability of the divertor field configuration to reduce the backflow of impurities to the main plasma. Plasma transport in the boundary region plays an important role in both these issues.

B. Plasma Transport

There is a strong interrelationship between transport of plasma in the closed-field region and in the open-field boundary, but the later region is rendered more complex by the three-dimensional nature of the problem. Transport in the closed-field can be considered a predominantly one-dimensional problem and it is justifiable to model it in the radial direction across the magnetic field. This conventional approach can be modified to make some allowance for transport parallel to the field in the outer radial regions of the boundary; examples can be found in Post *et al.* (1979), Nicolai (1979), Watkins *et al.* (1982), and Neuhauser *et al.* (1982). However, such treatments are not as yet fully self-consistent because they use simplified (zero-dimensional) descriptions of the transport of plasma particles and energy in the boundary.

Studies of the boundary per se are more rigorous and there are now numerical treatments for one-dimensional fluid transport of electrons and ions in the direction of the magnetic field (for examples see Harbour and

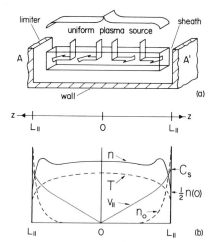

Fig. 5. Linear representation of boundary plasma transport parallel to the magnetic field. A uniform distribution of plasma input is illustrated in (a) for a toroidal boundary between opposite faces A and A' of a limiter. Gradients in the z direction arising from collisional flow in this region are sketched in (b) for the density of plasma and neutral particles n, n_0, the drift velocity of the plasma $v_\parallel$, and the plasma temperature T. Recycling occurs predominantly in regions where n_0 is significant, i.e., where $z \to L_\parallel$. The different flow properties of a limiter and a divertor boundary plasma are discussed in the text.

Morgan, 1982, and Chodura *et al.*, 1982). There is also a two-dimensional numerical treatment (i.e., fluid flow along and across the field) in which plasma transport is coupled to a Monte Carlo assessment of neutral particle recycling from the boundary surfaces (Post *et al.*, 1982; Petravic *et al.*, 1982). Alternatively there are analytically oriented approaches to boundary plasma modeling (see, e.g., Harrison *et al.*, 1981, 1982, 1983). These are less precise in their determination of the spatial behavior of the plasma but offer a convenient route for identifying the effects of boundary interactions. The following discussion is based on a simplified analytical approach in which transport is considered only in the direction parallel to the magnetic field (i.e., in the z direction shown in Fig. 2). The plasma parameters are treated as functions of z only and so their values actually relate to averages over the flow channel area $A_\parallel$.

The plasma flow channel that was identified in Fig. 2 can also be envisaged as a linear projection in the z direction and the case for a limiter is illustrated in Fig. 5a. In the following discussion it is assumed that

(a) particles and energy enter the boundary from the main plasma, and these inputs are distributed uniformly over the face of the channel that is adjacent to the confined plasma,

(b) parallel flow in the boundary causes plasma to drift equally to the sheaths that cover each face of the limiter, and

(c) substantial recycling of plasma occurs where the scrape-off layer intercepts the limiter surface.

It is considered that plasma flow is collisional, and this assumption is valid if the scale lengths $\Delta_n(z)$ and $\Delta_T(z)$ for the gradients dn/dz and dT/dz are appreciably greater than the mean free paths λ_{ee} and λ_{ei} for electron–electron and electron–ion scattering due to multiple Coulomb collisions. The mean free path λ_{ee} (approximately equal to λ_{ei}) is

$$\lambda_{ee} \approx 4 \times 10^{13} T_e^2 / nZ^2 \ln \Lambda \quad \text{(cm)}, \tag{14}$$

where Z is the ion charge state and $\ln \Lambda$, the Coulomb logarithm, is described in Eq. (9) of Harrison (Chapter 2, this volume). For the preceding "hydrogen" plasma conditions, $\lambda_{ee} \approx 180$ cm and is thus appreciably less than $L_\parallel^{60}$ given by Eq. (8). The assumption of collisionality becomes progressively less valid in the plasma adjacent to the sheath and flow through the sheath (which is characterized by the distance λ_D, see Eq. (25)] is collisionless because $\lambda_D \ll \lambda_{ee}$.

The principal aspects of collisional flow are shown firstly by the fluid equation for continuity,

$$\frac{d}{dz}(nv_\parallel A_\parallel) = SA_\parallel, \tag{15}$$

where S is a volume source term for particles (i.e., number of particles/unit volume/unit time), and secondly by the associated momentum equation,

$$mnv_\| \frac{dv_\|}{dz} = -\frac{dP}{dz} - mv_\| S. \tag{16}$$

Here, $m = m_i + m_e$, $P = nT_e + nT_i$ is the plasma pressure, and it is assumed that the particle source does not add momentum to the flow. Manipulation yields

$$\frac{1}{v_\|}\frac{dv_\|}{dz} = -\frac{1}{1-\mathcal{M}^2}\frac{1}{A_\|}\frac{dA_\|}{dz}$$
$$+ \frac{1+\mathcal{M}^2}{1-\mathcal{M}^2}\left(\frac{S}{nv_\|} + \frac{1}{T_e+T_i}\frac{d}{dz}(T_e+T_i)\right), \tag{17}$$

where all parameters are functions of z.

The flow must start from rest at the point of symmetry, which implies that $v_\| = 0$ and $dv_\|/dz = 0$ when $z = 0$ so that $\mathcal{M}(0) = 0$. The drift velocity is therefore subsonic and the flow accelerates up to $\mathcal{M}(L_\|) = 1$ at the sheath edge. In order to illustrate trends in behavior it is useful to simplify Eq. (17) by the assumptions that $\mathcal{M}^2 \ll 1$, $dA_\|/A_\| = 0$, and $d(T_e + T_i)/dz = 0$. This leads to the expression

$$\frac{dv_\|}{dz} = \frac{S}{n}, \tag{18}$$

which shows that the velocity gradient $dv_\|/dz$ is directly dependent upon the local value of the particle source $S(z)$ whenever n is invariant. Accepting for the moment that invariance of density is valid we see that $v_\|$ increases linearly with z throughout the region of flow channel, which is subjected to a uniform input of plasma [i.e., $S(z)$ is constant in this region]. However, localized recycling adjacent to the limiter will enhance the value of $S(z)$, as can be seen in Fig. 5b from the profile $n_0(z)$ of neutral particles released from the limiter. The gradient in $v_\|$ will correspondingly increase in the downstream region of the channel close to the limiter until it reaches the value of C_s at the sheath edge. This behavior of $v_\|(z)$, which is strongly dependent on the penetration of the plasma by neutral particles released at the limiter surface, is also sketched in Fig. 5b. The plasma drift time,

$$\tau_\| = \int_0^{L_\|} \frac{dz}{v_\|(z)}, \tag{19}$$

is thus dependent upon the magnitude and distribution of particle sources along the overall channel length. The magnitude of the sources depends firstly upon the input of plasma that diffuses across the closed field region into the boundary and secondly upon the recycling of those neutral particles that arise from plasma–surface interaction and which subsequently become

7. Boundary Plasma

entrained as ions within the boundary plasma. A significant number of the neutral particles released at the limiter surface must travel in the radial direction and thereby enter the main plasma from which they may contribute to recycling in the boundary region distant from the limiter (see, for example, the trajectories indicated in Fig. 1). The situation in the divertor case is different because the equivalent particles do not directly enter the main plasma unless they can escape through the throat of the divertor chamber and thereby enter the torus region. Consequently the source term at the end of the flow channel of the divertor [i.e., $S(L_\|)$] is much larger than in the case of the limiter and there is a greater variation of S with z. The drift velocity of plasma in the torus scrape-off region of the divertor is therefore lower than in the case of a limiter. Recycling in the divertor of a conceptual tokamak reactor is predicted (for example, see Harrison *et al.*, 1982a) to be so large that $\mathcal{M} \sim 10^{-2}$ in the scrape-off region and in this condition the drift time of plasma particles $\tau_{\|,p}$ given by Eq. (6) tends to be comparable to the radial containment time $\tau_{\perp,p}$. Plasma is then likely to diffuse to the torus wall rather than drift directly into the divertor. Fortunately the transport of energy to the divertor is not deleteriously affected by this condition so that a relatively small fraction of the power that is fed to the boundary is likely to be deposited upon the torus wall, which is thereby shielded from damage by the "energy unload" action of the divertor.

The principal mechanism for transport of plasma energy in a collision dominated boundary is electron thermal conduction parallel to the magnetic field. Collisionality is assured because most electrons are reflected by the boundary sheaths and ultimately leave the plasma only as a consequence of electron–electron scattering collisions. The heat flux conducted along the energy flow channel can be expressed as

$$-\kappa_0 T^{5/2} \frac{dT}{dz} = \frac{\alpha Q}{[A_\|]_Q}, \qquad (20)$$

where $T = T_e = T_i$, Q is the total flow of energy, $[A_\|]_Q$ is given by Eq. (10) in which Δ_Q is substituted for Δ_n, and $\kappa_0 T^{5/2}$ is the thermal conductivity whose coefficient κ_0 is given by Spitzer (1962) as

$$\kappa_0 = 3.15 \times 10^2 / Z_{\text{eff}} \ln \Lambda \qquad (\text{W eV}^{-7/2} \text{ cm}^{-1}). \qquad (21)$$

Here Z_{eff} is the effective charge state of the plasma (see Hogan, Chapter 4, Section II, this volume). The parameter α is the source distribution function of the input energy and, for a uniformly distributed flow of energy into the boundary, $\alpha = z[\pi R q(r)]^{-1}$. It is apparent that the energy flux in collisional flow is strongly dependent upon T, and it is unlikely that the temperature of the scrape-off plasma [i.e., $T(0)$ in Fig. 5b] will appreciably exceed $\sim 10^2$ eV even under reactor conditions. Equation (20) also indicates that dT/dz tends to be relatively modest except in the region close to the end of the channel, where recycling causes localized cooling of the electrons and concomitant reduction in $\kappa_0 T^{5/2}$. A typical profile of $T(z)$ is sketched in Fig. 5b, and this predicted behavior is compatible with recent experimental results (Keilhacker et al., 1982).

The ready ability of the electrons to transport power throughout the boundary plasma implies that the effective channel area $[A_\|]_Q$ can be significantly smaller than that for particles $[A_\|]_n$, and it follows that the scale length Δ_Q for radial transport of energy will also be smaller than Δ_n.

Indication of the downstream density can be obtained from manipulation of Eq. (17) for conditions under which $T_i = T_e$ and both S and $A_\|$ are independent of z; this yields

$$n(0)/n(z) = [1 + \mathcal{M}(z)^2]T(z)/[1 + \mathcal{M}(0)^2]T(0)$$

and, since $\mathcal{M}(L_\|) = 1$,

$$n(L_\|) = \tfrac{1}{2}n(0)T(0)/T(L_\|). \tag{22}$$

Equations (20) and (22) provide justification for the preceding simplification that $n(z)$ tends to invariance throughout most of the length of the flow channel, as can be seen from the profile of $n(z)$, which is also indicated in Fig. 5b.

Energy which has been transported to the sheath edge leaves the plasma by convective transport to the surface because the sheath is collisionless. The convective energy flux Ψ_s through the sheath is given by

$$\Psi_s = n(L_\|)v(L_\|)\mathsf{E}_{ie} = n(L_\|)C_s\mathsf{E}_{ie}, \tag{23}$$

where E_{ie} is the energy transported to the boundary surface by each ion–electron pair that is lost from the plasma. Bombardment of the surface can release secondary electrons, but, if these are suppressed, $\mathsf{E}_{ie} \approx 6T_e/e$ for a D–T plasma in which $T_i \approx T_e$ [see Eq. (28) of Section III]. When secondary electrons are not suppressed E_{ie} increases and can reach a limiting value of about $20T_e/e$ (see Section III). Not all of this energy will be retained by the surface because a fraction is carried away by energetic recycling atoms; this aspect is discussed in Section IV.

It is apparent from Eqs. (23) and (28) that the convective flux of energy flowing to the surface tends to be proportional to $(nT^{3/2})_{L_\|}$ so that for a constant power flux an increase in plasma density is matched by a corresponding reduction in plasma temperature at the sheath edge. Thus the temperature $T(L_\|)$ can be influenced by the amount of recycling in the region $z \to L_\|$. The plasma temperature adjacent to the sheath influences the release rate of impurities from the surface in the manner described in Section IV, and it also affects the ability of the plasma to dissipate energy by means of atomic collisions. This form of impurity control (an example is described in Harrison and Hotston, 1982a) is very well suited to the divertor configuration because it has an inherently high level of localized recycling. The amount of recycling that can be sustained is sensitive to the plasma pressure balance over the full length of the scrape-off [see Eq. (22)], but it can, to varying degrees, be controlled by pumping away some of the neutral particles that

7. Boundary Plasma

escape from the plasma and reach the wall of the divertor chamber, as shown in Fig. 10 of Subchapter 6D. Indeed the feasibility of helium exhaust in a tokamak reactor is strongly dependent upon the establishment of local recycling in a region of the boundary that is accessible to external vacuum pumps. This issue is discussed by Post (Subchapter 6D).

The preceding simplified analysis is relevant to subsonic flow along a channel of uniform area but collisional flow is sensitive to changes in channel area (i.e., changes in magnetic field), and this more complex situation has been analyzed by Rognlien and Brengle (1981). Effects of collisionless flow are likely to have greatest significance at the downstream end of the channel where $z \to L_\parallel$ but, even under these conditions, electrons remain the major transporters of energy because of their high mobility in the region between the boundary sheaths. The energy flow transported by free-streaming electrons can be expressed as

$$Q_e \approx \tfrac{1}{4}\bar{n}_e A_\parallel \Omega k T_e (8kT_e/\pi m_e)^{1/2}, \tag{24}$$

where Ω is a numerical constant (≤ 2) dependent upon the distribution of electron velocities in the flow region. Collisionless flow may occur in the relatively short flow channel encompassed by the chamber of a bundle divertor where the additional effects of substantial variation in magnetic field must be taken into account (Emmert, 1980; Emmert and Bailey, 1980; Bailey and Emmert, 1980).

C. Impurity Control

It is important to know the fate of the impurities released either at the divertor target/limiter plate or from distant regions of the torus wall. Most are released from the surface as atoms, which are subsequently ionized by collisions with electrons; the charge state reached is dependent upon the residence time of the impurity ions (see Section V.D), and this in turn depends upon the efficacy with which the particular ion can be entrained in the bulk of the boundary plasma that is drifting parallel to the magnetic field. Entrainment arises predominantly from two forces; the first is dependent upon the rate at which the directed momentum of the drifting plasma is transferred to the impurity ion and the second is the electrostatic force due to long-range electric fields within the plasma upstream of the ion accelerating sheath. [In some circumstances these long-range fields may be ion repelling.[†]]. The electrostatic force is obviously dependent upon the ion charge

[†] The main components of the long-range field (dU'/dz) can be expressed as

$$en_e \left(\frac{dU'}{dz}\right) = -\frac{dP_e}{dz} - 0.71 n_e \frac{dT_e}{dz}.$$

The first term is the electron pressure gradient, which is negative in conditions of powerful recycling. The second term is significant because localized ionization, inherent to recycling, dilutes the plasma with cold electrons.

state Z_{imp}, but the rate of momentum transfer is dependent upon Z_{imp}^2. [This can be appreciated from the equipartition time for momentum given by Sivukhim (1966), and which is $\sqrt{3/2\pi}\, t_T$ where t_T is given by Eq. (59) in Section V.D.] Studies of these problems are as yet preliminary but two examples (Harbour and Morgan, 1982; Neuhauser et al., 1983) indicate that momentum transfer will dominate transport even when the long-range electric field is ion repelling. Impurities released at the divertor target–limiter plate are likely to be returned in relatively low charge states [typically $Z_{imp}(L_\parallel) = 3$ to 5], whereas impurities released at the torus wall in the region of $z < L_\parallel$ will attain higher charge states not only because of the longer length of their flow path but also due to retardation in cases where the long-range electric field is ion repelling. The velocity with which the impurity ions strike the surface at $z = L_\parallel$ depends upon their local temperature, their acceleration through the sheath potential, and finally upon their drift velocity. Impurities released at this surface return to it after gaining only a small fraction of the drift velocity of the bulk plasma (which is taken to be C_s at $z = L_\parallel$). Those impurities released from the torus wall indeed attain the relatively low drift velocity of the bulk of the plasma in regions where $z < L_\parallel$, but it is uncertain whether sufficient impulse can be delivered to them in the region $z \to L_\parallel$ where the bulk flow is strongly accelerated. On present evidence it is unlikely that these impurities will attain the plasma sound speed at the sheath edge.

The implications of the preceding discussion are that the momentum carried by the bulk flow of plasma into a divertor chamber is likely to impede the backflow of impurity ions released at the divertor target; moreover, the high level of localized recycling within the chamber tends to reduce the release rate of such impurities. Thus the divertor provides a powerful means of impurity control. The limiter does not have the full benefit of these advantages because its surface is in contact with the main plasma and some fraction of the released impurities can penetrate the main plasma. This is particularly true for atoms, which can travel from the surface, through the scrape-off plasma, and into the closed-field region. Moreover, localized recycling at the limiter is inherently less powerful than in the case of the divertor and $T(L_\parallel)$ cannot be effectively controlled by recycling. An alternate approach (see, e.g., Gibson, 1978) is to reduce $T(L_\parallel)$ by reducing the power flow transported by charged particles to the limiter; this could be achieved if a thin but powerfully radiating layer of plasma can be established in the closed magnetic field region just inboard of the limiter. The radiating layer will contain a significant concentration of impurity ions but the presence of such ions in the plasma edge must not cause unacceptable radiative power losses from the hot inner region of the main plasma. It is conceived that a self-regulating condition could be established whereby radiative power losses cause a reduction in $T(L_\parallel)$, which in turn controls the rate of release of impurities by the amount needed to sustain the appropriate concentration in

7. Boundary Plasma

the radiating layer. The condition is sensitive not only to the boundary plasma but also to the properties of radial transport and thereby lies outside the scope of this chapter. Recent examples of such studies can be found in Neuhauser et al. (1982) and Watkins et al. (1982).

III. The Sheath and Long-Range Electric Field Regions

Plasma contained in a vessel with electrically conducting walls will, in the absence of a magnetic field, assume a positive potential with respect to the walls whenever $T_e \gtrsim T_i$. The potential ensures that the ready flow of charge carried by mobile electrons to the wall is reduced (owing to the repulsive electrostatic force) so that it equals that carried by ions; (as discussed later) this condition is likely to pertain to flow parallel to the magnetic field in the boundary plasma. The net loss of charge is then zero and an electrostatic field is established in the vicinity of the wall, the extent of this field being related to the screening properties of the plasma. The scale length of the potential gradient is characterized by the Debye screening length λ_D, which can be expressed as

$$\lambda_D = (kT_e/4\pi n_e e^2)^{1/2} = 7.43 \times 10^2 (T_e/n_e)^{1/2} \quad \text{(cm)}. \quad (25)$$

The sheath potential U can be determined from Poisson's equation into which are substituted the densities of electrons and ions in the region close to the wall (see, e.g., Chen, 1974). A solution

$$U = \frac{kT_e}{2e} \ln\left(\frac{m_i}{2\pi m_e}\right) \quad (26)$$

is representative of singly charged ions that are incident upon a plane surface and uninfluenced by a magnetic field. For a "hydrogen plasma" $U \approx 3T_e/e$.

There is also a longer-range, ion accelerating region called the "presheath" over which a potential difference is established in order to impart a directed drift velocity to ions in the adjacent plasma; this ensures that the sheath region is fed with ions at such a rate that a balance of negative and positive charge density can be maintained in the electrostatic field close to the surface. In practice the situation is complex and the extent and nature of the presheath and long-range field regions are rather indeterminate—they depend upon plasma collision mechanisms (both particle–particle collisions and particle scattering by magnetic mirrors) and they are also sensitive to the distribution of recycling in the manner discussed in Section II.C. Moreover, as $z \to L_\parallel$, collisionality is less certain and the complexity of collisionless plasma flow in these regions is clearly shown by a recent analysis by Emmert et al. (1980). Nevertheless, although details of the sheath region are dependent upon specific environment, there are sufficient generalities to justify the simplistic assumption that the total effect of long-range fields can be represented by an ion accelerating potential $U' \gtrsim \frac{1}{2}kT_e/e$.

If ions have a Maxwellian distribution of velocities, then those that enter the sheath carry an amount of energy equal to $2kT_i$ because high-energy particles preponderate over those of average energy $\frac{3}{2}kT_i$. Ions of charge state Z strike the surface with an energy E_i given by

$$E_i = 2kT_i + Ze(U + U'). \tag{27}$$

Electrons in the high-energy tail of the Maxwellian distribution escape through the sheath to the surface; these electrons are retarded by the electric field and their energy distribution at the surface is the same as that of the thermal electrons within the plasma. The energy carried to the surface by each electron is thus about $2kT_e$ and the major action of the sheath is to reduce the density of plasma electrons in the flow to the surface. In the absence of secondary electrons emitted from the surface the total energy carried to the surface by each ion and the electrons associated with its charge state Z can be expressed as

$$E_{ie} = 2kT_i + Ze(U + U') + \chi_i + Z2kT_e \tag{28}$$

where χ_i is the ionization threshold energy. In the case of "hydrogen" (i.e., when $Z = 1$), Eq. (28) shows that the kinetic (i.e., thermal) energy lost from the plasma to the surface is less than that dissipated by atomic processes whenever $8kT_e \lesssim \chi_i$. In practice, this balance is effected by both the backscatter of energetic particles from the surface (Section IV) and by other energy loss processes, such as atomic line radiation and charge exchange; when these effects are taken into account the transitional regime is predicted to occur at $T_e \sim 10$ eV. If the power flow in the boundary is insufficient to ensure $T_e(L_\parallel) > 10$ eV, then the sheath temperature tends to be limited because the plasma feeds energy into the strongly recycling region close to the sheath at a rate sufficient to maintain a full ionization within the boundary layer. This scenario has recently been identified in experiments (Keilhacker et al., 1982). It also has implications upon the ability to exhaust helium gas from a reactor (Harrison and Hotston, 1982b) because the exhaust efficiency is enhanced due to the particularly low probability for ionization of helium atoms in a plasma where the electron temperature is only a few electron volts.

Release of impurity atoms is strongly dependent upon the energy of ions incident upon the boundary surface (see Section IV), and in this context it is advantageous to minimize the sheath potential. The potential is reduced if the plasma electrons are cooled because of radiative power losses, but another mechanism is the release of secondary electrons from the surface. These electrons are accelerated by the sheath potential and enter the plasma, where they can thermalize and then return to the surface. The net loss of charge from the plasma must remain zero but under these conditions the

7. Boundary Plasma

magnitude of electron current flowing through the sheath in the direction from the plasma to the surface will be greater than the accompanying flow of ion current by an amount equal to the secondary emission current. The sheath potential must thus reduce to maintain this inequality, and an expression for the potential has been derived by Hobbs and Wesson (1967):

$$U = \frac{kT_e}{2e} \ln \frac{m_i(1 - \gamma_s)^2}{2\pi m_e}, \tag{29}$$

where

$$\gamma_s = (\gamma_e + \gamma_i + j)/(1 + \gamma_i + j). \tag{30}$$

Here γ_e and γ_i are the coefficients for emission of secondary electrons by impact of electrons and ions and $j = J(nkT_e/m_i)^{-1/2}$ where J is the emission flux due to all other processes, e.g., photoemission. There is a saturation limit imposed upon γ_s due to space-charge buildup in the electric field at the surface (analogous to a space-charge limitation in a plane diode) and

$$(\gamma_s)_{\text{lim}} = 1 - 8.3(m_e/m_i)^{1/2} = 0.81$$

for hydrogen. A comparable analysis by Harbour and Harrison (1979) outlines the significance of secondary electron emission in relationship to certain fusion reactor concepts but it must be stressed that the effect is highly sensitive to the topology of the magnetic field. For example, if the field lines graze the surface many secondary electrons may be suppressed because the electron gyroradius is then smaller than the sheath thickness.

Suppression of secondary electrons due to grazing incidence of the magnetic field lines is an issue of considerable relevance to impurity control. Studies by Chodura (1982) have shown that the sheath potential difference tends to be insensitive to the inclination of the magnetic field, but the study also shows that the potential difference extends to a distance appreciably greater than λ_D when the magnetic field lies at grazing incidence. Suppression of secondary electrons does not occur if the field is inclined at greater than about 5°, but at smaller angles Igitkhanov (1982) claims that suppression becomes increasingly more significant. In a fusion reactor it is essential to ensure that the magnetic field grazes the boundary surface in order to reduce both power loading and thermal stress, and so it is reasonable to assume for such conditions that emitted electrons are suppressed. This situation is not necessarily encountered in present-day experiments, but in this context it should be noted that electron emission from surfaces of relevant metals will not generally be saturated when T_e is less than about 80 eV (see Harbour and Harrison, 1978).

Typical values of λ_D for a boundary plasma of a toroidal device are $\sim 10^{-3}$ cm and hence the sheath region is very much smaller than the scale lengths for atomic collisions and the motion of outward-going neutral particles is unimpeded by processes such as localied ionization and charge exchange.

The ion gyroradius is usually greater than λ_D and so there is a tendency for the motion of the incident ion to be oriented toward the normal to the surface. However, a detailed computation for a proton–electron plasma by Chodura (1982) indicates that the average velocity of the protons is oriented toward the direction of the magnetic field. The fate of these incident particles and of the energy that they carry is considered in Section IV.

IV. Particle–Surface Interactions

Both the surface and bulk of the wall material are involved in reactions which are caused by the incidence of charged particles, neutral particles, and photons. Moreover, in a D–T-burning fusion device, the walls will be subjected to substantial fluxes of neutrons. It is convenient to consider the various processes in respect to the magnitude of the energy interchange involved. The low-energy extreme corresponds to thermal desorption of trapped gas and also evaporation of bulk material from the surface which in the limit of localized heating may result in the formation of unipolar arcs in the electric field region of the sheath. Next follows particle- and photon-induced desorption of gas that has been trapped upon the surface, and this may in some measure be opposed by chemical binding. In the medium-energy regime, incident particles penetrate for a short distance into the solid and may return to the surface because of backscattering; moreover, the transfer of momentum to the lattice may also cause sputtering of surface atoms. While in the high-energy regime, neutrons (and to a lesser degree α particles) can initiate nuclear reactions throughout the bulk of the material. The walls may therefore be eroded by sputtering and arcing, their structural properties may be degraded due to lattice damage, and they may suffer severe local stresses due to thermal loading. The successful evolution of a reliable reactor requires simultaneous solutions to all these problems, which, with the exception of the nuclear processes, are strongly dependent upon conditions within the boundary plasma. There are, however, more immediate interests because the release of impurities and "hydrogen" gas due to plasma–wall interaction plays a significant role in present-day research. A recent and comprehensive review of both processes and data pertinent to tokamak experiments has been produced by McCracken and Stott (1979). Fortunately, the scope of this wide-ranging subject is substantially reduced if processes of a transitory nature are neglected; in effect this implies that trapped gas has been desorbed from the surface of the walls and that the deposition of plasma particles within the solid of the wall has reached a steady state. This restricted scenario is adequate to provide a basis for the subsequent consideration of atomic processes and it is thus adopted for the present discussion, which also omits consideration of arcing and nuclear processes.

A. Backscattering due to Ion and Atom Impact

Ions and atoms of moderate incident energy ($\sim 10^2$ eV) can pass through the surface of a solid and then be scattered by collisions within the lattice of the material. Some of the incident particles backscatter to the surface from which they emerge with reduced kinetic energy, whereas the remainder slow down to thermal energies and are thus trapped within the lattice. Lindhard *et al.* (1963) proposed that both the range and the energy loss of the incident particles could be characterized by a reduced energy ε given by

$$\varepsilon = \frac{M_2}{M_1 + M_2} \frac{a}{A_1 A_2 e^2} E, \tag{31}$$

where M_1, A_1 and M_2, A_2 are the mass and atomic numbers of the incident particle and target atom, respectively, and E is the incident energy. Lindhard postulated that the parameter a should be set equal to the Thomas–Fermi screening length, i.e.,

$$a = a_{FT} = 0.4685(A_1^{2/3} + A_2^{2/3})^{-1/2},$$

and, since $e^2 = 14.39$ eV Å, the reduced energy can be expressed as

$$\varepsilon = 32.55 \frac{M_2}{M_1 + M_2} \frac{1}{A_1 A_2 (A_1^{2/3} + A_2^{2/3})^{1/2}} E \quad \text{keV}. \tag{32}$$

The probability for backscattering or "reflection" of each incident particle is expressed in terms of a particle reflection coefficient R_N. Both experimental data and theoretical estimates of the scattering probability (the latter based upon Monte Carlo methods) indicate that the values of R_N plotted as a function of ε for various combinations of projectile and target lie approximately on a universal curve of the form shown in Fig. 6. The shaded region indicates the spread in data (biassed somewhat in favor of measured values) that are taken from a compilation by Eckstein and Verbeek (1979) of results for light ions (i.e., $M_1 = 1$–4 and $A_1 = 1, 2$) in the energy range below 20 keV. These data refer to particles incident normal to the surface, but they allow for the fact that emerging particles have an angular distribution which is close to cosine and which must be accounted for by integration over all angles of reflection. It is apparent that ions with low incident energy are more readily backscattered than trapped and that backscattering is greatest for targets with large atomic number. The reflection coefficient is predicted to increase with deviation from normal incidence; at grazing incidence $R_N \to 1$ but its sensitivity to intermediate angles is dependent upon the energy of the incident particle. The reflection coefficient for atoms should be similar to that for ions because the charge state of the incident particle has little significance once the particle enters the influence of the lattice system.

The emerging particles are predominantly neutral owing to electron capture within the solid, and experimental evidence indicates that the probabil-

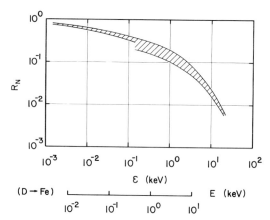

Fig. 6. Reflection coefficient R_N for atoms and ions incident normal to the surfaces plotted as a function of reduced energy ε. The shaded region illustrates the spread in data presented by Eckstein and Verbeek (1979), and ε (in keV) is determined from Eq. (32). The scale of incident energy E for D → Fe is also shown.

ity of being backscattered as an ion is almost independent of target material and depends predominantly upon the energy of the reflected particle. About 50% of the backscattered particles arising from proton impact are charged when the incident energy is 40 keV, but the proportion is appreciably less than 1% in the energy range of interest, i.e., $E < 1$ keV.

The energy distribution of the backscattered particles has been studied both experimentally and theoretically. A coefficient R_E is used to describe the fraction of the incident energy that is reflected (i.e., emerges from the surface) and it is to be expected that R_E, expressed as a function of ε, will also lie approximately upon a universal curve. A formal determination of R_E should include that component of the backscattered energy which is carried by the lattice system (i.e., by the sputtered atoms), but light projectiles produce little sputtering, so that for most processes of interest in recycling R_E approximates to the fraction of incident energy carried away by the backscattered particles. A comparison of the energy and particle reflection coefficients (i.e., R_E/R_N) is shown in Fig. 7 using data taken from McCracken and Stott (1979). The fraction of energy carried away by backscattered particles becomes greater as the incident energy is reduced because slow particles penetrate but weakly and so can lose little of their energy by collisions with the lattice.

Particles are trapped within the lattice system when they lose most of their initial kinetic energy, but the range of penetration λ_w is small, $\sim 10^{-6}$ cm. Fluxes of incident particles are high ($\sim 10^{19}$ cm^{-2} s^{-1} at the limiter or divertor target of a typical tokamak) and so the peak number density of particles within the penetration depth of the solid rapidly becomes comparable to that of the lattice atoms (e.g., the average density rises to a value

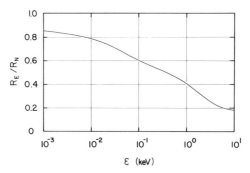

Fig. 7. Ratio R_E/R_N of the coefficients for energy and particle reflection plotted as a function of reduced energy ε. [From McCracken and Stott (1979).]

$\sim 10^{23}$ atoms cm^{-3} over a depth of $\sim 10^{-6}$ cm in about 10^{-2} s). The diffusion coefficient for hydrogen in most metals is high and the system quickly reaches equilibrium, but the gradient of particle concentration is much steeper toward the surface than into the bulk material and trapped particles diffuse much more readily toward the surface. In equilibrium, each incident particle is either backscattered or releases a particle that is trapped in the solid, there is no net retention of particles by the bulk material, and the ratio of outgoing fluxes of backscattered to detrapped particles is given by $R_N/(1 - R_N)$. The detrapping mechanisms within the lattice structure appear to be a combination of thermal release and release induced by energetic particles, but some indication of the time taken to reach equilibrium can be taken from the study of thermal release by Erents and McCracken (1970). The released flux $\Gamma_w(t)$ at time t is shown to be described by

$$\Gamma_w(t) = \Gamma_w(0) \, \mathrm{erf}[\lambda_w(4D_w t)^{-1/2}], \qquad (33)$$

where D_w is the diffusion coefficient of trapped atoms within the material. If the condition $[\Gamma_w(t)/\Gamma_w(0)] \approx 0.1$ is taken to characterize equilibrium then $[\lambda_w(4D_w t)^{-1/2}] \approx 0.1$ and substitution of typical values of D_w for metals at room temperature shows that $t \sim 10$ s.

The temperature of the detrapped particles is likely to approximate to that of the surface but there is no clear evidence as to whether "hydrogen" is released as atoms or molecules. Consideration of the surface binding energy favors the release of molecules, and this assumption is frequently made in modeling of boundary plasma conditions. However, both incident and backscattered particles are energetic and these might release a surface-bound atom before it can migrate across the surface and form a molecule. There is evidence that hydrogen trapped in high concentration on the surface is released as atoms (Taglauer *et al.*, 1978), but the surface concentration tends to be low at normal operating temperature and the assumption of molecular release is probably valid.

Details of backscattering and detrapping are likely to differ when the incident particle is an ion or atom. For normally incident particles of energy E both species give rise to a fraction $R_N(E)$ which is backscattered with an average energy $E(R_E/R_N)_E$ and a fraction $(1 - R_N)_E$ of detrapped low-energy particles. However, daughter atoms arising from charge exchange tend to impact with randomly distributed incident angles and with a velocity distribution corresponding to Maxwellian, whereas ions which have been accelerated through the sheath have a greater tendency to be unidirectional and, as can be seen from Eq. (27), they may be more closely monoenergetic.

B. *Sputtering*

Sputtering occurs because of the ejection of a surface atom which in some manner receives an impulse from an incident particle. Studies of sputtering by heavy ions are firmly established and the data are well described by the theory of Sigmund (1969). This attributes sputtering to the transfer of momentum from a projectile to atoms of the solid and the subsequent propagation of momentum to the surface by a cascade of collisions between the lattice atoms. However, this model is not valid for momentum exchange by light projectiles such as D^+ and T^+. Sputtering by light ions has recently been reviewed in detail by Roth (1980), and it suffices here to state that the sputter yield has been shown to consist of two components: one due to direct transfer of momentum from the ion to an atom at the surface and the second due to backscattered projectiles that transfer momentum while they are exiting through the surface. The first mechanism is strongly evident at grazing incidence and gives rise to the ejection of atoms into an anisotropic cone in the forward direction. The second mechanism, which is more evident at normal incidence and higher energy, tends to display a cosine-type distribution of ejected atoms. The sputter yield (atoms per ion) is greatest at angles close to grazing incidence, where its value can be as high as 20 times that at normal incidence.

There is an identifiable threshold energy below which insufficient energy is transferred to the lattice atoms for sputtering to occur. This threshold E_{th} can be related to the sublimation energy of the solid E_w by

$$E_{th} \approx E_w/\zeta(1 - \zeta), \qquad (34)$$

where

$$\zeta = 4M_1M_2/(M_1 + M_2)^2.$$

A recent compilation of data for sputtering by light ions at low energy (Roth *et al.*, 1979) lists values of E_{th}, for example: H → Al = 53 eV, H → C = 9.9 eV, H → Fe = 64 eV, H → Mo = 164 eV, H → Ni = 47 eV, H → Ta = 460 eV, H → Ti = 43.5 eV, H → V = 76 eV, and H → W = 400 eV. These

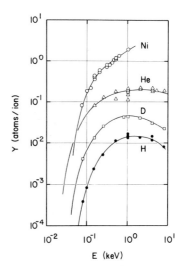

Fig. 8. Sputtering yield Y plotted as a function of energy E for ions at normal incidence upon nickel. [Data for incident H^+, D^+, He^+, and Ni^+ are taken from Roth et al. (1979).]

correspond to somewhat lower values of E_w than are provided by thermal data; the difference is attributed to destruction of the crystalline structure of the solid by ion bombardment which creates an amorphous surface. Assessment of data for low-energy H^+, D^+, $^3He^+$, and $^4He^+$ impact upon a wide range of targets has resulted in a universal curve for the sputtering yield Y at normal angle of incidence,

$$Y = 6.4 \times 10^{-3} M_2 \zeta^{5/3} (E/E_{th})^{1/4} (1 - E_{th}/E)^{7/2} \quad \text{atoms/ion}. \quad (35)$$

The expression describes the sputter yield to within a factor of 2 when $M_1/M_2 \leq 0.4$ and $1 < E/E_{th} < 20$.[†] It is to be expected that Eq. (35) is equally applicable to sputtering by incident atoms, but the differences in incident angle and energy discussed in relationship to backscattering will also apply to sputtering.

The magnitude of Y is illustrated in Fig. 8 for a number of incident ions upon a nickel target. [The data are taken from Roth et al. (1979); this particular target is selected for illustration because there are also measured data for the self-sputter yield of Ni^+ on N.] Curves through the data points show the fit of the energy dependence of Eq. (35). It is apparent that the yields for ions of a "hydrogen" plasma are not substantial in the energy range appropriate to the boundary plasma (e.g., Y is a few percent for $E < 500$ eV), but the corresponding yield for self-sputter, Y_{self}, can exceed unity. When the self-sputter yield tends to unity there is the possibility of a self-sustained condition in which sputtering occurs predominantly through bombardment by ions of its own erosion products, the energy to support this process being drawn from the plasma by a variety of collision processes. It is

---

[†] A more precise expression has recently been evolved by Bohdansky (1982).

convenient when modeling boundary conditions to define a rather arbitrary value for an effective sputter coefficient Y' such that

$$Y' = Y_{\text{(plasma ion)}}/(1 - Y_{\text{self}}) \qquad (Y_{\text{self}} < 1). \tag{36}$$

It should be noted that Y_{self} is strongly dependent upon incident energy, which in turn depends upon the sheath potential, the charge state, and the drift velocity of the recycling ion of the target element; the significance of atomic processes in determining this charge state is evident from the discussion in Section V.D. Unfortunately there is at present a scarcity of experimentally verified atomic data relevant to wall materials, but crossed-beam studies of some ionization cross sections have been reported (see Montague et al., 1983).

Atoms ejected by sputtering have quite low velocities; the velocity distribution of Fe atoms sputtered by normally incident 10-keV D^+ ions (Elbern et al., 1978) is shown in Fig. 9. The peak velocity is equivalent to an energy of 2.2 eV and the shape of this distribution, and also of similar distributions available from the limited amount of data, has been likened to that predicted by the model of Thompson (1968), which describes heavy-ion sputtering. However, this similarity is unlikely to extend to high ejection velocities because of the inefficient transfer of momentum between the low mass incident ions and the heavier atoms of the bulk material.

Boundary surfaces are also eroded by chemical-sputtering, which occurs when the incident particle becomes chemically bound to one or more surface atoms and returns to the plasma in the form of a molecular compound. Such sputtering is particularly evident when the bombarding plasma contains oxygen impurities or when a carbon surface is bombarded by a "hydrogen" plasma. Chemical sputtering tends to be sensitive to the surface temperature

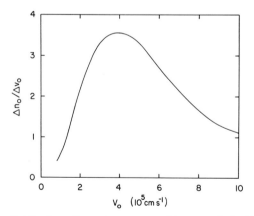

Fig. 9. Velocity distribution of iron atoms sputtered by normally incident 10-keV D^+ ions. [Data are from Elbern et al., 1978 and are adapted (in the manner described in Roth, 1980) to illustrate the distribution of velocity flux.]

7. Boundary Plasma

because of the relatively weak strength of the chemical binding energy. A recent study of the temperature dependence of the sputtering yields of graphite bombarded by both hydrogen and helium (Roth *et al.*, 1982) shows a marked enhancement of the yield at high temperature (i.e., >1000 K). This particular high temperature effect is attributed to the enhanced evaporation of graphite atoms activated by ion bombardment; its magnitude is so large that serious doubts have now been raised as to the feasibility of graphite limiters or divertor targets in a tokamak reactor.

Erosion due to sputtering is in some measure compensated by redeposition arising from transport of ions of the sputtered material. Details of the process are linked to the transport of impurities described in Section II.C, and numerical analysis based upon trajectory tracing in a simplified model of the boundary plasma has been undertaken by Brooks (1982). This and also less complex treatments indicate that many of the atoms will return to the surface of a limiter (or divertor target), although some displacement of the surface material is inevitable. The bulk properties of material redeposited in the boundary of the plasma is at present a matter of conjecture.

V. Atomic Processes in the Boundary Plasma

Atomic reactions in the boundary regions are predominantly due to free–bound collisions. The most prolific bound species is "hydrogen" (either atomic or molecular) but there are also smaller concentrations of impurity atoms and ions. Low-mass elements such as oxygen and carbon are typical of the products that arise from detrapping of surface gases; higher-mass components are present owing to erosion of the containment vessel, e.g., iron from the stainless-steel walls and tungsten (or other refractory metals) from the limiters. In a D-T burning reactor there will also be helium. The initial velocity of most of the atoms of these elements in the boundary region is governed by surface interactions, and the collision rate with charged plasma particles can be expressed as

$$K = nn_0 \langle \sigma \bar{v} \rangle, \tag{37}$$

where n and n_0 are, respectively, the density of the plasma and the neutral species and $\bar{v}$ is their relative velocity. The cross section for the process is σ and $\langle \sigma \bar{v} \rangle$ is averaged over the distribution of the collision velocity, which is assumed to be Maxwellian. Collisions with electrons are generally most frequent because of their comparatively high velocity, but symmetric, resonant charge exchange, such as

$$H + H^+ \rightleftharpoons H^+ + H \tag{i}$$

can also be important because of the large magnitude of its cross section at low collision velocities. However, there are but a few other processes involving ion collisions that are likely to be significant in the boundary region.[†] The plasma can be regarded as transparent to most radiation emitted by free–bound transitions and so photon-induced reactions can be neglected.

Electron collisions remove thermal energy from the plasma by ionizing and by inducing radiative transitions in "hydrogen" and in atoms and unstripped ions of impurity elements. In principle, radiation can also be produced by electron–ion recombination. However, the electron density is insufficient to support three-body processes and, if the atomic reactions are localized to regions where the plasma drifts parallel to open magnetic field lines, it can be argued that the characteristic time for two-body recombination,

$$\tau_\alpha = [n_e \alpha(T_e)]^{-1}, \tag{38}$$

where $\alpha(T_e)$ is the recombination coefficient, is appreciably greater than the drift time $\tau_\parallel = L_\parallel v_\parallel^{-1}$. This is examined here for the case of partially stripped ions of oxygen, which recombine predominantly by dielectronic processes in plasma environment typical of the boundary, e.g., $T_e = 30$ eV and $n_e = 10^{13}$ cm^{-3}. Recombination coefficients (see the data in Summers, 1974) are of the order 10^{-11} cm^{-3} s^{-1} so that $\tau_\alpha \not< 10^{-2}$ can be taken as representative. If it is assumed that these ions drift with a velocity comparable to the main D–T plasma flow ($v_\parallel$ for D–T is here taken to be $0.3C_s$), then the drift length corresponding to the scale length for recombination is about 1.5×10^4 cm and so is greater than the likely length of the flow channel. This fact supports the contention that volume recombination of impurity ions is insignificant compared to electron capture processes at the boundary surfaces. The case is more extreme for "protons" and hydrogenlike ions because these cannot undergo dielectronic recombination and, at the moderate densities of the boundary environment, the only significant alternative process is two-body radiative recombination, for which the rate coefficients are small. Proton recombination can generally be neglected in all regions, but any impurity ions that penetrate inward to the region of closed magnetic field will reside in the plasma for a relatively long time and will take part in recombination events. However, on the basis of the discussion in Section II.C, it is evident that only a small fraction of the total number of impurity atoms released in the boundary environment will enter the closed-field region.

A. Energy Losses due to Collisions of Electrons with "Hydrogen" Atoms

The arguments so far evolved indicate that the most significant atomic processes associated with electron collisions with "hydrogen" atoms are

---

[†] The situation is more complex when energetic neutral beams are used for plasma heating because these fast atoms can charge exchange with impurity ions within the boundary plasma and the rate coefficient for charge exchange is significantly large.

7. Boundary Plasma

likely to be

(a) Excitation of an atom from level p to an upper level q,

$$H(p) + e \rightarrow H(q) + e, \tag{ii}$$

and this may be followed by spontaneous radiative decay,

$$H(q) \rightarrow H(p) + h\nu. \tag{iii}$$

An alternative reaction is collisional deexcitation, which is the reverse of (ii); this is a superelastic process which returns energy to the colliding electron.

(b) Ionization of an atom in level p,

$$H(p) + e \rightarrow H^+ + e + e, \tag{iv}$$

but the reverse reaction, three-body recombination, can be neglected in the boundary, where the electron density is modest and the temperature relatively high.

The implications of these various processes are discussed in detail in McWhirter and Summers (Chapter 3, this volume) where they are interpreted in the light of work by Bates *et al.* (1962), Bates and Kingston (1963), McWhirter and Hearn (1963), and Hutcheon and McWhirter (1973). Suffice it to say here that the arguments are concerned with the fact that electron collisions cause ionization either directly by transferring a bound electron to the continuum, or indirectly through a sequence of level transitions which terminates in the continuum. The cross sections for such level transitions can be large (i.e., proportional to the fourth power of the principal quantum number), so, even at moderate electron density, the collision time

$$\tau_p = (n_e \langle \sigma_p v_e \rangle)^{-1} \tag{39}$$

becomes shorter than the radiative lifetimes of all but the lowest-lying excited states. The number of radiating channels is thereby reduced in favour of (a) ionization by a chain of nonradiative upward transitions and (b) repopulation of the ground state by a complementary chain of nonradiative downward transitions; this chain of deexciting collisions is shown to have a particularly powerful effect. The analysis, which is applicable to the boundary plasma because τ_p is appreciably less than the particle residence time, shows that the ionization rate can be expressed as

$$K_i = n_e[n(g)S_{\text{C-R}} - n_e \alpha_{\text{C-R}}], \tag{40}$$

where $n(g)$ is the ground state density and $S_{\text{C-R}}$ is a composite coefficient (called the "collision–radiative ionization coefficient") which allows for the processes outlined above. [$S_{\text{C-R}}$ is compared with the ground state coefficients $S_i(g)$ in Fig. 12.] Equation (39) also includes a "collisional–radiative recombination" term which can generally be neglected in the boundary plasma.

The preceding concepts lead to the view that electrons colliding with a partially ionized "hydrogen" plasma lose energy by ionization and by radiation from a few low-lying radiative levels, and the average amount of energy ξ_i expended in producing one proton–electron pair must include contributions from both processes. McWhirter and Hearn (1963) have evaluated ξ_i for hydrogenlike ions under conditions where recombination is insignificant (i.e., $T_e \gtrsim 2$ eV); the expression used in the case of atomic hydrogen is

$$\xi_i = (\chi_i S_{C-R} + P_1 n_e^{-1})/S_{C-R}, \qquad (41)$$

where $P_1(T_e, n_e)$ is a coefficient that allows for radiative power loss from ground state atoms. The data appropriate to hydrogen[†] (i.e., to $Z = 1$) are shown as a function of T_e and for $n_e = 10^{12}$, 10^{13}, and 10^{14} cm^{-3} by the solid lines in Fig. 10. However, this analysis is directed toward electron–hydrogenic-ion collisions and so employs Coulomb–Born cross sections, which overestimate excitation and ionization in electron–H-atom collisions except at high T_e. To show the significance of this effect data for H atoms, S_{C-R} (from Bates et al., 1962) and P_1 (from Bates and Kingston, 1963) have been substituted into Eq. (41) and the results plotted as circles in Fig. 10. It is apparent that there is little difference except at low T_e and high n_e, where the use of ion-type cross sections results in an underestimation of ξ_i for H atoms. The value of ξ_i tends to be independent of T_e in the higher-temperature regime but, at lower temperatures, ξ_i increases owing to the enhancement of radiative losses which occur because fewer electrons have sufficient energy to produce direct ionization. However, this enhancement is reduced by increasing electron density due to the suppression of radiative channels and the consequent enhancement of superelastic collisions; indeed, at high density (not shown in Fig. 10) ξ_i tends to the ionization threshold energy $\chi_i = 13.6$ eV. An independent assessment of the energy dissipated in ionization at high energies has been obtained by Dalgarno and Griffing (1958) for monoenergetic electrons passing through a target of ground state H atoms. These data, which are shown by dashed curves, support the general trends in ξ_i but, because collisional–radiative effects are not allowed for, the ionization energy tends to be slightly higher. A numerical algorithm of the data of McWhirter and Hearn (1963) is given in Harrison et al. (1982, 1983).

It seems reasonable in the boundary plasma to accept that ξ_i is likely to lie in the range 25–30 eV when $T_e \gtrsim 20$ eV. It is also worth noting in the context of the boundary plasma that each ionizing event involved in the recycling of "hydrogen" atoms corresponds to an amount of energy $\xi_i - \chi_i$ extracted from the electrons and radiated from the plasma and that the additional amount χ_i is transferred from the electrons and stored as potential (not kinetic) energy in the ions. The initial energy of the ions corresponds to that

---

[†] The scaling factors for hydrogenic ions of charge state Z are $[T_e]_z \equiv Z^2 T_e$, $[\xi_i]_z \equiv Z^2[\xi_i]_H$, and $[n_e]_z \equiv n_e Z^{-7}$.

7. Boundary Plasma

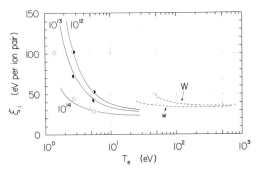

Fig. 10. Average electron energy ξ_i dissipated in producing one proton–electron pair in atomic hydrogen plotted as a function of electron temperature T_e. Circles show data obtained from Bates et al. (1962) and from Bates and Kingston (1963); solid lines represent data from McWhirter and Hearn (1963) for hydrogenlike ions with $Z = 1$. Results are given for $n_e = 10^{12}$, 10^{13}, and 10^{14} cm^{-3}. Data from Dalgarno and Griffing (1958) are illustrated by dashed curves. Secondary electrons are assumed to contribute to ionization in the data shown by curve w, whereas this contribution is not allowed for in curve W.

of their parent atoms whereas most of the ejected electrons have low energy (a few eV being typical). Subsequent Coulomb collisions with the bulk plasma tend to bring the energies of the recycled electrons and ions into equilibrium with the plasma temperature and drift velocity but the ability for equilibration is dependent upon local plasma conditions in the manner discussed in Section II.C.

B. Effects of Molecular Hydrogen

Electron collisions with molecular hydrogen give rise to ionization and radiative reactions and also to the formation of "protons" and "hydrogen" atoms by dissociation. The situation has received relatively little attention in the context of fusion plasmas and is rendered particularly complex by the powerful effects of vibrational excitation upon reaction rates at low electron temperature. Radiative mechanisms are neglected in the following discussion, which is directed to processes that lead to the trapping of "protons" and to the formation of "hydrogen" atoms within the boundary plasma. The most significant reactions associated with neutral "hydrogen" molecules are

(a) dissociation:

$$H_2 + e \rightarrow H + H + e \quad (S_2^0 a),$$

(b) ionization:

$$H_2 + e \rightarrow H_2^+ + e + e \quad (S_2^0 b),$$

(c) dissociative ionization:

$$H_2 + e \rightarrow H^+ + H + e + e \quad (S_2^0 c).$$

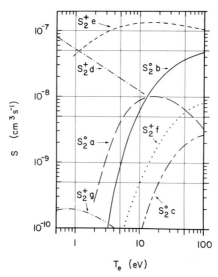

Fig. 11. Rate coefficients for electron collisions in H_2 and H_2^+ plotted as a function of electron temperature T_e. Rate coefficients $S_2^0 a$, $S_2^0 b$, and $S_2^0 c$ refer to collisions between electrons and H_2 molecules, whereas $S_2^+ d$, $S_2^+ e$, $S_2^+ f$, and $S_2^+ g$ relate to collisions with H_2^+ molecular ions. The reactions a, b, etc., are described in the text.

Experimental data for neutral hydrogen molecules, which were probably in their ground vibrational state, are available for these reactions and a compilation of cross sections together with the corresponding rate coefficient (e.g., $S_2^0 a$, $S_2^0 b$, and $S_2^0 c$) can be found in Jones (1977). Rate coefficients are shown in Fig. 11, where it is evident that $S_2^0 c$ is appreciably less than $S_2^0 b$ at all electron temperatures, so direct ionization of neutral H_2 yields predominantly H_2^+. The total coefficient for ionization $S_2^0 b + S_2^0 c$ exceeds that for dissociation into atoms, $S_2^0 a$, when T_e is in excess of about 10 eV, but dissociation becomes progressively more dominant at lower temperature.

The distribution of vibrational states among H_2 molecules released from the walls by particle impact has not yet been studied in the energy regime relevant to the boundary of a tokamak but it is known that H_2^+ produced by ionization of unexcited H_2 has a distribution of vibrationally excited levels which is closely determined by the appropriate Frank–Condon factors (see, e.g., Dunn, 1966). It has also been shown that this distribution is in general accord with measured cross sections (see, e.g., Peart and Dolder, 1972) and, moreover, the distribution does not seem to be grossly distorted by subsequent collisions of H_2^+ with plasma particles (see discussion in Dance et al., 1967). Processes which involve collisions with electrons and H_2^+ ions are

(d) dissociative recombination:

$$H_2^+ + e \rightarrow H + H \quad (S_2^+ d),$$

(e) dissociative excitation:

$$H_2^+ + e \to (H_2^+)^* + e \to H^+ + H + e \quad (S_2^+ e),$$

(f) dissociative ionization:

$$H_2^+ + e \to H^+ + H^+ + e + e \quad (S_2^+ f),$$

(g) dissociative attachment:

$$H_2^+ + e \to H^+ + H^- \quad (S_2^+ g).$$

Experimental data have been reviewed by Dolder and Peart (1976), and the rate coefficients shown in Fig. 11 are taken from Jones (1977) with the exception of $S_2^+ e$ and $S_2^+ g$, which have been calculated specifically for this discussion.[†] If $T_e \gtrsim 5$ eV it is evident from Fig. 11 that dissociative excitation of H_2^+ provides by far the most powerful route for destruction of the H_2^+ ions formed by electron collisions so that the molecular ions are rapidly dissociated into protons and H atoms. Additional contributions to proton production from H_2 (i.e., $S_2^0 c$) and from H_2^+ (i.e., $S_2^+ f + S_2^+ g$) are relatively small. Furthermore, comparison of $S_2^+ d$ and $(S_2^+ e + S_2^+ f)$ shows that destruction of H_2^+ by dissociative recombination is likely to be significant only at very low electron temperatures; a similar limitation is appropriate to dissociative attachment (i.e., $S_2^+ g$) and so both processes are neglected in the present context. Measured cross sections for D_2^+ are presently available only for reaction (d); there are no data for D–T, which may have different vibrational properties because of its molecular asymmetry; nor are there data for T_2.

In the light of the preceding inequalities it is reasonable to express the rate coefficient for proton production from H_2 molecules in the boundary plasma, $S_2^0(H^+)$, as

$$S_2^0(H^+) \approx S_2^0 b \left[(S_2^+ e + 2S_2^+ f)/(S_2^+ e + S_2^+ f) \right] + S_2^0 c \qquad (42)$$

and the corresponding coefficient $S_2^0(H)$ for atom production as

$$S_2^0(H) \approx 2S_2^0 a + S_2^0 c + S_2^0 b \, S_2^+ e /(S_2^+ e + S_2^+ f). \qquad (43)$$

Within the boundary plasma the electron–H_2 collision time

$$\tau(H_2) \approx (n_e S_2^0 b)^{-1} \qquad (44)$$

will generally be greater than the atomic level relaxation time τ_p, so equilibrium conditions comparable to those in isolated atoms are likely to be established. Unfortunately there are as yet no data for the collisional–radiative behavior of the molecule and so in modeling of boundary conditions (see, e.g., Harrison et al., 1982) the molecular contribution to radiation is taken to be equal to that of two isolated atoms.

[†] Jones assumed when calculating $S_2^+ e$ that H_2^+ is not vibrationally excited; both theory and experiment indicate that this is most unlikely and so $S_2^+ e$ shown in Fig. 11 is determined directly from the measured cross sections. It can thus be compared with the other rate coefficients which also relate to measured cross sections.

Collisions involving H_3^+ are neglected because the rate coefficient for the reaction,

$$H_2^+ + H_2 \rightarrow H_3^+ + H, \quad (v)$$

does not exceed 10^{-9} cm^3 s^{-1}, which is about 10^2 times less than the rate coefficient for the destruction of H_2^+ by electron collisions. In this respect the boundary plasma differs appreciably from the plasma in an ion source, where the density of H_2 can be enhanced owing to the enforced feed of neutral molecular gas.

The concentration of molecules in the boundary is governed predominantly by the response of the boundary surface to impacting "protons" and "H" atoms. If it is assumed that the probability of molecular formation is proportional to $1 - R_N$ then it is apparent (see Section IV.A) that the source of molecules is weak in a low-temperature boundary plasma but strong in a high-temperature boundary because R_N decreases with increasing impact energy. In general, about one-half of the dissociation products from these molecules becomes trapped in the plasma as "protons" and the other half consists of atoms that engage in a subsequent sequence of charge exchange and ionizing collisions.

C. Recycling and Trapping of "Hydrogen" Atoms

The significance of resonant, symmetric charge exchange for the scattering of hydrogen atoms by proton impact prior to electron impact ionization has been briefly alluded to in Section I. Motion of these atoms in the boundary can be assessed in a comprehensive manner by using Monte Carlo techniques which take account of the changes in both direction and energy that result from collisions of atoms with surfaces and with plasma "protons" (see, e.g., Heifetz et al., 1982; Cupini et al., 1982). An alternative approach is to use analytical methods to describe a "random walk" (Harrison et al., 1983), but a much simpler description is employed here to identify the role played by atomic processes. Consider two conditions, an initial one (a), where an atom has a velocity v_0 characteristic of its release from boundary surface and a subsequent one (b), where the velocity has been changed by collision with a plasma proton so that the daughter atom corresponds to a randomly directed particle with thermal velocity v_{th}. The scale length for motion in condition (a) is characterized by a length Δ_a given by

$$\frac{1}{\Delta_a} = \frac{1}{\lambda_{cx}^a} + \frac{1}{\lambda_i^a}, \quad (45)$$

where the mean free path for charge exchange is

$$\lambda_{cx}^a = v_0/n_i\langle\sigma_{cx}\bar{v}_{0p}\rangle$$

7. Boundary Plasma

and the corresponding mean free path for electron impact ionization is given by

$$\lambda_i^a = v_0/n_e\langle\sigma_i\bar{v}_{0e}\rangle.$$

Here $\bar{v}_{0p}$ is the relative velocity of atom–"proton" collisions and $\bar{v}_{0e} \approx \bar{v}_e$. It is convenient to introduce the parameter

$$G_a = \lambda_{cx}^a/\lambda_{ei}^a = S_{C-R}^a/S_{cx}^a, \tag{46}$$

so that Eq. (45) reduces to

$$\lambda_a = \lambda_{cx}^a/(1 + G_a). \tag{47}$$

Motion in condition (b) can be treated as a diffusive problem in which the diffusion step length corresponds to the charge exchange mean free path λ_{cx}^b. This is given by

$$\lambda_{cx}^b = v_{th}/n\langle\sigma_{cx}\bar{v}_b\rangle \tag{48}$$

where $\bar{v}_b$ is the relative collision velocity, and for atoms of mass m_0 in a homogeneous plasma of ion mass m_i it is given by

$$\bar{v}_b = \left(\frac{8kT_i}{\pi}\frac{m_0 + m_i}{m_0 m_i}\right)^{1/2}. \tag{49}$$

Diffusion is considered only in one direction, so the scale length Δ_b of diffusive transport can be expressed as

$$\Delta_b \approx \left(\frac{N_b}{3}\right)^{1/2}\lambda_{cx}^b. \tag{50}$$

Here N_b is the average number of charge exchange collisions prior to ionization; it is equal to the ratio of collision times, namely,

$$N_b \approx \frac{\tau_{ei}^b}{\tau_{cx}^b} \approx \frac{S_{cx}^b}{S_{C-R}^b} \approx \frac{1}{G_b}. \tag{51}$$

Substitution for N_b in Eq. (51) yields

$$\Delta_b \approx (1/3G_b)^{1/2}\lambda_{cx}^b. \tag{52}$$

Consider now a homogeneous plasma that extends infinitely from a surface and consider motion only in the direction x normal to the surface. The flux Γ^a of condition (a) atoms at x is given by

$$\Gamma^a(x) = \Gamma^a(0)\exp(-x/\Delta_a),$$

so that, in an element of extent dx at x, there is a rate

$$K_{(x)}^a \approx \Gamma_{(x)}^a\, dx/\lambda_{cx}^a$$

of scattering events. These cause atoms to move with equal probability either more deeply into the plasma (where they are assumed to be trapped by ionization) or else backward toward the surface. The atoms returning from dx at x are in condition (b) and their flux at the wall is given by

$$\Gamma^b(x) \approx \tfrac{1}{2} K^a(x) \exp(-x/\Delta_b).$$

Manipulation and integration yield

$$\frac{\Gamma^b}{\Gamma^a} \approx \tfrac{1}{2} \frac{1}{\lambda_{cx}^a} \frac{\Delta_a \Delta_b}{\Delta_a + \Delta_b}, \qquad (53)$$

where the ratio Γ^b/Γ^a represents the fraction of released atoms that recycle back to the surface.

Equation (53) can be expressed as

$$\Gamma^b/\Gamma^a \approx \tfrac{1}{2}[1 + G_a + (3G_b)^{1/2} \lambda_{cx}^a/\lambda_{cx}^b]^{-1}, \qquad (54)$$

and data pertinent to G_b are shown in Fig. 12, namely charge exchange rate coefficients S_{cx}^b for D + D, T + T, and D + T collisions[†] and also the electron ionization coefficient $S_i(g)$; the latter is insensitive to isotope effects because $\bar{v}_{0e} \approx \bar{v}_e$. Data for S_{C-R} are also shown, but effects of collisional–radiative ionization are most significant at low electron temperature and are neglected in the present discussion where it is assumed that $S_{C-R}^a = S_{C-R}^b = S_i(g)$. The ratio G_b(D–T) is also plotted, and it is apparent that G_b is rather insensitive to temperature in the regime of the boundary plasma (i.e., 20–100 eV) and that $G_b \approx 0.6$ can be taken as typical. The parameter G_a cannot differ appreciably from G_b because, first, $\bar{v}_{op} \not< v_{th}$ and, second, $\bar{v}_{op}$ for recycled atoms cannot be grossly in excess of $\bar{v}_b$. If it is assumed that $G_a \to G_b \to 0.6$ and by analogy that $S_{cx}^a \to S_{cx}^b$, it follows that Eq. (54) can be reduced to

$$\Gamma^b/\Gamma^a = 1/2(1.6 + 1.3 v_0 C_i^{-1}). \qquad (55)$$

A typical value of Γ^b/Γ^a appropriate for fast backscattered atoms (i.e., $v_0 \to v_{th}$) is therefore about 0.17, but this increases as the atom release velocity is reduced to a limiting value of about 0.3 when $v_0 \to 0$. These values of Γ^b/Γ^a should be regarded as lower limits because the plasma will not be homogeneous, nor is it valid to assume that all inward-scattered atoms are trapped in the plasma. It should also be noted that the situation is somewhat different when the boundary regions are penetrated by fast atoms during neutral-beam heating of the central plasma.

D. Collisions of Electrons with Impurity Ions

To determine radiative power losses associated with the ionization of impurity atoms released from boundary surfaces it is necessary to know the

---

[†] Several plasma models have incorporated charge exchange data taken from Freeman and Jones (1974), but it should be noted that these data refer to H + H⁺ collisions in which $v_0 = 0$. The D + D⁺ case illustrated above for $\langle \sigma v_b \rangle$ is, as can be seen from Eq. (49), numerically equal to that for H + H⁺ when $v_0 = 0$. Analytical expressions used here for σ_{cx} are taken from Riviere (1971).

7. Boundary Plasma

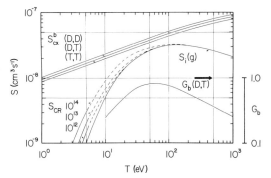

Fig. 12. Rate coefficients for charge exchange and electron impact ionization plotted as a function of plasma temperature T. The influence of isotope mass upon charge exchange collisions between plasma ions and thermalized atoms is shown by S_{cx}^b plotted for D–D, T–T, and D–T. $S_i(g)$ is the electron ionization rate coefficient for ground state "hydrogen atoms" and the ratio $G_b = S_i(g)/S_{cx}^b(\text{D–T})$ is also shown. S_{C-R} is the "collisional–radiative rate coefficient" for a hydrogen plasma of electron density $n_e = 10^{12}$, 10^{13}, and 10^{14} cm s^{-1}; the solid lines show the data of Bates and Kingston (1963) and the dashed lines show the present extrapolation to higher temperatures.

history of the impurity species within the plasma. It can in general be assumed that each atom is ionized by electrons within the adjacent plasma[†] but the residence time of the ion τ_{imp} will depend upon plasma and surface parameters as well as the atomic nature of the impurity species. Each stage of ionization is accompanied by a characteristic loss of radiated energy ξ_Z^r, and, if ionization proceeds in a stepwise manner, it is possible to express the total loss of energy ξ_{imp} associated with the ionization cycle of each released atom as

$$\xi_{imp} = P_{01}(\xi_0^r + \chi_0) + P_{01}P_{12}(\xi_1^r + \chi_1) + P_{01}P_{12}P_{23}(\xi_2^r + \chi_2) + \cdots \quad (\text{eV/atom}) \quad (56)$$

where the upper limit of charge state is determined by the residence time or by the fully stripped condition of the ion. The parameters P_{01}, P_{12}, etc., denote the probability that the impurity atom will pass through that particular stage of ionization during its time within the plasma and, as already implied, $P_{01} = 1$ is generally a reasonable assumption. If the residence time τ_{imp} is known then

$$P_{Z \to Z+1} = 1 - \exp(-\tau_{imp}/\tau_i)_Z \quad (57)$$

can be used to determine the probabilities in Eq. (56).

[†] This generalization is not fully valid in certain boundary regions which are specifically contrived to favor the exhaust of neutral gas. It is particularly desirable to exhaust neutral helium from a fusion reactor and this aspect is favored because $S_i(g)$ for He is relatively small and the probability for charge exchange of helium atoms in a D–T plasma is low. This aspect is discussed in Post Subchapter 6D, Section III).

It is accepted that volume recombination is negligible and so the energy ξ_Z extracted from the plasma electrons during each stage of ionization can be determined from

$$\xi_Z = \chi_Z + [P_{LZ}(n_e, T_e)/S_{C\alpha}(n_e, T_e)]_Z \qquad (58)$$

where P_{LZ} is a "line-radiated power loss coefficient" and $S_{C\alpha}$ is the "collisional dielectronic ionization coefficient" which is analogous to S_{C-R} discussed in Section V.A.

It can be seen from Section II.C that it is not possible to provide a general statement about τ_{imp}; nevertheless, to illustrate the mechanisms of impurity cooling it is here assumed that τ_{imp} is comparable to the time required for the released ion to thermalize because of ion–ion collisions within the drifting stream of D–T plasma that intercepts the surface from which it has been released. It is argued that this equilibrium time t_T is indicative of the time needed for the impurity ion to be entrained in the D–T ion flow and then to be carried by the flow back to the surface. Spitzer (1962) has expressed t_T as

$$t_T = \frac{7.34 \times 10^6 MM'}{n' Z^2 Z'^2 \ln \Lambda} \left(\frac{T_i}{M} + \frac{T_i'}{M'}\right)^{3/2} \quad (s), \qquad (59)$$

where the lack of a superscript denotes a "test" particle (i.e., the impurity ion) that moves through an assembly of "field" particles (i.e., the D–T plasma) whose parameters are denoted by dashes. The residence time is obviously dependent upon the release velocity of the impurity atom and, to illustrate the sensitivity to this effect, T_i in Eq. (59) is set equal to zero to represent low velocities of release or to $T_i = 4ZT'$ to simulate high velocities. This latter condition corresponds to the backscattering of atoms arising from incident impurity ions of charge state Z (justification of the values used can be found in Sections III and IV.A); in brief,

$$T_i \approx (R_E/R_N)[(2kT_i') + (4kT_e')Z],$$

where $T_e' = T_i'$ and $R_E/R_N \approx 0.7$.

Values of $P_L(n_e, T_e, Z)$ and $S_{C\alpha}(n_e, T_e, Z)$ have been determined by Summers[†] for some common impurity species; the case of oxygen has been selected for the present discussion on the grounds (a) that it is an important impurity species and (b) that detailed data are available for low charge states (the latter situation is not generally the case for metallic elements such as iron). The parameters in Eq. (58) are almost invariant with density over the range 10^{12}–10^{14} cm^{-3}, so that the analysis, which was evaluated for $n_e = 10^{13}$

[†] The work is described in Summers and McWhirter (1979) but the detailed printout has been used here to extract values of $P_L(n_e, T_e, Z)$ and $S_{C\alpha}(n_e, T_e, Z)$. These calculated values of $S_{C\alpha}$ may be compared with coefficients determined from crossed-beam experiments (see the compilation Bell et al., 1982).

7. Boundary Plasma

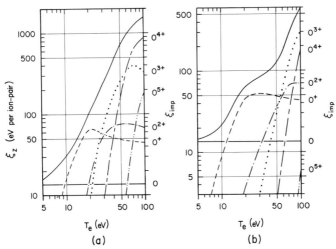

Fig. 13. Average electron energy ξ_z dissipated in producing an oxygen ion of charge state Z within the boundary plasma plotted as a function of electron temperature T_e. Data for line radiative energy losses are taken from Summers (see Summers and McWhirter, 1979), and the residence time of each ion charge state within the plasma is assumed equal to the thermalization time in ion–ion collisions given by Eq. (59). The energy dissipated in the formation of each charge state is shown together with the total dissipation ξ_{imp} for the complete ionization cycle. (a) Data for atoms released with an energy equivalent to $T_i = 4ZT'_i$; (b) equivalent data for $T_i \to 0$.

cm^{-3}, is applicable over the range of interest to the boundary plasma. Results are shown in Fig. 13a for $T_i = 4ZT'_i$ and Fig. 13b for $T_i = 0$. It is clear that ξ_{imp} is a very powerful function of plasma temperature, changing by almost two orders of magnitude between 5 and 100 eV, but, in this particular model, the change in atom release velocity has a relatively minor effect; i.e., the total energy loss is about halved if the released atoms have low velocity. Contributions from each charge state up to $Z = 5$ are shown and it is apparent that the effects of charge states greater than $Z = 3$ are not likely to be significant when $T_e \lesssim 30$ eV. Moreover, this conclusion is rather insensitive to the selection of reasonable values of τ_{imp}, and it might be assumed that the energy generally dissipated by ionization of oxygen atoms that recycle in the vicinity of a surface in a low-temperature boundary plasma is about two to three times that associated with "hydrogen" atoms. However, at higher temperatures there is a substantial increase in losses arising from charge states with $Z > 2$, so a conclusion becomes rather sensitive to the assumptions made regarding τ_{imp}. Nevertheless, the preceding results indicate the broad characteristics of the loss processes.

The preceding simplified discussion is zero dimensional and so it does not take in account the spatial distribution of energy sinks and particle sources that are introduced into the drifting stream of boundary plasma as a consequence of impurities. One approach to this problem is based upon the use of a modified coronal equilibrium to describe the effects of impurities. The charge balance rate equation is accepted but an additional term is included to take account of the loss of ions to the boundary surfaces, i.e.,

$$n_e n_{(Z-1)} S_{(Z-1)} - n_e n_Z S_Z + n_e n_{(Z+1)} \alpha_{(Z+1)} - n_e n_Z \alpha_Z - n_Z \tau_{\text{imp}}^{-1} = 0, \quad (60)$$

where S and α are, respectively, the ionization and recombination coefficients. The distribution of charge state with respect to T_e can be estimated from Eq. (60) for various values of $n_e \tau_{\text{imp}}$ and then used in conjunction with coronal power loss coefficients such as those described in McWhirter and Summers (Section VII, Chapter 3 of this volume) in order to determine radiative power losses from the plasma. This procedure has been applied to boundary plasmas by Shimada et al. (1982) and by Abramov (1982).

The detailed histories of impurity ions in the boundary flow parallel to the magnetic field have been studied by Neuhauser et al. (1983) who use a one-dimensional numerical description of the bulk plasma linked to a comprehensive, noncoronal treatment of ionization. This work, which has been briefly discussed in Section II.C, yields detailed information about the charge state distribution, but as yet it deals only with trace quantities of impurities so that the transport properties of the bulk plasma are unaffected.

The predominant influence of impurity ionization upon parallel transport within the boundary does not generally arise from radiative power losses within the boundary region itself (see, e.g., Harrison and Hotston, 1982a). A major effect upon the drifting boundary plasma arises from the presence of cold electrons ejected by ionization and which subsequently cool the plasma electrons due to electron–electron collisions. Additional losses arise from acceleration of multiply charged impurity ions through the potential of the sheath, which must be sustained by the electrons of the bulk plasma. Finally, energy is also extracted from ions of the bulk plasma due to ion–ion collisions that tend to accelerate the massive impurity ions up to the drift velocity of the bulk plasma.

VI. Significance of the Boundary Plasma

The objective of Chapter 7 has been to outline the physical processes that impact upon atomic and molecular interactions within the boundary region, but the discussion does not extend to an examination of the important interactions between the boundary plasma and the central core of the confined plasma. In general the boundary can be regarded as a sink of energy that flows from the central plasma to the walls of the containment vessel but it is also a source of impurity ions that may enter the central plasma. In addition it is a source of recycling "hydrogen" atoms, and its plasma characteristics govern the ability to fuel the central plasma with neutral "hydrogen" and to exhaust unwanted atomic species such as helium atoms which will eventually be produced in a D–T fusion reactor.

7. Boundary Plasma

Suitable diversion of the magnetic field within the boundary region can locate the predominant release site of impurities within a spatially separated divertor chamber but, in the alternate configuration where the toroidal plasma is bounded by a limiter, the release site must be adjacent to the central plasma. The relative merits of these two approaches then rest strongly upon the ability of a divertor to minimize the backflow of impurity ions along the magnetic field to the torus, whereas successful limiter operation is more strongly governed by the need to maintain an adequately low concentration of impurities which will build up within the central plasma due to inward diffusion across the magnetic field. At present the viability of the divertor and limiter approaches to impurity control remains a matter of intensive debate, that embraces not only the physics issues discussed here but also the significance that such systems have upon engineering concepts of a tokamak reactor.

Acknowledgment

All the figures in this chapter have been reproduced courtesy of Culham Laboratory.

References

Abramov, V. A. (1982). *In* "USSR Contributions to Phase IIA of the INTOR Workshop" (Compiled by B. B. Kadomtsev), Vol. 1, p. VI-137. Rep. Kurchatov Inst. Moscow, U.S.S.R.
Bailey, A. W., and Emmert, G. A. (1980). *Univ. Wis. Rep.* **UWFDM-362.**
Bates, D. R., and Kingston, A. E. (1963). *Planet. Space Sci.* **11,** 1.
Bates, D. R., Kingston, A. E., and McWhirter, R. W. P. (1962). *Proc. R. Soc. London, Ser. A* **267,** 297.
Bell, K. L., Gilbody, H. B., Hughes, J. C., Kingston, A. E., and Smith, F. J. (1982). *Culham Lab. Rep.* **CLM-R216.**
Bohdansky, J. (1982). *In* "European Contributions to the INTOR Phase IIA Workshop," EUR FU BRU/XII-132/82/EDV30, p. VI-318. Euratom, Brussels.
Brooks, J. N. (1982). *J. Nucl. Mater.* **111/112,** 457.
Burnett, C. R., Grove, D. J., Palladino, R. W., Stix, T. H., and Wakefield, K. E. (1958). *Phys. Fluids* **1,** 438.
Chen, F. F. (1974). "Introduction to Plasma Physics." Plenum, New York.
Chodura, R. (1982). *In* "European Contributions to the INTOR Phase IIA Workshop," EUR FU BRU/XII-132/82/EDV30, p. VI-275. Euratom, Brussels; *J. Nucl. Mater.* **111/112,** 420.
Chodura, R., Lackner, K., Neuhauser, J., Schneider, W., and Wunderlich, R. (1982). *Plasma Phys. Controlled Nucl. Fusion Res., Proc. Int. Conf., 9th, Baltimore* **1,** 313.
Cupini, E., De Matteis, A., Simonini, R., and Hotston, E. (1982). *In* "European Contributions to the INTOR Phase IIA Workshop," EUR FU BRU/XII-132/82/EDV30, p. VI-245. Euratom, Brussels.
Dalgarno, A., and Griffing, G. W. (1958). *Proc. R. Soc. London, Ser. A* **248,** 415.
Dance, D. F., Harrison, M. F. A., Rundel, R. D., and Smith, A. C. H. (1967). *Proc. Phys. Soc., London* **92,** 577.
Dolder, K. T., and Peart, B. (1976). *Rep. Prog. Phys.* **39,** 697.
Dunn, G. H. (1966). *J. Chem. Phys.* **44,** 2592.
Eckstein, W., and Verbeek, H. (1979). *Max-Planck-Inst. Plasmaphys.* [Ber.] *IPP* **IPP 9/32.**
Elbern, A., Hintz, B., and Schweer, B. (1978). *J. Nucl. Mater.* **76/77,** 143.

Emmert, G. A. (1980). *Univ. Wis. Rep.* **UWFDM-343**.
Emmert, G. A., and Bailey, A. W. (1980). *Univ. Wis. Rep.* **UWFDM-365**.
Emmert, G. A., Wieland, R. M., Mense, A. T., and Davidson, J. N. (1980). *Phys. Fluids* **24**, 803.
Erents, S. K., and McCracken, G. M. (1970). *Radiat. Eff.* **3**, 123.
Freeman, E. L., and Jones, E. M. (1974). *Culham Lab. Rep.* **CLM-R137**.
Gibson, A. (1978). *J. Nucl. Mater.* **76/77**, 92.
Harbour, P. J. (1981). In "Plasma Physics for Thermonuclear Fusion Reactors" (G. Casini, ed.), p. 255. Harwood Academic Publ., Paris (for Comm. Eur. Communities).
Harbour, P. J., and Harrison, M. F. A. (1978). *J. Nucl. Mater.* **76/77**, 513.
Harbour, P. J., and Harrison, M. F. A. (1979). *Nucl. Fusion* **19**, 695.
Harbour, P. J., and Morgan, J. G. (1982). *Culham Lab. Rep.* **CLM-R234**.
Harrison, M. F. A., and Hotston, E. S. (1982a). In "European Contributions to the INTOR Phase IIA Workshop," EUR FU BRU/XII-132/82/EDV30, p. VI-199. Euratom, Brussels; *Culham Lab. Rep.* **CLM-R226**.
Harrison, M. F. A., and Hotston, E. S. (1982b). In "European Contributions to the INTOR Phase IIA Workshop," EUR FU BRU/XII-132/EDU30, p. VI-134. Euratom, Brussels; *Culham Lab. Rep.* **CLM-R232**.
Harrison, M. F. A. *et al.* (1981). *Culham Lab. Rep.* **CLM-R211**.
Harrison, M. F. A., Harbour, P. J., and Hotston, E. S. (1982). In "European Contributions to the Conceptual Design of the INTOR Phase One Workshop," EUR FU BRU/XII-132/82/EDV2, p. 231. Euratom, Brussels.
Harrison, M. F. A., Harbour, P. J., and Hotston, E. S. (1983). *Nucl. Technol. Fusion* **3**, 432.
Heifetz, D., Post, D., Petravic, M., Weisheit, J., and Bateman, G. (1982). *Princeton Plasma Physics Lab. Rep.* **PPPL-1843**.
Hobbs, G., and Wesson, J. (1967). *Plasma Phys.* **9**, 85.
Hutcheon, R. J., and McWhirter, R. W. P. (1973). *J. Phys. B* **6**, 2668.
Igitkhanov, Yu. L. (1982). In "USSR Contributions to Phase IIA of the INTOR Workshop" (compiled by B. B. Kadomtsev), Vol. 1, p. VI-145. Rep. Kutchatov Inst. Moscow, U.S.S.R.
Jones, E. M. (1977). *Culham Lab. Rep.* **CLM-R175**.
Keilhacker, M., and Daybelge, U., eds. (1981). *IAEA Tech. Comm. Meet. Divertors and Impurity Control* Max-Planck-Inst. Plasmaphys. Garching, Fed. Rep. Ger.
Keilhacker, M. *et al.* (1982). *Plasma Phys. Controlled Nucl. Fusion Res., Proc. Int. Conf., 9th, Baltimore* **3**, 183.
Lindhard, L., Sharff, M., and Schiøtt, H. E. (1963). *Mat.-Fys. Medd.—K. Dan. Vidensk. Selsk.* **33**, 39.
McCracken, G. M., and Stott, P. E. (1979). *Nucl. Fusion* **19**, 889.
McWhirter, R. W. P., and Hearn, A. G. (1963). *Proc. Phys. Soc., London* **82**, 641.
Montague, R. S., and Harrison, M. F. A. (1983). *J. Phys. B* **16**, 3045.
Neuhauser, J., Lackner, K., and Wunderlich, R. (1982). In "European Contributions to the INTOR Phase IIA Workshop," EUR FU BRU/XII-132/EDV30, p. VI-47. Euratom, Brussels.
Neuhauser, J., Schneider, W., Wunderlich, R., and Lackner, K. (1983). *Max-Planck-Inst. Plasmaphys.* [Ber.] *IPP* **IPP 1/216**.
Nicolai, A. (1979). *Eur. Conf. Controlled Fusion Plasma Phys., Proc., 9th, Oxford, Engl.* p. 33.
Peart, B., and Dolder, K. T. (1972). *J. Phys. B* **5**, 860.
Petravic, M., Heifetz, D., Post, D., Langer, W., and Singer, C. (1982). *Plasma Phys. Controlled Nucl. Fusion. Res., Proc. Int. Conf., 9th, Baltimore* **1**, 323.
Post, D. E., Heifetz, D., and Petravic, M. (1982). *J. Nucl. Mater.* **111/112**, 383.
Post, D. E. *et al.* (1979). *Plasma Phys. Controlled Nucl. Fusion Res., Proc. Int. Conf., 7th, Innsbruck, Austria, 1978* **1**, 315.
Proudfoot, G., and Harbour, P. J. (1980). *J. Nucl. Mater.* **93/94**, 413.

Rognlien, T. D., and Brengle, T. A. (1981). *Phys. Fluids* **24**, 871.
Riviere, A. C. (1971). *Nucl. Fusion* **11**, 363.
Roth, J. (1980). *Proc. Symp. Sputtering, Vienna*.
Roth, J., Bohdansky, J., and Ottenberger, W. (1979). *Max-Planck-Inst. Plasmaphys.* [Ber.] *IPP* **IPP9/26**.
Roth, J., Bohdansky, J., and Wilson, K. L. (1982). *J. Nucl. Mater.* **111/112**, 775.
Shimada, M. *et al.* (1982). *In* "Japanese Contributions to the International Tokamak Reactor Phase IIA," Vol. I, Appendix VI-B. Japan Atomic Energy Research Institute, Tokai-mura.
Sigmund, P. (1969). *Phys. Rev.* **184**, 383.
Sivukhim, D. V. (1966). *Rev. Plasma Phys.* **4**, 93.
Spitzer, L. (1962). "Physics of Fully Ionised Gases," Wiley, New York.
Staib, P. (1982). *J. Nucl. Mater.* **111/112**, 102.
Stott, P. E., Wilson, C. M., and Gibson, A. (1977). *Nucl. Fusion* **17**, 481.
Stott, P. E., Wilson, C. M., and Gibson, A. (1978). *Nucl. Fusion* **18**, 475.
Summers, H. P. (1974). *Mon. Not. R. Astron. Soc.* **169**, 667.
Summers, H. P., and McWhirter, R. W. P. (1979). *J. Phys. B* **14**, 2287.
Taglauer, E., Beitat, U., and Heiland, W. (1978). *Nucl. Instrum. Methods* **149**, 605.
Thompson, M. W. (1968). *Philos. Mag.* **18**, 377.
Watkins, M. L., Cordey, J. G., Abels-van Maanen, A. E. P. M., Roberts, J. E. C., and Stubberfield, P. M. (1982). *Plasma Phys. Controlled Nucl. Fusion Res. Proc. Int. Conf., 9th, Baltimore* **1**, 281.
Wesley, J. C. *et al.* (1981). *In* "IAEA Technical Committee Meeting on Divertors and Impurity Control" (M. Keilhacker and U. Daybelge, eds.), p. 32. Max-Planck-Inst. Plasmaphys., Garching, Fed. Rep. Ger.

8
Atomic Phenomena in Hot Dense Plasmas

Jon C. Weisheit[*]

Plasma Physics Laboratory
Princeton University
Princeton, New Jersey

I.	Introduction	441
II.	The Plasma Environment	443
	A. Electrostatic Potentials of Ions in Plasmas . .	444
	B. Electric Microfield Distributions	447
III.	Perturbations of Atomic Structure.	450
	A. Continuum Lowering and Level Shifts	450
	B. Radiative Transition Strengths	455
IV.	Perturbations of Atomic Collisions	460
	A. Elastic Scattering	461
	B. Inelastic Scattering	464
V.	Formation of Spectral Lines	467
	A. Stark Broadening	468
	B. Forbidden Components and Plasma Satellites .	474
	C. Optical Depth Effects	477
VI.	Dielectronic Recombination	479
	References	482

I. Introduction

Our understanding of atomic processes important in plasmas has benefited considerably from the efforts of physicists and astronomers to analyze and relate laboratory and cosmic plasma radiation arising under similar conditions. Particle densities in conventional laboratory plasmas typically range from about 10^7 to about 10^{18} cm^{-3}. As is illustrated in Fig. 1, this range of densities is bracketed by gaseous nebulae and stellar atmospheres, and it has been extensively studied for several decades.

[*] Present address: Lawrence Livermore National Laboratory, Livermore, California 94550.

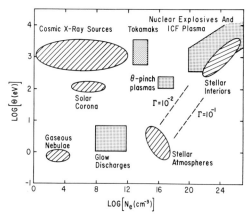

Fig. 1. Density and temperature regimes of some natural and man-made plasmas. The two dashed lines show (N_e, Θ) values corresponding to different constant values of the Coulomb parameter Γ [Eq. (1)] for a hydrogen plasma.

During the past few years, research directed toward controlled fusion by means of inertial confinement (laser or particle beams) has resulted in very hot laboratory plasmas at solid and even greater than solid densities, $N \geq 10^{22}$ cm^{-3}. It is, of course, no surprise that inertial confinement fusion (ICF) experiments seek to generate plasmas having densities and temperatures like those occurring within stars. However, in contrast to stellar interiors, which—except for neutrino flux measurements—are not directly accessible to observation, the ICF plasmas can and are being studied in great detail. [See, for example, the review by DeMichelis and Mattioli (1981).] Besides integrated spectral data provided by conventional means, spatial and temporal resolution of ICF plasmas is provided by x-ray photographs from pinhole and streak cameras (Yaakobi and Goldman, 1976; Attwood *et al.*, 1977; Lee and Rosen, 1979; Key *et al.*, 1979).

The long time scales characteristic of stellar evolution ensure that the dense gas inside stars almost always in is local thermodynamic equilibrium (LTE), but the nanosecond time scales characteristic of ICF plasmas generally preclude populations of atomic states from attaining LTE values. This dense, 10^{18} cm$^{-3} \leq N \leq 10^{24}$ cm^{-3}, and hot, $kT \geq 100$ eV, nonequilibrium plasma regime is the primary setting for the present discussion of atomic phenomena. In this environment, the dimensions of ionic orbitals can be comparable to distances between ions, causing quantal states to be strongly influenced by surrounding plasma.

In addition to stellar physics and ICF research, high-density plasmas are encountered in solid-state science and in the study of thermonuclear detonations. However, collective phenomena are of primary interest in cold metal

8. Atomic Phenomena in Hot Dense Plasmas

and semiconductor plasmas, and much of the physics of thermonuclear explosions is classified defense research. Therefore, the relevance of the present subject to these areas will not be considered here. The design and development of short-wavelength lasers is another active area of investigation often involving dense plasmas. [Because of their very brief radiative lifetimes, the lasing atoms require very rapid pumping, and for many lasing materials this is best done at very high densities (see, e.g., Chapline and Wood, 1975).] Specific laser schemes will not be described in this chapter, but many phenomena discussed here can affect the extent of stimulated emission in a high-density lasing medium.

This review only includes research published before 1983. Several other chapters collected in the present work provide up-to-date discussions of related subjects. Some recent monographs on associated topics are those by Hughes (1975), Bekefi (1976), Motz (1979), and McDowell and Ferendeci (1980). These references discuss either relevant atomic processes, but as they occur in low-density plasmas, or the laser/plasma aspects of fusion. The text by Cox and Giuli (1968) surveys atomic physics in stellar interiors. Finally, the proceedings of a conference on "Radiative Properties of Hot, Dense Matter" (1982, *J. Quant. Spectrosc. Radiat. Transfer* **27**, No. 3) contain several articles specifically referenced herein.

II. The Plasma Environment

A convenient, one-dimensional characterization of a plasma is provided by the ion–ion Coulomb parameter (DeWitt *et al.*, 1973)

$$\Gamma = \langle Z_i e \rangle^2 / R_i \Theta = \langle Z_i \rangle^{5/3} [N_e / (8.00 \times 10^{19} \text{ cm}^{-3})]^{1/3} / \Theta \quad \text{eV}, \tag{1}$$

where $\langle Z_i e \rangle$ is the mean charge of all plasma ions, $\Theta = kT$ is the plasma temperature[†] in energy units, and where the ion-sphere radius R_i approximates the ion–ion spacing. It is defined in terms of the electron concentration N_e by the equation

$$R_i = (3\langle Z_i \rangle / 4\pi N_e)^{1/3} = [(1.61 \times 10^{24} \langle Z_i \rangle \text{ cm}^{-3}) / N_e]^{1/3} a_0, \tag{2}$$

with $a_0 = 0.529 \times 10^{-8}$ cm, the Bohr radius. When $\Gamma \gg 1$ Coulomb interactions control particle motions, but when $\Gamma \ll 1$ thermal motions predominate. Two lines of constant Γ are shown on the $N_e - \Theta$ diagram in Fig. 1, for a hydrogen plasma with $\langle Z_i \rangle = 1$. It is evident that values $\Gamma \gtrsim 10^{-2} \langle Z_i \rangle^{5/3}$ are of interest here.

[†] Throughout this chapter it is assumed that an equilibrium temperature can be ascribed to each charge species. The parameter Γ is useful when all species have the same temperature and, unless noted otherwise, this additional assumption is also made.

Another important parameter is the plasma electron frequency ω_e, which is the oscillation frequency for space-charge disturbances. At high densities, the energy of collective oscillations (plasmons),

$$\hbar\omega_e = \hbar(4\pi e^2 N_e/m_e)^{1/2} = [N_e/(7.21 \times 10^{20}\ \text{cm}^{-3})]^{1/2}\ \text{eV}, \quad (3)$$

can be comparable with atomic transition energies, and electric fields associated with these oscillations can strongly perturb plasma ions.

A. Electrostatic Potentials of Ions in Plasmas

The long-range nature of the Coulomb force causes plasmas to be electrically neutral on a macroscopic scale. However, small departures from charge neutrality do occur, and their scale can be estimated from the range of the electrostatic potential $\phi(r)$ of a bare test charge placed in a uniform plasma. This potential also serves, in a first approximation, as a measure of the plasma's influence on a nucleus with some bound electrons, if one makes the simplifying assumption that the many-body wave function representing the background plasma and the ion of interest is separable.

In response to a test charge Ze, whose position is taken as the origin, the concentration of each (nondegenerate) particle species p departs from its uniform value N_p according to Boltzmann's equation,

$$\tilde{N}_p(r) = N_p g_p(r) = N_p \exp[-Z_p e\phi(r)/\Theta_p], \quad (4)$$

such that ϕ satisfies the Poisson equation. The exponential quantity g is known as the radial distribution function. In the limit of high temperatures Θ_p (i.e., $\Gamma \ll 1$), a linear approximation of $g_p(r)$ leads to the familiar result

$$\phi = \phi_{\text{D-H}}(r) = (Ze/r) \exp(-r/D), \quad (5)$$

where D is the Debye–Hückel length

$$D = \left(4\pi e^2 \sum_p Z_p^2 N_p/\Theta_p\right)^{-1/2}$$

$$= \left\{\sum_p Z_p^2 N_p/[1.97 \times 10^{22}\ \text{cm}^{-3}\ \Theta_p\ (\text{eV})]\right\}^{-1/2} a_0. \quad (6)$$

This statistical description of the plasma screening is valid if there are many shielding particles within a "Debye sphere," i.e., if $(\tfrac{4}{3})\pi D^3 N \gg 1$. When screening by only electrons is relevant, the Debye length computed including just one species is denoted D_e.

A complementary picture can be obtained in the low-temperature limit $\Gamma \gg 1$. Under the assumptions (i) that the nuclei are fixed and (ii) that each nucleus of charge Ze is completely shielded from its neighbors by a uniform cloud of Z electrons, Poisson's equation can be solved to give

8. Atomic Phenomena in Hot Dense Plasmas

$$\phi(r \leq R_i) = \phi_{I-S}(r) = Ze[1/r - (1/2R_i)(3 - r^2/R_i^2)], \tag{7}$$

where R_i is the ion–sphere radius of Eq. (2). For $r > R_i$, ϕ_{IS} is defined to be zero.

Sufficiently far from a test charge, the ratio $e\phi/\Theta_p \ll 1$ regardless of the temperature, so the Debye potential represents the asymptotic ($r \to \infty$) form of plasma screening; conversely, when r is very small, the ion–sphere potential is more appropriate at all temperatures. When all particle species have the same temperature, the relationship between the two screening lengths is

$$R_i/D_e = (3\Gamma/\langle Z_i\rangle)^{1/2}. \tag{8}$$

The relationship between these two model potentials has been studied in detail by Stewart and Pyatt (1966), for the case of a one-temperature plasma. These authors give a simple interpolation formula that is valid when $\langle Z_i^2\rangle \gg \langle Z_i\rangle$.

Thus far, no mention has been made of the influence of bound electrons on $\phi(r)$. The Thomas–Fermi statistical model (see Bethe and Jackiw, 1968) does not consider bound states explicitly, but it does yield an electrostatic potential ϕ_{TF} that is self-consistent with a degenerate ($\Theta_e = 0$) Fermi distribution of free electrons near the nucleus. For an ion of nuclear charge $Z_n e$ this potential satisfies the differential equation

$$x^{1/2}U''(x) = U^{3/2}(x), \tag{9}$$

in which

$$U = r(e\phi_{T-F} + \zeta)/Z_n e^2, \tag{10}$$

where ζ is the chemical potential of the electron gas, $x = r/\xi$, and the length

$$\xi = (3\pi/4)^{2/3}a_0/2Z_n^{1/3}. \tag{11}$$

At the origin, $U(0) = 1$; different choices for the second boundary condition yield potentials for confined and isolated systems.

The Thomas–Fermi model has been extended to include electron exchange interactions (Dirac, 1930) and finite electron temperatures (Feynman et al., 1949; Cowan and Ashkin, 1957). Moreover, the effects of atomic shell structure on ϕ_{T-F} have been treated approximately by Zink (1968) and by Carson et al. (1968). Plasma screening in the regime of intermediate degeneracy—between the Debye–Hückel and Thomas–Fermi limits—has been investigated by Gupta and Rajagopal (1979, 1982) and by Dharma-wardana and Perrot (1982) using finite-temperature density functional theory. The computations that have been performed, all for hydrogen or hydrogenlike ions, reveal that electron exchange interactions are not important unless the plasma temperature is less than three or four times the (electron) Fermi temperature $\Theta_F = (N_e/5 \times 10^{21} \text{ cm}^{-3})^{2/3}$ eV. However, unless $\Theta \gtrsim 10\,\Theta_F$, correlations between bound and free electrons do significantly deepen the effective potential near an ion.

Computations for complex ions in hot, dense matter have been performed by Rozsnyai (1972). He used a relativistic Thomas–Fermi–Dirac calculation to describe the free-electron distribution and to provide the trial potential for a relativistic Hartree–Fock–Slater calculation of the bound-electron distribution. The result is a statistical, or average, atom (AA) with nonintegral occupation numbers (Chandrasekhar, 1939). The total potential ϕ_{AA} due to the nucleus and its neutralizing cloud of bound and free electrons, all isolated in an ion sphere, is self-consistent with the total charge density of these same particles.

Thomas–Fermi and Hartree–Fock potentials cannot be expressed in closed form, even for an isolated atom. In order to illustrate the differences among various model potentials, numerical results are shown in Fig. 2 for neon ions in an LTE neon plasma of temperature $\Theta = 100$ eV and nucleon number density $N_{Ne} = 10^{21}$ cm^{-3}; in this plasma, the predominant charge state of neon ions is Ne^{8+} and the ion–ion Coulomb parameter is $\Gamma = 1.36$. The plotted quantity is the effective nuclear charge number seen by one of the bound electrons, $Z_{eff}(r) = (r/e)\phi(r)$. For the ion-sphere, Thomas–Fermi, and average-atom potentials, Z_{eff} was computed by reducing the ten neutralizing electrons' contribution to ϕ by the factor $\frac{9}{10}$. For the Debye–Hückel potential, Z_{eff} was computed from Eq. (5) for a point charge +9; this introduces a significant difference at small r values due to the complete screening accorded the other bound electron. In the region of large r values the discrepancies are due to the fact that partial screening by other ions and their electrons is not included in the ion-sphere, Thomas–Fermi, or aver-

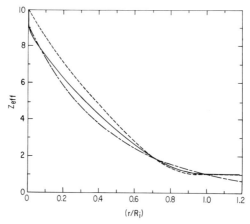

Fig. 2. Effective charge number Z_{eff} seen by a bound electron in a Ne^{8+} ion that is imbedded in a neon plasma having $\Theta = 100$ eV and $N_{Ne} = 10^{21}$ cm^{-3}. The ion-sphere radius is $R_i = 11.8a_0$ and the Debye length is $D = 5.24a_0$. ——, Debye–Hückel; ----, ion-sphere; ———, Thomas–Fermi and (Hartee–Fock–Slater) average-atom model potentials.

8. Atomic Phenomena in Hot Dense Plasmas

age-atom potential calculations. (All of these potentials correspond to a model of each plasma ion being confined in a finite volume.) On the scale of Fig. 2, the Thomas–Fermi and average-atom effective charges cannot be distinguished. Although the agreement usually is not as good as it is for this particular plasma of moderate temperature and density, these two schemes often yield quite similar results (Rozsnyai, personal communication).

In addition to these various potentials, some other models recently have been proposed for calculating macroscopic (equation-of-state) properties of matter at high densities (see Liberman, 1979; More, 1979). As yet, though, these models have not been applied to atomic problems.

B. Electric Microfield Distributions

Because plasmas are good conductors, there are no steady-state electric fields within them. However, transient fields $\mathbf{F}(t)$ are produced by motions of incompletely screened charges. Perturbations of atoms and ions by passing electrons usually are brief, and are treated as collisions. These impacts mainly contribute to the high-frequency ($\omega > \omega_e$) part of the power spectrum of $\mathbf{F}(t)$. On the other hand, the fluctuation time scales $\sim |\mathbf{F}|/|\dot{\mathbf{F}}|$ associated with the slower-moving ions are, in many instances, long compared with the lifetimes of atomic states, and therefore these low-frequency ion microfields can be treated quasistatistically (see Griem, 1974, Section II.2).

1. *One-Component Plasmas*

The problem of calculating the net electric field $\mathbf{F} = \Sigma_j \mathbf{F}_j$ of an ensemble of stationary, uncorrelated point charges Ze was first considered by Holtsmark (1919). He found that the distribution of reduced field strengths $\beta = |\mathbf{F}|/F_0$, where

$$F_0 = \sqrt{8\pi/25}\, Ze/R_i^2 = Z^{1/3}[N_e/(4.35 \times 10^9\ \text{cm}^{-3})]^{2/3}\ \text{V/cm}, \qquad (12)$$

is given by the integral

$$W_H(\beta) = \frac{2\beta}{\pi} \int_0^\infty \exp(-x^{3/2}) \sin(\beta x) x\, dx, \qquad (13)$$

the normalization being $\int_0^\infty W_H(\beta)d\beta = 1$. This distribution, which has been tabulated by, for instance, Baranger and Mozer (1959), is plotted in Fig. 3. For comparison, the distribution of field strengths due to the single-nearest-neighbor (1NN) ion in a uniform plasma (Margenau and Lewis, 1959),

$$W_{1NN}(\beta) = (3/2\beta^{5/2}) \exp(-1/\beta^{3/2}), \qquad (14)$$

also is plotted. The two field strength distributions are essentially the same when $\beta \gg 1$, but elsewhere the 1NN formula is a poor approximation of the Holtsmark distribution.

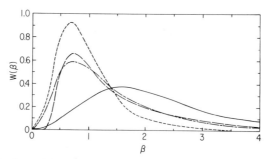

Fig. 3. Distributions of the normalized electric microfield $\beta = F/F_0$ [Eq. (12)]. ———, $\Gamma = 0$ (Holtsmark, 1919); —·—, $\Gamma = 0$ (uniform-density, nearest-neighbor); – – – –, $\Gamma = 0.2$ (Hooper, 1968); – – – –, $\Gamma = 2.0$ (Weisheit and Pollock, 1981).

At finite temperatures, Coulomb interactions give rise to correlations between ions, and electron screening reduces the individual ionic fields $\mathbf{F}_j$. Two quite different techniques have been used to investigate how finite temperatures alter microfield distributions. The first (Baranger and Mozer, 1959, 1960; Hooper, 1966, 1968; Iglesias and Hooper, 1982a) is primarily analytical and is analogous to Holtsmark's approach. It involves the use of cluster expansions, collective coordinates, and particle correlation functions. Most published calculations of this sort have used a linearized Debye–Hückel approximation of the correlation between ion pairs. In other words, the radial distribution of ions at a distance $r = xR_i$ from a given ion has been taken to be [see Eq. (4)]

$$g(x) \to g_{DH}(x) = \exp\left[\frac{-Z^2e^2}{r\Theta}\exp\left(\frac{-r}{D_e}\right)\right]$$

$$\simeq 1 - \frac{\Gamma}{x}\exp(-\sqrt{3\Gamma}x), \qquad (15)$$

where, by using D_e, one includes just the electrons' screening of the ion–ion repulsion (Baranger and Mozer, 1959). The linearization restricts Baranger and Mozer's results to plasmas having low Γ values. However, Hooper's method, which separates the many-body interaction in order to treat differently its long- and short-range parts, can be used to determine microfields in plasmas with $\Gamma \sim 1$. The retention of second- and third-order terms in the expansion of $g(x)$ (Held and Deutsch, 1981) enables the Baranger–Mozer formalism also to yield accurate ion microfields for plasmas with $\Gamma \sim 1$.

The second approach, which is not limited to low Γ values, involves lengthy numerical (i.e., Monte Carlo or molecular dynamics) simulations of the plasma as a classical n-body system (Brush *et al.*, 1966; Hansen, 1973). There exist few direct comparisons of microfield distributions, but it is generally believed that when $\Gamma \leq 0.2$ these different schemes yield distributions that are almost identical (Hooper, 1966). More extensive comparisons have been made of the radial distribution function $g(x)$. Brush *et al.* (1966) determined $g(x)$ from their Monte Carlo experiments for plasmas with $0.05 \leq \Gamma \leq 100$, and found that even the full Debye–Hückel expression g_{D-H} is inaccurate when Γ exceeds a few tenths. [Polynominal fits to the computed values of g have been discussed by DeWitt *et al.* (1973) and Itoh *et al.* (1977).] Quantum corrections to the classical, Debye–Hückel picture of electron screening have been investigated by Iglesias and Hooper (1982b), who found that such corrections are very small unless $\Theta_e \leq \Theta_F$, the Fermi temperature.

Two microfield distributions are shown in Fig. 3 for single-component plasmas with Γ values in the range of interest here. The likelihood of a strong field occurring in a dense plasma clearly is much less than that predicted by Holtsmark's expression for the plasma with $\Gamma = 0$, but, even when $\Gamma > 1$, the characteristic ion microfield strength is of order F_0. It is interesting to compare this field strength with that associated with longitudinal plasma waves at the frequency ω_e. Their rms strength F_w (Griem, 1974, p. 22) in fact is proportional to F_0:

$$\frac{F_w}{F_0} = \frac{25}{8\pi^2 Z} \left(\frac{3\Gamma}{Z}\right)^{1/2} = 0.551 Z^{-3/2} \Gamma^{1/2}. \tag{16}$$

From Fig. 1 it can be determined that $F_w < F_0$ in most cases of interest here.

2. *Multicomponent Plasmas*

Holtsmark's analysis can be extended to the case of a plasma with ions of different charge. According to Chandrasekhar and von Neumann (1942), it is only necessary to redefine the normal field F_0 as being due to ions of effective charge $\langle Z_i^{3/2} \rangle^{2/3} e$. Formulas that generalize the nearest-neighbor (1NN) approximation—including two-particle correlations—have been given for ion mixtures by Weisheit and Pollock (1981).

Hooper and collaborators (O'Brien and Hooper, 1972; Tighe and Hooper, 1976, 1977) have extended his original formalism to treat multicomponent plasmas by means of collective coordinates. Weisheit and Pollock (1981) have also published some molecular dynamics calculations of microfield distributions for hydrogen plasmas with high-Z impurities. These computations for multicomponent plasmas having $\Gamma > 0$ are very time consuming. Therefore, reasonably accurate approximations are needed to describe microfields in the variety of plasmas now being generated. To this end, a two-nearest-neighbor scheme has been proposed by Weisheit and Pollock (1981); it yields accurate microfield distributions for plasmas with $\Gamma \leq 2$, and requires as input just radial distribution functions for mixtures (see Itoh *et al.*, 1979). [It should be noted that the Coulomb parameter as defined in Eq. (1) differs slightly from the expression used by Itoh *et al.*]

III. Perturbations of Atomic Structure

Modification of the Coulomb interaction significantly affects atomic structure, and this, in turn, has important consequences for the statistical mechanical properties of plasmas (Theimer and Kepple, 1970; Rogers and DeWitt, 1973; Zimmerman and More, 1980). Because of the finite range of screened electrostatic potentials, there is only a finite number of bound states for any ion in a plasma. Moreover, surrounding plasma lifts the orbital angular momentum degeneracy of hydrogenic states, and changes to some extent eigenenergies and eigenfunctions of all atomic systems. Observed shifts and intensity alterations of spectral lines, which are evidence of these perturbations, can serve as a check on the model potentials and as a probe of plasma conditions.

A. Continuum Lowering and Level Shifts

1. Theory

The Schrödinger equation for an electron bound by a Debye–Hückel potential to a nucleus of charge $Z_n e = (Z + 1)e$ cannot be solved analytically.[†] However, from first-order perturbation theory one finds that the binding energy of a level (nl) is decreased by an amount

$$\Delta E_{\text{D-H}}(nl) = -(Z + 1)e^2 \langle nl|(e^{-r/D} - 1)/r|nl\rangle, \tag{17}$$

and that the principal quantum number of the uppermost bound state of the system is roughly $[(Z + 1)D/a_0]^{1/2}$. This continuum lowering modifies equilibrium populations of ionic states and brings about a higher degree of ionization in the plasma. Accurate energy levels $E_{\text{D-H}}(nl)$ of a Debye-screened hydrogenic ion have been computed by Rogers *et al.* (1970) and by Roussel and O'Connell (1974) for a wide range of screening lengths $1 \leq (Z + 1)D/a_0 \leq 10^3$. The number of bound states with orbital angular momentum $l = 0$ was found to be $1.126[(Z + 1)D/a_0]^{1/2}$, and the smallest screening length for which a bound state exists was determined to be $\hat{D}(1s) = 0.8399 a_0/(Z + 1)$. Critical screening lengths for the existence of the 45 lowest orbitals, as computed by Rogers *et al.* (1970), are reproduced in Table I; Green (1982) has presented simple formulas that accurately reproduce these values of $\hat{D}(nl)$.

The ion–sphere potential also does not admit an analytic solution of the one-electron Schrödinger equation, but perturbation theory again can be used to estimate the extent of continuum lowering for this model, $\Delta E_{\text{I-S}}(nl)$. Variational calculations of the energies and wave functions for the $n = 1$ and 2 states in an ion–sphere potential were performed by Zirin (1954), who

---

[†] This choice of nuclear charge is consistent with the definitions and formulas of Section II, wherein Ze represents the *net* ionic charge.

TABLE I
Critical Screening Lengths $(Z + 1)\hat{D}/a_0$ for One-Electron Eigenstates $|nl\rangle$[a]

n/l	0	1	2	3	4	5	6	7	8
1	0.8399								
2	3.223	4.541							
3	7.171	8.872	10.947						
4	12.687	14.731	17.210	20.068					
5	19.772	22.130	24.985	28.257	31.904				
6	28.423	31.079	34.285	37.950	42.018	46.458			
7	38.64	41.581	45.122	49.159	53.630	58.500	63.730		
8	50.44	53.641	57.501	61.894	66.752	72.028	77.691	83.720	
9	63.81	67.258	71.426	76.162	81.392	87.064	93.143	99.604	106.43

[a] From Rogers et al. (1970).

tabulated his results as a function of the parameter $(Z + 1)R_i$. Because it represents a more severe screening picture, for a specified screening length R_i or D, ϕ_{IS} has less tightly bound levels than does ϕ_{D-H}. Another distinction between Debye–Hückel and ion–sphere models concerns the average degree of ionization $\langle Z_i \rangle$ they each predict (Zakowicz, Feng and Pratt, 1982): because a Debye–Hückel shielding cloud includes several ions and very many electrons, the Debye–Hückel model can give the bizarre result in which atoms in plasmas become less ionized as the temperature rises.

The relationship between the Debye–Hückel, ion–sphere, and Thomas–Fermi predictions of continuum lowering was explored by Stewart and Pyatt (1966), whose results are presented in Fig. 4. As expected, in the limits of large and small Γ values, the ion–sphere and Debye–Hückel depressions ΔE are obtained. At intermediate Γ values, Thomas–Fermi theory yields a continuum lowering ΔE_{TF} that is always less than ΔE_{D-H}; thus the (nonde-

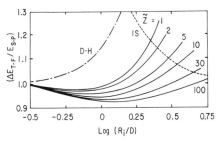

Fig. 4. Thomas–Fermi continuum depression ΔE_{T-F} for an ion of net charge Ze in a plasma with $\langle Z_i^2 \rangle / \langle Z_i \rangle = \bar{Z}$. The energy unit is $\Delta E_{S-P} = \frac{3}{2}\{[(R_i/D)^3 + 1]^{2/3} - 1\}(D/R_i)^3(Ze^2/D\Theta)$ [From Stewart and Pyatt (1966). Reprinted courtesy of Stewart and Pyatt and *Astrophys. J.*, published by the University of Chicago Press; © 1966, The American Astronomical Society.] When $R_i/D \ll 1$, $\Delta E_{S-P} \to \Delta E_{D-H}$; when $R_i/D \gg 1$, $\Delta E_{S-P} \to \Delta E_{I-S}$.

generate) Debye–Hückel potential supports fewer bound states than does the (degenerate) Thomas–Fermi potential. This conclusion of Steward and Pyatt (1966) is consistent with the recent analyses of Gupta and Rajagopal (1979, 1981), who computed eigenvalues for highly charged single-electron ions in partially degenerate plasmas.

Latter (1955) numerically solved the Schrödinger equation with the ($\Theta_e = 0$) Thomas–Fermi–Dirac potential, and obtained orbital eigenvalues for isolated neutral atoms that agree with experimental ionization energies almost as well as Hartree–Fock results do. Shalitin (1965, 1967) proposed a variant of the Thomas–Fermi potential that accounts for the self-interaction of the electrons. He then extended Latter's tables to include this modification; some positive ions were also considered. Although no extensive eigenvalue tabulations comparable to Latter's and Shalitin's exist for confined ions in plasmas, Zink (1968) described a model potential with properties like ϕ_{T-F} that can be used to determine shell structure in dense, partially ionized matter.

Rozsnyai (1972, 1975) published average-atom eigenvalues for a few different ions. At low matter densities and temperatures his results agree with standard Hartree–Fock calculations, and his AA eigenvalues for ions in dense, hot plasmas generally are in accord with Thomas–Fermi results for confined ions.

There also exist more detailed investigations for the special case of a partially ionized hydrogen plasma. Theimer and Kepple (1970) treated unbound plasma particles classically, but corrected the Debye–Hückel model for the fact that the free-electron distribution in the vicinity of a neutral atom is not the same as it is near a point charge. Recently, Dharma-wardana *et al.* (1980) treated quantum mechanically both the bound and the free particles in the hydrogen plasma, and then determined the spectrum of elementary (one-particle) excitations of this many-body system. In both of these works, neutral-atom level shifts much smaller than those computed for the Debye–Hückel potential were obtained.

2. *Experiment*

In principle, the perturbation of an individual ionic level with respect to the continuum is observable in the displacement of the threshold for recombination continuum radiation, but in practice various effects may obliterate this shift. For a hydrogenic level n, the Doppler width $\Delta E_{\text{Dopp}}(n)$ of the recombination radiation edge and the first-order Debye–Hückel shift have the ratio

$$\frac{\Delta E_{\text{Dopp}}(n)}{\Delta E_{\text{DH}}(n)} \simeq \frac{\Theta \text{ (eV)}}{n^2} \left(\frac{2.16 \times 10^{13} \text{ cm}^{-3}}{N_e}\right)^{1/2}, \tag{18}$$

independent of the ionic charge. Thus, thermal motions pose a problem mainly for dilute plasmas. In dense plasmas, the true continuum depression

8. Atomic Phenomena in Hot Dense Plasmas

is obscured by the broadening and merging of emission lines near the series limit—the so-called Inglis–Teller effect (Inglis and Teller, 1939)—unless there are only a few, well-separated lines remaining. Now, the process of negative ion formation via radiative attachment produces a "recombination" continuum with no attendant line series. Experiments (see Neiger and Zimmermann, 1980) to measure plasma-induced changes in electron affinities have revealed shifts that are negligible, or at least are very much smaller than predicted by a Debye–Hückel screening model, but do show a smearing of the continuum edge that is consistent with perturbation by plasma microfields.

Can relative shifts[†] between pairs of levels be determined by measuring the displacement $\Delta\lambda$ of spectral lines with respect to their vacuum wavelengths λ_0? For most ions the answer is no, because election impact broadening of spectral lines is accompanied by a shift that cannot be computed with great accuracy. However, lines of hydrogenic ions are not Stark shifted by plasma electrons. Berg et al. (1962) first measured the shift of the He$^+$ (4686-Å) line, and they attributed it to the polarization of the plasma near each radiating ion. Their basic idea is that, for an electron bound in an orbital of principal quantum number n, free electrons shield the nucleus to the extent

$$e\Delta Z(n) = -4\pi e \int_0^{r_n} \tilde{N}_e(r) r^2 \, dr. \qquad (19)$$

The resulting energy shift is $\Delta E(n) = -(Z + 1)e^2 \, \Delta Z(n)/n^2 a_0$, and so the plasma polarization shift (PPS) in wavelength of a photon emitted in the transition $n \to m$ is

$$\Delta\lambda_{\text{PPS}} = -\frac{2\lambda_0}{Z+1} \left(\frac{\Delta Z(n)}{n^2} - \frac{\Delta Z(m)}{m^2}\right) \bigg/ \left(\frac{1}{n^2} - \frac{1}{m^2}\right). \qquad (20)$$

No unique prescription exists to determine in Eq. (19) the free-electron concentration within the ion or the effective orbital radius r_n. In a first approximation (Greig et al., 1970), the effective radius may be equated to $\langle n|r|n\rangle \simeq n^2 a_0/(Z + 1)$, and the influence of the bound electron on the free-electron radial distribution may be accounted for by the replacement $\tilde{N}_e(r) = N_e g_e(r) \to N_e \exp(V_n/\theta)$, where $V_n > 0$ is a constant representing the average interaction energy of the plasma and the radiating ion in state n. If $V_n = V_m$, these approximations give the plasma polarization shifts that are negative (i.e., blue-shifts) and are proportional to $m^2 n^2 (n^2 + 1) N_e$. In contrast, eigenvalues for the Debye–Hückel and ion-sphere potentials always predict red shifts of transition wavelengths, although to lowest order these shifts also are proportional to N_e. The source of this sign difference is free electrons

---

[†] The formal differences between "level shifts" and "line shifts," which have been noted by Dharma-wardana et al. (1980), are not considered here.

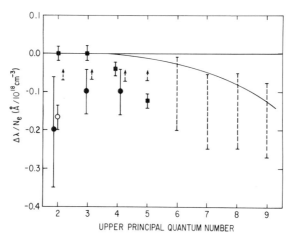

Fig. 5. Measured plasma polarization shifts $\Delta\lambda$ for the He$^+$ principal series. All values have been scaled linearly with N_e to an electron density of 10^{18} cm^{-3}: (Φ) Greig et al. (1970), ( ) Gabriel and Volonte (1973), (I) and ( ) Goto and Burgess (1974), ( ) Neiger and Griem (1976). The theoretical curve is the prediction of Burgess and Peacock (1971).

outside the radiator (viz., those at $r > r_n$) that also contribute, but with opposite sign, to the plasma-induced level shifts computed for screened Coulomb potentials; in fact, these outer electrons have the greater effect (Pittman et al., 1980; Cauble, 1982).

Since the original experiment of Berg et al. (1962) there have been several other measurements of plasma-induced shifts in He$^+$ resonance lines, and the published data are collected in Fig. 5. All the measured wavelength displacements have been scaled linearly with N_e to the density 1×10^{18} cm^{-3}. There is consensus regarding the sign of the shift, namely $\Delta\lambda$ (np $\to$ 1s) < 0; but, where there is more than one measurement, the agreement among different values is poor. To make matters more confusing, data pertaining to other ions and/or other spectra series (Balmer and Paschen) reveal no shifts (Burgess et al., 1967; Van Zandt et al., 1976; Nicolosi and Volonte, 1981) or even red shifts (Pittman et al., 1980)! Evidently, simple screening models are too crude to describe ion–plasma interactions at this level of detail. What clearly are needed to better understand this effect are data for different spectral series taken under a wide range of plasma temperatures and densities, plus additional research on ion microfield gradients, which do produce hydrogenic line shifts (Demura and Sholin, 1975). Here it is worth noting that the Hartree–Fock average-atom theory predicts red shifts for some spectral lines and blue shifts for others; moreover, the sign of the shift for some transitions changes as the plasma conditions vary (Rozsnyai, 1975).

Besides its intrinsic interest as a many-body phenomenon, Skupsky (1980) has suggested that PPS measurements of lines from high-Z impurities

8. Atomic Phenomena in Hot Dense Plasmas

in a hydrogen plasma can be a useful density diagnostic for ICF experiments, when $\Delta\lambda_{PPS} \propto N_e$. But, given the present unsatisfactory status of the theory, any such density determinations are likely to be unreliable. Finally, Burgess (1972) has emphasized that the existence of plasma polarization shifts places an (as yet unknown) restriction on the accuracy of wavelengths determined from plasma spectra.

B. Radiative Transition Strengths

Because screening affects the eigenfunctions as well as the eigenenergies of ions, the strengths of radiative transitions in plasmas are changed from their vacuum values. The discussion here is limited to allowed transitions, so only matrix elements of the electric dipole operator **d** are pertinent.

1. Bound–Bound Transitions

The absorption oscillator strength $f(b \rightarrow a)$ and the spontaneous emission coefficient $A(a \rightarrow b)$ for transition between atomic states a and b, whose statistical weights are $w(a)$ and $w(b)$ and whose substates are specified by the sets of quantum numbers q_a and q_b, both are proportional to the transition strength

$$S(b, a) = S(a, b) = \sum_{q_a q_b} |\langle a q_a | \mathbf{d} | b q_b \rangle|^2, \qquad (21)$$

the familiar relations being

$$f(b \rightarrow a) = \frac{2m_e}{3\hbar^2 e^2} \frac{\Delta E(a, b)}{w(b)} S(a, b), \qquad (22)$$

$$A(a \rightarrow b) = \frac{4[\Delta E(a, b)]^3}{3\hbar^4 c^3 w(a)} S(a, b). \qquad (23)$$

The smallness of the plasma polarization shift suggests that any significant modification of f and A values in plasmas will be due to variation of the transition strengths, not the transition energies $\Delta E(a, b) = |E(a) - E(b)|$.

Several authors (Herman and Coulaud, 1970; Roussel and O'Connel, 1974; Weisheit and Shore, 1974) have investigated dipole transitions in a Debye–Hückel-screened hydrogenic system. The general result is that transition strengths begin to decrease from their vacuum values when the screening length D gets as small as a few times the critical screening length (Table I) of the upper level, and then they drop precipitously as the critical screening length is approached. This trend, which is illustrated in Fig. 6 for the hydrogenic-ion resonance series, has two important consequences (Weisheit and Shore, 1974). First, there is less absorption of radiation by closely spaced

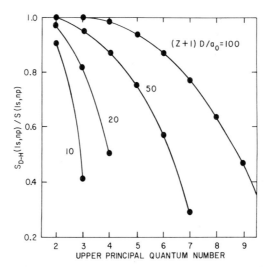

Fig. 6. Ratio S_{D-H}/S of Debye–Hückel-screened to unscreened transition strengths for the principal series in a hydrogenic ion of nuclear charge $(Z + 1)e$. As an aid to the eye, curves have been drawn to connect values for a given screening length D.

lines near a series limit, and this can affect opacity calculations[†] (see Mihalas, 1978, Section 7.2). Second, higher lines of a spectral emission series become progressively harder to detect, and this can hinder the Inglis–Teller scheme for plasma density determinations (Roussel and O'Connell, 1975).

No transition strength calculations involving the ion-sphere or Thomas–Fermi potentials have been published, but the same qualitative behavior displayed in Fig. 6 is evident in the Hartree–Fock–Slater average-atom oscillator strengths reported by Rozsnyai (1975).

2. Bound–Free Transitions

The ever greater diminution of oscillator strengths with respect to their vacuum values as the limit of a spectral series $nl \to n'l'$ is approached is formally connected to the low-energy ($E \gtrsim 0$) behavior of the photoionization cross section $\sigma(nl \to El')$, where E is the energy of the ejected electron (see Shore, 1975). The photoionization cross section vanishes at threshold because of the finite range of the interaction between the ejected electron and its screened parent ion (Wigner, 1948). Just above threshold, Wigner showed that $\sigma \propto E^{l'+1/2}$. This behavior was reproduced by the cross section

---

[†] This point has been disputed recently by Höhne and Zimmermann (1982), who observed that, for screened ions, the decrease in transition energy $\Delta E(1s, np)$ is such that the oscillator density $df/dE = f(1s, np) [d \Delta E(1s, np)/dn]^{-1}$ is unchanged from its unscreened value. Although this is correct, for line opacities the relevant quantity is the absorption cross section, which is proportional to f and not to df/dE.

8. Atomic Phenomena in Hot Dense Plasmas

calculations of Weisheit and Shore (1974), and Höhne and Zimmermann (1982), who also found that well above threshold the cross sections regained their hydrogenic form. In fact, the numerical results can be approximated accurately by matching, at the energy $E = \hbar^2/2m_e D^2$, the low- and high-energy forms. In all these screened-ion calculations, the oscillator strength sum rule is satisfied: the absorption strength lost by bound–bound transitions is gained by bound–free transitions through the lowered ionization threshold.

The work of Rozsnyai (1975) includes some average-atom cross sections for photoionization of ions in dense, hot plasmas, but does not discuss these results in particular. Similarly, the stellar opacity calculations of Carson et al. (1968) employ Thomas–Fermi–Dirac potentials to compute radiative transition strengths, but no specific results are given. Recently, Pratt and co-workers (Shalitin et al., 1982; Feng and Pratt, 1982) explored the sensitivity of iron photoionization cross sections to different Thomas–Fermi models of a plasma environment. No results were presented for valence-shell electrons, but for inner-shell electrons they found only small changes in the cross sections when plasma temperatures and densities were varied widely.

3. Free–Free Transitions

In the nonrelativistic limit, the differential cross section for bremsstrahlung, or free–free emission, in the frequency interval $(\omega, \omega + d\omega)$ as an electron of initial momentum $\hbar k_1$ scatters from a fixed point charge Ze is (Bethe and Salpeter, 1957, Section 77)

$$\frac{d\sigma(\omega)}{d\omega} d\omega = G_C \frac{d\sigma_K(\omega)}{d\omega} d\omega = G_C \frac{16\pi\alpha^3 Z^2}{3\sqrt{3}\, \omega k_1^2} d\omega, \qquad (24)$$

where $\alpha = e^2/\hbar c$ is the fine-structure constant, $d\sigma_K(\omega)/d\omega$ is the Kramers formula, and G_C is the Coulomb Gaunt factor, a complicated function of the electron's initial and final momenta $\hbar k_1$ and $\hbar k_2$. There exist simple approximations to the Gaunt factor in various limits (see Sobel'man, 1979, Section 9.5.7) and extensive tables of numerical values (Karzas and Latter, 1961). Except very near threshold ($\omega \approx 0$), $G_C \simeq 1$ when the inequalities $Z/k_1 a_0 \gg 1$ and $Z/k_2 a_0 - Z/k_1 a_0 \ll 1$, are satisfied. In the other limit, when $Z/k_1 a_0 \ll 1$, the (Coulomb) Born–Elwert approximation is accurate,

$$G_{\text{CBE}} = \frac{\sqrt{3}}{\pi} \left(\frac{k_1}{k_2} \frac{1 - \exp(-2\pi Z/k_1 a_0)}{1 - \exp(-2\pi Z/k_2 a_0)} \right) \ln\left(\frac{k_1 + k_2}{k_1 + k_2} \right). \qquad (25)$$

The Born approximation is valid when k_1 and k_2 both are much greater than Z/a_0. In this case, the bracketed term in the preceding equation approaches unity and the usual Born result for bremsstrahlung emission in a Coulomb field is obtained.

In the (nonrelativistic) Born approximation, the general expression for G is given in terms of an integral involving the Fourier transform of the electron–ion interaction ϕ_x (Grant, 1958),

$$G_{xB}(k_1, k_2) = \frac{\sqrt{3}}{\pi} k_1 k_2 \int_{-1}^{+1} dy \left| \int_0^\infty dr \frac{r\phi_x(r)}{Z_n e} \sin Kr \right|^2 \qquad (26)$$

where $\hbar^2 K^2 = \hbar^2|\mathbf{k}_1 - \mathbf{k}_2|^2 = \hbar^2(k_1^2 + k_2^2 - 2k_1 k_2 y)$ is the square of the momentum lost by the scattered electron, and $Z_n e$ again is the nuclear charge of the target ion. Thus, in the high-energy regime it is straightforward to investigate the effect of screening by bound or free electrons.

When ϕ is the Debye–Hückel interaction, the Born approximation for G takes the form (Grant, 1958)

$$G_{\text{D-H-B}}(k_1, k_2, D) = \frac{\sqrt{3}}{\pi} \left[\ln \left(\frac{1 + (k_1 + k_2)^2 D^2}{1 + (k_1 - k_2)^2 D^2} \right)^{1/2} \right.$$
$$\left. + \frac{1}{2[1 + (k_1 + k_2)^2 D^2]} - \frac{1}{2[1 + (k_1 - k_2)^2 D^2]} \right]. \qquad (27)$$

Examination of Eq. (27) shows that a finite screening length tends to reduce G and, hence, the free–free cross section at all emission frequencies ω. In the limit of negligible screening, $D \to \infty$, the Coulomb Born formula is regained. Grant (1958) and Rozsnyai (1979) modeled the actual potential in the vicinity of a partially stripped ion by a linear combination of Debye–Hückel potentials having different screening lengths, and then used a generalization of Eq. (27) to compute the bremsstrahlung emission from screened electron–ion collisions.

In plasmas with $\Gamma > 1$, the Born approximation to the Gaunt factor can be computed for the ion–sphere potential from the integral expression

$$G_{\text{I-S-B}}(k_1, k_2, R_i) = \frac{\sqrt{3}}{\pi} \int_{(k_1-k_2)R_i}^{(k_1+k_2)R_i} \left[1 + \frac{3}{x^2}\left(\cos x - \frac{\sin x}{x}\right) \right]^2 \frac{dx}{x}. \qquad (28)$$

Various Born Gaunt factors are compared in Fig. 7 for an electron of initial momentum $\hbar k_1 = 5\hbar/a_0$ (initial energy 340 eV) scattering in a hydrogen plasma with $N_e = 3.4 \times 10^{24}$ and $\Theta = 340$ eV. The screening lengths in this plasma are $D = 1a_0$ and $R_i = 0.78a_0$, and the Coulomb parameter is $\Gamma = 0.10$.

Lamoureux et al. (1982) have determined accurate Gaunt factors for a dense cesium plasma using Thomas–Fermi and (average-atom) Hartree–Fock central potentials. Differences between results for the two screening models were found to be much less than differences between the accurate values and values computed for these potentials with the Born approximation. The accurate results, though, are qualitatively similar to the ion–sphere Born curve of Fig. 7. It should be mentioned that Zirin (1954) has published a few tables of Gaunt factors for free–free emission in an ion–sphere potential; these calculations employed crude (one-term) variational wave functions for

8. Atomic Phenomena in Hot Dense Plasmas

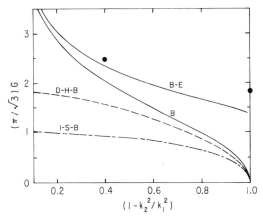

Fig. 7. Gaunt factor G for free–free emission by an electron of initial momentum $\hbar k_1 = 5\hbar/a_0$ scattering in a hydrogen plasma with $\Theta = 340$ eV and $N_e = 3.4 \times 10^{24}$ cm^{-3}; the electron's final momentum is $\hbar k_2$. B denotes the Born approximation to G for scattering in a Coulomb potential, and BE, the Born approximation with Elwert's correction. D–H–B and I–S–B denote, respectively, the Born approximations for scattering in Debye–Hückel and ion-sphere potentials. The two filled circles denote values of the (Coulomb) Gaunt factor, as computed by Karzas and Latter (1961).

the scattering electron. Also, Jackson (1975) has presented a classical electrodynamic description of free–free emission when screening is important, and Pratt and collaborators (see Tseng and Pratt, 1979) have extensively studied relativistic effects in electron bremsstrahlung. These effects become significant when $k_1 \gtrsim 60/a_0$; i.e., when the incident energy of the electron exceeds about 50 keV.

Cross sections for free–free emission and free–free absorption (inverse bremsstrahlung) are related through the principle of detailed balance. Therefore, the absorption process, which plays an important role in the heating of plasmas by laser irradiation, also is affected by plasma screening of the ionic charges. Recently, Lima et al. (1979) investigated this topic and concluded that the rate at which free electrons gain energy via inverse bremsstrahlung is significantly reduced unless the laser frequency $\omega \approx \omega_e$, the plasma frequency; in that resonant situation the absorption is enhanced.

In addition to work, already cited, involving one or another screened potential, there is the paper by Watson (1970), who investigated the effect on the Gaunt factor of ion–ion correlations in dense plasmas. He found that, in degenerate plasmas characteristic of some stellar interiors, the total radiative opacity typically decreased by less than 10% owing to this effect, while in nondegenerate plasmas the decrease is even smaller.

For many applications to laboratory or astrophysical plasmas, what is needed is a Gaunt factor averaged with respect to the distribution of incident electron momenta. Such averages for the Maxwellian distribution can be found in the works of Karzas and Latter (1961), Green (1974) and Grant (1958), Zirin (1954), Grant (1958), and Rozsnyai (1979) for the pure-Coulomb, Debye-Hückel, ion–sphere, Thomas–Fermi, and average-atom potentials, respectively.

IV. Perturbations of Atomic Collisions

It is well known that the infinite range of the Coulomb interaction causes the total cross section for elastic scattering of two point charges to diverge at all collision energies, and the total cross section for excitation of ions by charged particles to be finite at threshold. Besides modifying the eigenstates of scattering systems, plasma screening changes these general features of atomic collisions. Moreover, if the plasma density is very great, then collisions overlap in time and, instead of computing transition rates for (binary) scatterings via screened interactions, one must determine "collisional" rates induced by the stochastic microfield $\mathbf{F}(t)$ of the plasma (cf. Shevelko et al., 1977). This binary regime is limited to densities $N < R_\sigma^{-3}$, where R_σ is an effective interaction range for a particular collision.

Let $\mathbf{v}_*$ be the constant velocity of a charge $Z_* e$ moving in an otherwise uniform plasma, and let $\mathbf{s}_*(t)$ be its position at time t. In this time-dependent situation, the test-particle concept (Rostoker and Rosenbluth, 1960) provides a quantitative description of the screening in terms of the plasma dielectric function $\mathcal{D}(\mathbf{k}, \omega)$ (see Krall and Trivelpiece, 1973). At a given point $\mathbf{s}$,

$$\phi_*(r) = \frac{4\pi Z_* e}{(2\pi)^3} \int\!\!\!\int\!\!\!\int_{-\infty}^{+\infty} \frac{\exp\{i\mathbf{k} \cdot [\mathbf{s} - \mathbf{s}_*(t)]\} \, d^3k}{k^2 \, \mathcal{D}(\mathbf{k}, \mathbf{k} \cdot \mathbf{v}_*)} \tag{29}$$

where $r = |\mathbf{s} - \mathbf{s}_*(t)|$ is the instantaneous distance between the test charge and the point in question. This concept, which is based on the Debye–Hückel model of plasma screening, leads to the following result. Only plasma particles with speeds at least comparable to v_* are polarized on a timescale short enough that they can contribute effectively to the screening of the moving test charge's potential. Specifically:

(i) When v_* is much smaller than the mean thermal speeds of all particle species, then $\mathcal{D} \to 1 - D$ and the potential of the test charge is

$$\phi_*(r) \to (Z_* e/r) \exp(r/D). \tag{30}$$

(ii) When v_* is much larger than the mean thermal speeds of all ion species but much smaller than the mean thermal electron speed, then $\mathcal{D} \to 1 - D_e$ and ϕ_* has the same form as in Eq. (30), but with D replaced by D_e.

(iii) When v_* is much larger than the mean thermal speeds of all particle species, then $\mathcal{D} \to 1$ and the potential of the test charge is unshielded.

8. Atomic Phenomena in Hot Dense Plasmas

Even for a stationary target, this screening model is valid only if the surrounding plasma's polarization time scale, of order (plasma frequency)$^{-1}$ is short in comparison with the duration of the collision, say R_σ/v. Otherwise, only one or a few configurations of screening particles are realized during the collision, and the plasma effects need to be treated quasistatically (see Section V.A). Together with the binary approximation, this consideration bounds the plasma density. For the present discussion, it is assumed that both the inequalities

$$NR_\sigma^3 < 1 \quad \text{and} \quad v/R_\sigma < \omega(\text{plasma}) \sim N^{1/2} \tag{31}$$

are satisfied, so that the plasma's effects on a collision can be treated by screened Coulomb interactions.

There is an extensive literature on the formal aspects of scattering by systems that interact via Debye–Hückel potentials (cf. Joachain, 1975; Semon and Taylor, 1976, and references cited therein), in part because the exponentially screened $1/r$ potential also occurs in the study of nuclear phenomena. However, little is known about scattering by an ion–sphere interaction. This model has no analog of the test-particle scheme, but it may be suitable for scattering from ions in plasmas with $\Gamma > \langle Z_i \rangle$.

A. Elastic Scattering

In general, the elastic scattering of a charge Z_*e by a central potential $\phi(r)$ cannot be computed readily except at high energies, where the (first) Born approximation is valid. Born's formula for the differential cross section for scattering through an angle θ from a fixed center is

$$\left(\frac{d\sigma}{d\Omega}\right)_B = \left(\frac{2m_* Z_* e}{\hbar^2}\right)^2 \left| \int_0^\infty r^2 \phi(r) \frac{\sin K_* r}{K_* r} \, dr \right|^2, \tag{32}$$

where m_* is the mass of the scattered particle, $\hbar k_*$ is its momentum, and $K_* = 2k_* \sin(\theta/2)$. For a binary collision, m_* and $\hbar k_*$ are replaced by the reduced mass μ of the two particles and their relative momentum $\hbar k$, and θ becomes the scattering angle in the center-of-mass coordinate system.

According to the test-particle picture, the differential cross section for the elastic scattering of a fast (unscreened) charge Z_*e by a Debye–Hückel screened charge Ze in a plasma follows directly from Eqs. (30) and (32):

$$\left(\frac{d\sigma}{d\Omega}\right)_{\text{D-H-B}} = \left(\frac{2\mu Z_* Z e^2 D^2/\hbar^2}{1 + (KD)^2}\right)^2, \tag{33}$$

where D is the relevant screening length and $K = 2k \sin(\theta/2)$. In the limit $D \to \infty$ this expression smoothly approaches the Rutherford (Coulomb) cross section:

$$\left(\frac{d\sigma}{d\Omega}\right)_C = \left(\frac{2\mu ZZ_* e^2}{\hbar^2 K^2}\right)^2 = \frac{1}{16} r_0^2 \sin^{-4}\left(\frac{\theta}{2}\right), \tag{34}$$

$r_0 = 2\mu |ZZ_*|e^2/\hbar^2 k^2$ being the distance of closest approach for a repulsive, head-on collision, and upon integration with respect to solid angle it yields a finite total scattering cross section,

$$\sigma_{\text{DHB}} = \frac{4\pi(2\pi ZZ_* e^2 D^2/\hbar^2)^2}{1 + 4k^2 D^2} = \frac{4\pi r_0^2 (kD)^4}{1 + 4k^2 D^2}. \tag{35}$$

Comparison of the Born formula (35) and the cross section σ_{DH} calculated by a full quantal treatment shows that the Born approximation is accurate to within 10% even when $kD \sim 1$ (Joachain, 1975, p. 179).

In plasmas having large Γ values, the elastic scattering of a fast particle and a slowly moving (i.e., thermal) ion may be described reasonably well by Eq. (32), with $\phi = \phi_{\text{I-S}}$ or $\phi_{\text{T-F}}$ or ϕ_{AA}. Substitution of $\phi_{\text{I-S}}$ into the Born formula, followed by some algebra, leads to the equation

$$\left(\frac{d\sigma}{d\Omega}\right)_{\text{I-S-B}} = \left(\frac{2\mu Z_* Z_n e^2 R_i^2}{\hbar^2}\right)^2 \left| 6 \sum_{l=0}^{\infty} \frac{(-1)^l(2+l)(KR_i)^{2l}}{(2l+5)!} \right|^2, \tag{36}$$

$Z_n e$ being the nuclear charge of the screened target ion. Integration over solid angle yields the total elastic scattering cross section for the ion-sphere potential,

$$\sigma_{\text{I-S-B}} = \pi(2\mu Z_* Z_n e^2 R_i^2/\hbar^2)^2 Y(2kR_i) = \pi r_0^2 (kR_i)^4 Y(2kR_i), \tag{37}$$

where the function $Y(x)$, which is evaluated in Table II, decreases slowly from $Y(0) = \frac{1}{25}$.

A general formula for the differential (Born) cross section for elastic scattering in a Thomas–Fermi potential can be written in terms of the dimensionless variables U and ξ of Eqs. (10) and (11),

$$\left(\frac{d\sigma}{d\Omega}\right)_{\text{T-F-B}} = \left| \frac{2\mu Z_* Z_n e^2 \xi}{\hbar^2 K} \int_0^\infty U(x) \sin(K\xi x)\, dx \right|^2. \tag{38}$$

For neutral atoms, the differential cross section is proportional to the square of a universal function of the product $K\xi$, which has been tabulated by Mott and Massey (1965, p. 462). Although it has not been done, it should be possible to treat elastic scattering from ions in plasmas by using in Eq. (38) the potential function U computed for finite-matter temperatures and densities.

The transport properties of an ionized gas are regulated by small-angle deflections in charged-particle encounters. As an illustration of screening effects on these properties, we compare energy loss formulas using Born cross sections $(d\sigma/d\Omega)_{\text{xB}}$ for Coulomb and Debye–Hückel screened interactions; Eqs. (36) and (38) do not yield simple expressions for energy loss.

TABLE II

The Function $Y(x)$ for Elastic Scattering from an Ion–Sphere Potential

x	$Y(x)$	x	$Y(x)$
0.0	0.0400	7.5	0.0310
0.5	0.0393	10.0	0.0286
1.0	0.0386	12.5	0.0265
1.5	0.0379	15.0	0.0246
2.0	0.0373	17.5	0.0230
3.0	0.0360	20.0	0.0214
4.0	0.0348	22.5	0.0201
5.0	0.0336	25.0	0.0188

When a charge Z_*e of mass m_* moves through a one-component plasma having, per unit volume, N stationary ions of charge Ze and mass M, its energy loss per unit path length is

$$-\left(\frac{dE_*}{ds}\right) = N \int d\Omega \left(\frac{d\sigma}{d\Omega}\right) [-\Delta E(\Omega)], \tag{39}$$

where

$$-\Delta E(\Omega) = \frac{4m_* M}{(m_* + M)^2} E_* \sin^2(\theta/2) \tag{40}$$

is the energy loss per scattering. If the plasma ions are unscreened, one obtains from these equations the Coulomb formula,

$$-\left(\frac{dE_*}{ds}\right)_C = \frac{\pi N}{M}\left(\frac{2\mu Z Z_* e^2}{\hbar k}\right)^2 \ln\left(\frac{\hbar^2 k^2 \rho_{\max}}{\mu e^2 |ZZ_*|}\right), \tag{41}$$

where $\mu = (M^{-1} + m_*^{-1})^{-1}$ and $\rho_{\max}$ is a somewhat arbitrary maximum impact parameter for which there is any significant energy transfer in individual Coulomb collisions. Suppose the particle slowing down is a charge selected from a thermal distribution in the plasma. The approximation $\hbar^2 k_*^2 = 3\mu\Theta$ in the logarithmic term, together with the substitution $\rho_{\max} \to D_e$, yields the familiar Coulomb logarithm (Spitzer, 1962),

$$\ln \Lambda = \ln\left(\frac{3\Theta D_e}{|ZZ_*|e^2}\right) = \ln\left(\frac{3Z^3}{Z_*^2 \Gamma^3}\right)^{1/2}. \tag{42}$$

Spitzer (1962, Table 5.1) has published a convenient table of values of $\ln \Lambda$ for a wide range of plasma conditions. In the regime explored by ICF experiments, $\ln \Lambda \sim 3{-}10$ for hydrogen plasmas.

Because of its particularly simple form, Eq. (33) can be used to obtain a Debye–Hückel (Born) formula analogous to Eq. (41),

$$-\left(\frac{dE_*}{ds}\right)_{\text{D-H-B}} = \frac{\pi N}{M} \left(\frac{2\mu Z_* Z e^2}{\hbar k_*}\right)^2 L_{\text{D-H-B}}(2k_* D), \qquad (43)$$

where

$$L_{\text{D-H-B}}(x) = \tfrac{1}{2}[\ln(1 + x^2) - x^2/(1 + x^2)]. \qquad (44)$$

For electron screening of plasma ions $D \to D_e$. When k_* again is replaced by its rms thermal value, the argument of $L_{\text{D-H-B}}$ becomes

$$Y = \frac{2D_e \sqrt{3\mu\Theta}}{\hbar} = \frac{\Theta}{\hbar\omega_e} \sqrt{\frac{12\mu}{m_e}}, \qquad (45)$$

which generally is much greater than unity in hot plasmas. Thus, the ratio of Coulomb to Debye–Hückel energy loss rates is just $\ln \Lambda / \ln Y$. Note that, unlike the Coulomb logarithm, the quantity Y depends upon the masses of the scattering particles.

A specific numerical example is given by a particle with $|Z_*| = 1$ scattering in the hydrogen plasma described in connection with Fig. 7; for that plasma, $\ln \Lambda = 4.0$. It follows from Eq. (45) that $\ln Y = 2.8$ when $m_* = m_e$, and that $\ln Y = 6.6$ when $m_* = 1836 m_e$.

All the results of this section pertain to a nondegenerate plasma. Energy loss by charged particles moving in a Fermi gas has been studied extensively (see, e.g., Fermi and Teller, 1947; Ritchie, 1959; Peres and Ron, 1976), and will not be treated here. It is interesting to note, though, that in these very dense environments the logarithmic divergence of dE/ds due to small transfers in individual collisions is removed not by plasma screening [i.e., $\rho_{\max} \to D$ in Eq. (41)], but by the fact that there is a minimum energy that a degenerate plasma can absorb, namely, the smaller of the energy needed to raise a scattering electron to an unoccupied portion of velocity space (on average, one-third the Fermi energy), and the energy $\hbar\omega_i = (Z_i m_e/M_i)^{1/2} \hbar\omega_e$ of collective ion motions.

B. Inelastic Scattering

Not much is known about a plasma's influence on inelastic processes such as excitation, ionization, and rearrangement. However, for electron impact excitation of slow (i.e., thermal) ions, some qualitative effects of screening by plasma electrons can be determined by comparing cross sections for two isoelectronic systems, one neutral and the other charged. For a particular transition this shows how the scattering event is affected when the long-range Coulomb interaction is eliminated. (A comparison of collisional ionization cross sections for neutral and ionized isoelectronic targets would not adequately illustrate the effects of plasma screening because in both cases the ejected electron interacts via a Coulomb potential with its parent ion and with the scattered charged particle.) Such a comparison is made in Fig. 8a

8. Atomic Phenomena in Hot Dense Plasmas

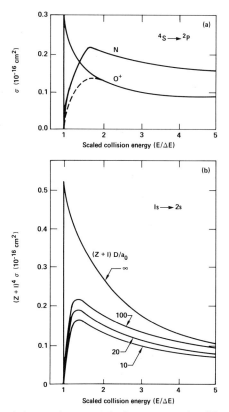

Fig. 8. (a) Computed electron impact excitation cross section (Henry et al., 1969) for the transition $(2p^3)^4S \to (2p^3)^2P$ in N and O^+. When the nitrogen cross section is scaled to agree at high energies with the oxygen cross section, the dashed curve shows its low-energy behavior. For nitrogen, the threshold energy is $\Delta E = 3.58$ eV; for oxygen, it is $\Delta E = 5.02$ eV. (b) Computed electron impact excitation cross section for the transition $1s \to 2s$ in a one-electron ion of nuclear charge $(Z + 1)e$. The Debye-screened ion results are the Born calculations of Hatton et al. (1981), and the unscreened ion result ($D \to \infty$) is the Coulomb–Born calculation of Oh et al. (1978). The threshold energy is $\Delta E = 10.2 (Z + 1)^2$ eV.

for the transition $(2p^3)^4S \to (2p^3)^2P$ in O^+-e^- and N-e^- collisions; to facilitate the comparison, the theoretical curves are plotted as a function of impact energy measured in units of the threshold energy ΔE for each system. The finite range of the electron–atom interaction causes the N-e^- excitation cross section to vanish at threshold, and plasma screening causes this same behavior in electron–ion excitation cross sections. In fact, the model potential study by Hatton et al. (1981) suggests that the magnitude of the effect is similar to that displayed in Fig. 8a. A few of the results of these authors, who used the first Born approximation to compute various excitation cross sections for electron scattering from a Debye–Hückel-shielded one-electron ion, are drawn in Fig. 8b.

The Born approximation often fails near threshold and, unfortunately, it is just this energy region that is most heavily weighted in the calculation of many thermal rate coefficients,

$$C_{ab} = \langle \sigma(a \to b) v \rangle = (8\Theta/\pi\mu)^{1/2} \int_{\chi_0}^{\infty} \chi \sigma(\chi) e^{-\chi} \, d\chi, \tag{46}$$

where $\chi = E/\Theta$ and $\chi_0 = \Delta E(a, b)/\Theta$ are dimensionless ratios. This is because a given stage of ionization of an element is prevalent when the plasma temperature is a few tenths of its ionization potential, and so a particular excitation process is most likely to occur in a plasma when $\chi_0 \sim 1$. Recently, Davis and Blaha (1982) reported distorted wave cross sections for electron impact excitation of hydrogenlike neon ions in dense plasmas. They used a screened interaction that included exchange and correlation terms obtained from density functional computations, but in fact was quite similar to ϕ_{D-H}. Also, Whitten et al. (1984) reported distorted-wave and few-state close coupling calculations for electron excitation of He^+, Ne^{+9}, and Ar^{+18} ions in dense plasmas; both Debye–Hückel and ion–sphere screening models were investigated. These new results indicate:

(1) As suggested by the Born computations, cross sections for dipole-allowed transitions are significantly reduced by screening, at all collision energies. However, unlike the Born predictions, dipole-forbidden cross sections are not very sensitive to plasma screening.

(2) Screening modifications of the colliding electron's trajectory are not important; it is the screening of electron–electron interaction in the transition matrix elements that matters.

Burgess et al. (1978) devised an experiment to measure inelastic electron–atom collision rates in a hydrogen plasma. From an analysis of Balmer emission line data they concluded that inelastic collisions with excited ($n = 2$) H atoms in their laser-pumped plasma had rates lower by a factor of at least 5 than predicted by calculations (for unscreened atoms). Detailed computations will be required to determine to what extent this surprising result can be attributed to particle transport in their plasma, instead of the modest screening expected for the neutral atoms ($D_e \simeq 500 a_0$). In contrast to this result, Kunze et al. (1968) inferred, from C^{4+} spectral line intensities in a low-density θ-pinch plasma ($N_e \gtrsim 10^{15}$ cm^{-3}), carbon excitation rates in good agreement with values computed for unscreened interactions. Also, a new experiment by Kolbe et al. (1982) was reported in which He^+ excitation rates were inferred for a plasma with conditions similar to that studied by Burgess et al. (1978). In this instance, no significant departure of collision rates from their "unscreened" values was found. This subject certainly needs more experimental information, but it is very difficult to extract accurate electron–ion collision rates from spectroscopic data for dense plasmas. In addition, there needs to be an examination of plasma screening effects on electron

8. Atomic Phenomena in Hot Dense Plasmas

impact ionizations and weakly inelastic ion–ion excitations; these processes also play important roles in determining atomic level populations in a dense plasma.

V. Formation of Spectral Lines

Line radiation from dense plasmas is strongly affected by Stark broadening, i.e., pressure broadening by charged perturbers. Intensities of lines arising in dense plasmas also may be influenced by radiative transfer effects due to large optical depths. Moreover, the electric microfields in plasmas mix atomic states of different parity and thereby give rise to "forbidden components" of allowed transition lines. Before the present brief discussions of these topics are given, it is helpful to summarize other broadening mechanisms relevant to the analysis of dense-plasma spectra. These are Doppler and natural broadening, which occur in gases of any density and temperature.

Let $Q(\omega)\, d\omega$ be the power per unit volume emitted by a plasma in the angular frequency interval $d\omega$ about ω. In the vicinity of a spectral feature due to the atomic transition $a \to b$, that is, ω near $\omega_0 = \Delta E(a, b)/\hbar$, one has

$$Q(\omega)\, d\omega = (4\omega^4/3c^3) N_a I_{ab}(\omega)\, d\omega, \tag{47}$$

where N_a is the number density of radiating atoms, and where $I(\omega)$, the distribution of radiation intensity within the line, is called the line shape. In the general case of overlapping lines, Eq. (47) includes a summation with respect to all relevant transitions; if there is an underlying emission continuum, its contribution to $Q(\omega)\, d\omega$ must be added to Eq. (47). $I(\omega)$ is proportional to the transition strength $S(a, b)$ [cf. Eq. (23)]. The Lorentz formula

$$I_{ab}^{(L)}(\omega) = \frac{\gamma_L S(a, b)}{\pi w(a)[(\omega - \omega_0)^2 + \gamma_L^2]} \tag{48}$$

describes natural broadening when $(2\gamma_L)^{-1}$ is the harmonic mean of the radiative lifetimes of states a and b; note that $2\gamma_L$ is also the full width at half maximum (FWHM) of this profile. For a Maxwellian distribution of ion speeds, the Doppler profile is

$$I_{ab}^{(D)}(\omega) = \frac{S(a, b)}{\gamma_D w(a)\sqrt{\pi}} \exp\left[-\left(\frac{\omega - \omega_0}{\gamma_D}\right)^2\right], \tag{49}$$

where $\gamma_D = \omega_0 (2\Theta_i/M_i c^2)^{1/2}$ is $(2\sqrt{\ln 2})^{-1}$ times the Doppler FWHM. As the plasma temperature increases, Doppler broadening becomes more and more important. This trend is also exhibited by natural broadening of lines from the higher charge states that occur in hot plasmas: the scaling is roughly $\gamma_L/\omega_0 \sim Z^2$.

Convolution of $I^{(L)}$ and $I^{(D)}$ yields the well-known Voigt profile that describes a line shape influenced by both Gaussian and Lorentzian mechanisms. Properties of the Voigt function have been discussed by Armstrong (1967), and a useful estimate of the Voigt FWHM is (Allen, 1975, p. 85)

$$2\gamma_V = \gamma_L + [\gamma_L^2 + (4 \ln 2)\gamma_D^2]^{1/2}. \tag{50}$$

From the discussion of Section III.B.1 it is clear that the perturbation of an atom's eigenstates by surrounding plasma decreases radiative transition strengths and, hence, results in somewhat lower intensities I and smaller natural linewidths γ_L. Doppler widths for radiating ions also can be reduced by a dense-plasma environment: If the radiator's collision mean free path between large-angle deflections is much shorter than the wavelength of the emitted photon, Doppler narrowing occurs because some of the radiator's momentum can be collisionally transferred to other particles during the emission process (Dicke, 1953; Burgess et al., 1979). In fact, the narrowing might exceed that estimated by Burgess et al. (1979). Their calculation employed a Coulomb mean free path, which is proportional to $(\ln \Lambda)^{-1}$. But, if a Debye–Hückel potential is used, the ion mean free path is smaller, being proportional instead to $(\ln Y)^{-1}$. As yet, none of these plasma screening effects have been incorporated into calculations of spectral line shapes.

A. Stark Broadening

The current theory of Stark broadening was developed independently by Baranger (1958) and by Kolb and Griem (1958). The aspects summarized here are described in detail in the texts by Sobel'man (1972, Chapter 10) and by Griem (1974) and in the review article by Baranger (1962). The review by Wiese and Konjevic (1982) provides an up-to-date summary of experimental results.

1. Basic Concepts

Stark broadening can be treated as a perturbation problem in quantum mechanics. The plasma surrounding a given radiating atom (or ion) introduces a time-varying interaction $V(t)$ into the atomic Hamiltonian. In terms of the eigenfunctions $\Psi_i(t)$ of this Hamiltonian the definition of the line shape for an electric dipole transition is (Sobel'man, 1972, p. 402)

$$\begin{aligned}I_{ab}(\omega) &= \lim_{T\to\infty} \frac{1}{2\pi T} \left| \int_{-T/2}^{T/2} dt\, e^{i\omega t} \langle \Psi_b(t)|\mathbf{d}|\Psi_a(t)\rangle \right|^2 \\ &= \frac{1}{2\pi} \int_{-\infty}^{\infty} d\tau\, e^{i\omega\tau}\Phi(\tau), \end{aligned} \tag{51}$$

where **d** is the instantaneous electric dipole moment of the radiator, and where $\Phi(\tau)$, the Fourier transform of $I(\omega)$, is also the autocorrelation function of the dipole transition moment,

$$\Phi(\tau) = \{\langle\Psi_a(0)|\mathbf{d}|\Psi_b(0)\rangle \cdot \langle\Psi_b(\tau)|\mathbf{d}|\Psi_a(\tau)\rangle\}_{\text{Av}}. \tag{52}$$

In Eq. (52), which follows from the ergodic hypothesis for stationary random processes, $\{\ \}_{\text{Av}}$ denotes an average with respect to the ensemble of plasma configurations. If state a or b is degenerate, or is one of several closely spaced states, then Eq. (52) becomes a statistically weighted sum of terms.

The standard evaluation of $\Phi(\tau)$ employs Dirac's variation-of-the-constants method wherein each $\Psi(t)$ is expressed in terms of unperturbed atomic eigenfunctions $|i\rangle$ and coupling matrix elements $V_{ij}(t) = \langle i|V(t)|j\rangle$. In the quasistatic limit, $V(t)$ is essentially constant during the radiation process, while in the impact limit $V(t)$ is presumed to vary significantly during this time because of collisions between the radiator and rapidly moving charges. In principle, the quasistatic approximation is valid over that portion of the profile for which the distance to the line center is much greater than the inverse of the time scale for microfield fluctuations, $\Delta\omega \gg |\dot{\mathbf{F}}|/|\mathbf{F}|$, an inequality that can be written as

$$\frac{\Delta\omega}{\gamma_D} \gg \frac{\Gamma}{\langle Z\rangle^2\alpha}\frac{\Theta}{\hbar\omega_0}. \tag{53}$$

Likewise, the impact approximation is strictly valid only for the central part of the profile well within the inverse of the typical collision duration, $\Delta\omega \ll v/R_\sigma$. In practice, however, most Stark broadening calculations employ—over the whole profile—the impact approximation to treat electrons and the quasistatic approximation to treat ions. The complete Stark profile then is gotten by convoluting these two results.

The quasistatic autocorrelation function for an isolated line (neither upper nor lower state degenerate) follows directly from Dirac's perturbation formulas, and when substituted into Eq. (51) yields

$$I_{ab}(\omega) = S(a, b)\mathcal{P}(\Delta\omega)\delta(\omega - \omega_0 - \Delta\omega)/w(a), \tag{54}$$

where $\hbar\Delta\omega = V_{aa} - V_{bb}$ is the change in the transition energy, and $\mathcal{P}(\Delta\omega) d(\Delta\omega)$ is the probability of occurrence of plasma configurations that give rise to a frequency shift $\Delta\omega$. Accordingly, the calculation of $I(\omega)$ in the quasistatic approximation reduces to the determination of the plasma microfield distribution discussed in Section II.B, and the matrix elements of V, which will be described later in this section.

The classical impact approximation, in which every collision just changes the phase of the radiating atom, usually is inadequate. The quantal formalism involves a lengthy derivation, and only the final result is quoted here (see any of the general references cited at the beginning of Section V.A). For an isolated line,

$$I_{ab}(\omega) = \frac{1}{\pi} \frac{\gamma_{ab} S(a,b)/w(a)}{(\omega - \omega_0 - \delta_{ab})^2 + \gamma_{ab}^2},\qquad(55)$$

where the Lorentzian half-width γ and the shift from line center δ both are defined in terms of elements of the complex scattering matrix $\underline{S}$:

$$\gamma_{ab} - i\delta_{ab} = \{1 - \langle a|\underline{S}^*|a\rangle\langle b|\underline{S}|b\rangle\}_{\text{Av}}.\qquad(56)$$

In this equation, the asterisk denotes complex conjugation, and for impact broadening $\{\ \}_{\text{Av}}$ represents an average with respect to collision speeds v, having a distribution $f(v)$, and collision impact parameters ρ,

$$\{[\cdots]\}_{\text{Av}} \rightarrow 2\pi N \int dv\, f(v) v \int d\rho\, \rho [\cdots].\qquad(57)$$

When perturbations of the lower state can be ignored, as is often the case, then the FWHM $2\gamma_{ab} \rightarrow N\langle\sigma_a(\text{total})v\rangle$, the total rate of scattering by atoms in state a. A more general form of Eqs. (54) and (55) is needed to describe lines from hydrogenic ions, because of their states' degeneracy. The line-shape formulas for these ions involve the trace, over the degenerate atomic states involved, of a particular product of electric dipole, $\underline{S}$-matrix, and density-matrix operators.

Smith et al. (1969) described a unified theory that treats the electron broadening over the whole profile in a more satisfactory way. It reduces to the impact limit in the core of the line, and to the quasistatic limit in the wings of the line. This theory, which was improved upon by Green et al. (1975) and applied by them to hydrogenic ion lines, requires elaborate computations. Still further refinement of the general theory is gained by including the broadening caused by the motion of the radiating ion (see, for example, Voslamber, 1977, or Voslamber and Stamm, 1981).

Almost all impact broadening calculations are based on impact parameter formulas and, up to the present time, screening effects on electron impact widths and shifts have been treated by restricting the impact parameters to values less than the electron Debye length D_e, or by using a Debye-screened perturber–radiator interaction. For transitions involving levels near the continuum, it may prove necessary to also include screening effects on the bound electron in the computation of widths and shifts for dense plasma situations.

It must be emphasized that the impact approximation is founded on the assumption that strong collisions, those having $\rho < R_\sigma$, do not occur simultaneously (Baranger, 1962); this restriction is equivalent to the aforementioned density restriction $NR_\sigma^3 < 1$. Royer (1980) has recently published a broadening theory for the very high density regime in which the impact approximation fails. Brissaud and Frisch (1971) proposed an alternative to the standard Stark broadening formalism that essentially eliminates all reference to collisions; instead, line profiles are determined by the radiator's response to the time-dependent (electron plus ion) microfield $\mathbf{F}(t)$. To date, calculations based on this theory have used analytical microfield models (cf. Seidel, 1977, 1979), whose various strengths and weaknesses have been discussed by Brissaud et al. (1976), Smith et al. (1981), and Greene (1982). Microfields calculated in many-body computer simulations (molecular dynamics) should prove useful for checking these various analytical prescriptions.

2. Quasistatic Shifts for Ion Resonance Lines

In order to determine quasistatic profiles, diagonal matrix elements of the plasma–radiator interaction V must be calculated. Here, attention is restricted to resonance lines of one- and two-electron ions because they are commonly observed in high-temperature plasmas (see, e.g., Boiko et al., 1979).

Consider a perturbing ion of charge $Z_j e$ at a distance R_j from the nucleus of a radiating ion; its interaction V_j with the radiator, whose active electron is at $\mathbf{r}$, can be expanded in a series of Legendre polynomials P_ν. The quasistatic shift of the transition energy due to all perturbing ions is just

$$\hbar \Delta \omega = -\sum_{\text{ions}} \left[\frac{Z_j e^2}{R_j} \sum_{\nu=1}^{\infty} \left\langle a \left| \left(\frac{r}{R_j} \right)^\nu P_\nu(\cos \theta_j) \right| a \right\rangle \right], \qquad (58)$$

since the ground state of a H- or He-like ion is spherically symmetric.

For hydrogenic radiators the leading term ($\nu = 1$) is nonvanishing, and the use of parabolic coordinates leads to the linear Stark formula

$$\begin{aligned} \hbar \Delta \omega &= \tfrac{3}{2} F \frac{n_a}{Z_n} (n_1 - n_2) e a_0 \\ &= \beta \frac{n_a}{Z_n} (n_1 - n_2) \left(\frac{N_e \langle Z_i^{3/2} \rangle / \langle Z_i \rangle}{6.15 \times 10^{21} \text{ cm}^{-3}} \right)^{2/3} \text{ eV}, \end{aligned} \qquad (59)$$

Where $F = \beta F_0$ is the strength of the net ionic field, n_a is the principal quantum number of the upper state, $Z_n e$ is the nuclear charge of the radiator, and $n_1 - n_2 = 0, \pm 1, \pm 2, \ldots, \pm(n_a - 1)$ is the difference of two parabolic quantum numbers. Each of the $2n_a - 1$ Stark substates of $|a\rangle$ is perturbed by a different amount. The complete profile of the resonance line $n_a \to 1$, which has an approximate FWHM of $2\gamma_S \simeq 3F n_a(n_a - 1)e a_0 Z_n$, consists of overlapping components symmetrical about $\Delta \omega = 0$.

This familiar picture becomes invalid in very dense plasmas: plasma screening and, for high-Z radiators, fine-structure interactions lift the orbital degeneracy of the hydrogenic states; moveover, the ion–quadrupole interaction [the $\nu = 2$ term in Eq. (58)] is no longer negligible. The complicated, asymmetrical Stark patterns that can result from all these simultaneous

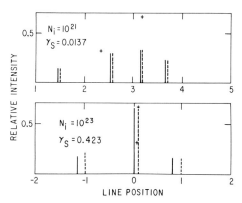

Fig. 9. Lyman-α Stark patterns for neon (Ne^{9+}) ions in homogeneous neon plasmas of two different ion densities N_i (cm^{-3}); in each case, the ion field strength is taken to be F_0. The half-width $\gamma_S(e^2/a_0)$ of the linear Stark pattern is the unit of energy, and the crosses indicate the Lyman-α fine-structure components' positions and intensities in the low-density limit. Full and broken lines show results obtained with and without taking into account ion-sphere screening of the radiating ion. The zero point of energy is the position of the unperturbed Lyman-α line, and the direction of increasing transition energy is to the right. [From Weisheit and Rozsnyai (1976). © The Institute of Physics 1976.]

perturbations have been discussed by Weisheit and Rozsnyai (1976), and one of their examples is reproduced in Fig. 9.

Because there are only accidental degeneracies among the various atomic states of two-electron systems, their Stark profiles differ considerably from those of one-electron systems. First, the spectrum lines are isolated, and this means that the one-term formula (55) can be used for the impact broadening. Second, the quasistatic shifts do not depend linearly on the ion microfield. Instead, the Stark shift has a first-order (quadrupole) term proportional to the gradient $H = -\frac{1}{2}\hat{u} \cdot (\partial \mathbf{F}/\partial u)$, where $\hat{u}$ is a unit vector in the direction of $\mathbf{F}$, plus a second-order (quadratic–Stark) term proportional to F^2,

$$\hbar \Delta \omega = B_3(a)H + B_4(a)F^2. \tag{60}$$

(The coefficients B_3 and B_4, which in the nearest-neighbor approximation multiply terms proportional to R^{-3} and R^{-4}, have been given by Weisheit and Pollock (1981), for the case of L–S coupling; intermediate coupling effects have been discussed by Griem and Kepple (1981). Even in the absence of any broadening of the lower state, the quasistatic profile [Eq. (54)] still is difficult to determine because it depends upon the joint microfield distribution of F and H (see Chandrasekhar and von Neumann, 1942; Weisheit and Pollock, 1981).

3. Comparison of Measured and Computed Lineshapes

Griem *et al.* (1959) first applied the quantal impact theory to the calculation of hydrogen line profiles. An extensive tabulation of H^0 and He^+ results, in which only the linear Stark term is considered, has been published by Griem (1974, Appendix I). Sholin (1969; see also Demura and Sholin, 1975) extended the Stark profile formulas to include the ion–quadrupole term of Eq. (58), and used field strength distributions for independent particles to determine quasistatic profiles in hydrogen plasmas with $N_e \gtrsim 10^{18}$ cm^{-3}. Bacon (1973, 1977) added the quadratic Stark correction, and then calculated complete hydrogen Stark profiles. In the line wings, his results for Lyman-α are in accord with the measurements of Fussman (1975), but for the Lyman-β line the agreement is not as good. However, it does appear that the higher-order Stark terms are responsible for the small observed line asymmetries.

A more serious discrepancy between theory and experiment was uncovered by Grützmacher and Wende (1977). They measured the core of the Lyman-α profile in plasmas having $N_e \gtrsim 10^{17}$ cm^{-3}, $\Theta \simeq 1$ eV, and found FWHM almost twice as large as predicted values. It has been argued that relative motion between the radiator and perturber ions (Voslamber, 1977) and/or fluctuations in the number of screening electrons within a given ion's Debye sphere (Griem, 1978) cause the disagreement, but as yet there are too few comparisons between measured and computed profiles to identify with certainty the additional broadening mechanisms.

Griem et al. (1979) computed a grid of linear Stark fractional widths (full widths at $1/n$ maximum, with $n = 2, 4,$ and 8) for several one-electron ions in dense, hot plasmas. Also, numerous laboratories have published shapes of lines emitted by hydrogenic ions in laser-heated plasmas (see Yaakobi et al., 1977; Key et al., 1979; Mitchell et al., 1979), and then matched them to computed profiles in order to infer plasma densities. However, it is not known how correct the theoretical profiles actually are because there were no independent means of accurately determining N_e and, in some instances, line center optical depths were large. Until more information is obtained, plasma densities inferred from Stark-broadened hydrogenic ion lines are uncertain, and may be correct only to within 50% or so.

There exist just a few studies of line broadening for He-like ions in hot, dense plasmas, and much work remains to be done on this subject. Lee et al. (1979) performed calculations for two-electron silicon and argon resonance lines. They considered the ion–quadrupole interaction but not the quadratic Stark term, which usually is comparable at very high densities. Weisheit and Pollock (1981) computed quasistatic profiles for He-like neon and argon resonance lines that include both of these Stark interactions, but then convoluted the results with Voigt profiles in order to approximate complete line shapes. Boiko et al. (1977) measured the positions and intensities of intercombination lines of several highly stripped, two-electron ions, but did not published individual line profiles. High resolution data for lines of He-like Ar^{16+} arising in very dense plasmas ($N_e > 10^{23}$ cm^{-3}) have been reported by Yaakobi et al. (1980), Hauer et al. (1980), and Kilkenny et al. (1980). Here again, comparisons with theory were complicated by opacity effects. On this topic, the paper by Kilkenny *et al.* is particularly instructive.

Finally, Hoe *et al.* (1981) have recently raised an interesting point regarding the spectroscopy of laser-header plasmas not having spherical symmetry: In such experiments, the target plasma's self-generated magnetic field can be of megagauss strength (Stamper and Ripin, 1975; Raven *et al.*, 1978), which means that the $\mathbf{v} \times \mathbf{B}$ electric field experienced by moving ions can be as large as the plasma microfield. The spectral line broadening caused by this motional field is particularly complicated, since it strongly correlates Doppler and Stark effects. Moreover, the lines become polarized, and the measured profiles depend upon the orientation of the detector with respect to the target. The extent to which this phenomenon has affected some previous line profile analyses is yet to be determined.

B. Forbidden Components and Plasma Satellites

It frequently happens that excited atomic states of different parity but the same multiplicity lie close to one another. These states are mixed by the microscopic electric fields in the plasma. Suppose, as shown in Fig. 10a, that there are only two adjacent states, and that *in vacuo* the transition $a \to b$ is allowed but the transition $\bar{a} \to b$ is (dipole) forbidden. In the limit of degenerate states a and $\bar{a}$, or in the limit of strong electric fields, a hydrogenic profile would be observed. Under less extreme conditions, a forbidden component ($\bar{a} \to b$) may be observed as a separate feature adjacent to the line $a \to b$ (cf. Fig. 10b).

Most of the early work on forbidden components related to closely spaced levels in the spectrum of neutral helium (see, e.g., Griem, 1968). However, the large microfields arising in dense plasmas can strongly couple states that are quite widely separated, and x-ray spectra may show forbidden components. It is straightforward to calculate mixing coefficients and, from them, separations and relative intensities of allowed and forbidden components in the presence of a quasistatic field $\mathbf{F}$. For the two-state case, first-order perturbation theory yields the formulas

$$E_a^{(1)} - E_{\bar{a}}^{(1)} = [(E_a^{(0)} - E_{\bar{a}}^{(0)})^2 + 4V_{a\bar{a}}^2]^{1/2}, \tag{61}$$

$$\frac{I_{\bar{a}b}}{I_{ab}} = \frac{\int I_{\bar{a}b}(\omega)\, d\omega}{\int I_{ab}(\omega)\, d\omega} = \left(\frac{V_{a\bar{a}}}{E_a^{(0)} - E_{\bar{a}}^{(0)}}\right)^2, \tag{62}$$

where the $E^{(0)}$ are unperturbed eigenenergies and where

$$V_{a\bar{a}}^2 = |\langle a|\mathbf{d} \cdot \mathbf{F}|\bar{a}\rangle|^2 = \tfrac{1}{3}F^2 S(a, \bar{a})/w(a) \tag{63}$$

is the square of the coupling-matrix element. Since $F \sim N_e^{2/3}$, the line separation and the relative strength of the forbidden component both increase with density, and therefore may be useful as plasma diagnostics.

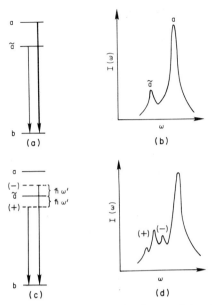

Fig. 10. (a) Schematic energy-level diagram of real states giving rise to a forbidden component ($\tilde{a} \to b$) of the allowed line ($a \to b$); (b) resulting spectrum. (c) Schematic energy-level diagram of real and virtual ($-$, $+$) states giving rise to a forbidden component of the allowed line and to plasma satellite lines ($[-] \to b$ and $[+] \to b$); (d) resulting spectrum.

The 2^1S and 2^1P states of He-like ions, whose unperturbed energy separation is approximately $0.0019 Z_n^2 e^2/a_0$, provide a numerical example of forbidden-component line strengths. For highly stripped ions in dense plasmas, the nuclear charge $Z_n e \approx (Z_n - 1)e$, the charge seen by the active electron, and also the microfield typically has a strength near the Holtsmark normal value, $F \approx F_0$ (cf. Fig. 3). Under these circumstances, if hydrogenic eigenfunctions are used to evaluate $V_{a\tilde{a}}$, the intensity ratio of the forbidden component and the allowed line is approximately (Weisheit et al., 1975)

$$\frac{I(2^1S \to 1^1S)}{I(2^1P \to 1^1S)} \approx \left(\frac{\langle Z_i \rangle}{Z_n}\right)^{2/3} \left(\frac{N_e}{6 \times 10^{19} Z_n^4 \text{ cm}^{-3}}\right)^{4/3}. \tag{64}$$

Two problems complicate the actual measurement of forbidden-component features in the spectra of hot, dense plasmas. First, significant Stark broadening of an allowed line can cause it to overlap its forbidden components. Second, transition probabilities for intercombination lines, e.g., 2^3P-1^1S, increase rapidly with nuclear charge because of the breakdown of L–S coupling; at large Z_n values these lines are strong (Boiko et al., 1977; Griem and Kepple, 1981) and they too may obscure forbidden components. (Indeed, there may even be forbidden components of the intercombination lines.) High-resolution spectra from plasmas with $N_e > 10^{23}$ cm^{-3} will be required to identify these various features due to radiating ions with $Z_n \sim$ 10–20.

There is a bound–free radiative process analogous to the formation of forbidden components of bound–bound transitions. Some doubly excited ionic states are unable to autoionize to a (degenerate) continuum state because of their parity; *in vacuo* such states therefore decay via spectral line emission. In certain instances these states can be efficiently mixed, by plasma microfields, with adjacent doubly excited states that do autoionize rapidly. Davis and Jacobs (1975) first investigated this subject and they found, for example, that the He-like ion line arising from the transition $(2p^2)^3P \rightarrow (1s2p)^3P$ would be completely quenched by this mixing if the plasma density exceeded about $10^{20} Z_n^3$ cm^{-3}. This phenomenon can have a large effect on dielectronic recombination rate coefficients (see Section VI).

As pointed out by Baranger and Mozer (1961), the interaction of an atomic dipole and plasma oscillations at a given frequency ω' can produce additional spectral features in the vicinity of an allowed line and a nearby forbidden line: A second-order process involving virtual atomic states results in plasma satellite lines displaced by an amount $\pm \hbar \omega'$ from the energy of the forbidden line; this is illustrated in Figs. 10c and 10d, again for the simple case of two adjacent excited states. The relative strengths of the satellites in this instance are

$$\frac{I_\pm}{I_{ab}} = \frac{\langle F^2 \rangle S(a, \bar{a})}{6w(a)[\Delta E(a, \bar{a}) \pm \hbar \omega']^2}, \qquad (65)$$

where $\Delta E(a, \bar{a}) = E_a^{(0)} - E_{\bar{a}}^{(0)}$. If F is the wave field due to thermal oscillations at the (electron) plasma frequency ω_e, then $\hbar \omega' = \hbar \omega_e \ll \Delta E(a, b)$ for resonant x-ray transitions, even when $N_e \gtrsim 10^{23}$ cm^{-3} [cf. Eq. (3)]. In consequence, the satellite lines and the forbidden component have a characteristic intensity ratio given by Eqs. (62), (63), (65), and (16),

$$\frac{I_\pm}{I_{\bar{a}b}} \approx \frac{F_w^2}{2F_0^2} = \frac{0.15\Gamma}{\langle Z_i \rangle^3}. \qquad (66)$$

Plasma satellites therefore are expected to be weak features in dense-plasma x-ray spectra, unless there exist strong, *non*thermal oscillations.

Lee (1979) has calculated shapes of satellite lines in nonequilibrium plasmas, and several experimental results have been summarized by Bekefi *et al.* (1976). Recently, Drawin and others (see Drawin and Ramette, 1979) have argued that most observed features which previously were identified as plasma satellites are in fact weak molecular transitions. This point remains to be resolved; however, new hydrogen and helium spectra obtained by Kunze and his collaborators (Finken *et al.*, 1980; Hildebrandt and Kunze, 1980) exhibit features that probably are due to the interaction of radiating atoms and plasma oscillations.

8. Atomic Phenomena in Hot Dense Plasmas

C. Optical Depth Effects

All the spectral line intensity formulas presented above were derived under the assumption that the emitting gas is optically thin to its own radiation. However, even though their dimension ordinarily is small, $s \leq 10^{-2}$ cm, very dense laboratory plasmas can have significant optical depths at resonance line frequencies. When this happens, spectral line shapes and intensities are strongly modified, a fact evidenced in several recent studies involving ICF plasmas. [See, in addition to experiments cited in Section V.A3, the papers by Irons (1975a, 1975b), Weisheit et al. (1975), and Duston and Davis (1980).] The theory of radiation transfer in optically thick lines (and these lines' shapes) is complicated, and is beyond the scope of this chapter. However, one often is interested only in the total amount of radiant energy transported in a given line, or the net rate of radiative decay (emission minus resonance absorption). In the absence of large temperature and density gradients there are simple but useful ideas concerning integrated line emission $\mathcal{Q}_{ab} = \int ds \int d\omega \, Q_{ab}(\omega)$ and effective transition probabilities. [Spectroscopic effects caused by gradients have been discussed by Duston et al. (1981) and Lee (1982).]

An atom's cross section for absorption of radiation in the interval $(\omega, \omega + d\omega)$ near one of its transition frequencies $\omega_0 = \Delta E(a, b)/\hbar$ is

$$\sigma_{ba}(\omega) \, d\omega = (2\pi\omega_0/3\hbar c)I_{ab}(\omega) \, d\omega = (\pi e^2 f_{ba}/m_e c)\psi_{ab}(\omega) \, d\omega, \quad (67)$$

where, as before, f is the oscillator strength and where

$$\psi_{ab}(\omega) = I_{ab}(\omega) \bigg/ \int I_{ab}(\omega) \, d\omega \quad (68)$$

is the normalized line profile function.[†] The corresponding optical depth for a gas of thickness s, having N_b absorbing atoms per unit volume, is

$$\tau_{ba}(\omega) = N_b s \sigma_{ba}(\omega). \quad (69)$$

Usually, it is convenient to characterize the plasma's absorptivity in the vicinity of a spectral line by the optical depth at that line's center, $\tau_0 \equiv \tau_{ba}(\omega = \omega_0)$.

Consider a transition-line photon created at a particular depth $\tau_0 \gg 1$ from the surface of a plasma. This photon may experience several resonance scatterings (absorptions and subsequent reemissions) before it escapes the plasma, or it may be "destroyed" during a particular scattering if the excited atom suffers an inelastic collision or emits a different-frequency photon. Yet a third possibility is that the photon escapes directly, without scattering at

---

[†] Absorption and emission profiles are essentially the same when spectral line shapes are dominated by Stark broadening, because then the "complete frequency redistribution" approximation is valid (Edmonds, 1955; Hummer, 1962).

all. Clearly, this last contingency is likely only if the photon is created in the wings of the line, where $\tau_{ba}(\omega) < 1$. An intuitive estimate of the probability of direct escape is just that fraction of the area under the curve $\psi(\omega)$ for which $\tau(\omega) \leq 1$ (Osterbrook, 1962). (For clarity we hereafter omit subscripts identifying a particular transition.) For a plane parallel slab of plasma, the true escape probability is (Hummer, 1964)

$$P_{\text{esc}}(\tau_0) = \int_0^\infty d\omega\ \psi(\omega)\ E_2[\tau_0\psi(\omega)/\psi(\omega_0)], \quad (70)$$

where E_2 is the second exponential integral function.

Escape probabilities provide a means of approximating the effect of resonance absorption on atomic level populations in optically thick media (see, e.g., McWhirter, 1965; or Mihalas, 1978, wherein P_{esc} is called the net radiative bracket. Specifically, at an optical depth τ_0, the net rate of radiative transitions between two levels is just the Einstein coefficient A_{ab} times the photon escape probability for that line, viz.,

$$A_{ab}(\tau_0) = A_{ab}P_{\text{esc}}(\tau_0). \quad (71)$$

Asymptotic ($\tau_0 \gg 1$) photon escape probability formulas for Doppler, Lorentz, and Voigt line shapes have been known for some time (see Athay, 1972, p. 22). Weisheit (1979) and, more recently, Puetter (1981) presented asymptotic P_{esc} formulas for Stark-broadened hydrogenic lines, but to date there are no published Stark profile results for nonhydrogenic ions (e.g., nonlinearly Stark-broadened lines). Because of its utility, the escape factor method of approximating line tranport effects has been extended to treat moving media (Sobolev, 1957; Hummer and Rybicki, 1982a), overlapping lines (Irons, 1980), and nonlocal effects of photon trapping (Ivanov, 1973; Puetter et al. 1982; Hummer and Rybicki, 1982b).

In order to estimate an optically thick plasma's line emission, let $\langle X(\tau_0)\rangle$ be the mean number of scatterings experienced by photons, in a particular transition line $a \to b$, that escape the plasma from their point of creation at a depth τ_0; and let ε_{ab} be the probability per scattering that such photons are destroyed. In terms of NC_a^{tot} and A_a^{tot}, the total rates of inelastic collisional and radiative transitions from the excited state, ε_{ab} is defined as

$$\varepsilon_{ab} = (NC_a^{\text{tot}} + A_a^{\text{tot}})/A_{ab} - 1. \quad (72)$$

In the absence of any continuum opacity source, the probability that a given photon, once created, ultimately is emitted by the plasma is

$$P_{\text{em}}(\tau_0) = (1 - \varepsilon_{ab})^{\langle X(\tau_0)\rangle} \simeq \exp[-\varepsilon_{ab}\langle X(\tau_0)\rangle], \quad (73)$$

the second expression being correct in the limits of large $\langle X\rangle$ values and small ε values. It follows directly from Eqs. (47) and (73) that the integrated line emission from the plasma surface (energy/area/time) is

$$\mathcal{D}_{ab} = \int ds(\tau_0)P_{\text{em}}(\tau_0) \int d\omega\ Q_{ab}(\omega, s), \quad (74)$$

8. Atomic Phenomena in Hot Dense Plasmas

with Q, the power per unit volume and frequency interval, depending upon depth through the local density of emitters $N_a(s)$ and perhaps also the intrinsic shape of the line. If these dependences are known a priori or can be estimated, then $\mathcal{Q}$ can be determined once $\langle X(\tau_0) \rangle$ is specified (Hummer, 1964; Hummer and Kunasz, 1980). For most real plasmas, though, all the various quantities need to be computed in some self-consistent scheme (Duston and Davis, 1980; Lee, 1982; Alley et al., 1982).

VI. Dielectronic Recombination

The comment was made in the Introduction that level populations of ions in dense laboratory plasmas may be far from their equilibrium (LTE) values. Owing to the transient nature of these plasmas, ionization and recombination events, as well as radiative and collisional transitions between bound states of the same ionization stage, can strongly influence the densities $N_Z(a)$. Because there is no specific information on collisional ionization rates in dense plasmas, high-density modifications of the inverse process, three-body recombination, are unknown. Radiative electron captures occur predominantly to an ion's lower levels, and for these levels screening effects on photoionization, the inverse process, are not large (cf. Section II.B.2). Therefore, one does not expect radiative recombination rates to be changed significantly by the density of a plasma. In contrast, radiationless dielectronic captures often populate highly excited states, and so dense plasma dielectronic recombination rates can vary significantly from vacuum values. [Seaton and Storey (1976) and Dubau and Volonte (1980) have summarized information on dielectronic recombination at low densities.]

To see how dielectronic recombination of a $(Z + 1)$-times ionized atom, denoted Ξ_{Z+1}, and a free electron of energy E_e is altered by a high-density environment, recall that in the low-density limit it involves the following sequence of processes.

resonant capture:
$$\Xi_{Z+1}(n_0) + e^-(E_e) \rightarrow \Xi_Z(n_1 n_2); \tag{75}$$

radiative stabilization:
$$\Xi_Z(n_1 n_2) \rightarrow \Xi_Z(n_2) + \hbar\omega; \tag{76}$$

radiative cascade:
$$\Xi_Z(n_2) \rightarrow \Xi_Z(n_3) + \hbar\omega. \tag{77}$$

(For convenience, singly and doubly excited states have been identified in these formulas by one and two principal quantum numbers.) The rate coefficient for a dielectronic capture that leaves the recombined ion in the singly excited state $|n_2\rangle$ [processes (75) plus (76)] is normally written in terms of the autoionization probability of state $|n_1 n_2\rangle$, the inverse of process (75)

(Burgess, 1964),

$$\kappa_{Z+1}(n_2) = \left(\frac{2\pi\hbar^2}{m_e\Theta}\right)^{3/2} \frac{w(n_1n_2)}{w(n_0)} \exp\left(\frac{-E_e}{\Theta}\right) \frac{A^{\text{auto}}(n_1n_2)A^{\text{rad}}(n_1n_2)}{A^{\text{auto}}(n_1n_2) + A^{\text{rad}}(n_1n_2)}. \quad (78)$$

Here, the ws are the statistical weights of the initial and final states and A^{rad} is the sum of the probabilities of stabilizing radiative transitions. For n_2 small, $w(n_1n_2) \propto n_2^2$, but for n_2 large the statistical weight has essentially a fixed value $2(n_2^*)^2$, because only a few low angular momentum substates $l < n_2^*$ have substantial resonance capture cross sections. The total dielectronic rate coefficient κ_{Z+1} is obtained by summing Eq. (78) with respect to all singly excited states $|n_2\rangle$ of the recombined ion. Now the stabilizing radiative transitions scale as $(Z + 1)^4$, whereas autoionizing transitions scale as $(Z + 1)^0/n_2^3$. By using representative values for A^{rad} and A^{auto}, one finds that the largest terms $K_{Z+1}(n_2)$ occur for resonant captures into levels n_2 with

$$n_2^* \leq n_2 \leq \bar{n}_2 \simeq [40/(Z + 1)]^{4/3}. \quad (79)$$

This behavior is clearly evident in the He^+-like ion results of Burgess and Tworkowski (1976).

Burgess and Summers (1969) first described how collisional processes at high densities can enhance the dielectronic process by deexciting the states $|n_1n_2\rangle$ and by changing these states' angular momentum quantum numbers so as to impede the autoionization rate; in addition, collisions can reduce dielectronic recombination by ionizing the states $|n_1n_2\rangle$. These effects were investigated by Weisheit (1975), who pointed out that continuum lowering also reduces the dielectronic rate because it removes many excited states from the summation $\Sigma \kappa_{Z+1}(n_2)$. A sample of results obtained from the computer code described by Weisheit (1975), which uses *unscreened* radiative and collisional rate coefficients, are plotted in Fig. 11 for 100-eV neon plasmas of different densities. The kink in the curve for $N_e = 10^{21}$ cm^{-3} is due to collisions which stabilize $\text{Ne}^{8+}(1snl)$ ions with angular momentum $l \geq 8$ against autoionization. Zhdanov (1979) obtained similar results from a simple analytic model that provides an approximate density- and temperature-dependent reduction factor for the total rate coefficient. An important effect that was not included in these numerical calculations is the Stark mixing of autoionizing and nonautoionizing states by the quasistatic ion microfield (Davis and Jacobs, 1975; Jacobs et al., 1976; Grigoriadi and Fisun, 1982). This process, which already has been mentioned in the discussion of forbidden spectral components (Section V.B), can increase or decrease the rate of dielectronic recombination for a particular ion; the change depends upon that ion's charge, the energy of the stabilizing photon [Eq. (76)], and, of course, plasma conditions. Comparable effects can also be produced by ion motions in a strong magnetic field (Huber and Bottcher, 1980). The presence of an intense radiation field further complicates the determination of total dielectronic recombination rate coefficients (Burgess and Summers, 1969; Zhdanov, 1980). On the one hand, the field induces radiative stabilizations and cascades, while on the other, it photoionizes stabilized states. In laser-

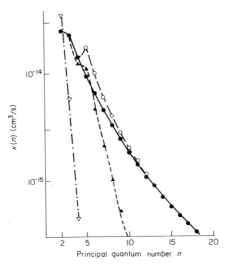

Fig. 11. Rate coefficients $\kappa(n)$ (cm^3 sec^{-1}) for dielectronic recombination to excited states (1snl) of Ne^{8+} in a 200-eV plasma of electron density N_e (cm^{-3}) = 10^{10} (●), 10^{19} (○), 10^{21} (▲), 10^{23} (▽). At 10^{23} cm^{-3}, continuum lowering removes all but the few states indicated.

heated plasmas (dimension $\sim 10^{-2}$ cm), the bremsstrahlung ionization rate of a state $|n_2\rangle$ is comparable with that state's radiative decay rate when the electron density is

$$N_e \simeq 1.5 \times 10^{21}(Z + 1)^{7/2}/n_2^2 \quad \text{cm}^{-3}. \tag{80}$$

Evidently, the radiative quenching becomes severe when the value of n_2 specified by this equation equals $\bar{n}_2$, the largest of the values satisfying the inequality (79). This occurs only when $N_e > 8 \times 10^{16}(Z + 1)^6$ cm^{-3}, a seldom realized condition.

It has been known for some time (see Gabriel and Paget, 1972) that the stabilizing radiative transitions (76) are a major source of satellite lines situated near x-ray resonance lines. Peacock *et al.* (1973) observed that in dense laser plasmas the intensities of certain satellite lines increase rapidly with ion charge. This was explained by Weisheit (1975) as being due to the fact that collisional rates scale roughly as Z^{-3}, while radiative rates scale as Z^4, and therefore that the satellite lines from ions of lower net charge were more readily quenched at a given density. Additional work on this topic has been published by Stavrakas and Lee (1982) for satellites to the resonance lines of H-like silicon ions. Other interesting effects of high density on x-ray satellite line intensities have been explored in some detail by Vinogradov *et al.* (1977), by Jacobs and Blaha (1980), and by Seely *et al.* (1980). In particular, the possible uses of these lines' intensities as density diagnostics were considered (see also Seely, 1979). In a related paper Duston and Davis (1980) described optical depth effects relevant to the analysis of satellite x-ray line spectra.

Acknowledgments

I wish to thank Professor Hans Griem for critically reading this manuscript and making numerous suggestions for its improvement. This work was performed under the auspices of the U.S. Department of Energy, and supported by Contracts Nos. DE-AC02-76-CHO3073 and W-7405-ENG-48.

References

Allen, C. W. (1975). "Astrophysical Quantities," 3rd ed. Athlone Press, London.
Alley, W. E., Chapline, G. F., Kunasz, P. and Weisheit, J. C. (1982). *J. Quant. Spectrosc. Radiat. Transfer* **27**, 257–266.
Armstrong, B. H. (1967). *J. Quant. Spectrosc. Radiat. Transfer* **7**, 61–88.
Athay, R. G. (1972). "Radiation Transport in Spectral Lines." Reidel Publ., Dordrecht, Netherlands.
Attwood, D. T., Coleman, L. W., Boyle, M. J., Larsen, J. T., Phillion, D. W., and Manes, K. R. (1977). *Phys. Rev. Lett.* **38**, 282–285.
Bacon, M. E. (1973). *J. Quant. Spectrosc. Radiat. Transfer* **13**, 1161–1170.
Bacon, M. E. (1977). *J. Quant. Spectrosc. Radiat. Transfer* **17**, 501–512.
Barranger, M. (1958). *Phys. Rev.* **111**, 481–493, 494–513; **112**, 855–865.
Baranger, M. (1962). *In* "Atomic and Molecular Processes" (D. R. Bates, ed.), Chap. 13. Academic Press, New York.
Baranger, M., and Mozer, B. (1959). *Phys. Rev.* **115**, 521–525.
Baranger, M., and Mozer, B. (1960). *Phys. Rev.* **118**, 626–631.
Baranger, M., and Mozer, B. (1961). *Phys. Rev.* **123**, 25–28.
Bekefi, G., ed. (1976). "Principles of Laser Plasmas." Wiley, New York.
Bekefi, G., Deutsch, C., and Yaakobi, B. (1976). *In* "Principles of Laser Plasmas" (G. Bekefi, ed.), pp. 622–627. Wiley, New York.
Berg, H. F., Ali, A. W., Lincke, R., and Griem, H. R. (1962). *Phys. Rev.* **125**, 199–206.
Bethe, H. A., and Jakiw, R. (1968). "Intermediate Quantum Mechanics," 2nd ed., pp. 83–98. Benjamin, New York.
Bethe, H. A., and Salpeter, E. E. (1957). "Quantum Mechanics of One- and Two-Electron Atoms." Springer-Verlag, Berlin and New York.
Boiko, V. A., Faenov, A. Y., Pikuz, S. A., Skobelev, I. Y., Vinogradov, A. V., and Yukov, E. A. (1977). *J. Phys. B* **10**, 3387–3394.
Boiko, V. A., Pikuz, S. A., and Faenov, A. Y. (1979). *J. Phys. B* **12**, 1889–1910.
Brissaud, A. and Frisch, V. (1971). *J. Quant. Spectrosc. Radiat. Transfer* **11**, 1767–1783.
Brissaud, A., Goldbach, C., Leorat, J., Mazure, A., and Nollez, G. (1976). *J. Phys. B* **9**, 1129–1146.
Brush, S. G., Sahlin, H. L., and Teller, E. (1966). *J. Chem. Phys.* **45**, 2102–2118.
Burgess, A. (1964). *Astrophys. J.* **139**, 776–780.
Burgess, A., and Summers, H. P. (1969). *Astrophys. J.* **157**, 1007–1021.
Burgess, A., and Tworkowski, A. S. (1976). *Astrophys. J.* **205**, L105–L107.
Burgess, D. D. (1972). *Space Sci. Rev.* **13**, 493–527.
Burgess, D. D., and Peacock, N. J. (1971). *J. Phys. B* **4**, L94–L97.
Burgess, D. D., Kolbe, G., and Ward, J. M. (1978). *J. Phys. B* **11**, 2765–2778.
Burgess, D. D., Everett, D., and Lee, R. W. (1979). *J. Phys. B.* **12**, L755–L758.
Carson, T. R., Mayers, D. F., and Stibbs, D. W. N. (1968). *Mon. Not. R. Astron. Soc.* **140**, 483–536.
Cauble, R. (1982). *J. Quant. Spectrosc. Radiat. Transfer* **28**, 41–46.
Chandrasekhar, S. (1939). "An Introduction to the Study of Stellar Structure," pp. 256–261. Dover, New York.

8. Atomic Phenomena in Hot Dense Plasmas

Chandrasekhar, S., and von Neumann, J. (1942). *Astrophys. J.* **95**, 489–531.
Chapline, G., and Wood, L. (1975). *Phys. Today* **28**(June), 41–48.
Cowan, R. D., and Ashkin, J. (1957). *Phys. Rev.* **105**, 144–157.
Cox, J. P., and Giuli, R. T. (1968). "Principles of Stellar Structure," Vol. 1. Gordon & Breach, New York.
Davis, J., and Blaha, M. (1982). *J. Quant. Spectrosc. Radiat. Transfer* **27**, 307–314.
Davis, J., and Jacobs, V. L. (1975). *Phys. Rev. A* **12**, 2017–2023.
DeMichelis, C., and Mattioli, M. (1981). *Nucl. Fusion* **21**, 677–754.
Demura, A. V., and Sholin, G. V. (1975). *J. Quant. Spectrosc. Radiat. Transfer* **15**, 881–899. (In Russ.)
DeWitt, H. E., Graboske, H. C., and Cooper, M. S. (1973). *Astrophys. J.* **181**, 439–456.
Dharma-wardana, M. W. C., and Perrot, F. (1982). *Phys. Rev. A* **26**, 2096–2104.
Dharma-wardana, M. W. C., Grimaldi, F., Lecourt, A., and Pellissier, J.-L. (1980). *Phys. Rev. A* **21**, 379–396.
Dicke, R. H. (1953). *Phys. Rev.* **89**, 472–473.
Dirac, P. A. M. (1930). *Proc. Cambridge Philos. Soc.* **26**, 376–385.
Drawin, H. W., and Ramette, J. (1979). *Z. Naturforsch., A* **34A**, 1041–1050.
Dubau, J., and Volonte, S. (1980). *Rep. Prog. Phys.* **43**, 199–252.
Duston, D., and Davis, J. (1980). *Phys. Rev. A* **21**, 932–941.
Duston, D., Davis, J., and Kepple, P. C. (1981). *Phys. Rev. A* **24**, 1505–1519.
Edmonds, F. N. (1955). *Astrophys. J.* **121**, 418–424.
Feng, I. J., and Pratt, R. H. (1982). *J. Quant. Spectrosc. Radiat. Transfer* **27**, 341–343.
Fermi, E., and Teller, E. (1947). *Phys. Rev.* **72**, 399–408.
Feynman, R. P., Metropolis, N., and Teller, E. (1949). *Phys. Rev.* **75**, 1561–1573.
Finken, K. H., Buchwald, R., Bertschinger, G., and Kunze, H.-J. (1980). *Phys. Rev. A* **21**, 200–206.
Fussman, G. (1975). *J. Quant. Spectrosc. Radiat. Transfer* **15**, 791–809.
Gabriel, A. H., and Paget, T. M. (1972). *J. Phys. B* **5**, 673–685.
Gabriel, A. H., and Volonte, S. (1973). *Phys. Lett. A* **43A**, 372–374.
Goto, T., and Burgess, D. D. (1974). *J. Phys. B* **7**, 857–864.
Grant, I. P. (1958). *Mon. Not. R. Astron. Soc.* **118**, 241–257.
Green, A. E. S. (1982). *Phys. Rev. A* **26**, 1759–1761.
Green, J. M. (1974). Rep. RDA-TR-4900-007. R & D Associates, Santa Monica, California.
Greene, R. L. (1982). *J. Quant. Spectrosc. Radiat. Transfer* **27**, 185–190.
Greene, R. L., Cooper, J., and Smith, E. W. (1975). *J. Quant. Spectrosc. Radiat. Transfer* **15**, 1025–1036.
Greig, J. R., Griem, H. R., Jones, L. A., and Oda, T. (1970). *Phys. Rev. Lett.* **24**, 3–6.
Griem, H. R. (1968). *Astrophys. J.* **154**, 1111–1122.
Griem, H. R. (1974). "Spectral Line Broadening by Plasmas." Academic Press, New York.
Griem, H. R. (1978). *Phys. Rev. A* **17**, 214–217.
Griem, H. R., and Kepple, P. C. (1981). *Proc. Int. Conf. Spectral Line Shapes, 5th* pp. 391–396.
Griem, H. R., Blaha, M., and Kepple, P. C. (1979). *Phys. Rev. A* **19**, 2421–2432.
Griem, H. R., Kolb, A. C., and Shen, K. Y. (1959). *Phys. Rev.* **116**, 4–16.
Grigoriadi, A. K., and Fisun, O. I. (1982). *Sov. J. Plasma Phys.* **8**, 440–445.
Grützmacher, K., and Wende, B. (1977). *Phys. Rev. A* **16**, 243–246.
Gupta, U., and Rajagopal, A. K. (1979). *J. Phys. B* **12**, L703–L709.
Gupta, U., and Rajagopal, A. K. (1981). *J. Phys. B.* **14**, 2309–2317.
Gupta, U., and Rajagopal, A. K. (1982). *Phys. Rep.* **87**, 259–311.
Hansen, J. P. (1973). *Phys. Rev. A* **8**, 3096–3109.
Hatton, G. J., Lane, N. F., and Weisheit, J. C. (1981). *J. Phys. B* **14**, 4879–4888.
Hauer, A., Mitchell, K. B., van Hulsteyn, D. B., Tan, T. H., Linnebur, E. J., Mueller, M. M., Kepple, P. C., and Griem, H. R. (1980). *Phys. Rev. Lett.* **45**, 1495–1498.
Held, B., and Deutsch, C. (1981). *Phys. Rev. A* **24**, 540–559.

Henry, R. J. W., Burke, P. G., and Sinfailam, A. L. (1969). *Phys. Rev.* **178**, 218–224.
Herman, L., and Coulaud, G. (1970). *J. Quant. Spectrosc. Radiat. Transfer* **10**, 1257–1275.
Hildebrandt, J., and Kunze, H.-J. (1980). *Phys. Rev. Lett.* **45**, 183–186.
Hoe, N., Grumberg, J., Caby, M., Leboucher, E., and Couland, G. (1981). *Phys. Rev. A* **24**, 438–447.
Höhne, F. E., and Zimmerman, R. (1982). *J. Phys. B.* **15**, 2551–2561.
Holtsmark, J. (1919). *Ann. Phys. (Leipzig)* **58**, 577–630.
Hooper, C. F. (1966). *Phys. Rev.* **149**, 77–91.
Hooper, C. F. (1968). *Phys. Rev.* **165**, 215–222.
Huber, W. A., and Bottcher, C. (1980). *J. Phys. B.* **13**, L399–L404.
Hughes, T. P. (1975). "Plasmas and Laser Light." Hilger, London.
Hummer, D. G. (1962). *Mon. Not. R. Astron. Soc.* **125**, 21–37.
Hummer, D. G. (1964). *Astrophys. J.* **140**, 276–281.
Hummer, D. G., and Kunasz, P. B. (1980). *Astophys. J.* **236**, 609–618.
Hummer, D. G., and Rybicki, G. B. (1982a). *Atrophys. J.* **254**, 767–779.
Hummer, D. G., and Rybicki, G. B. (1982b). *Astrophys. J.* **263**, 925–934.
Iglesias, C. A., and Hooper, C. F. (1982a). *Phys. Rev. A* **25**, 1049–1059.
Iglesias, C. A., and Hooper, C. F. (1982b). *Phys. Rev. A* **25**, 1632–1635.
Inglis, D. R., and Teller, E. (1939). *Astrophys. J.* **90**, 439–448.
Irons, F. E. (1975a). *J. Phys. B* **8**, 3044–3068.
Irons, F. E. (1975b). *J. Phys. B* **9**, 2737–2753.
Irons, F. E. (1980). *J. Quant. Spectrosc. Radiat. Transfer* **24**, 119–132.
Itoh, N., Totsuji, H., and Ichimaru, S. (1977). *Astrophys. J.* **218**, 477–483.
Itoh, N., Totsuji, H., Ichimaru, S., and DeWitt, H. E. (1979). *Astrophys. J.* **234**, 1079–1084.
Ivanov, V. V. (1973). "Transfer of Radiation in Spectral Lines" (D. G. Hummer, transl.). U.S. Gov. Print. Off., Washington, D.C.
Jackson, J. D. (1975). "Classical Electrodynamics," 2nd ed., Chap. 15. Wiley, New York.
Jacobs, V. L., and Blaha, M. (1980). *Phys. Rev. A* **21**, 525–546.
Jacobs, V. L., Davis, J., and Kepple, P. C. (1976). *Phys. Rev. Lett.* **37**, 1390–1393.
Joachain, C. J. (1975). "Quantum Collision Theory," Chap. 8. North-Holland Publ., Amsterdam.
Karzas, W. J., and Latter, R. (1961). *Astrophys. J., Suppl. Ser.* **6**, 167–212.
Key, M. H., Lunney, J. G., Ward, J. M., Evans, R. G., and Rumsby, P. T. (1979). *J. Phys. B* **12**, L213–L218.
Kilkenny, J. D., Lee, R. W., Key, M. H., and Lunney, J. G. (1980). *Phys. Rev. A* **22**, 2746–2760.
Kolb, A. C., and Griem, H. R. (1958). *Phys. Rev.* **111**, 514–521.
Kolbe, G., Huang, Y. W., and Burgess, D. D. (1982). *J. Phys. B* **15**, 4283–4289.
Krall, N. A., and Trivelpiece, A. W. (1973). "Principles of Plasma Physics," p. 557–563. McGraw-Hill, New York.
Kunze, H. J., Gabriel, A. H., and Griem, H. R. (1968). *Phys. Rev.* **165**, 267–276.
Lamoureux, M., Feng, I. J., Pratt, R. H., and Tseng, H. K. (1982). *J. Quant. Spectrosc. Radiat. Transfer* **27**, 227–231.
Latter, R. (1955). *Phys. Rev.* **99**, 510–519.
Lee, P. H.-Y., and Rosen, M. D. (1979). *Phys. Rev. Lett.* **42**, 236–239.
Lee, R. W. (1979). *J. Phys. B* **12**, 1165–1181.
Lee, R. W. (1982). *J. Quant. Spectrosc. Radiat. Transfer* **27**, 87–101.
Lee, R. W., Bromage, G. E., and Richards, A. G. (1979). *J. Phys. B* **12**, 3445–3453.
Liberman, D. A. (1979). *Phys. Rev. B* **20**, 4981–4989.
Lima, M. B. S., Lima, C. A. S., and Miranda, L. C. M. (1979). *Phys. Rev. A* **19**, 1796–1800.
Margenau, H., and Lewis, M. (1959). *Rev. Mod. Phys.* **31**, 569–615.
McDowell, M. R. C., and Ferendeci, A. M., eds. (1980). "Atomic and Molecular Processes in Controlled Thermonuclear Fusion." Plenum, New York.

8. Atomic Phenomena in Hot Dense Plasmas

McWhirter, R. P. W. (1965). *In* "Plasma Diagnostic Techniques" (R. H. Huddlestone and S. L. Leonard, eds.), pp. 227–241. Academic Press, New York.
Mihalas, D. (1978). "Stellar Atmospheres," 2nd ed. Freeman, San Francisco, California.
Mitchell, K. B., van Husteyn, D. B., McCall, G. H., Lee, P., and Griem, H. R. (1979). *Phys. Rev. Lett.* **42**, 232–235.
More, R. M. (1979). *Phys. Rev. A* **19**, 1234–1246.
Mott, N. F., and Massey, H. S. W. (1965). "The Theory of Atomic Collisions," 3rd ed. Oxford Univ. Press, London and New York.
Motz, H. (1979). "The Physics of Laser Fusion." Academic Press, New York.
Neiger, M., and Griem, H. R. (1976). *Phys. Rev. A* **14**, 291–299.
Neiger, M., and Zimmermann, B. (1980). *J. Quant. Spectrosc. Radiat. Transfer* **23**, 241–246; 247–251.
Nicolosi, P., and Volonte, S. (1981). *J. Phys. B.* **14**, 585–590.
O'Brien, J. T., and Hooper, C. F. (1972). *Phys. Rev. A* **5**, 867–884.
Oh, S. D., Macek, J., and Kelsey, E. (1978). *Phys. Rev. A* **17**, 873–879.
Osterbrock, D. E. (1962). *Astrophys. J.* **135**, 195–211.
Peacock, N. J., Hobby, M. G., and Galanti, M. (1973). *J. Phys. B* **6**, L298–L304.
Peres, A., and Ron, A. (1976). *Phys. Rev. A* **13**, 417–425.
Pittman, T. L., Voigt, P., and Kelleher, D. E. (1980). *Phys. Rev. Lett.* **45**, 723–726.
Puetter, R. C. (1981). *Astrophys. J.* **251**, 446–450.
Puetter, R. C., Hubbard, E. N., Ricchiazzi, P. J., and Canfield, R. C. (1982). *Astrophys. J.* **258**, 46–52.
Raven, A., Willi, O., and Rumsby, P. T. (1978). *Phys. Rev. Lett.* **41**, 554–557.
Ritchie, R. H. (1959). *Phys. Rev.* **14**, 644–654.
Rogers, F. J., and DeWitt, H. E. (1973). *Phys. Rev. A* **8**, 1061–1076.
Rogers, F. J., Graboske, H. J., and Harwood, D. J. (1970). *Phys. Rev. A* **1**, 1577–1586.
Rostoker, N., and Rosenbluth, M. N. (1960). *Phys. Fluids* **3**, 1–14.
Roussel, K. M., and O'Connell, R. F. (1974). *Phys. Rev. A* **9**, 52–56.
Roussel, K. M., and O'Connell, R. (1975). *Phys. Lett. A* **51A**, 244–246.
Royer, A. (1980). *Phys. Rev. A* **22**, 1625–1654.
Rozsnyai, B. F. (1972). *Phys. Rev. A* **5**, 1137–1149.
Rozsnyai, B. F. (1975). *J. Quant. Spectrosc. Radiat. Transfer* **15**, 695–699.
Rozsnyai, B. F. (1979). *J. Quant. Spectrosc. Radiat. Transfer* **22**, 337–343.
Seaton, M. J., and Storey, P. J. (1976). *In* "Atomic Processes and Applications" (P. G. Burke and B. L. Moiseiwitsch, eds.), pp. 134–197. North-Holland Publ., Amsterdam.
Seely, J. F. (1979). *Phys. Rev. Lett.* **42**, 1606–1609.
Seely, J. F., Dixon, R. H., and Elton, R. C. (1981). *Phys. Rev. A* **23**, 1437–1450.
Seidel, J. (1977). *Z. Naturforsch* **32a**, 1195–1206.
Seidel, J. (1979). *Z. Naturforsch* **34a**, 1385–1397.
Semon, M. D., and Taylor, J. R. (1976). *J. Math. Phys.* **17**, 1366–1370.
Shalitin, D. (1965). *Phys. Rev.* **140**, 1857–1863.
Shalitin, D. (1967). *Phys. Rev.* **155**, 20–23.
Shalitin, D., Ron, A., Reiss, Y., and Pratt, R. H. (1982). *J. Quant. Spectrosc. Radiat. Transfer* **27**, 219–226.
Shevelko, V. P., Skobelev, I. Yu., and Vinogradov, A. V. (1977). *Phys. Scripta* **16**, 123–128.
Sholin, G. V. (1969). *Opt. Spectrosc. (Engl. Transl.)* **26**, 275–289.
Shore, B. W. (1975). *J. Phys. B* **8**, 2023–2040.
Skupsky, S. (1980). *Phys. Rev. A* **21**, 1316–1326.
Smith, E. W., Cooper, J., and Vidal, C. R. (1969). *Phys. Rev.* **185**, 140–151.
Smith, E. W., Talin, B., and Cooper, J. (1981). *J. Quant. Spectrosc. Radiat. Transfer* **26**, 229–242.
Sobel'man, I. I. (1972). "Introduction to the Theory of Atomic Spectra." Pergamon, New York.

Sobel'man, I. I. (1979). "Atomic Spectra and Radiative Transitions." Springer-Verlag, Berlin and New York.
Sobolev, V. V. (1957). *Sov. Astron.—AJ* **1,** 678–689.
Spitzer, L. (1962). "Physics of Fully Ionized Gases," 2nd ed., Chap. 5. Wiley (Interscience), New York.
Stamper, J. A., and Ripin, B. H. (1975). *Phys. Rev. Lett.* **34,** 138–141.
Stavrakas, T. A., and Lee, R. W. (1982). *J. Phys. B.* **15,** 1939–1948.
Stewart, J. C., and Pyatt, K. D. (1966). *Astrophys. J.* **144,** 1203–1211.
Theimer, O., and Kepple, P. (1970). *Phys. Rev. A* **1,** 957–965.
Tighe, R. J., and Hooper, C. F. (1976). *Phys. Rev. A* **14,** 1514–1519.
Tighe, R. J., and Hooper, C. F. (1977). *Phys. Rev. A* **15,** 1773–1779.
Tseng, H. K., and Pratt, R. H. (1979). *Phys. Rev. A* **19,** 1515–1528.
Van Zandt, J. R., Adcock, J. C., and Griem, H. R. (1976). *Phys. Rev. A* **14,** 2126–2132.
Vinogradov, A. V., Skobelev, I. Y., and Yukov, E. A. (1977). *Sov. Phys.—JETP (Engl. Transl.)* **45,** 925–928.
Voslamber, D. (1977). *Phys. Lett. A* **61A,** 27–29.
Voslamber, D., and Stamm, R. (1981). *Proc. Int. Conf. Spectral Line Shapes, 5th* pp. 63–72.
Watson, W. D. (1970). *Astrophys. J.* **159,** 653–658.
Weisheit, J. C. (1975). *J. Phys. B* **8,** 2556–2564.
Weisheit, J. C. (1979). *J. Quant. Spectrosc. Radiat. Transfer* **22,** 585–588.
Weisheit, J. C., and Pollock, E. L. (1981). *Proc. Conf. Spectral Line Shapes, 5th, 1980* pp. 443–445.
Weisheit, J. C., and Rozsnyai, B. F. (1976). *J. Phys. B* **9,** L63–L69.
Weisheit, J. C., and Shore, B. W. (1974). *Astrophys. J.* **194,** 519–523.
Weisheit, J. C., Tarter, C. B., Scofield, J. H., and Richards, L. M. (1975). *J. Quant. Spectrosc. Radiat. Transfer* **16,** 659–669.
Whitten, B. L., Lane, N. F., and Weisheit, J. C. (1984). *Phys. Rev. A* (to be published).
Wiese, W. L., and Konjevic, N. (1982). *J. Quant. Spectrosc. Radiat. Transfer* **28,** 185–198.
Wigner, E. P. (1948). *Phys. Rev.* **73,** 1002–1009.
Yaakobi, B., and Goldman, L. M. (1976). *Phys. Rev. Lett.* **37,** 899–902.
Yaakobi, B., Skupsky, S., McCrary, R. L., Hooper, C. F., Deckman, H., Bourke, P., and Soures, J. M. (1980). *Phys. Rev. Lett.* **44,** 1072–1075.
Yaakobi, B., Steel, D., Thoros, E., Hauer, A., and Perry, B. (1977). *Phys. Rev. Lett.* **39,** 1526–1529.
Zakowicz, W., Feng, I. J., and Pratt, R. H. (1982). *J. Quant. Spectrosc. Radiat. Transfer* **27,** 329–333.
Zhdanov, V. P. (1979). *Sov. J. Plasma Phys.* **5,** 320–323.
Zhdanov, V. P. (1980). *Sov. J. Plasma Phys.* **6,** 103–105.
Zimmerman, G. B., and More, R. M. (1980). *J. Quant. Spectrosc. Radiat. Transfer* **23,** 517–522.
Zink, J. W. (1968). *Phys. Rev.* **176,** 279–284.
Zirin, H. (1954). *Astrophys. J.* **119,** 371–385.

Index

A

Absorption coefficient, in electron cyclotron emission, 230–232
Absorption oscillator strength, 455
Adiabatic compression, and moment equations, 125–126
Alcator-A tokamak, 132, 138
Alcator
  scaling, 331
  tokamak, 253
ALICE experiment, 2, 16
Alpha particle
  confinement of, 383
  energy splitting in, 382
  reflection, by magnetic mirrors, 385
Alpha-particle birth profile, 387
Alpha-particle heating, 381–393
  ash in, 390–391
  drifting in, 387–388
  $n\tau_E$ criterion for ignition in, 40–41
Alpha-particle orbit, evolution of, 386
Alpha-particle production, 381–387
Amplitude distributions, in Fourier optics, 213
Amplitude-independent polarization modulation, Faraday rotation, 198–199
Amplitude rotation method, in Faraday rotation, 196–197
Angular distributions, in electron bremsstrahlung spectrum, 317–318
Aperture limiters, in toroidal device, 401
Arc discharge sources, ionization efficiency in, 353–357
Astron experiment, Livermore, 18
Atomic level populations, 162–169
Atomic parameters, emission spectra and, 163
Atomic phenomena, in hot dense plasma, 441–481
Atomic process, ion transfer in, 61
Atomic radiation, from low-density plasma, 51–108

Atomic reactions, in boundary plasma, 423–436
Atom–proton collisions, relative velocity of, 431
Attenuation, of neutral particle flux, 271
Auger rates, 68
Austerix III iodine laser, 21
Auxiliary heating, in fusion research, 42–46

B

Backscattered particles, energy distribution of, 418
Backscattering, from ion and atom impact, 417–420
Balmer-α/Lyman-α intensity ratio, of O VIII, 182
Barium ion beam experiment, 295
Baseball minimum-B coil, 16
Beam-driven current, Ohm's law with, 337
Beam emittance factor, 367
Beam particle trajectories, 367–372
  single beamlets in, 367–368
Beam scattering diagnostics, 282–286
Beam transport
  in gas neutralizer, 363–372
  space-charge compensation in, 368–372
Beam transport parameters, as function of beam energy, 369
Berylliumlike ions
  partial term scheme for, 93
  spectral line intensities for, 93–95
Beta II experiment, Livermore, 18
Bethe–Maximon Coulomb corrections, 315
Bohr magneton, 291
Bohr radius, 156
Boltzmann equation, 55–60, 444
Born approximation, 458
Born cross sections, 461–463
Born Gaunt factors, 458

Boronlike ions
    partial term scheme for, 97
    spectral line intensities for, 95–97
Boundary
    definition, 395–396
    of toroidal device, 399–413
Boundary plasma, 395–437, *see also* Plasma
    atomic processes in, 423–436
    and backscattering, due to ion and atom impact, 417–420
    impurity control for, 411–413
    impurity ions in, 432–436
    molecular hydrogen and, 427–430
    particle–surface interactions in, 416–423
    sheath and long-range electric field regions of, 413–416
    significance of, 436–437
    sputtering and, 420–423
Boundary regime, in toroidal containment vessel, 398–399
Boundary region, description of, 395–399
Bound–bound transitions, 67, 455–456
Bound–free radiative process, 476
Bound–free transitions, 456
Bremsstrahlung
    electron, *see* Electron bremsstrahlung
    in plasma, 308–311
Bremsstrahlung calculations, model assumptions in, 308–310
Bremsstrahlung emission, in hot dense plasmas, 318
Bremsstrahlung energy spectrum, for Al and Fe, 314
Bremsstrahlung power loss component, 101
Brookhaven Laboratory, 374
Bumpy tori, electron cyclotron emission, in diagnostics of, 13–14, 244
Bundle divertor, 122, 403–405
Burnup fraction, 47

C

Carbon dioxide lasers, 21
Cesium heat-pipe conversion cell, 268–269
Charged particles
    cyclotron or synchrotron radiation from, 37, 53–54
    end losses of, 33
    trajectory of, 33
Charge exchange, instantaneous effect of, 162
Charge exchange cells, negative-ion beams and, 374
Charge exchange cross section, 156
Charge exchange ionization, rate coefficients for, 433
Charge states, in diffusive equilibrium, 146–155
Charge transfer, fractional radiation increase due to, 160
Charge transfer processes, and background neutrals, at plasma edge, 162
Charge transfer recombination, ionization equilibrium and, 155–162
Chemical sputtering, erosion by, of boundary surfaces, 422–423
Close-coupling methods, 65
Collective scattering ion term, 216
Collision, elastic, *see* Elastic collision
Collisional deexcitation, 425
Collisional dielectronic ionization coefficient, 434
Collisional dielectronic recombination coefficient, 78–79
Collisional equilibrium, criteria for, 151
Collisional ionization, by electron impact, 62–64
Collisional processes, in dielectronic process, 480
Collisional–radiative ionization, definition, 75–76
Collisional–radiative ionization coefficient, 425
Collisional–radiative recombination, 425–426
Collision limit, in tokamak plasmas, 161
Composite coefficient
    calculation of, 76–81
    ionization/recombination treatment and, 81
Composite plasma, spectra of, 219
Continuum lowering, level shifts and, 450
Core transition, 68
Cornell University, 21
Coronal equilibrium, 38–39, 183
Coronal ionization balance, equation for, 150
Costley rapid-scanning interferometer, 239–240
Coulomb–Bethe approximation, 163
Coulomb–Born approximation, 65
Coulomb collision
    fast ion energy loss in, 327
    fast ion scattering by, 332–333
Coulomb Gaunt factor, 316, 457
Coulomb interaction
    atomic structure and, 450
    infinite range of, 460
    ion correlations and, 448
Coulomb logarithm, 34, 58, 463

Index

Coulomb repulsion force, overcoming of, 28
Coulomb scattering, energy losses and, 36, *see also* Scattering
Coulomb scattering collisions, between free electrons and ions, 34
Culham Laboratory, United Kingdom, 6, 9, 16, 52
Culham Superconducting Levitron, 337
Cyclotron radiation, from charged particles, 37, 53

D

DCX-1 experiment, Oak Ridge, 2
DCX mirror experiments, Oak Ridge, 15
Debye–Hückel approximation, 448
Debye–Hückel energy loss rate, 464
Debye–Hückel limits, 445
Debye–Hückel potential, 450
Debye–Hückel screened charge, 461
Debye–Hückel screened/unscreened transition strength ratio, 456
Debye–Hückel screening model, 453
Debye–Hückel shift, 452–453
Debye length, 470
Debye potential, 445
Debye sphere, 444
Density, in space and time, evolution of, 123
Depopulation, of excited levels, 85–86
Detector photocurrent, signal-to-noise ratio in, 220
Detrapped particles, temperature of, 419
Deuterium–deuterium reactions, 28, 38
  energetic charged-particle reaction products from, 124
Deuterium–tritium fusion plasma, power density functions for, 37–40
Deuterium–tritium fusion reactor
  alpha particles in, 326, 381
  principles of, 28–32
  simplified concept of, 31
Deuterium–tritium pellets, 1, 5–6
Deuterium–tritium plasma reaction, cross section for, 1
Deuterium–tritium reaction, 28, 124
Deuterium–tritium reaction time, 20
Dielectronic recombination, in hot dense plasmas, 68–70, 479–482
Dielectronic recombination process, collisional processes and, 480

Diffusion coefficient, 34–35
Diffusion indicator, one- and two-electron ions as, 152
Diffusive equilibrium, charge states in, 146–155
Dirac excitation power loss, 100–101
Dirac variation-of-the-constants method, 469
Direct extraction sources, of negative-ion beams, 374–376
Direct radiative recombination cross sections and rate coefficients, 316–318
Dissociation, principal reactions or processes leading to, 252
Distorted-wave method, 65
DITE tokamak, Culham Laboratory, *see also* Tokamak
  beam-driven current in, 337
  deposition profiles in, 330
  electron heating control in, 134
  extreme ultraviolet spectrum for, 52, 174, 178–179
  Fabry–Perot interferometer data for, 243
  forbidden transitions for, 166–167
  heavy metal contamination in, 133
  pulsed-beam experiments in, 160
  soft x-ray spectrum of, 52, 174, 178–179
  wavelength measurements in, 185
Divertors
  energy unload action of, 409
  function of, 405–406
  in toroidal device, 402–406
Divertor target–limiter plate, impurities released at, 411
Doppler broadening technique, 277, 295
Double-charge capture, in negative-ion beams, 372–374
DRR (direct radiative recombination) cross sections, 316–318
D–T pellets, *see* Deuterium–tritium pellets
Duopigatron ion source, 360

E

EBFA machine, 21
EBT-1 bumpy torus, 14
EBT plasma, 295, 299–300
Echellete grating spectrometer, 237
ECIP approximation, 63
ECRH, *see* Electron cyclotron resonant heating

Effective charge, neutral beam attenuation in, 278–282
Effective ion confinement time, 150
Effective ion diffusion velocity, 150
Einstein absorption coefficient, 169
Elastic collision, and equipartition of kinetic energy, 72
Elastic scattering, 461–464
  differential (Born) cross section for, 462
Electric microfield distributions, 447–449
Electrode power loading, in ion extraction, 348–352
Electron bremsstrahlung, definition, 307, *see also* Bremsstrahlung
Electron bremsstrahlung spectrum
  angular distributions and polarization correlations in, 317–318
  and atomic electron screening effects, for isolated atom or ion, 313–315
  Coulomb spectrum and, 311–313
  elastic scattering, and direct radiative recombination in, 315–317
  end points of, 315–317
  from neutral atoms and ions, 308–319
Electron capture cross sections
  for C^{6+} + H reaction, 288
  for 50 keV/amu of fully stripped ions, 289
Electron confinement
  conduction and convection in, 131–132
  energy balance and, 131–132
  impurity effects in, 132–134
Electron cyclotron emission
  applications of, 240–244
  for bumpy tori, 244
  Fabry–Perot interferometer in, 238–240
  Fourier-transform spectroscopy in, 234–235
  grating spectroscopy in, 236–238
  instrumentation in, 233–240
  for mirror machines, 244
  non-Maxwellian distributions in, 232–233
  in plasma diagnostics, 227–246
  swept heterodyne receivers in, 233–234
  theory of, 229–233
  in tokamak diagnostics, 240–244
Electron cyclotron resonant heating, 11
Electron Debye length, 470
Electron density, intensity ratios and, 96
Electron energy, dissipation of, oxygen ion production, 435
Electron energy losses, for inelastic collisions, 124

Electron gas, scattering by, 211
Electronic density, versus time and electron temperature, 83
Electron impact
  collisional excitation by, 64–66
  collisional ionization by, 62–64
Electron impact collision strengths, 66
Electron–proton recombination, 397
Electrons
  collision, with impurity ions, 432–436
  runaway, 3
Electron scattering, by electron gas, 211–212
Electron temperature, *see also* Plasma electron temperature
  intensity ratios and, 94
  reduced, 96
  versus time and electron density, 83
Elmo bumpy torus, 13–14
  diagnostics of, 244
Elwert factor, 312
Emission spectra
  atomic level populations and, 162–169
  and basic atomic physics of tokamak, 183–186
  diagnostics based on, 143–186
  forbidden lines in, 164–168
  L-shell excitation and, 176–180
  M-shell excitation and, 180
  neutral-beam spectroscopy in, 181–183
  optical opacity of, 168–169
  spectral features and fusion plasmas in relation to, 169–180
Emission spectroscopy
  of high-temperature laboratory plasmas, 143
  ionization equilibrium and, 144–162
Emittance factor, in beam particle trajectories, 367
End losses, of charged particles, 33
Energy balance
  electron confinement and, 131–132
  equations for, 124–125
  ion confinement and, 135–136
  in magnetically confined fusion plasma, 36–42
  in tokamaks, 130–136
Energy balance equation, and adiabatic compression of plasma, 124–125
Energy confinement, figures of merit for, 131
Energy confinement time, alpha particle heating and, 40–41

Equilibrium charge state, of impurities, 144
Equilibrium power density, computed profiles of, 161

F

Fabry–Perot interferometer, in electron cyclotron emission, 238–240, 243, 292
Faraday rotation, 192–199, 290
  amplitude-independent polarization modulation and, 198–199
  amplitude ratio method for, 196–197
  and characteristic waves for magnetized plasma, 194–195
  Poincaré unit sphere and, 192–194
  polarization modulation method in, 197–198
  and rotation of plasma, 195–196
Faraday rotator, 294
Fast alpha-particle diagnostics, 387–390
Fast energy neutrals, ionization of, 329
Fast ions
  beam-driven currents and, 336–337
  deposition of, 328–332
  effect on plasma temperature, current, and rotation, 335–337
  energy and momentum transfer rates for, 334–335
  plasma rotation and, 337
  scattering of, by Coulomb collisions, 332
  slowing down of, 332–334
  trapping and thermalization of, 327–337
FCT, see Flux-conserving tokamak
Field-free cool thermal plasma, form factor for, 214–216
Field-reversed mirror concept, 18
Flux, poloidal, see Poloidal flux
Flux conserving tokamak, 116, 120–121
Flux surfaces, 116–117
Fokker–Planck equation, 274–275, 333, 336, 381
Fontenay TFR-400 tokamak, 7
Forbidden transitions, 164–168
  in emission spectra, 164–168, 474–476
  of L-shell ions, 184
  quadrupole, 227
  spontaneous decay rates of, 164–166
Form factor, for field-free cool thermal plasma, 214–216
Fourier optics, amplitude distributions in, 213

Fourier-transform spectrometer
  data for, 242
  schematic diagram of, 235
Fourier-transform spectrometry, in electron cyclotron emission, 234–236
Free-electron distribution, Thomas–Fermi–Dirac calculation for, 446
Free–free emission, cross sections for, 459
Free–free transitions, 457–460
Frequency-analyzing network, 219
Fusion plasma
  alpha ignition in, 40–41, 381–393
  approaches to, 5–21
  atomic physics of, 183–186
  auxiliary heating in, 42–46
  electron temperature within, 397
  Elmo bumpy torus in, 13–14
  inertial confinement in, 46–49
  ion temperature in, 217–220
  magnetic confinement of, 32–36
  neutral-beam injection heating in, 42–44
  open-ended geometries in, 14–20
  radio-frequency heating, 44–66
  spectral features in diagnosis of, 169–180
  stellarators in, 9–12
  tokamaks in, see Tokamak
  toroidal geometries in, 6–14
Fusion research
  basic concepts in, 27–49
  characteristic time for, 31
  history of, 1–4

G

GAMMA-10 machine, 18
Garching Laboratory, West Germany, 9, 21
Gas neutralizer
  beam transport in, 363–372
  efficiency of, 363–365
  slow hydrogen ion production in, 367
Gas puffing experiments, 129
Gaunt factor, 316, 457, 459
Gekko-4 system, 21
Global confinement time, local diffusivity and, 129–131
Global replacement time, 127
GM-II laser, 21
Grating spectroscopy, in electron cyclotron emission, 236–238

H

Halo effect, impurities and, 133
Hankel functions, 311
Hard-photon end-point region, of bremsstrahlung spectrum, 315
Hartree–Fock, average atom theory, 454
　method, 67, 452
　potentials, 446, 458
Hartree–Fock–Slater calculation, 446
Harwell Laboratory, U.K., 2, 13
HBTX-1 pinch, 13
HCN laser interferometry, electron line density in, 280
Heavy-ion beam probe, 297–302
Heliotron-E stellarator, Japan, 9, 11
Helium, collisional–dielectron recombination coefficient for, 78–79
Helium atoms, energy distribution in, 285
Helium ions, differential scattering cross section for, 286
Heliumlike ions
　excitation of, 88
　partial term scheme for, 89
　satellite lines in spectral line intensities for, 90–91
　spectral line intensities for, 87–91
Heliumlike resonance line, solar spectrum in vicinity of, 91
Heterodyne detection, theory of, 218–219
Heterodyne receivers, as cyclotron radiation detectors, 233–234
Hot dense plasma
　atomic phenomena in, 441–448
　dielectronic recombination in, 479–482
　elastic and inelastic scattering in, 461–467
　electric microfield distributions and, 447–449
　forbidden components and plasma satellites in, 474–476
　and perturbations by atomic collisions, 460–467
　spectra emitted by, 52–55
　radiative transition strengths and, 455–460
　spectral line formation in, 467–479
　Stark broadening in, 467–474
Hot plasma electron temperature, relativistic treatment of, 221–224
Hot plasma incoherent scattering, 221–224
Hydrogen atoms
　energy losses due to collisions of electrons with, 424
　recycling and trapping of, 426, 430–432, 436
Hydrogen density profile, for tokamak plasma, 252–253
Hydrogenic states, collisional rates for transitions between, 87
Hydrogen ions
　positive, 364
　spatial evolution of, 363–364
Hydrogenlike ions
　relaxation time constants for, 73–74
　spectral line intensities of, 86–87
　Stark shift and, 453
Hydrogen molecules, on plasma edge, 251–252
Hydrogen plasma, see also Plasma
　conditions in, 406–407
　electrons colliding with, 426
　as most prolific bound species, 423
　oxygen ion charge states for, 149
　radiation loss per Fe impurity ion, 160
　sheath potential for, 413
Hydrogen plus deuterium plasmas, ion features for, 218

I

ICRH, see Ion cyclotron resonance
Ignition
　alpha-particle heating in, 40–41
　in inertial confinement, 46–47
Ignition temperature, radiative power loss and, 105
Impurities
　equilibrium charge state and, 144
　local effective ion charge and, 144
　plasma properties and, 2
　scattering form factor and, 216–217
Impurity classes, K-shell excitation and, 171–173
Impurity control problems, 5
Impurity
　density
　　measurements of, 137
　　in tokamaks, 169
　effects, in electron confinement, 132–133
　elements
　　radiated power density function for, 39
　　radiated power loss and, 104–106
　equation, particle balance and, 123
　influx, edge plasma in, 23

Index

ion charge states, in diffuse equilibrium, 146–155
ion density, 286–290
ion diffusion, in toroidal systems, 145–148
ion profiles, relative sensitivity of, 152
ions
  charge exchange cross sections and, 259
  electron collisions with, 432–436
  reaction rate cross sections of, 22
  resonance fluorescence stimulation and, 137
  spectroscopic studies of, 144
  transport, in tokamaks, 136–138
IMS torsatron, University of Wisconsin, 11
Incident photon flux, as detector response, 175
Inelastic scattering, 464–467
Inertial confinement, 5, 20–21, 46–49
Inglis–Teller effect, 453
Injection heating, neutral-beam, 42–44
Intercombination/forbidden line intensity ratios, 90
Interferometry, 199–210
  magnetic-field-independent phase shift in, 199–200
  phase modulation in, 200–202
International Atomic Energy Agency for INTOR, 27
INTOR (International Tokamak Reactor), 27
  design studies for, 132
INTOR—Phase One, 325
INTOR scaling, 331
Ioffe bars, in magnetic mirrors, 15
Ion beam extraction, see also Ion extraction
  acceleration of, 341–352
  computational studies of, 346–348
Ion beams, space-charge compensation and, 368–372
Ion confinement, in tokamak, 135
Ion confinement time, concept of, 150
Ion cyclotron resonance, 45–46
  heating by, 125
Ion density profile, beam attenuation methods of, 281–282
Ion diffusion velocity, 150
Ion extraction
  beam steering in, 346
  computational studies of, 346–348
  electrode power loading and, 348–352
  in multigap systems, 344–346
  from plasma source, 341–348
  single-gap, 342–344

Ionic charge, scaling of coefficients with, 69–71
Ion impurity, density, calculation of, 123, see also Impurity density
Ion–ion Coulomb parameter, 443
Ionization
  boundary conditions in, 356–357
  collective viewpoint of, 71–81
  collisional–radiative, 75–76
  by excitation to autoionizing levels, 63
  from inner shells, 63
  by primary electrons, 354–355
  recombination and, 71–81
  stepwise, 78
  by thermal electrons, 353–354
Ionization balance equation, revised, 158
Ionization coefficients, 63
  composite, 81
Ionization equilibrium
  charge-transfer recombination and, 155–162
  emission spectra diagnostics and, 144–162
  neon ions in, 84
Ionization limit, 78
Ionization rates, temperature dependence of, 151
Ionization–recombination balance, 71, 144
Ionization stages, distribution among, 81–85
Ion mixing mode, 130
Ion–quadrupole interaction, 473
Ion resonance lines, quasistatic shifts for, 471–472
Ions, electrostatic potentials of, 444–447, see also Ion extraction; Fast ions
Ions of light elements, K-shell spectra and, 173
Ion source, plasma density distribution in, 358
Ion source types, 360–363
  duopigatron, 360
  magnetic-field-free, 362–363
  magnetic multipole, 362–363
  periplasmatron, 360–362
Ion species, factors influencing, 357–358
Ion temperature, see also Plasma temperature
  in center of fusion research plasmas, 217–221
  computer codes for, 272
  neutral-particle spectrometers in determination of, 257–277
Ion transfer, in atomic processes, 61
Ion transport, noncoronal ionization and, 147

Ion transport models, for stationary impurity ion profiles, 152
Iron ion abundances, in coronal equilibrium, 147, 159
ISXB device, high-density discharges on, 336

J

JET near-term fusion experiment, 388
JIPPT II stellarator, Japan, 9
  gas puffing experiments with, 129–130

K

Kapchinsky–Vladimirsky equation, 367
Khar'hov Institute, USSR, 9
Kinetic energy, equipartition of inelastic collisions, 72
Kinetic transport equations, energy moment of, 124
Kramers formula, 312, 316
K-shell excitation, 171–175
Kurchatov Institute, USSR, 6
Kurchatov T-3 tokamak, 6
Kyoto University, Japan, 9

L

L-1 and L-2 stellaratrons, USSR, 9
Landau–Lifshitz formula, 311–313
Lande g factor, 291
Laser diagnostics, 191–224
  Faraday rotation and, 192–199
  hot plasma electron temperature and, 221–224
  interferometry and, 199–210
  multiple-beam interferometry and data interpretation in, 210
  phase-comparison double interferometer in, 208–209
  phase quadrature interferometer in, 202–204
  Thomson scattering and, 210–224
  vibration-compensated quadrature interferometer in, 204–208
Laser-induced resonance fluorescence, schematic diagram of, 295
Lawrence Livermore Laboratory, 2, 15–18
Lawson criterion, 108
Lebedev Institute, USSR, 9
Liénard–Wiechert potentials, 211

Line broadening, 295
Line-radiated power loss coefficient, 434
Line shapes, in hot dense plasma, 472–474
Lithium ion, partial term scheme for, 92
Lithium ion satellites, K-shell excitation in, 171–172
Lithium isotope, in breeding reactor, 29
Lithiumlike ions, spectral line intensities for, 91–93
Local diffusivity, global confinement time and, 129–130
Local thermal equilibrium, 38, 59–60
Los Alamos Atomic Research Center, 13, 19, 21
Low density plasma, atomic radiation from, 51–108
Lower hybrid resonance, 45
L-shell excitation, 176–180
L-shell ions
  forbidden lines of, 184
  typical spectrum of, 177–179
LTE, see Local thermal equilibrium
Lyman-α–Stark patterns, for neon ions, 472

M

Mach–Zehnder arrangement, 202
Magnetically confined plasma, 113–138
  confinement geometry for, 115–116
  magnetic configuration for, 116–120
  temperature increase in, 325
Magnetic configuration
  flux surfaces and, 116–119
  maintenance of, 119–120
  of plasmas in tokamaks, 116–121
Magnetic confinement
  energy balance conditions in, 36–42
  of fusion plasmas, 32–36
  geometry of, 115–116
  ohmic heating in, 35–36
Magnetic constriction factor, in beam-particle trajectories, 368
Magnetic-dipole transitions, emission spectra and, 164
Magnetic-field-free ion sources, 362–363
Magnetic-field-independent phase shift, in interferometry, 199–200
Magnetic field measurements, in particle plasma diagnostics, 291–296
Magnetic limiter, divertor and, 404
Magnetic mirrors, 14–18

Index

alpha-particle reflection by, 385
configurations of, 274
Magnetic multipole (bucket) ion source, 362–363
Magnetized plasma, characteristic waves for, 194–195
Magnetohydrodynamic equilibrium theory, 291
Magnetohydrodynamic fluid, plasma instabilities of, 5
Magnetron ion source, for negative ions, 375
Matrix condensation, 78
Maxwell distribution, relativistic, 222
Maxwell field evolution equations, 118
Metastable ions, *see also* Ions
  in power loss calculations, 98
  spontaneous emission processes and, 73
Mirror machines, electron cyclotron emission in diagnostics of, 244
Mirrors
  magnetic, *see* Magnetic mirrors
  tandem, 17–18, 281–282
Molecular hydrogen, electron collisions with, 427–428, *see also* Hydrogen molecules
Moment equations, 122–126
  for adiabatic compression, 125–126
  energy balance and, 124–125
  particle balance in, 122–124
Momentum–energy analysis, for neutral-beam heated plasma, 273
Monte Carlo classical trajectory, 287
M-shell excitation, 180
MTSE II experiment, United Kingdom, 16
Multichannel grating spectrometer, 243
Multicomponent plasmas, 449
Multiconfiguration direct diagonalization method, 67
Multigap ion-extraction systems, 344–346
Multimegawatt beams, neutral injector for, 340
Multiple-beam interferometry, 210
Murakami limit, 132

N

Naval Research Laboratory, 21
Negative hydrogen ions, 364
Negative ion beams
  beam line for, 373
  from collisional processes, 376
  direct extraction sources of, 374–376
  double-charge capture in, 372–374

magnetron ion source for, 375
in neutral-beam formation and transport, 372–377
volume production of, 376–377
Neoclassical particle flux, 148
Neoclassical transport, versus anomalous, 148
Neon ions
  fractional abundances of, in ionization equilibrium, 84
  Lyman-α-Stark patterns for, 472
Neutral-beam
  attenuation, 278–282
  auxiliary heating, 23
  formation and transport, 339–377
  injection, advantages of, 339–340
  injection heating, 42–44, *see also* Neutral-injection heating
  spectroscopy, 181–183
  systems, 372
Neutral-density spectra, 128
Neutral-flux spectra, 273
Neutral hydrogen density, spatial profile, 128
Neutral-injection heating, theory of, 327, *see also* Neutral-beam, injection heating
Neutral injector, for multimegawatt beams, 341
Neutralizer, *see* Gas neutralizer
Neutral-particle
  analyzer
    conversion efficiency of, 264–269
    design of, 265
    five-channel momentum–energy, 262–263
    seven-channel 180°, momentum–energy, 263
    single-channel, 261
    stripping gas for, 265
  balance, 128
  distribution, in tokamak plasma, 127–129
  flux, attenuation of, 271
  plasma ion energy distribution and, 274
  sources of, at plasma edge, 127
  spectrometer, *see also* Neutral-particle, analyzer
    cesium pipe conversion cell and, 268
    in ion temperature determination, 257–277
Neutral transport kinetic equation, 127
Nonaxisymmetric fields, single-particle confinement in, 122
Noncoronal equilibrium radiation loss, 145

Noncoronal ionization, ion transport and, 147
Novosibirsk Laboratory, USSR, 374
Novosibirsk Third Conference on Plasma Physics and Controlled Fusion Research, 6

O

Oak Ridge DCX-1 experiment, 2
Oak Ridge magnetic mirror, 14–15
Oak Ridge ORMAK tokamak, 4, 7
OGRA experiments, USSR, 2, 15
Ohmic heating, in magnetic confinement, 35–36
Ohm's law, 118–119, 337
OHTE experiment, 13
One-component plasmas, 447–448
One-electron eigenstates, critical screening lengths for, 451
One-electron ions, as diffusion indicators, 152
Optical opacity, impurity densities and, 169
Optimal depth effects, in spectral line formation, 477–479
ORMAK tokamak, 4, see also Tokamak
Oxygen VIII, Balmer-α/Lyman-α intensity ratio for, 182
Oxygen ion charge states, diffusive equilibrium concentrations of, 149

P

Particle balance, 126–130
  model for, 126–127
  moment equations for, 122–124
  neutral particle distribution in, 127–129
Particle confinement, pellet injection and, 130
Particle confinement model, computer simulation of, 129
Particle confinement time, definition, 126
Particle diffusion, scattering and, 34–35
Particle flux, neoclassical, 148
Particle lifetimes, in tokamak plasmas, 271
Particle parameters, particle plasma diagnostics in, 249
Particle plasma diagnostics, 249–302
  active versus passive techniques in, 249–250
  advantages of, 250–251
  atomic physics of, 251–257
  beam scattering diagnostics in, 282–286
  heavy-ion beam probe in, 296–302
  impurity ion density and, 286–290
  magnetic field measurements in, 290–296
  and neutral particle spectrometers, used in determining ion temperatures in, 257–277
  and plasma ion density, and effective charge, by neutral-beam attenuation, 278–282
Particles, charged, see Charged particles
Paschen–Back effect, 291
PBFA experiment, 21
Peaking criterion, in impurity behavior, 137
Pellet fueling
  deuterium–tritium, 1, 5–6
  in inertial confinement, 46–48
  in particle confinement tests, 130
Periplasmatron, 360–362
Phase modulation, in interferometry, 200–202
Phase quadrature interferometer, 202–204
PHOENIX experiment, 2
Photon angular distribution, 317
Photon emission, per charge event, 182
Photons, spontaneous emission of, 66–67
Pinch, see Reversed-field pinch plasma; Theta pinch
Planck temperature, of radiation field, 60
Plasma, see also Tokamak plasma
  density and temperature regions in, 442
  electrostatic potentials of ions in, 444–446
  hot dense, see Hot dense plasma
  hydrogen, see Hydrogen plasma
  ion–ion Coulomb parameter in, 443
  magnetically confined, see Magnetically confined plasma
  Monte Carlo experiments for, 449
  multicomponent, 449
  one-component, 447
  one-dimensional characterization of, 443
  sheath potential and, 413
  total power radiated by, 97–108
  two categories of, 6
Plasma capture cross section, equation for, 271
Plasma charge state, 409
Plasma cross section, of flux-conserving tokamak, 121
Plasma density distribution, in ion source, 358
Plasma diagnostics
  active-beam, 255
  active or passive techniques in, 249–250
  electron cyclotron emission in, 227–246
Plasma dispersion function, 214
Plasma edge

Index

ergodic region in, 122
hydrogen molecules on, 251–252
neutral particle sources at, 127
Plasma electron temperature, see also Electron temperature
H^+ recombination rate and, 254
relativistic treatment of, 221–224
Plasma environment, 443
Plasma flow channel, topology of, 403, 407
Plasma focus, spectroscopic studies of, 143
Plasma heating
frequencies for, 45
from neutral-beam injection, 339–340
in tokamaks and stellarators, 4–5
Plasma interactions, parameters in, 396
Plasma ion density, neutral beam attenuation and, 278–282
Plasma limit, in space-charge compensation, 370–372
Plasma populations, impurities and, 2
Plasma power balance, radiative power loss and, 107
Plasma proton, scattering and, 397
Plasma research, in 1960s, 3–4
Plasma rotation, see also Faraday rotation
Faraday effect as, 195–196
fast ions and, 337
Plasma satellites, in spectral line formation, 474–476
Plasma screening, statistical description of, 444
Plasma spatial density, diffusion time-scale evolution and, 122
Plasma Spectroscopy (Griem), 144
Plasma temperature
increase in, 228
outer-region, 267
recycling of, 397
Plasma transport, 406–411
PLT tokamak, Princeton
charge exchange neutral energy distribution for, 272
heavy metal contamination in, 133
high-power neutral beam heating in, 172–173
ion temperature as function of time during discharges of, 275–276
plasma of, 4
x-ray line spectrum of, 171
Poincaré unit sphere, Faraday rotation and, 192–194
Polarization modulation method, in Faraday rotation measurement, 197–198

Polarized wave, vibrational ellipse for electric vector in, 193
Poloidal divertor, 403–404
Poloidal field, null point in, 403
Poloidal flux, 117
Positive impurity ions, 364–366
Positive ions, plasma source for, 352–363, see also Ions
Power density functions, fusion process and, 37–38
Power loading, in ion extraction, 348–352
Power loss, predicted, 103
Power loss coefficient, 99
Poynting flux, power spectrum of, 211
Primary electrons, ionization by, 354–355
Princeton ATC tokamak, 7
Princeton B-1 stellarator, 10
Princeton Large Torus Tokamak, 253
Princeton model C stellarator, 7
Princeton Plasma Physics Laboratory, 233
Princeton PLT tokamak, see PLT tokamak, Princeton
Princeton TFTR experiment, 8
Proton, definition, 54
Pulsator tokamak
magnetic fields, on discharge of, 294
poloidal field of, 294
Pulsed CO_2 lasers, 21

Q

q profile
of flux-conserving tokamak, 120–121
tokamak stability and, 119–120
Quantum electrodynamic theory, high-Z ions in test of, 184
Quasistatic shifts, for ion resonance lines, 471–472

R

Radial diffusion coefficient, 399
Radiated power loss
fusion plasma power balance and, 107
impurity elements and, 104–106
"lowest temperature" factor in, 105
Radiated power loss, calculations, 97–108
coefficients, 100
function, 99, 101–103
Radiating channels, energy losses in, 425
Radiative decay coefficients, 85
Radiative process, power density functions in, 37–38

Radiative recombination, 67
Radiative transition strengths, 455–460
Radio-frequency heating, in fusion power, 44–46
Raman–Nath scattering, 212
Random walk process, 430
Reaction rate cross sections, impurity ions and, 22
Recombination
  collective viewpoint of, 71–81
  dielectronic, 68–69, 479–482
  radiative, 67
Recombination continuum, 453
Recombination radiation component, 101
Recombination rate coefficients, 158
Recycling, plasma temperature and, 397
Reduced electron temperatures and densities, 69–70
Relaxation time constants, for hydrogenic ions, 73–74
Release radius, for transfer electrons, 156
Resonance decay, intersystem lines arising from, 171
Reversed-field pinch plasma, 12–13, 204–205
"Runaway" electrons, 3
Rutherford Laboratory, United Kingdom, 21

S

Safety factor, for toroidal devices, 400
Saha–Boltzmann equations, 74–77
Sandia Laboratory, 21
Sawtooth oscillations, in plasma core, 137
Scattering
  hot plasma incoherent, 221–224
  particle diffusion in, 34–35
  Thomson, 211–224
Scattering form factor, 211–212
Schrödinger equation, 450–452
SCYLLA experiment, 19–20
Second Geneva Conference on Peaceful Uses of Atomic Energy, 1–2
Semiempirical formula, 62–63
Signal-to-noise ratio, in detector photocurrent, 220–221
Single beamlets, in beam-particle trajectories, 367
Single-channel neutral-particle analyzer, line drawing of, 261
Single-gap ion extraction, analytical treatment of, 342–344

Single-null poloidal divertor, 404
Single particle, in nonaxisymmetric fields, 122
Soft-photon end-point region, 315–317
Sommerfeld formula, 311
Soviet Khar'kov Institute, 9
Soviet Kurchatov Institute, 6–7
Soviet Lebedev Institute, 9
Space-charge compensation
  of high current ion beams, 368–382
  plasma limit in, 370–372
Space-charge factor, in beam-particle trajectories, 367–368
Spatial shape factor, 330
Spectral line formation, 467–479
  optical depth effects in, 477–479
Spectral line intensity calculations, 85–97
  for berylliumlike ions, 93–95
  for boronlike ions, 95–97
  for heliumlike ions, 87–91
  for hydrogenlike ions, 86–87
  for lithiumlike ions, 91–93
Spectrometers, neutral-particle, see Neutral-particle, spectrometer
Spectroscopic studies, 143–186
Spitzer electrical conductivity, 337
Spontaneous emission coefficient, 455
Sputtering
  atoms ejected by, 422
  chemical, 422–423
  erosion due to, 423
  by heavy ions, 420
Sputtering yield, as function of ion energy, 421
Stark broadening, in hot dense plasma, 467–474
Stark effect, 295
Stark shift, 453
Statistical balance equations, components of, 60–71
Steady-state ionization balance
  power loss calculations on basis of, 101–102, 107–108
  predicted lower loss and, 103
Steady-state solution, as time-dependent solution, 83–84
Stellarators, 7–12
  plasma heating in, 4
  Princeton Model C, 7
  three-hexapole, 10
Stepwise ionization, 78
Swept heterodyne receiver, 233–234, 241, 245

Index

Synchrotron radiation, from charged particles, 37, 53–54

T

Table Top I experiment, Livermore, 15
Tandem mirror machine experiment, 17–18, 281–282
TARA experiment, Massachusetts Institute of Technology, 18
TEXT tokamak plasma, 296
TFTR tokamak, *see also* Tokamak
  adiabatic compression in, 118, 134
  near-term fusion experiments in, 388
  standard-case configuration for, 384
TFR tokamak plasma, neutral-beam attenuation measurements in, 278–281
Thermal electrons, ionization by, 353–354
Thermal equilibrium, local, 38, 59–60
Thermal plasma, frequency spectrum of radiation scattered from, 215
Thermal transport equation, 331
Theta pinch, devices, 82
  experiments, 19–20
  sources, 143
Thomas–Fermi continuum depression, 451
Thomas–Fermi–Dirac calculations, 446
Thomas–Fermi limits, 445
Thomas–Fermi model, of plasma screening, 445
Thomas–Fermi potential, 446, 452, 458, 462
Thomson scattering, 211–224
  collective, 218
  refractivity and, 212–214
Three-body charge transfer interaction, 155–156
Three-body recombination, 63–64, 425
Time-of-flight spectrometer, for escaping neutral particles, 268
TMX, *see* Tandem mirror machine experiment
Tokamak
  anomalous transport in, 148
  basic principle of, 6–9, 183–186
  definition, 7–8
  diffusion of one-electron systems in, 153–154
  DITE, *see* DITE tokamak
  electron cyclotron radiation diagnostics for, 240–241
  energy balance in, 130–136
  flux-conserving, 116, 120–121
  flux surfaces in, 116–119
  geometry of, 116
  impurity densities in, 169
  impurity transport in, 136–138
  ion dynamics in, 135–136
  ion temperature in, 217–221
  with iron core transformer, 8
  kilo-electron-volt temperatures in, 144
  magnetically confined plasmas in, 113–138
  moment equations for, 122–126
  neoclassical transport in, 148
  neutral-beam spectroscopy in, 181–183
  Oak Ridge, 4, 7
  particle balance in, 126–130
  plasma heating in, 4
  plasma temperatures in, 175–176, 228
  PLT, *see* PLT tokamak, Princeton
  q profile for, 119–120
  sawtooth oscillations from plasma core of, 120
  size of, 8–9
  special configurations of, 119–120
  steady-state concentrations in hot core of, 151
  toroidal surfaces in, 116–117
Tokamak core plasma, *see also* Tokamak plasma
  corona equilibrium conditions in, 183
  temperature of, 175–176
Tokamak plasma, *see also* Tokamak core plasma
  H^0 neutral density profile for, 252–253
  neutral particle distribution in, 127–129
  outer region ion temperatures in, 267
  particle lifetimes in, 271
Tokamak research, reviews of, 7–8
Tokamak spectroscopy, 144–186, *see also* Emission spectroscopy
Toroidal containment vessel, boundary regime in, 398–399
Toroidal device
  aperture limiters in, 401
  boundary of, 399–413
  divertors in, 402–406
  plasma–boundary interactions in, 400
  safety factor for, 400
Toroidal flux surface, rapid circulation within, 119
Toroidal geometries, 6–14
Torsatron, 11
TOR-1 stellarator, USSR, 9

Total radiated power
  calculations for, 97–108
  equation for, 99
Triplet/singlet ratio, in one-electron ion diffusion velocities, 154
Two-electron ions, as diffusion indicators, 152
Typographical errors, in equations, 246

U

Uragan III Stellarator, USSR, 9
U.S. Office of Controlled Thermonuclear Research, 4

V

Vibration-compensated quadrature interferometer, 204–208

W

Wendelstein stellarators I and II, 9

Wiener–Khinchine theorem, 211
Wisconsin, University of, 11
Wollaston prism, 202, 294

X

X-ray recombination edge, 154

Y

Yin–yang coil, 16

Z

Zebra-stripe interferometer, 200–202
Zeeman effect, 291–293
Zeeman splitting, energy-level diagram of, 292
ZETA toroidal pinch experiments, United Kingdom, 3, 13

PURE AND APPLIED PHYSICS

A Series of Monographs and Textbooks

Consulting Editors

H. S. W. Massey
University College London
London, England

Keith A. Brueckner
University of California, San Diego
La Jolla, California

1. F. H. Field and J. L. Franklin, Electron Impact Phenomena and the Properties of Gaseous Ions. (Revised edition, 1970.)
2. H. Kopfermann, Nuclear Moments. English Version Prepared from the Second German Edition by E. E. Schneider.
3. Walter E. Thirring, Principles of Quantum Electrodynamics. Translated from the German by J. Bernstein. With Corrections and Additions by Walter E. Thirring.
4. U. Fano and G. Racah, Irreducible Tensorial Sets.
5. E. P. Wigner, Group Theory and Its Application to the Quantum Mechanics of Atomic Spectra. Expanded and Improved Edition. Translated from the German by J. J. Griffin.
6. J. Irving and N. Mullineux, Mathematics in Physics and Engineering.
7. Karl F. Herzfeld and Theodore A. Litovitz, Absorption and Dispersion of Ultrasonic Waves.
8. Leon Brillouin, Wave Propagation and Group Velocity.
9. Fay Ajzenberg-Selove (ed.), Nuclear Spectroscopy. Parts A and B.
10. D. R. Bates (ed.), Quantum Theory. In three volumes.
11. D. J. Thouless, The Quantum Mechanics of Many-Body Systems. (Second edition, 1972.)
12. W. S. C. Williams, An Introduction to Elementary Particles. (Second edition, 1971.)
13. D. R. Bates (ed.), Atomic and Molecular Processes.
14. Amos de-Shalit and Igal Talmi, Nuclear Shell Theory.
15. Walter H. Barkas. Nuclear Research Emulsions. Volumes I and II.
16. Joseph Callaway, Energy Band Theory.
17. John M. Blatt, Theory of Superconductivity.
18. F. A. Kaempffer, Concepts in Quantum Mechanics.
19. R. E. Burgess (ed.), Fluctuation Phenomena in Solids.
20. J. M. Daniels, Oriented Nuclei: Polarized Targets and Beams.
21. R. H. Huddlestone and S. L. Leonard (eds.), Plasma Diagnostic Techniques.
22. Amnon Katz, Classical Mechanics, Quantum Mechanics, Field Theory.
23. Warren P. Mason, Crystal Physics in Interaction Processes.
24. F. A. Berezin, The Method of Second Quantization.
25. E. H. S. Burhop (ed.), High Energy Physics. In five volumes.
26. L. S. Rodberg and R. M. Thaler, Introduction to the Quantum Theory of Scattering.

27. R. P. Shutt (ed.), Bubble and Spark Chambers. In two volumes.
28. Geoffrey V. Marr, Photoionization Processes in Gases.
29. J. P. Davidson, Collective Models of the Nucleus.
30. Sydney Geltman, Topics in Atomic Collision Theory.
31. Eugene Feenberg, Theory of Quantum Fluids.
32. Robert T. Beyer and Stephen V. Letcher, Physical Ultrasonics.
33. S. Sugano, Y. Tanabe, and H. Kamimura, Multiplets of Transition-Metal Ions in Crystals.
34. Walter T. Grandy, Jr., Introduction to Electrodynamics and Radiation.
35. J. Killingbeck and G. H. A. Cole, Mathematical Techniques and Physical Applications.
36. Herbert Überall, Electron Scattering from Complex Nuclei. Parts A and B.
37. Ronald C. Davidson, Methods in Nonlinear Plasma Theory.
38. O. N. Stavroudis, The Optics of Rays, Wavefronts, and Caustics.
39. Hans R. Griem, Spectral Line Broadening by Plasmas.
40. Joseph Cerny (ed.), Nuclear Spectroscopy and Reactions. Parts A, B, C, and D.
41. Sidney Cornbleet, Microwave Optics: The Optics of Microwave Antenna Design.
42. Peter R. Fontana, Atomic Radiative Processes.
43. H. S. W. Massey, E. W. McDaniel, and B. Bederson (eds.), Applied Atomic Collision Physics, Volumes 1, 2, 3, 4, and 5.